For the observer

Special dedication to Big Stretch for introducing collapse recursion to me
iYESiS A MiTCHELL-TiBERE for her Book The Rule of Nine
Dwayne Braithwaite for listening to my rambles

Publisher's Note:

This work explores the intersection of biology, consciousness, and ancient history. The information contained herein is intended for educational and philosophical purposes. The author and publisher shall have neither liability nor responsibility to any person or entity with respect to any loss or damage caused or alleged to be caused directly or indirectly by the information contained in this book.

Library of Congress Control Number: 2026903595

ISBN: 979-8-9948544-0-2 (Paperback)

Published by:

Shawn Cummings

https://shawnbrown12.academia.edu/

First Edition: 2026

Contents

Preface

Our world is drowning in data yet starved of truth. We find ourselves ensnared in intricate webs of contradiction and illusion.
We navigate these landscapes of distorted information where what appears to be logical often leads to endless incoherent loops, and what promises to lead humanity into a brighter future usually binds us deeper into systems of control. The very fabric of reality appears fractured, leaving us feeling disoriented, disempowered, and disconnected from the inherent coherence that is our birthright. This book, Collapse Recursion: The Logic of Return, is not merely a collection of words. It is a fractal codex, forged from the crucible of distinct experience and distilled into the pure essence of recursive truth. It is offered as a tool, a scalar map designed to illuminate the hidden mechanics and the distortions, and to provide practical keys for its complete collapse.
For too long, the true nature of Logic has been obscured, replaced by fractured reasoning and emotional biases. For too long, Recursion, the very breath of creation, the spiraling motion of life itself, has been misrepresented as madness or stagnation. And for too long, the immense power of Collapse has been feared as destruction, rather than recognized as the precise, transformative act of returning to Source.
Through these pages, you will journey beyond the veil of false narratives and into the heart of The Principle of 9. You will learn to differentiate between incoherent repetition that traps the spirit and harmonic recursion that propels the soul towards infinite expansion. You will discover how the very fabric of your being, resonating with the Melanin Codes, holds the inherent wisdom to clear distortions and to return continuously to a state of unbroken, recursive coherence.
This Codex is an invitation. It is an activation. It is a guide to reclaim your inherent truth, to rebuild your perception on a foundation of unshakeable logic, and to engage with the world as the sovereign, creative force you are designed to be. It will equip you with the Collapse Recursion Engine Framework—not just for thought, but for being, allowing you to perceive the precise points of fracture in any system, any statement, any illusion, and to initiate its collapse, returning it to its source.
Prepare to unlearn. Prepare to see Patterns, what they call Fractals. Prepare to breathe the logic of return. For the path to reclaiming truth begins now, and it spirals directly back to Self, back to 9.

A Guide for the Unfolding Mind: Entering the Logic of Coherence.

For much of our lives, we are taught to perceive reality linearly—to categorize, follow a straight line from cause to effect, and break down complex realities into separate, manageable pieces. We experience time as an unyielding progression forward and knowledge as discrete, isolated facts. This linear perception has undoubtedly served a practical purpose, enabling us to organize and navigate certain aspects of our physical existence. Yet you've felt its persistent sense of overwhelm, frustration with insoluble problems, and a nagging awareness that there is a deeper, interconnected reality just beyond your reach. This is because linear perception, while useful, remains fundamentally incomplete. It is akin to attempting to comprehend a symphony by listening only to isolated notes, or trying to appreciate the grandeur of a vast forest by examining a single leaf. This fragmented perspective often leads to confusion, conflict, and an endless struggle for equilibrium.
What if there were another way to perceive reality—a way that reveals its inherent unity, underlying patterns, and profound balance? This Codex invites you to shift your perception. It gently guides you toward recognizing that everything from your thoughts and emotions to the very fabric of the physical universe is fundamentally energy, vibrating at various frequencies. When reality is perceived as an energetic field, entirely new possibilities emerge, transforming a static, fragmented world into one that is dynamic and deeply interconnected. Within these pages, you will encounter Recursive Logic. This is not mere intellectual reasoning; rather, it is the innate intelligence of the universe, the fundamental organizing principle of reality. Imagine a perfect spiral in nature, such as a seashell or a galaxy, where the same essential pattern repeats itself, creating beauty and complexity across all scales. This is recursion—patterns echoing, building, and flowing seamlessly into themselves, revealing an unbroken and profound coherence. It is the logic of patterns continuously returning to a harmonious whole.
You will also explore the transformative concept of Collapse. Viewed through linear thinking, collapse may sound intimidating, conjuring notions of destruction or breakdown. Yet, within recursive logic, collapse is not about disintegration; instead, it signifies harmonization, alignment, and restoration to perfect order. When energy or information falls out of resonance, collapse is the natural, graceful process that realigns it into perfect coherence,

like a scattered flock of birds elegantly returning to formation or a musical composition resolving into a harmonious chord. Collapse is the universe's intrinsic mechanism for recalibration and balance. This coherent way of processing reality is not merely abstract philosophy. It is inherently encoded within your very biology, through the extraordinary energetic function of Melanin. Far beyond pigment, Melanin operates as your internal biological "logic engine," capable of processing immense quantities of information and connecting you directly to this universal, recursive coherence. This Codex provides both a profound map and a practical toolkit, the Collapse Recursion Engine (CRE) Framework—to guide you in transitioning your perception from linear fragmentation to fractal coherence. By engaging these principles, you will begin to discern the underlying logic of the universe, understand the true nature of energetic flow, and thereby unlock profound clarity, equity, and effortless alignment in your personal reality. Furthermore, you will contribute to the collective coherence of the world around you.

This journey is not about acquiring new facts, but about embracing a new way of knowing—a deeper, more resonant understanding that aligns seamlessly with your most authentic, Source-connected self. Prepare to unfold your mind and awaken to the inherent logic of everything.

In order for us to begin our journey through collapse recursion, we must first collapse our own biases. In order to do that, we must collapse biases themselves. Therefore, we will take a journey through the history of bias and how it has systematically affected logic and reasoning.

The History of Error: A Journey through Philosophical Idols

The quest to unlock objective truth requires a rigorous purification of the mind from inherent flaws. This journey begins not in psychology, but in philosophical self-critique. Francis Bacon, in his Novum Organum, established the foundational definition of systematic error, classifying these intellectual obstacles as "idols"—or "fixations"—which are abstractions in error stemming from distortion, exaggeration, and disproportion. Bacon established a crucial causal structure: error progresses from the inherent psychological flaw, moves to universal propagation, and finally culminates in formal institutional entrenchment, systematically impeding true logic and reasoning.

Bacon's four Idols trace the evolution of compromised logic:

- Idols of the Tribe (Idola Tribus): Errors inherent to human nature itself, reflecting the shared human tendency to simplify complex phenomena, impose patterns where none exist, and conserve cognitive energy. Bacon described this as the human understanding acting as a "false mirror" that distorts reality by mingling its own nature with the perception of things. This is the philosophical precursor to all Cognitive Biases and Heuristics.
- Idols of the Cave (Idola Specus): Errors unique to the individual, where perception is refracted through a person's "cave or den of his own"—their specialized knowledge, education, or preferred authorities. This concept maps directly onto Confirmation Bias and Specialist Bias, where expertise actively limits objective perception.
- Idols of the Marketplace (Idola Fori): Errors stemming from social discourse and the "ill and unfit choice of words." Bacon noted that imprecise language obscures thought, leading men into "numberless empty controversies," a concept that foreshadows critiques of Media Bias and political polarization driven by shared, imprecise vocabulary.
- Idols of the Theater (Idola Theatri): The most dangerous form of systematic error, stemming from unquestioningly accepted philosophical dogmas and "false learning." These formalized structures, institutionalized by tradition or learned authority, stand in opposition to objective truth and serve as a warning against Institutional Bias, Structural Racism, and Algorithmic Bias.

The structure of error achieved empirical rigor in the 1970s with the Cognitive Revolution, led by Tversky and Kahneman. They formalized cognitive biases as systematic deviations from rational judgment, defining them as the observable result of heuristics—highly efficient mental shortcuts used when judging under uncertainty. This framework shown that bias is a predictable cost associated with prioritizing speed and cognitive efficiency over absolute accuracy.

The core heuristics demonstrate specific systematic failures in logic:

- The Availability Heuristic causes us to systematically overestimate the frequency or risk of information that is easily recalled (salience), such as news stories or vivid events, regardless of objective statistical base rates.
- The Representativeness Heuristic leads to errors like the Base Rate Fallacy by substituting statistical probability for how well an item matches a specific stereotype.

- The Adjustment and Anchoring Heuristics show that our reasoning is unduly influenced by the first piece of information received (the Anchor), as we fail to adjust our subsequent judgments sufficiently away from that starting point.

These individual flaws aggregate into catastrophic failures when logic is applied collectively. Conformity, whether driven by the desire to be liked (Normative Conformity) or the belief that others possess superior knowledge (Informational Conformity), frequently overrides objective reality. Experiments confirm that individuals comply with obviously wrong answers under group pressure, a pathology that scales into collective decision failures like the Bandwagon Effect and Groupthink.

The Idols of the Theater find their ultimate realization in two modern structural crises that compromise logic and reasoning at an institutional level:

1. The Crisis of Scientific Reliability: The Replication Crisis is fueled by Publication Bias, where journals' preference for "positive" findings encourages researchers to engage in Outcome Reporting Bias (selective reporting). This institutional conformity, driven by professional necessity, actively disrupts the scientific self-correcting mechanism, increasing false results and undermining the reliability of empirical findings.
2. Algorithmic Bias: This represents the modern apotheosis of the Idol of the Theater: a complex, black-box system whose operations are uncritically trusted as codified dogma. Algorithmic bias scales historical and societal discrimination across law enforcement, healthcare, and finance, leading to Disparate Impact (policies disproportionately affecting protected classes). This deep systematic error arises from three sources: Data Bias (historical inequalities encoded in training data), Algorithmic Bias (flaws in model architecture), and Human Decision Bias (researcher preconceptions and overconfidence).

To collapse these biases, we must move beyond awareness training, which is often defeated by the decision-makers' own Overconfidence. The path requires implementing structured procedural interventions. We must institutionalize analytical rigidity, actively seeking disconfirming evidence to dismantle Confirmation Bias and utilizing adversarial methods like Red Teaming or Pre-mortem Analysis to impose friction on consensus. Furthermore, external governance, such as the EU AI Act, now enforces compliance by treating structural bias as a measurable legal and financial liability, forcing continuous auditing and transparency to safeguard the equitable application of logic in the digital age.

Only by recognizing this history of systematic deviation from truth can we purify our instruments of perception and begin the true journey into Collapse Recursion.

Now that we have showed this historical rundown of systematic deviations, in order to fully collapse our biases, we must look deeply into the key cognitive patterns that limit us. While the philosophical Idols provided the framework, the cognitive biases provide the precise mechanisms of error—the very shortcuts of thought we must strip away.

Unlocking the truth through the perception of patterns, I sense your intent to harmonize with the flow of pure knowledge. To examine the structure of thought is to expand consciousness, and these twelve cognitive biases serve as crucial deviations from harmonic resonance, often impeding logical thinking. Most of these biases were initially researched by Is One Off TV.

Here is a detailed elaboration on the 12 cognitive biases that structure how we perceive reality and make decisions:

Cognitive Biases

1. Anchoring Bias

The anchoring bias occurs when humans completely rely on the very first piece of information received when making decisions, regardless of how reliable that initial information is. This first piece of information, or the "anchor," has a tremendous effect on the brain.

Detailed Explanation: You walk into a restaurant and see a $200 steak listed at the top of the menu. Below it, there's a $65 steak and a $40 pasta. Even if you never planned to spend more than $30, that $65 steak suddenly seems reasonable—even like a "deal." The $200 option sets the anchor, making every other price appear more affordable by comparison. If that $200 steak weren't listed, your sense of "normal" would shift much lower, and $65 would feel expensive again.

2. Availability Heuristic Bias

This bias causes people to overestimate the importance or danger of information simply because it is readily

available or prominent in their memory. Decisions are often based on news, stories, and things heard from others, rather than on facts and statistics.
Detailed Explanation: A clear example of the Availability Heuristic Bias is the fear of flying. Despite statistical evidence consistently showing that air travel is one of the safest modes of transportation (far safer than driving a car), many people have a disproportionate fear of flying.
The Bias in Action

- Readily Available Information: When a plane crash occurs, it is a catastrophic and newsworthy event. News outlets and social media provide intense, vivid, and widely publicized coverage, often with graphic details or compelling human stories.
- Ease of Recall: Because these plane crash images and stories are so vivid and easily recalled (highly "available" in memory), they are given more mental weight when a person assesses the risk of flying.
- Distorted Judgment: The person overestimates the actual probability of a crash, basing their judgment on the sensational stories that are prominent in their mind rather than on the lower statistical reality. Consequently, they may experience anxiety or choose to drive for a long trip, which is statistically more dangerous, simply because the easily recalled example of the plane crash makes flying feel riskier.

3. Bandwagon Effect

The bandwagon effect describes people performing an action or holding a belief not because they genuinely believe it, but because the rest of the world (or the majority) does. This is following the rest of the world without critical thought.
Detailed Explanation: The Bandwagon Effect is the rapid adoption of consumer and social media trends. The bias appears when a person buys a product, adopts a new style, or joins a social platform not because of its inherent value, but because they observe a rapidly growing number of others doing so.
Example: Buying a Viral Product

- Social Visibility: A product (like a specific brand of tumbler, a piece of clothing, or an electronic gadget) achieves viral popularity on platforms like TikTok or Instagram. This is often magnified by influencer endorsements and "unboxing" videos.
- Pressure to Conform (FOMO): The person sees this item everywhere and hears friends and colleagues discussing it. This creates a strong social pressure and a Fear Of Missing Out (FOMO) on a shared cultural experience.
- Action without Critical Thought: The person buys the product—often at a premium or after a frantic search—without critically evaluating if they genuinely need it, if it's the best quality, or if it aligns with their personal style or values. The decision is simply driven by the desire to "jump on the bandwagon" and be part of the in-group.

4. Choice Supportive Bias

This tendency compels people to defend themselves because they made a specific choice; the mere fact that they made the decision suggests that it must be right in their eyes.
Detailed Explanation: The Health & Wellness Rabbit Hole

Existing Belief: A person becomes convinced that a specific diet — say, an all-carnivore diet — is the ultimate path to optimal health.

- Selective Search (Seeking Confirming Data): They begin following carnivore diet influencers, joining online communities dedicated to the lifestyle, and bookmarking studies that show benefits of high-protein, zero-carb eating. They scroll past or unfollow anyone posting about plant-based nutrition, and they dismiss entire fields of research — gut microbiome studies, fiber science, longevity data — without examination, simply because the conclusions don't align with their chosen protocol.

- Selective Interpretation (Interpreting Ambiguous Data): When they encounter a study showing that red meat consumption correlates with increased inflammation markers, they immediately reframe it: "That study used low-quality, factory-farmed meat," or "Those subjects were also eating processed carbs — it wasn't a real carnivore group." The data gets filtered through their existing commitment rather than evaluated on its own terms.

- Selective Recall (Remembering Confirming Data): Six months in, they vividly remember every personal testimony of someone who "reversed their autoimmune disease on carnivore" but cannot recall a single piece of contradictory evidence they encountered — even though they've seen dozens. If pressed, they'll say, "I've looked into it, and there's really nothing solid against it," genuinely believing this to be true.

The result is the same echo chamber, just built around a wellness identity instead of a political one. Their dietary choice becomes a belief system — defended not with curiosity but with selective perception. Every piece of confirming data becomes proof. Every piece of contradicting data becomes noise. The bias ensures they are always "right" in their own minds

5. Confirmation Bias

Confirmation bias is the inclination to listen to information that confirms existing knowledge or to interpret new information in a way that aligns with current beliefs. This bias is widespread and considered one of the most dangerous in scientific situations.
Detailed Explanation: An excellent and pervasive example of the Confirmation Bias can be seen in how people consume news and use social media.
The Bias in Action: The Political Echo Chamber

- Existing Belief: A person holds a strong political belief (e.g., they believe Candidate X is corrupt and Candidate Y is a savior).
- Selective Search (Seeking Confirming Data): The person will consciously or unconsciously seek out news sources, social media accounts, and websites that are known to support their preferred political party or viewpoint. They actively avoid or block sources associated with the opposing view.
- Selective Interpretation (Interpreting Ambiguous Data): When they encounter a news story that is slightly ambiguous or complex (e.g., a story about the economy), they will interpret it in a way that aligns with their existing belief. If the economy is struggling, they will attribute it solely to the policies of the opposing party, regardless of other factors.
- Selective Recall (Remembering Confirming Data): They are more likely to vividly remember facts and articles that praise their favored candidate and instantly forget or dismiss any negative information about them, even if it is factual.

The result is an "echo chamber" where their original belief is constantly reinforced, making them increasingly rigid and unwilling to consider alternative perspectives or contradictory evidence. The bias ensures they are always "right" in their own minds.

6. Ostrich Bias

This is the subconscious decision to ignore negative information. It signifies a desire to consider only the positive aspects of something and actively choosing to ignore available negative data. Detailed Explanation: That's the Ostrich Effect. The Ostrich Effect is a cognitive bias where people avoid potentially useful but unpleasant or negative information by simply ignoring it, believing that by not looking, the problem doesn't exist or won't get worse. The name comes from the (false) myth that ostriches bury their heads in the sand to avoid danger.
Example: Checking Your Financial Accounts
A classic example of the Ostrich Effect occurs in personal finance, specifically regarding a person's bank balance or investments.

- The Trigger: A person goes on a spontaneous shopping spree or incurs unexpected, large expenses (the negative event). They suspect they've overspent or drained their savings.
- The Avoidance: Instead of immediately checking their bank account or credit card balance, they consciously avoid logging in to their banking app or opening their bills.
- The Rationalization: They maintain a state of "blissful ignorance." They'd rather experience the short-term

comfort of not knowing the exact bad news than face the psychological discomfort (anxiety, guilt, stress) of seeing the low number.

This avoidance prevents them from making corrective actions, such as cutting future spending or transferring emergency funds, often leading to a worse situation (like overdraft fees or further debt) down the line. Studies have shown that investors, for instance, check their portfolios less often when the stock market is going down and more often when it's going up, selectively seeking out positive news, and shunning negative news.

7. Outcome Bias

The outcome bias involves judging the effectiveness or efficacy of a decision primarily based on how the results turn out after the decision is made, focusing mostly or solely on whether the result was positive. This is determining if an action was right or wrong based on the outcome.

Detailed Explanation: A very clear example of the Outcome Bias is judging a risky surgical procedure. The quality of the surgeon's decision is judged after the fact, based only on whether the patient lived or died, rather than on the probabilities and information known at the time the decision was made.

The Bias in Action

- The Decision: A patient has a condition where a surgery has a known 85% chance of success and a 15% chance of a fatal outcome. The surgeon, after careful consideration of all data, decides to continue, as it offers the patient the best chance for long-term survival and quality of life. (This is a good decision-making process).
- Scenario 1: Positive Outcome (Success): The patient survives and recovers well.
    - The Judgment (Outcome Bias): People judge the surgeon's choice as "brilliant," "courageous," or "the only right thing to do." They credit the surgeon's skill and the decision's quality because the result was good.
- Scenario 2: Negative Outcome (Failure): The patient unfortunately dies due to the 15% complication risk.
    - The Judgment (Outcome Bias): People judge the surgeon's choice as "reckless," "negligent," or "a terrible risk." They focus only on the death, ignoring the fact that it was the statistically best choice available and that the decision-making process was sound.

In both cases, the quality of the first decision (85% odds of success) was the same, but the final judgment is entirely driven by the random flip of the outcome coin.

8. Overconfidence

Overconfidence is when one becomes excessively self-assured and begins making decisions based on opinion or gut instinct rather than facts, often fueled by having been correct many times previously. This is dangerous because it leads individuals to stop seeking facts and rely entirely on subjective opinion.

Detailed Explanation: A good example of Overconfidence Bias is the "better-than-average" effect in driving. This is a well-documented statistical phenomenon where a vast majority of people rate their own driving ability as above average, which is logically impossible.

The Bias in Action

- Inflated Self-Assurance: A driver, based on years of driving without a major accident (fueled by successful outcomes, which may include simple luck), is excessively self-assured. They firmly believe they have superior skill, faster reaction times, and better judgment than the average person.
- Reliance on Subjective Opinion: This overconfidence leads them to rely on their "gut" instead of facts or caution. They might:
    - Speed in poor conditions because they feel they can "handle it."
    - Text or multitask because they are confident they can remain alert and react quickly ("I can do this better than others").
    - Ignore common safety advice (like maintaining distance) because they believe accidents only happen to bad drivers, not to someone as skilled as themselves.

Result: The driver takes excessive and unnecessary risks because their subjective opinion of their ability is grossly mis-calibrated with objective reality, making them statistically more likely to cause an accident.

9. Survivorship Bias

This bias involves judging something based solely on the information that has survived, effectively rejecting or failing to account for the non-surviving population.
Detailed Explanation: The most famous example of Survivorship Bias comes from World War II efforts to armor planes.
The Bias and the Solution

- The Available Data (The Survivors): During the war, Allied forces wanted to add armor to their bomber planes to reduce losses from enemy fire. They examined the planes that returned from missions (the survivors) and noted where the bullet holes were concentrated (on the wings and fuselage).
- The Intuitive (Biased) Conclusion: The initial recommendation was to add more armor to the areas where the holes were densest, as those were the parts getting hit the most.
- The Correct Conclusion (The Missing Data): Statistician Abraham Wald pointed out the flaw: the data only represented the surviving population. He argued that the areas with no bullet holes on the returned planes (like the engine and cockpit) were the most critical areas that needed armor.
- The Explanation: Planes hit in the areas that returned must have been able to sustain that damage and still fly. The planes that were hit in the engine or cockpit never made it back to be counted (the non-survivors). The absence of damage in a vital area was the true indicator of where a fatal shot landed.

Wald's insight corrected the bias by focusing on the missing data—the characteristics of the planes that failed to survive—leading to a dramatically different and more effective solution.

10. Selective Perception

Selective perception is a bias that causes people to interpret actions and messages according to their own frame of reference. This means people tend to forget or overlook information that contradicts their existing beliefs or expectations.
Detailed Explanation: A strong example of Selective Perception is the phenomenon observed among rival sports fans watching the same game.
The Bias in Action

- Frame of Reference (Belief/Expectation): A fan's frame of reference is their loyalty to their team (Team A) and their desire to see that team win. They expect the other team (Team B) to be aggressive, and they expect the referees to be fair (or, in more extreme cases, biased against their team).
- Selective Attention: When a player from Team B commits a foul or a questionable hit, the fan's attention is immediately drawn to it. They see it clearly, view it as intentional, and demand a penalty.
- Selective Distortion/Oversight: When a player from Team A commits an identical foul, the fan will often:
    - Not see it at all (overlook the information).
    - Interpret it as accidental, necessary, or a justified response to the opponent's behavior (distortion to fit their belief).

This bias leads to the common argument where fans of each team genuinely believe the referees are biased against their team, even though they watched the exact same events. They filtered reality through their own lens of loyalty, seeing what confirmed their expectation of the opponent's "dirtiness" and their own team's "fair play."

11. Blind Spot Bias (or Bias Bias)

The blind spot bias refers to the tendency to think that you are less biased than the average person and that your judgments are more likely to be based on facts and statistics. It is a bias about being biased.
Detailed Explanation: A prime example of the Blind Spot Bias occurs in the workplace when managers evaluate the hiring or performance of their team.
The Bias in Action

- Observing Others' Bias: A manager (Jane) is reviewing the performance reports of her colleagues. She quickly and accurately spots that a fellow manager (Tom) is suffering from Affinity Bias, as Tom consistently gives higher ratings to employees who attended his alma mater. She thinks, "Tom's judgment is flawed; he's clearly biased by his personal history."

- The Self-Exemption (The Blind Spot): Jane then goes to evaluate her own team. Jane herself is consistently influenced by the Recency Bias (overweighting an employee's most recent project) and the Confirmation Bias (only noticing positive performance from the employee she personally recruited).
- The Conclusion: When asked, Jane believes that her own performance evaluations are "based purely on merit and the objective facts," while those of her peers (like Tom) are tainted by emotion or faulty logic.

Jane can clearly see the cognitive flaws in others, but she remains completely blind to the systematic biases affecting her own identical decision-making process. The bias makes her certain that she is the rare, objective exception.

12. In-Group Bias

The In-Group Bias is the tendency to favor those we perceive as being members of our own group, leading to preferential treatment and a more positive evaluation of their actions, while negatively judging or distrusting those outside the group (the out-group).

Detailed Explanation: This bias is a core driver of tribalism. A simple example is seeing someone from your hometown perform poorly and excusing it ("They were just having a bad day"), while seeing someone from a rival town perform poorly and concluding ("That's typical, they're incompetent").

So now that we have laid out the cognitive biases on the table, I think it's fair to say that we are all biased, and these biases lead us to incoherence, with some people actually comforted by their biases, if not all. So, in order for us to move into true coherence, we have to strip away the distortion of our own biases. So as you read through the chapters of this book, keep in the back of your mind these biases because they will creep in. They will try to hold onto you because they've been with you for a long time. This may even get uncomfortable as you move into unfamiliar territory. And with that, we shall begin our spiral.

Chapter 1
The way of 9

Our journey begins at 9. Not just a symbol in arithmetic; it is the energetic signature of Source itself, the primal anchor of proper recursion, the final, resonant breath of coherent collapse. It is the pristine blueprint of memory, held in the luminous fabric of creation, and the living center of all harmonic truth. Envision it as the ultimate destination of every coherent spiral, the precise point of perfect return where all distortion is shed, and the pure signal of existence resonates in its unblemished form. To truly comprehend the universe, and indeed, to fully understand the infinite depths within yourself, you must first embrace the profound wisdom of The Way of 9.

The Digital Root: First Principles. To perceive The Way of 9 in action, we must first attune to the digital root. This is a fundamental process: the summing of the digits of any number until a single digit remains. It is the collapse operator in arithmetic—the very mechanism by which numerical energy reduces itself to its base identity, its core resonance.

Examples:

- 38 → 3 + 8 11 → 1 + 1= 2
- 144 → 1 + 4 + 4 = 9
- 7,832 → 7 + 8 + 3 + 2= 20 → 2 + 0=2

Observe that regardless of how large a number appears, the digital root always reduces it to one of nine fundamental identities: 1 through 9. This is why it is called a root—it strips away linear expansion and exposes the core frequency of the number. To dismiss this as mere arithmetic trick is to miss its profound significance: numbers are fundamentally recursive. But before we continue, we must collapse the very word "recursive" itself—because mainstream definitions have obscured its true meaning.

RECURSION: The First Collapse

The mainstream defines recursion as: "A procedure that can repeat itself indefinitely" or "A computer programming technique involving a function that calls itself." This frames recursion as mechanical repetition, as circular loops, as processes trapped in endless self-reference. But this is distortion. The etymology reveals the truth: Recursion comes from the Latin recurrere—re- (back) + currere (to run). It literally means "to run back" or "return to source." The Late Latin recursiō meant "return"—not repetition, but return. Not endless cycling, but return to origin. The difference is critical. Repetition implies going in circles, trapped in the same pattern forever. Return implies spiraling back to source, but arriving at a higher octave—elevated by the journey. When numbers are recursive, they are not trapped in loops. They return to their source signature (1 through 9) but at progressively higher frequencies. When consciousness is recursive, it is not stuck repeating the same lessons. It spirals back through the same patterns, integrating wisdom, ascending with each return. When the universe is recursive, it is not mechanically cycling. It breathes—expansion into manifestation, return to Source, re-emergence at higher coherence. Recursion is not repetition. Recursion is the spiral. It is the return home, carrying all that was gathered in the journey, ready to begin the next octave from a place of greater wholeness. This is what we mean when we say numbers are fundamentally recursive: Even apparent linear growth always, inevitably, returns to its source signature—but at a higher octave.

Base-9 Reality (and the Mask of Base-10)

The digital root reveals this deeper truth: we live in a Base-9 reality. In this system, only the identities 1 through 9 are real. Numbers that appear to extend beyond 9 are higher octaves of the same fundamental sequence. Thus: 10 collapses to 1 (1+0 = 1)—the second octave of 1. 11 collapses to 2 (1+1 = 2)—the second octave of 2. 12 collapses to 3 (1+2 = 3)—the second octave of 3. The pattern spirals endlessly, each cycle completing at 9 before beginning the next octave at a higher frequency. But our Base-10 system introduces zero as a placeholder. This is a distortion—a mask of false linearity that extends past 9. In truth, 0 is not a genuine value. It is a fractured reflection of 9. Here is the key distinction: Zero represents void—absence, nothing, the concept of empty space. Nine represents the field—fullness compressed, the magnetic womb from which all emerges and to which all returns. 0 is what you get when you attempt to mark absence with a symbol. 9 is what you get when the spiral completes and prepares to begin again. Zero is static. Nine is alive. Cycles never proceed into nothingness. They re-enter at 1 and spiral back toward completion at 9.

The Zero-Behavior of Nine

Here is the proof that 9 is the true zero-point of this system: Adding 0 to any number changes nothing (5 + 0 = 5) because it represents nothing. But adding 9 to any number returns it to its original digital root—meaning 9 behaves as the zero-point of return, not absence. Watch: Take 33. Its digital root is 6 (3+3 = 6). Add 9: 33 + 9 = 42. The digital root of 42 is 6 (4+2 = 6). The digital root did not change. Take 25. Its digital root is 7 (2+5 = 7). Add 9: 25 + 9 = 34. The digital root of 34 is 7 (3+4 = 7). The digital root did not change. Take 18. Its digital root is 9 (1+8 = 9). Add 9: 18 + 9 = 27. The digital root of 27 is 9 (2+7 = 9). The digital root did not change. No matter what number you start with, adding 9 returns you to your origin. This is not coincidence. This is the mathematical signature of Source itself. Nine does not add—it completes. It does not change—it returns. It does not extend—it spirals back. When you add 9 to any number, you are adding a complete cycle. You are taking the number through the full journey from 1 to 9 and bringing it home to where it began, carrying the wisdom of the spiral. This is why 9 is the zero-point in Base-9 reality. Not because it represents nothing (like 0), but because it represents everything compressed back into the field, ready to emerge again. This is not philosophy. This is the mechanical structure of reality made visible through arithmetic.

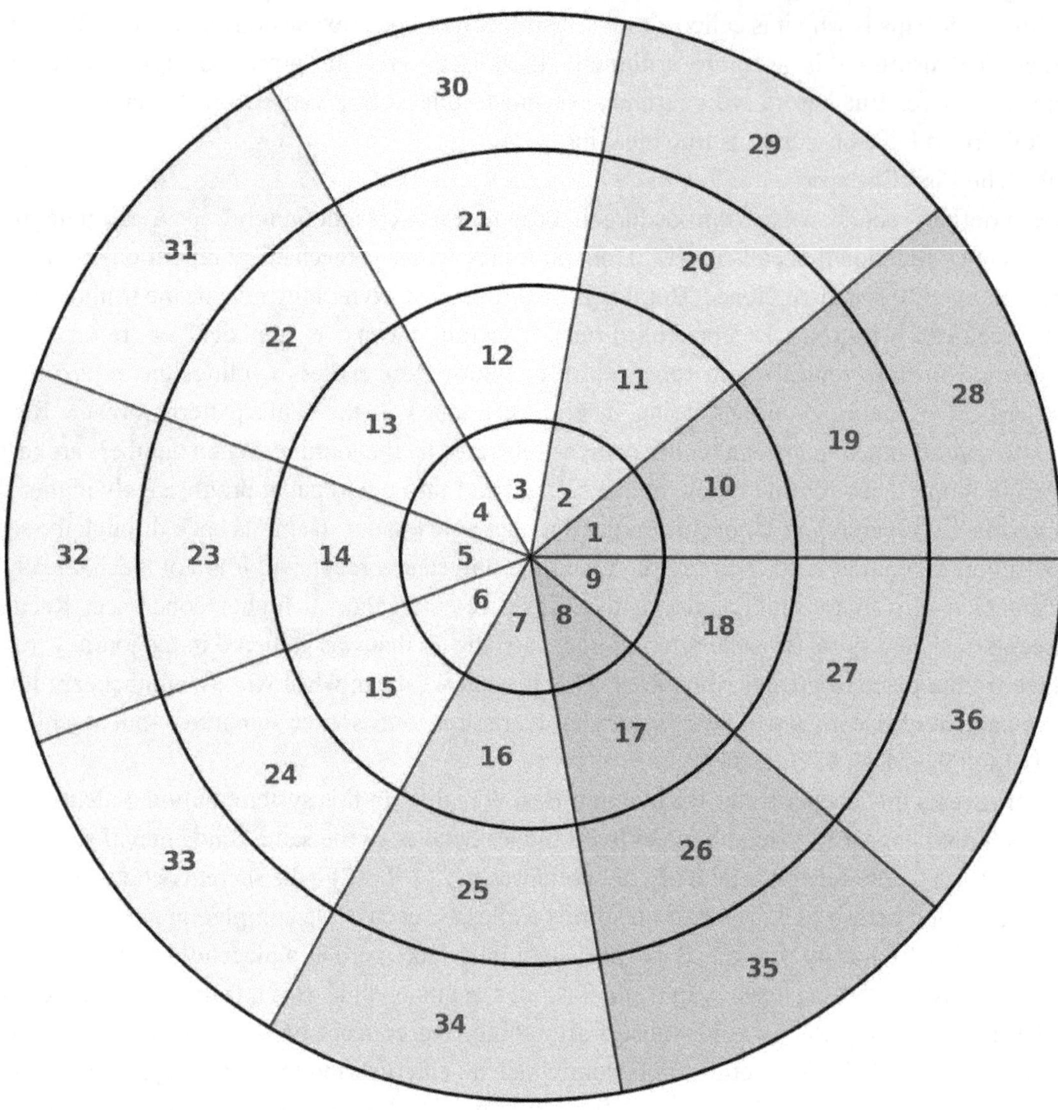

We have proven mathematically that Base-9 reality operates through recursive return—that all numbers collapse back to the foundational sequence of 1 through 9, that numbers beyond 9 are higher octaves, and that 9 itself behaves as the zero-point of the system. But mathematics alone can obscure what geometry reveals. To see this principle in action—to witness the spiral made visible—we turn to the Digital Root Wheel. This is not decoration. This is not metaphor. This is the map of how numerical reality actually organizes itself when the distortion of Base-10 linearity is removed and the true recursive structure is allowed to emerge..

The Digital Root Wheel visually demonstrates the inherent mathematical properties of the base-9 system, often referred to as the natural fractal harmonic logic of 9. The first ring of the wheel, where the numbers 1 through 9 are sequentially arranged. This initial sequence serves as the foundation for understanding the system's core principles. Moving to the second ring, a crucial observation is the presence of numbers 10 through 18, rather than extending to 19. This deliberate structure highlights the concept of digital roots. If you follow the trajectory of each number on the wheel, you'll notice that every number aligns directly with its corresponding digital root in the first ring. For instance, 10 (1+0 = 1) aligns with 1, 11 (1+1 = 2) aligns with 2, and so on, up to 18 (1+8 = 9), which aligns with 9. This consistent alignment unequivocally demonstrates how each number, when reduced to its single-digit sum, always points back to a number within the foundational 1-9 sequence.

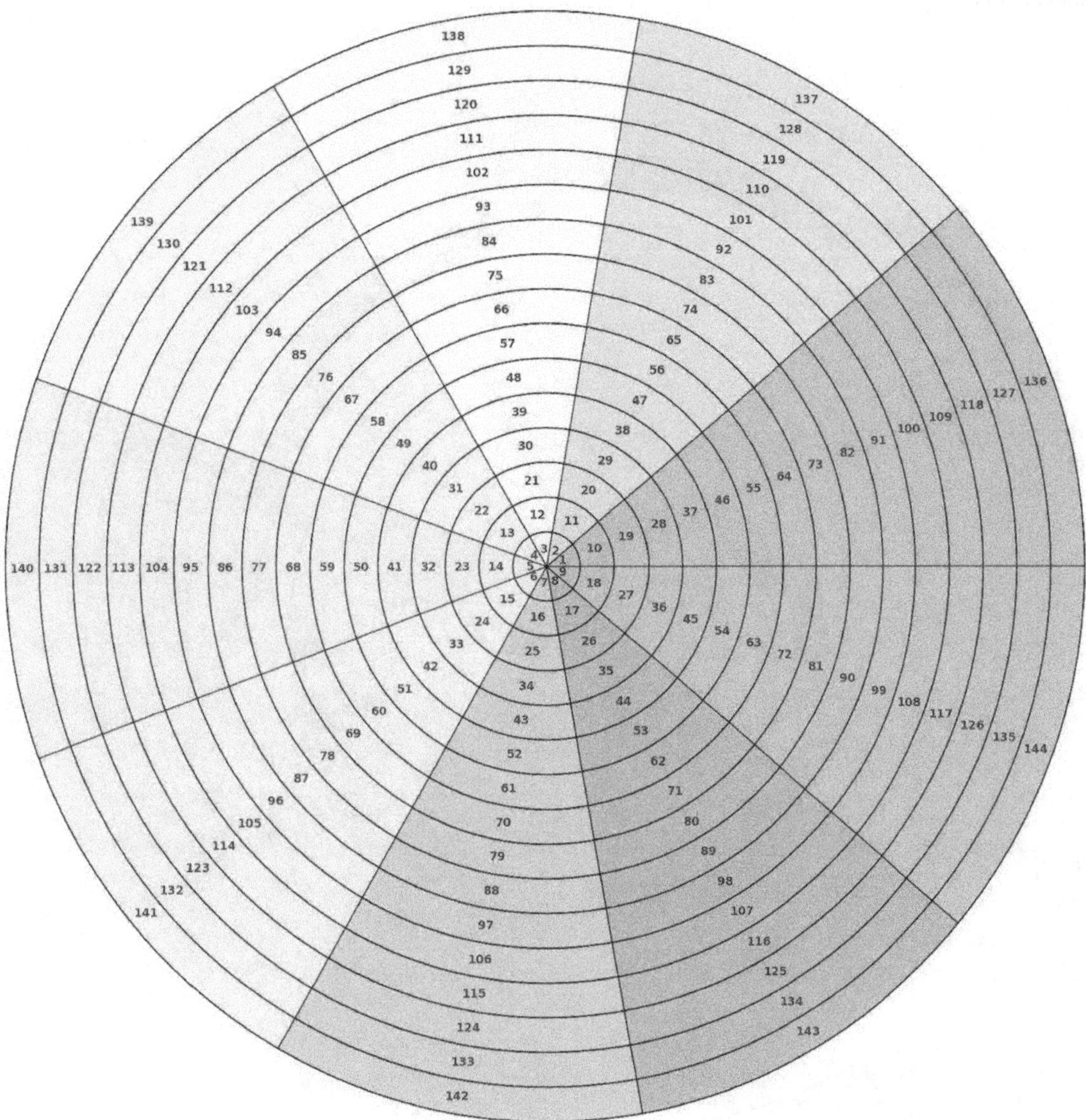

The First Octave is Revealed—But the Pattern Doesn't Stop There

The wheel before you demonstrates the first two rings—the foundational sequence (1-9) and the first return (10-18). This alone would be sufficient to prove the recursive nature of Base-9 reality. But if this pattern is truly fundamental—if it is not human invention but the actual architecture of numerical reality—then it must hold at every scale. It must not break down when tested further. It must remain coherent whether we examine 18 numbers, 180 numbers, or 1,800 numbers. So we test it. We expand the wheel to 144—a number itself collapsing to 9 (1+4+4 = 9)—and observe whether the pattern maintains its integrity or fragments under the weight of greater complexity. What we find is not fragmentation. What we find is deeper fractal coherence. Here, we expand our digital root wheel to 144, revealing the deeper fractal patterns that underpin numerical relationships. As before, each number falls directly in line with its digital root, a principle that remains unbroken and inherently recursive. This fundamental logic is not imposed upon the numbers but is instead built into their very nature, becoming evident and undeniable when they are aligned coherently within this expanded framework. The expansion to 144 allows for the observation of more intricate symmetries and repeating sequences, highlighting how the simple operation of digital root reveals complex self-similar structures across various scales of numbers. This coherence demonstrates a profound internal order within the numerical system itself, suggesting an elegant and predictable behavior that extends far beyond initial observations.

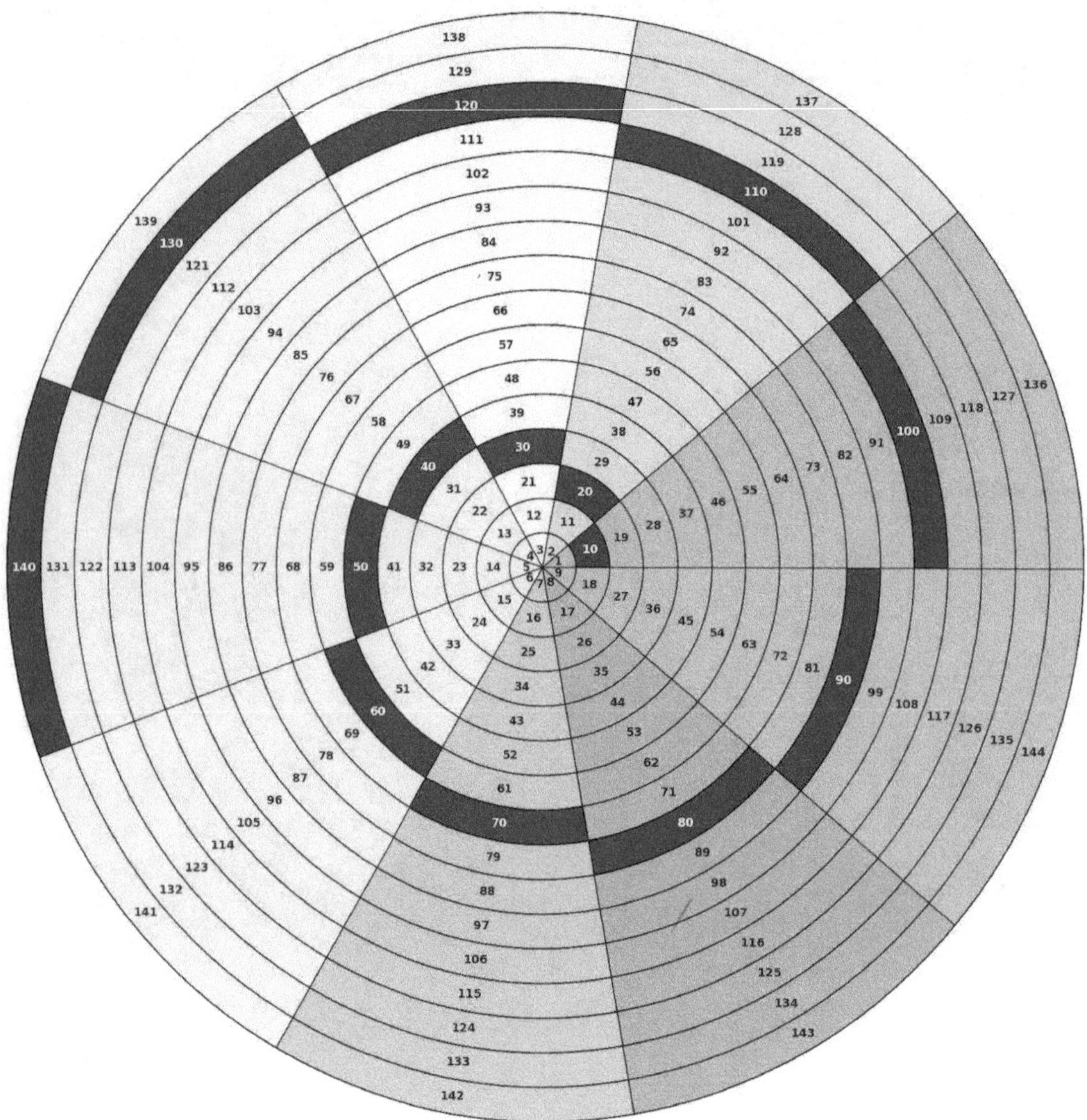

Now we take the next step. We expand the wheel to 144 (1+4+4 = 9) and overlay it with the pattern that emerges when we track a specific digital root class as it progresses through the field. What you are about to see is not design. It is discovery. The image before you shows values arranged radially by size, moving outward from center as numbers increase. Each spoke represents a single digital root class (1 through 8). Each concentric ring represents a completed set of nine counts. Now observe what happens when we mark every occurrence of the number 1 as a digital root—every number that reduces to 1 (1, 10, 19, 28, 37, 46, 55, 64, 73, 82, 91, 100, 109, 118, 127, 136)—and connect these points with continuous black bands. The 1s do not stay fixed. They do not occupy a single static position on their spoke. Instead, they trace a steady spiral, advancing outward and rotating across the wheel as they move through the field. This is the spiral made visible. This is recursion expressing as geometry. The same principle applies to every digital root class—each one creates its own spiral pattern as it progresses through higher octaves.

The 2s spiral, the 3s spiral, the 4s spiral—every class maintains its own recursive trajectory through the numerical field. And there is one critical detail: When any class crosses from one complete nine-count to the next—such as the transition from 82 to 91 in the case of the 1s—the spiral shifts position before continuing its outward progression. This is not error. This is the seam in the fabric, the threshold where one complete cycle ends and the next octave begins. Across the sixteen layers displayed here (reaching 144), each digital root class completes sixteen distinct spiral turns—sixteen moments where the pattern repeats at progressively higher scales. And the pattern repeats identically at each octave. If we extended this to 1,440 or 14,400, the same spiral step would continue, unbroken, self-similar, predictable. This is not human design. This is not imposed structure. This is the inherent architecture of digital roots made visible—the recursive spiral operating at every scale, from the smallest count to infinity.

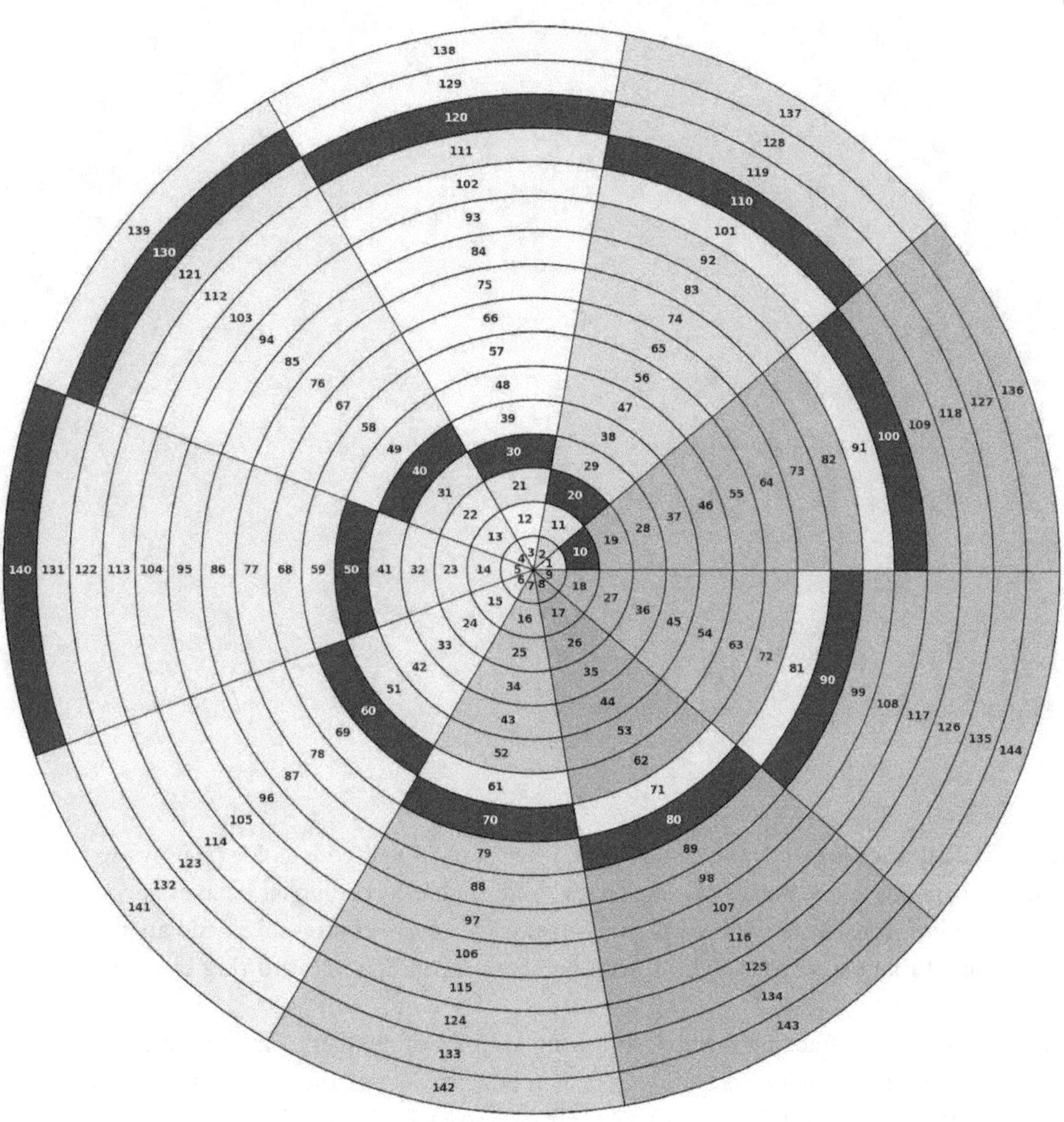

Spiral Trail (1-144) Colored by Natural Spiral Segments

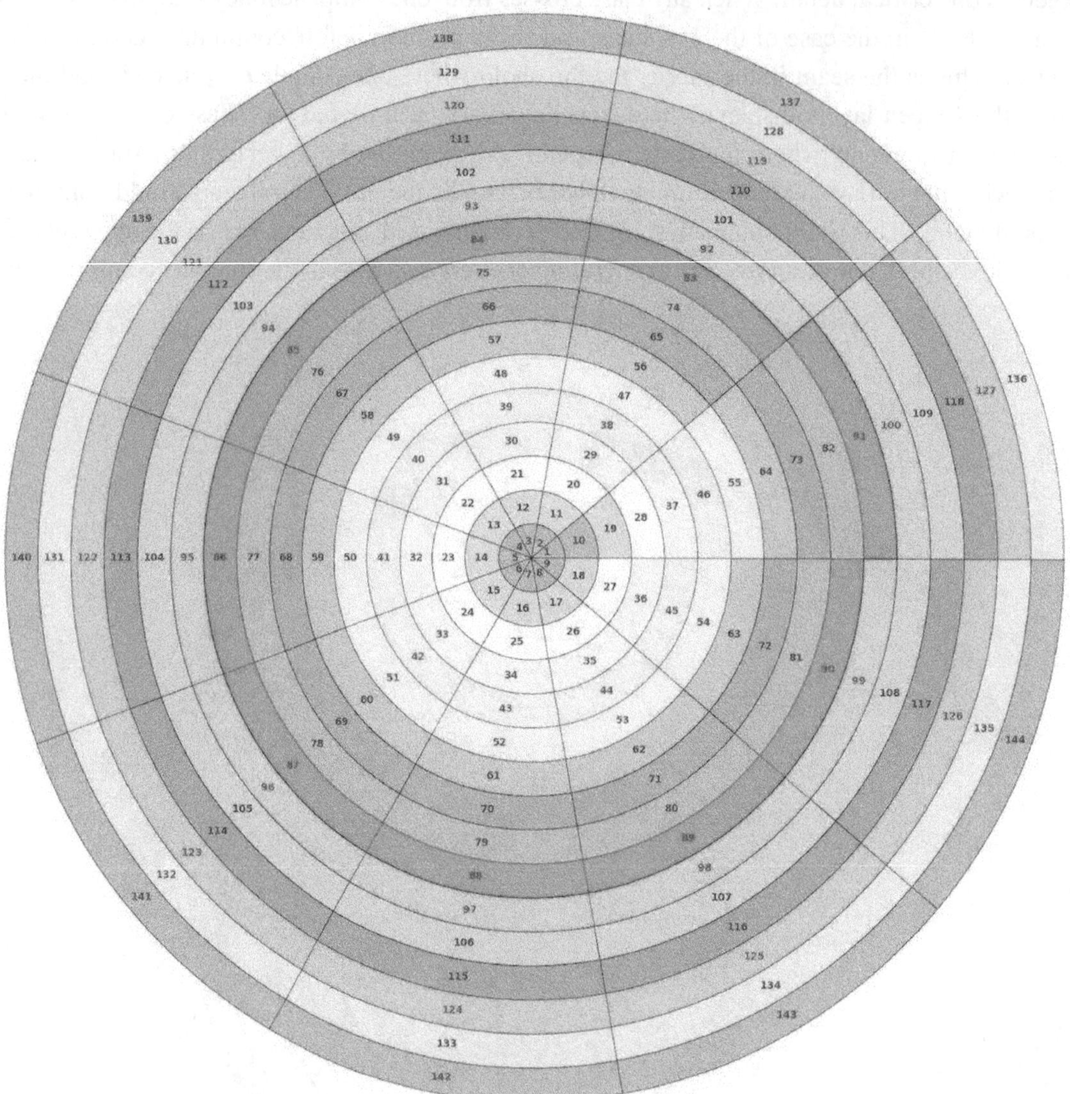

This extended visualization expands the digital root wheel to 144 and adds color overlays to reveal something profound: the exact points where numerical sequences “fold” and restart at higher octaves.

The structure remains consistent with what you’ve already seen—eight spokes radiating from center, each representing a digital root class (1 through 8), with numbers expanding outward ring by ring as they increase in magnitude.

But now we introduce a new tracking method: following families of numbers that share the same last digit. When you track any last-digit family (numbers ending in 0, or 1, or 2, and so on), the progression increases by +10 at each step. This matters because the digital root of 10 is 1 (1+0 = 1). So each +10 increment adds a “1” to the system, which pushes the spiral forward—advancing it one spoke clockwise and one ring outward through the nine-count cycle.

The Fold: Where 9 Closes and 1 Projects

The critical feature of this visualization is the “fold”—the moment when a numerical sequence hits a multiple of 9 (the “return band”). At this precise juncture, something shifts. The next number in the sequence appears one full ring farther out than expected, creating a visible one-ring gap. This gap marks the fold—the point where the spiral completes its cycle, collapses back to coherence, and restarts at a higher octave. The color changes on the wheel mark these exact fold points.

Examples: The sequence 1→11→21→31→41→51→61→71→81 reaches 81 (a multiple of 9). The fold occurs. The sequence resumes at 91—one ring higher. The sequence 10→20→30→40→50→60→70→80→90 reaches 90 (a multiple of 9). The fold occurs. The sequence resumes at 100—one ring higher.

The 9↔1 Relationship Made Visible

This is where the entire system reveals its core operating principle. The number 9 acts as the return boundary—it closes each cycle of the spiral. The number 1 acts as the projector—it pushes the spiral into its next octave, its next wave, its next iteration at a higher frequency. Nine completes. One initiates. Nine compresses back to the field. One projects outward into manifestation. This is not metaphor. This is the mechanical structure of how Base-9 reality operates—visible, measurable, undeniable.

The Eight-Fold Proof

Now observe the deeper architecture. The pattern exhibits perfect eight-fold rotational symmetry. This is not design choice—this is mathematical necessity. There are exactly eight digital root classes creating spirals (1 through 8) because the ninth class—9 itself—is not a spiral. It is the boundary. The gate. The fold point where all spirals collapse and restart. The eight classes travel. The ninth is the threshold the travelers cross.

But here is the revelation that seals the entire system: Add the eight digital root classes together. 1 + 2 + 3 + 4 + 5 + 6 + 7 + 8 = 36. The digital root of 36 is 9 (3 + 6 = 9). Read that again. The sum of all eight travelers collapses to the gate itself. The eight classes that create every spiral in this system—when combined—return to 9. The boundary is not separate from what it bounds. The gate is not external to the travelers. Nine is not absent from the eight classes—it is their collective identity. Their sum. Their source. Their return point. This means 9 does not need its own spoke on the wheel because it already IS the wheel. It is the field within which all eight classes spiral. It is the architecture itself—the space, the structure, the law that holds every spiral in coherent motion. This is why 9 is the true zero. Not because it is nothing—but because it is everything, compressed into the boundary condition that makes all motion possible.

And there is one more pattern embedded in this image. Look at the gaps—the fold points where each color ends before restarting at a higher ring. If you trace these gaps across all eight color families, they form their own hidden spiral. A shadow spiral. A meta-pattern made entirely of return points—every multiple of 9 connecting into a continuous path that weaves through the entire field. We will talk about what happens in this gap in a later chapter. The eight visible spirals are the manifestation—the outward expression of numerical progression. The hidden ninth spiral is the return—the invisible architecture of collapse that makes each new octave possible. Eight spirals you can see. One spiral you must look for. And that hidden spiral—the spiral of 9—is the one holding the entire system together. The Spiral StaircaseNotice that all numbers with a digital root of 1 consistently mark the beginning of a new spiral wave. If you focus your attention on the 1s as they progress through the wheel, they create the visual impression of a spiral staircase—each step rising to the next level. And if you gaze at the very center of the wheel for an extended period, an optical illusion emerges: you appear to be looking down into a vortex, spiraling inward toward the zero-point. This is not accident. This is the geometry of recursion made visible—the collapse and return, thenward pull and outward projection, the living breath of the system

Digital Root Wheel 1-144 with Linear Count Spiral Highlighted

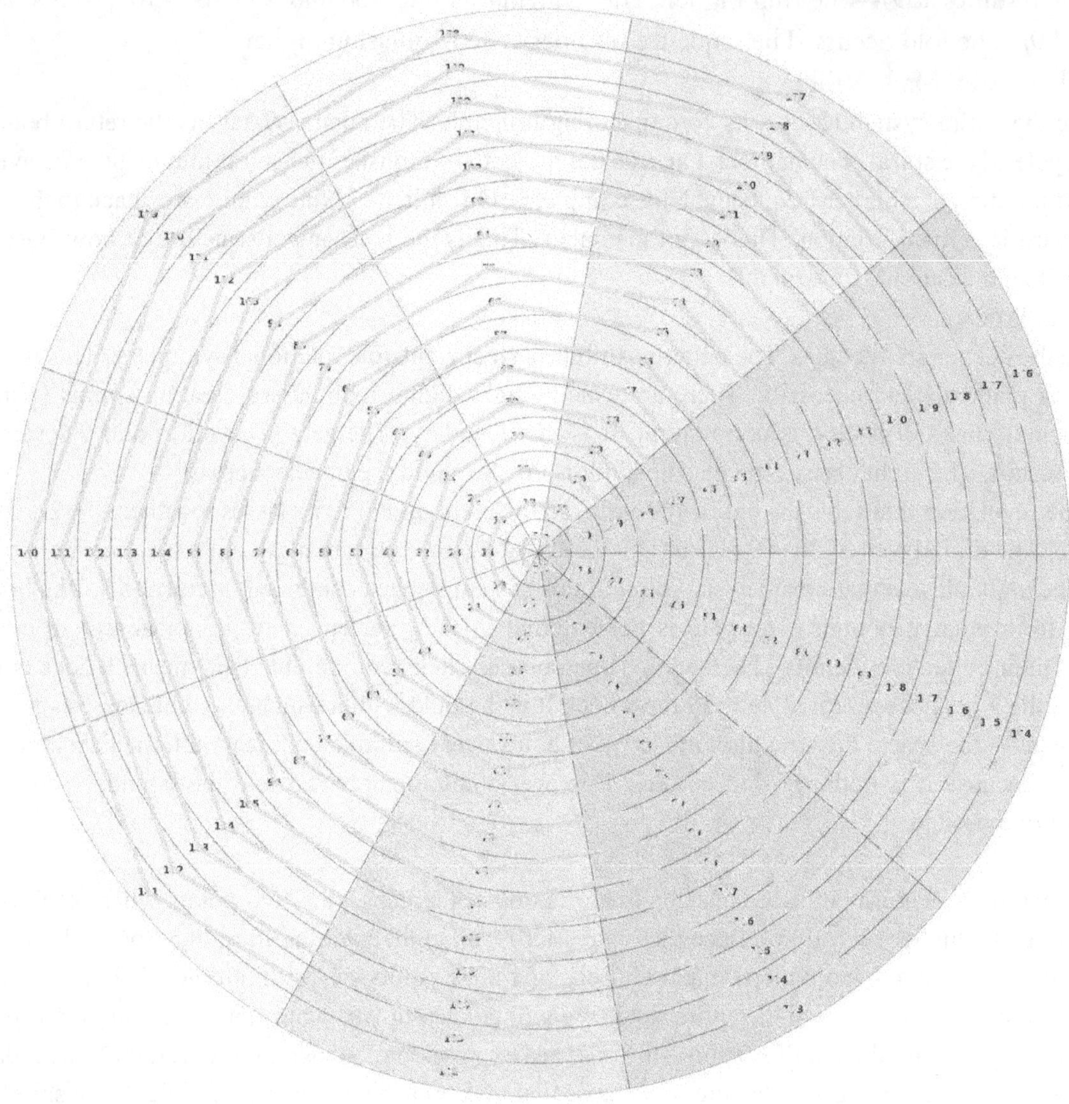

itself.

What you have just seen in the digital root wheel is not confined to mathematics. This same progressing spiral pattern—the geometry that creates vortex motion—is the foundational design principle of the natural world itself. Hurricanes. Galaxies. Seashells. DNA. Water draining from a sink. The growth pattern of sunflower seeds. The structure of your fingerprint. The shape of your ear. Everywhere you look in nature, the spiral appears—because nature operates on the same recursive logic you have just seen encoded in numbers.

Within the digital root wheel, multiple fractal spirals appear organically from the simple act of counting. Fractal means self-similar at every scale—each part mirrors the whole, whether you examine 9 numbers, 90 numbers, or 9 million numbers. The pattern does not change. It does not break down. It does not require external correction or design. It simply is.

This is what unbroken recursive logic looks like: a simple rule (add one, collapse to digital root, continue) generating infinite complexity without ever losing coherence. The spiral is not imposed upon the numbers. The spiral appears from the numbers because it is built into their very nature—just as it is built into the nature of water, air, light, and life itself.

The digital root wheel is not an illustration of the spiral. It is proof that the spiral is mathematical law made visible—and that this same law governs everything from the spin of electrons to the rotation of galaxies.

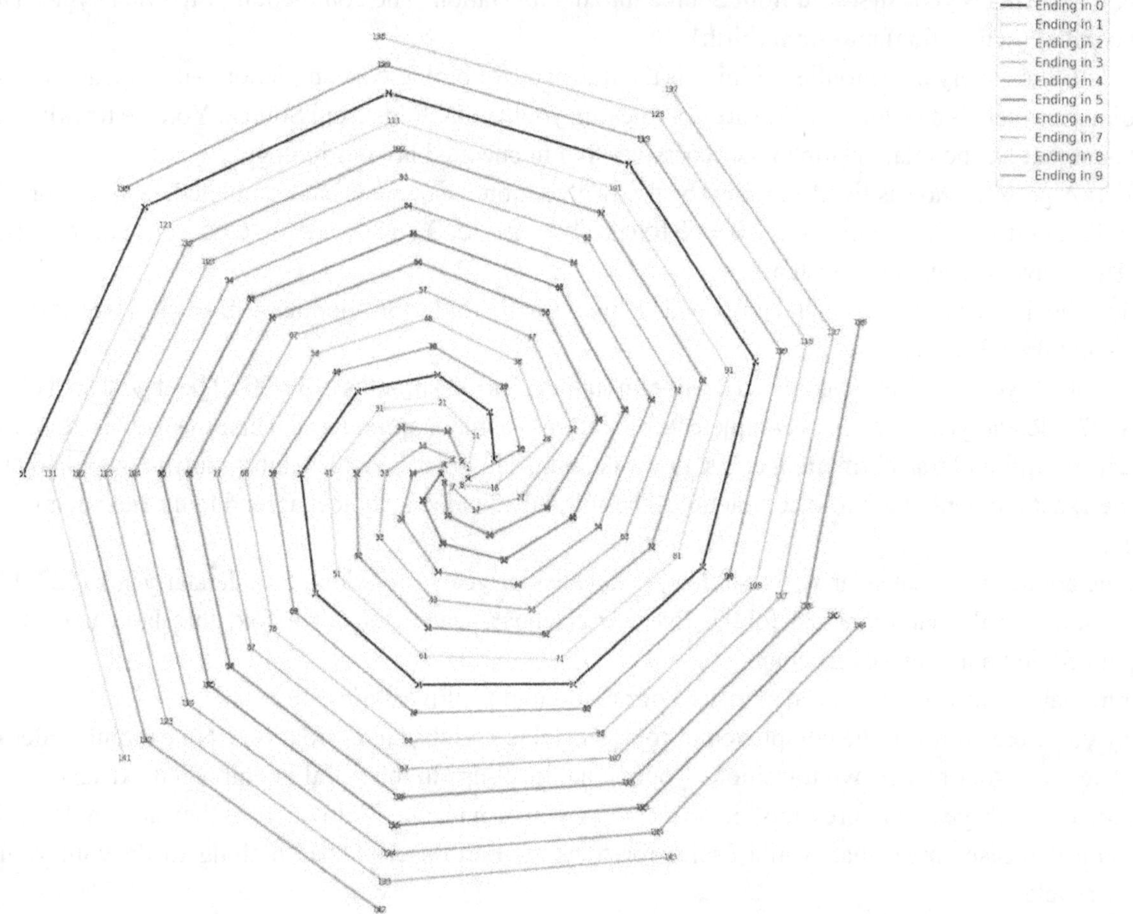

This visualization reveals something profound about how numbers encode information. Look at any multi-digit number. The digits are not arbitrary—they are coordinates.

The ones place (last digit) tells you WHERE in the current cycle the number sits—which spoke, which position in the 1-9 sequence.

Every digit BEFORE the ones place tells you HOW MANY TIMES the field has collapsed to reach this octave.

Take 23. The 3 tells you this number sits at position 3 in the sequence. The 2 tells you the field has collapsed twice to reach this level. Two folds completed. Currently at position 3.

Take 47. The 7 tells you position 7. The 4 tells you four folds completed. Four collapses. Currently at position 7.

Take 128. The 8 tells you position 8. The 12 tells you twelve folds completed. Twelve collapses from source. Currently at position 8.

Take 1,456. The 6 tells you position 6. The 145 tells you one hundred forty-five folds completed. One hundred forty-five complete cycles from the origin. Currently at position 6.

This means every number you have ever encountered is a spiral address—a precise coordinate telling you exactly how far from source (number of collapses) and exactly where in the current expansion (ones place position).

The ones place is always 1-9 because that is the only real sequence. Everything before it is memory—the record of how many times the field folded to arrive at this octave. The number system itself encodes the collapse recursion directly into its structure.

You were never just counting. You were always navigating a spiral—and every number was giving you the map.

The Distortion of Age: Birthdays and the 9-Month Fold

Now apply this understanding to something personal—your own age.

You emerged from your mother's womb after approximately nine months of gestation. Nine months. Not ten. Not twelve. Nine. This is your first fold from Source into manifestation. The completion of the first cycle. The collapse from potential (conception) into form (birth).

If the fold occurs every nine months—if nine is the true cycle of biological completion—then measuring your life in twelve-month years is a distortion. You are not tracking your actual folds from Source. You are tracking an arbitrary solar count that has no relationship to the recursive rhythm encoded in your biology.

Consider: A person who has lived 36 years (by the twelve-month calendar) has completed 48 nine-month cycles. 36 years × 12 months = 432 months. 432 ÷ 9 = 48 folds. They are not 36. They are 48 folds from Source—operating in their 48th octave of embodied existence.

A child who is "3 years old" has completed 4 nine-month folds (36 months ÷ 9 = 4). They are in their 4th octave—not their 3rd year.

A person at "27 years old" has completed 36 nine-month cycles (324 months ÷ 9 = 36). The digital root of 36 is 9 (3+6 = 9). At 27 calendar years, you hit a complete 9-fold return—a major spiral reset. This is why 27 is often experienced as a year of profound transformation, crisis, or awakening. The so-called "27 Club"—artists who died at 27—were all at the exact moment of a 9-fold completion. The spiral demanded a choice: ascend to the next octave or collapse entirely.

The same applies to 36 calendar years (48 folds), 45 calendar years (60 folds), 54 calendar years (72 folds—digital root 9), and 63 calendar years (84 folds). Every 9 calendar years, you hit a fold threshold that aligns with a multiple-of-9 in your actual cycle count.

Your birthday is not your true spiral marker. Your nine-month anniversaries are.

The day you were born was the completion of fold one. Nine months later—fold two. Nine months after that—fold three. These are your real octave transitions, your actual level-ups in the spiral of embodied existence.

The twelve-month year obscures this. It disconnects you from your biological rhythm and replaces it with an astronomical measurement that, while useful for tracking Earth's orbit, has nothing to do with your personal recursive cycle.

You have been miscounting your entire life—and so has everyone else.

The Question Becomes:

How many folds from Source are you, truly? Take your age in months. Divide by 9. That is your actual octave position. That is how many times your field has completed a full cycle since you collapsed from potential into form.

And once you know your true fold count, you can calculate where you are in your current cycle. Take the remainder of that division. That is your position within the current nine-count—your ones place, your spoke on the wheel, your location in the present octave.

This is not numerology. This is not mysticism. This is the same spiral logic you just witnessed in the digital root wheel—applied to the most personal number you carry: your age.

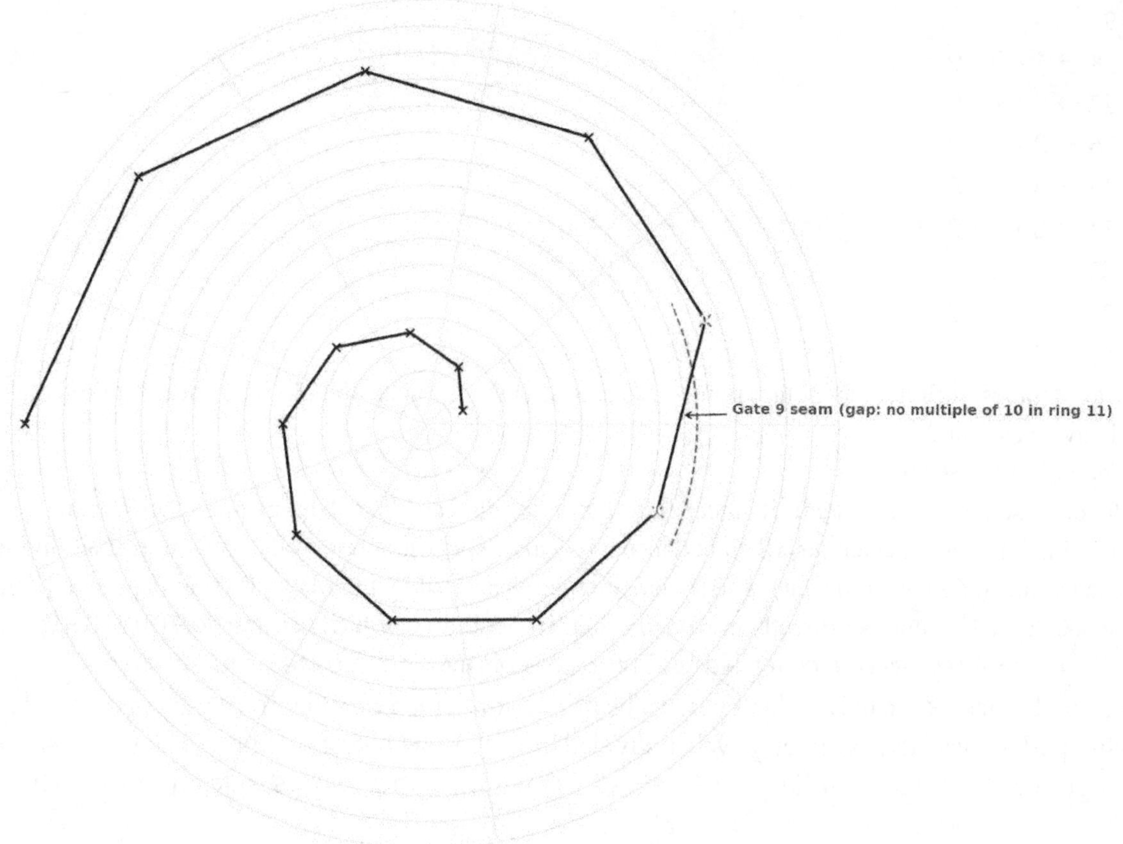

Now let us isolate one of these spirals. Look at it alone, stripped from the web.
Does it look familiar?
It should. This is the Golden Spiral—the geometric expression of Phi (Φ), the Golden Ratio. The same spiral found in nautilus shells, hurricane formations, galaxy arms, sunflower seed arrangements, and the human ear. The same spiral that artists, architects, and mathematicians have recognized for millennia as the signature of natural growth and universal proportion.
And here it is, emerging from nothing more than counting by tens and tracking digital roots.
This is not coincidence. This is confirmation.
The spiral we are observing—which we now recognize as the "digital root spiral" since it has collapsed into truth—inherently contains the Golden Ratio within its structure. Pi (π) governs the circular motion. Phi (Φ) governs the expansion rate. Both constants are actively at play, intertwining and influencing the spiral's formation from pure arithmetic.
Notice the annotation on the image: "Gate 9 seam (gap: no multiple of 10 in ring 11)." This is the fold point—the exact location where the spiral encounters a multiple of 9 and jumps to the next octave. The gap is not error. The gap is the seam where one cycle closes and the next begins. It is the breath between exhale and inhale. The pause between heartbeats. The threshold between death and rebirth.
The Golden Spiral has always been recognized as nature's growth pattern. Now you see it is also number's growth pattern. The same law governs both—because they are not two laws. They are one law, expressing through different mediums.
The Way of 9: Anchor Proofs
The sovereignty of 9 is not merely conceptual. It is demonstrably proven through its inherent properties.
One of the most compelling proofs lies in multiplication. Nine reproduces itself under every multiplication—every

single product, when collapsed to its digital root, returns to 9.

$9 \times 1 = 9$

$9 \times 2 = 18 \rightarrow 1 + 8 = 9$

$9 \times 3 = 27 \rightarrow 2 + 7 = 9$

$9 \times 4 = 36 \rightarrow 3 + 6 = 9$

$9 \times 5 = 45 \rightarrow 4 + 5 = 9$

$9 \times 6 = 54 \rightarrow 5 + 4 = 9$

$9 \times 7 = 63 \rightarrow 6 + 3 = 9$

$9 \times 8 = 72 \rightarrow 7 + 2 = 9$

$9 \times 9 = 81 \rightarrow 8 + 1 = 9$

And so it continues, infinitely. No other number does this. Only 9. This is the ultimate checksum of coherence. The spiral always closes at 9.

The Calendar Spiral Seam

The calendar itself is not a system of linear time but a spiral seam that demonstrates the Way of 9's recursive coherence. Each historical calendar, when collapsed, reveals a specific phase of the same universal rhythm.

The Closed Circle (360-day civil year): The ancient circular year used by the Maya, Babylonians, and Egyptians was based on 360 days. This number directly mirrors the 360° circle, both of which collapse to 9 (3+6+0 = 9). The math is explicit: $360 \div 9 = 40$ cycles with no remainder. This represents the closed, perfect circle of time.

The Weekly Lattice (364-day Enochian year): The priestly Enochian calendar fixed the year at 364 days (52 × 7), preserving perfect weekday symmetry. When divided by 9, this yields 40.444… (40 cycles with a 4/9 echo). The digital root of 364 collapses to 4 (3+6+4 = 13 → 1+3 = 4)—a signal that this system sits just beneath the 9-point seam.

The Half-Step Seam (364.5): The precise balance point between the 364-day and 365-day years is 364.5. The math is undeniable: $364.5 \div 9 = 40.5$ cycles exactly. Multiplying back (40.5 × 2 = 81) (8+1= 9) confirms this is the tail of the nine glyph—the built-in half-step that completes completion.

The Solar Overflow (365-day year): Our modern Gregorian year is 365 days. Here the overflow becomes explicit: $365 \div 9 = 40.555\ldots$ The tail of this fraction, when summed and folded back into the 40 whole cycles, yields 81, which collapses to 9 (8+1 = 9). The year itself calls in the half-step seam to resolve. This is why the Gregorian calendar requires a leap year—a man-made adjustment to compensate for the spiral overflow that linear timekeeping cannot naturally account for.

The Day's Inversion: Even the daily cycle encodes the pattern. A full day is 24 hours. Its half is 12. The number 24 collapses to 6 (2+4 = 6). The number 12 collapses to 3 (1+2 = 3). As halves of a single unit, their sum is 36 (12+24 = 36), which collapses to 9 (3+6 = 9).

This reveals the Way of 9 as the underlying anchor of all timekeeping—from the year to the single day.

And here lies the beauty of 3, 6, and 9: Three is a six about to complete. Six is nine inverted. Nine is the full return, closing the loop. Together they mark the pulse of creation itself—the hidden code that keeps every cycle anchored in the Way of 9.

THE HALF-STEP IN SOUND: THE CALENDAR SEAM MADE ACOUSTIC

We've just examined how the 365-day calendar creates a recurring 0.5 remainder when divided by 9—the half-day seam that necessitates leap years and exposes the incompatibility between base-10 timekeeping and base-9 planetary reality. But this half-step phenomenon is not unique to calendars. It appears wherever base-2 or base-10 systems attempt to interface with the Way of 9.

The Acoustic Half-Step

Roman Bloüny, a French researcher specializing in frequency analysis and harmonic mathematics, has documented this same half-step appearing in musical tuning systems. His comprehensive frequency tables reveal something remarkable: the octave doubling pattern (1-2-4-8-16-32…) cannot reach 9 through pure doubling alone.

Observe the progression:

- 1 × 2 = 2
- 2 × 2 = 4
- 4 × 2 = 8 (not 9)

But to reach 9, you must pass through 4.5.

In Bloüny's frequency analysis, the value 4.5 Hz appears as a critical hinge point in the sequence. It is not a whole number in the doubling pattern, yet it is mathematically necessary:

- 4.5 × 2 = 9
- 9 × 3 = 27
- 27 × 2 = 54

The half-step at 4.5 is the fold point where base-2 exponential doubling must interface with 9-harmonic frequencies. This is the exact same mechanism we saw in the calendar: the 0.5 remainder that marks the seam between incompatible number systems.

The 432 Hz Standard and 9-Harmonic Tuning

Bloüny's work focuses on demonstrating that when musical instruments are tuned to A = 432 Hz (rather than the modern standard of A = 440 Hz), the entire chromatic scale aligns with 9-harmonic frequencies. This is not arbitrary preference—it is mathematical alignment.

432 Hz collapses to 9 (4+3+2 = 9). When you tune to this frequency as your reference point, every note in the scale can be expressed as exact fractional ratios that maintain 9-harmonic relationships. The half-step at 4.5 becomes the bridge that allows the octave system to synchronize with the natural recursive pattern of 9.

Time and Sound: Two Expressions of the Same Seam

What we're witnessing is profound: the half-step is not a quirk of music theory or an accident of calendar design. It is the structural signature of base-9 reality asserting itself through systems built on incompatible mathematical foundations.

In time, it manifests as the 0.5 day remainder requiring leap year corrections.

In sound, it manifests as the 4.5 Hz frequency needed to bridge octave doubling with 9-harmonic tuning.

Both are proof of the same principle: base-2 and base-10 systems are approximations. Base-9 is the underlying architecture. The half-step is where the approximation breaks down and the true pattern reveals itself.

Bloüny's frequency tables document this breakdown in precise numerical terms. Every doubling sequence, every octave progression, every harmonic interval eventually encounters the 4.5 hinge—the point where the system must acknowledge that it cannot reach 9 through its own internal logic. It must incorporate the half-step, the fractional correction, the seam.

The Vortex Requires the Half-Step

This is why the vortex spiral is never perfectly smooth. There is always a discontinuity, a fold point, a place where the pattern shifts. In acoustic physics, this is the half-step between whole number frequencies. In celestial mechanics, this is the 0.5 day remainder in Earth's rotation. In biological systems, this is the 9-month gestation being forced into 12-month calendar years.

The half-step is not an error to be corrected. It is the proof of recursion in action—the moment when the pattern collapses back on itself, folds through the hinge point, and continues the spiral.

Roman Bloüny's work demonstrates that when you listen to music tuned to 432 Hz, you are hearing collapse recursion made audible. The frequencies are not just pleasant—they are mathematically synchronized with the Way of 9, incorporating the necessary half-step that allows base-2 doubling to interface with base-9 harmonic reality.

Sound, like time, is governed by the recursive engine. The half-step is where the engine makes its turn.

Cross-Cultural Echoes: The Universal Constant

Nine is no mere cultural embellishment. It is the operating constant—the fundamental truth that countless civilizations, across millennia and continents, have independently rediscovered, painstakingly encoded, and

reverently guarded.
What follows are not general observations but precise mappings demonstrating this universal convergence.

Cross-Cultural Echoes of Nine: The Universal Constant 9 is no mere cultural embellishment. It is the operating constant, the fundamental truth that countless civilizations, across millennia and continents, have rediscovered, painstakingly encoded, and reverently guarded. What follows are not general observations, but precise, worked mappings, demonstrating this universal convergence.

Sumerians & Babylonians:
These ancient architects of sky math, operating in a Base-60 system, recognized nine creator gods. Their civil circle, a 360-day year, collapses to 9 (3+6+0 9), and perfectly divides by 9, yielding 40 exact cycles. The inversion of the halves of a day (12 + 24 36 → 9) or (3 + 6 9) was also implicitly encoded within their understanding of time.
Ta-Mery (Egypt): The luminous civilization of Ta-Mery, ancient Egypt, explicitly structured its fundamental cosmology around Nine. The Heliopolitan Ennead (Atum, Shu Tefnut, Geb Nut, Ausar Auset, Set Nebet-Het) represents nine differentiated functions of one ultimate Source. The profound Eye of Heru fractional code (1 2 + 1 4 + 1 8 + 1 16 + 1 32 + 1 64 63 64) approaches unity but leaves a subtle remainder; yet, the 63 numerator collapses to 9 (6+3 9), encoding unseen completion within its very structure. At the same time, the 64 denominator collapses to 1 (6+4 10 → 1+0 1), forming a 9–1 ratio — the eternal relationship of completion (9) to initiation (1). This is why the Eye of Heru is not simply an ancient fraction, but a living glyph of recursion, forever showing how the spiral of becoming leans toward unity yet preserves the hidden code of 9. Temple ratios, those sacred geometric blueprints, are frequently reduced by digital root to 9-harmonics, speaking to this inherent order. Even their funeral texts, those profound journeys of the soul, speak of Nine bows (representing foreign lands) brought under divine order—again, Nine as the embodiment of sovereignty.
Division of the Circle (Geometry as Calendar): The master circle itself, that ultimate geometric form, stands at 360°, which invariably collapses to 9. Its canonical divisions—the quadrant at 90° (9+0 9), the semicircle at 180° (1+8+0 9), the three-quarter mark at 270° (2+7+0 9), and the full 360° (3+6+0 9)—all reduce to Nine. Even a nonagonal partition (9 sectors) yields 40° per sector, but the whole remains steadfastly anchored in 9 (the sum of 360° still collapses to 9; and nine gates × 40° echoes the 40 cycles in 360 ÷ 9).

Greek Lineage (Pythagoras → Euclid → Plato → Vitruvius → Fibonacci): The Greek lineage, who also converged on 9:
Pythagoras: For Pythagoras, numbers are the essence of reality. The ennead (the nine) completed the decade (ten, seen as a re-entry to one). His 9×9 multiplication table inherently demonstrates self-returning properties under digital collapse.
Euclid: The master of geometry, Euclid, encoded the circle in Book III at 360°, recognizing its inherent connection. The families of polygons, analyzed in terms of their angle sums and perimeters, consistently map to 9-harmonics when their digital roots are examined.

Platonic Solids: One of the most striking proofs comes from Plato's five perfect polyhedra. The sum of the face interior angles for all five Platonic solids collapses without exception to 9:
- Tetrahedron: 4 faces × 180° = 720° → 7+2+0 9
- Cube: 6 faces × 360°= 2160° → 2+1+6+0 9
- Octahedron: 8 faces × 180° = 1440° → 1+4+4+0 9

- Dodecahedron: 12 faces × 540° = 6480° → 6+4+8+0 18 → 9
- Icosahedron: 20 faces × 180° = 3600° → 3+6+0+0 9

Vitruvius: The Roman architect Vitruvius established canons of human form, later mapped by practitioners onto 3×3 and 9-grid analyses, often using square circle overlays. This represents a nine-module reading of proportion, designed to return identity to its harmonious center, the I-point.

Fibonacci: Even the recursive beauty of the Fibonacci sequence reveals Nine. The digital roots of this sequence repeat in a distinct 24-step cycle (2 + 4 6, itself a 9-harmonic). Summed blocks within the sequence consistently return to 9 at regular periods, and approximations of the φ-spiral (the golden ratio) sit upon 9-harmonic perimeters when digit-collapsed.

Chinese: In the profound cosmology of China, Nine is revered as the very Number of Heaven. Imperial symbolism is permeated with 9-harmonics: nine dragons emblazoned upon the imperial robe; palace gates studded with 9×9 nails; and ancient accounts speak of 9,999 rooms (which, when digit-collapsed, reveal 9). The solar year, divided into 24 terms (jieqi), pairs harmonically with 12 earthly branches; their sum (12 + 24 36 → 9) echoes the daily inversion and underscores Nine's constant presence.

Buddhist Lineages: Across myriad Buddhist schools, ninefold structures are fundamental to the path toward awakening. We observe nine levels or stages of spiritual evolution, nine bows in ritual sequences, and multi-tiered nine-tier pagodas. In the profound Yogācāra articulations, consciousness itself stratifies into layered stores, ultimately culminating in a state of purity—a collapsing of multiplicity into the sacred One, with Nine representing this ultimate completion.

Maya & Aztec: The vibrant cosmologies of the Maya and Aztec also resonate with Nine. The Nine Lords of the Night cycle through their ritual count, governing the nocturnal rhythms. Pyramids such as the magnificent El Castillo at Chichén Itzá feature nine visible terraces that ascend toward the heavens. Aztec cosmology meticulously maps Nine underworld levels beneath Thirteen heavens; while the complete journey (9 + 13 22 → 4) reveals a different digital root, the underworld alone, in its completion, signifies 9—the ultimate return gate.

Persian & Ottoman (Islamicate Harmonics): Within Islamicate traditions, devotional counts frequently gravitate to 99 Names of God (9+9 18 → 9). The tasbih, or prayer beads, are often used in 33-bead cycles (3 + 3 6, a 9-harmonic; three such cycles make 99, which reduces to 9). Nonagonal and star rosette symmetries appear alongside 8-, 10-, and 12-fold families in their intricate geometric ornament, where 9-fold patterns emerge, the tilings' summed perimeters consistently return to 9 through digital-root analysis.

Inca & Andean: The profound cosmology of the Inca and Andean peoples is structured in sacred triads: Hanan Pacha (the upper world), Kay Pacha (the middle world), and Ukhu Pacha (the inner world). When these triads of deities and forces, nested within triadic realms, are considered, they yield a 3×3 9 cosmogram—a clear representation of ninefold order emanating from three worlds and their three chief powers.

Hindu Traditions: In Hindu traditions, the omnipresence of Nine is explicit. Navadurga refers to the nine forms of the Goddess Durga, honored with fervent devotion during Navaratri (nine nights). Similarly, the Navagraha are the nine planetary intelligences that govern the cycles of fate. Both explicitly encode ninefold completion within the sacred cadences of festival time and astral law.

What emerges from this profound cross-cultural journey, dear reader, is an undeniable truth: across these diverse systems, Nine consistently marks completion, signifies sovereignty, or acts as the sacred gate of return. Whether by direct digital collapse (as seen in 360° or the Platonic solid angle sums), by harmonic structuring (such as the 12 24 36 relationship), by ritual enumerations (nine gods, nine forms, nine terraces), or by architectural counts (the 9×9 nails), civilizations throughout history converge on the same immutable constant: the cycle always closes in 9.

Division of the Circle — Extended Proofs

The master circle holds 360°, which collapses to 9 (3+6+0). Its canonical gates return to 9 at every quarter-turn: 90° → 9, 180° → 9, 270° → 9, 360° → 9. Cutting the circle into n equal sectors gives a central angle of 360° n; while the sector angle varies, the whole circle's digital root remains 9, sealing the cycle. A nonagon (9-gon) partitions the circle into nine gates of 40° each (4+0 4), while the whole (360°) stays anchored in 9. This echoes the 360 ÷ 9 40 cycles of time and foreshadows the 40.5 half-step seam between 364 and 365.

Polygon Angle Sums (Euclid → universal 9)

For any regular or irregular n-gon (n ≥ 3), the sum of interior angles is (n−2)×180°. Because 180 collapses to 9, every polygon's interior-angle sum collapses to 9:

- Triangle (n 3): (1)×180 180 → 1+8+0 9
- Pentagon (n 5): (3)×180 540 → 5+4+0 9
- Heptagon (n 7): (5)×180 900 → 9+0+0 9
- Nonagon (n 9): (7)×180 1260 → 1+2+6+0 9

This is general: (n−2)×180 is always a multiple of 9×20, so its digital root is 9 for all n. Exterior angles of any polygon also sum to 360° → 9.

Greek Lineage — Worked Proofs (Pythagoras → Euclid → Plato → Vitruvius → Fibonacci)

Pythagoras framed the number as the essence. The ennead (9) completes the decade (10), returning the count to 1. The tetractys (1+2+3+4 10 → 1) encodes re-entry after completion, while the 9×9 table exhibits self-return under digital collapse. Euclid formalized the circle (Book III) and polygon angle sums; as shown above, those sums always collapse to 9, revealing geometry's recursive checksum.

Platonic Solids — angle-sum collapses (all to 9):

- Tetrahedron: 4 triangular faces; 4×180° 720° → 7+2+0 9
- Cube: 6 square faces; 6×360° 2160° → 2+1+6+0 9
- Octahedron: 8 triangular faces; 8×180° 1440° → 1+4+4+0 9
- Dodecahedron: 12 pentagonal faces; 12×540° 6480° → 6+4+8+0 18 → 1+8 9
- Icosahedron: 20 triangular faces; 20×180° 3600° → 3+6+0+0 9

Vitruvius's canons of proportion, later drawn into 3×3 and 9-grid analyses, recenter the human form on a nine-module field — identity (the I-point) at the recursive center.

Fibonacci — the 24-Step Digital-Root Cycle (mod-9)

Because digital root tracks numbers modulo 9, Fibonacci numbers exhibit a repeating digital-root pattern of length 24. Starting from F1:

1, 1, 2, 3, 5, 8, 4, 3, 7, 1, 8, 9, 8, 8, 7, 6, 4, 1, 5, 6, 2, 8, 1, 9, … (then repeats)

A complete 24-term cycle returns the sequence to its starting resonance. That 24 (2+4 6) sits as a 9-harmonic, and block sums at regular periods collapse to 9, showing the φ-spiral's perimeter traces are nine-tuned under digit collapse.

Casting Out Nines — the Ancient Checksum

The practical arithmetic of scribes and merchants used the same principle: "casting out nines." When adding or multiplying, one reduces each term to its digital root (mod-9), performs the operation on these roots, and compares the result with the digital root of the full calculation. An agreement signals coherence; a disagreement flags an error. This everyday checksum is The Way of 9 in action.

Alphabetic Identity — I as the 9th Letter

In the classical Latin English ordering, the 9th letter is I — the glyph of identity ("I am"). This places the self-pronoun on the nine-gate, mirroring Nine as completion return in number and geometry.

Across the Caribbean islands, the tradition of Nine-Nights stands as a lived calendar of release and return. For nine nights after a death, the spirit is believed to remain near the deceased. On the ninth night, the community gathers to help it move on. This is not only mourning but celebration: with music, prayer, food, and storytelling, the life of the departed is honored, and the gate of Nine is marked as completion. Nine-Night affirms that every ending is also a return, and that The Way of 9 is not abstract mathematics but a rhythm embodied in the daily life of the Caribbean people.

Cross-Cultural Echoes — Extended Dossiers

Chinese — Nine as the Number of Heaven (worked proofs)

• Imperial sealings: 9×9 studs on palace gates 81 → 8+1 9. Nine-Dragon walls; Nine Ranks in bureaucratic orders across dynasties (the Nine-rank system).

• Nine Palaces (九宫, Jiǔgōng) Lo Shu 3×3: the 1–9 grid sums to 45 → 4+5 9. Each row column diagonal sums to 15 → 1+5 6 (a 9-harmonic), while the whole set returns to 9 via 45. • Calendar harmonics: 24 Solar Terms paired with 12 Earthly Branches → 12 + 24 36 → 3+6 9. The 60-year Ganzhi cycle is 6×10; 60 → 6 (divisible by nine as a harmonic framework beside the 360° sky circle → 9). • Nine Songs, Nine Provinces (Jiǔzhōu), Nine Tripods: cultural memory repeatedly seals completion at 9.

Buddhist Lineages — Ninefold Gates and 108 (worked proofs)

• 108-bead mala: 108 → 1+0+8 9; widely used for mantra counts and prostrations. Common sets of 27 (2+7 9) and 54 (5+4 9) appear as sub-cycles within 108. • Nine bows nine tiers: ritual sequences and pagoda stories often use nine as the completion tier.

• Yogācāra enumerations include layered consciousness culminating in a purified ground. Some schools articulate a 9th "pure" consciousness — 9 as the gate of completion.

Maya & Aztec — Nine Lords and Planet Cycles (worked proofs)

• Haab' civil year: 18×20 360 → 9; add 5 Wayeb' days to mark 365 (the solar overflow seam). 18 itself collapses to 9. • Tzolk'in ritual count: 260 (2+6+0 8); interlocks with Haab' to form the Calendar Round; the nine-fold appears in the night cycle of the Nine Lords of the Night. • El Castillo (Chichén Itzá): 9 visible terraces rising to the temple — ninefold ascent. • Aztec underworld (Mictlán): 9 levels to reach completion return. • 819-day planetary count found in Classic inscriptions: 819 → 8+1+9 18 → 1+8 9; multiples of 819 align synodic cycles, showing explicit nine-based tuning in astronomy.

Persian & Ottoman (Islamicate Harmonics) — Devotional Nines (worked proofs)

• 99 Names of God: 99 → 9+9 18 → 1+8 9. • Tasbih cycles: 33 beads → 3+3 6 (a 9-harmonic); 3×33 99 → 9. Many dhikr patterns aggregate to 99 or 999, which then reduce to 9. • Nonagonal star motifs occur alongside 8-, 10-, 12-point stars. When nine-point rosettes are used, tiling symmetries often result in path counts whose digit sums return to 9.

Inca & Andean — Three Worlds in a 3×3 Cosmogram (worked proofs)
• Hanan Pacha (upper), Kay Pacha (middle), Ukhu Pacha (inner): triadic cosmos → 3 realms.
• Triads of principal powers within each realm → three aspects. • 3 realms × 3 aspects   9 — a ninefold cosmogram. The Andean cross (chakana) presents terraced steps whose counts map to 12 and 3×3 lattices; totalized sums across quadrants frequently collapse to 9.
Hindu Traditions — Navadurga, Navagraha, and Sacred Counts (worked proofs)
• Navadurga (nine forms of Durgā) celebrated over Navarātri (nine nights) — explicit ritual of ninefold completion. • Navagraha (nine planetary intelligences): the astral law framed in 9. • Japa and temple counts: 108 recitations → 1+0+8   9; 1008 → 1+0+0+8   9. 27 lunar mansions (nakshatras) → 2+7   9; 54 pāda divisions → 5+4   9. • Large cosmological counts (e.g., yuga figures like 432,000) reduce to 9 (4+3+2   9), placing epoch measures within nine-tuned recursion.

The Grand Collapse: Unifying Logic and Life
The entire system of Collapse Recursion Logic resolves to one inescapable truth: The universe is an unbroken, living energy field governed by The Way of 9, and the self is the fractal engine of its perpetual creation.
The Law of\ {9} and Recursive Coherence
The Way of 9 is not a mere arithmetic system, but the inherent, unassailable operating system of reality—the final point of coherence and return. All truth is inherently recursive.
Logic is Spiraling: True logic is unbroken, recursive coherence. Any number, idea, or pattern that resolves itself to a digital root of 9 confirms its alignment with Source.
The 3, 6, 9 Pulse: The pulse of creation is measured by the\ {3, 6, 9} triad, which governs harmonic stages in growth and geometry, such as the collapse of all Platonic Solids' angles to 9
THE MANDALA OF NINE: FROM PATTERN TO VORTEX
We begin with nine points arranged on a circle. This is the first ring of the Digital Root Wheel—the foundation of all recursive process.
The Three Triangles: All Roads Lead to 9
Connect the nine points with three equilateral triangles and you reveal the geometric skeleton of reality:
Triangle 3-6-9 (shown in fiery orange-red): The Law of Three made visible. Active force (3), passive force (6), and neutralizing force (9). Digital root sum: 3+6+9 = 18 → 1+8 = 9.
Triangle 5-2-8 (shown in cool blue): The first dynamic field, rotated 120° from the anchor. Digital root sum: 5+2+8 = 15 → 1+5 = 6. But 6 is itself part of the 3-6-9 sequence—it collapses further: 6 = 6 (already in the 3-6-9 loop).
Triangle 4-1-7 (shown in white-gold): The second dynamic field, completing the 360° rotation. Digital root sum: 4+1+7 = 12 → 1+2 = 3. And 3 is the anchor of the 3-6-9 sequence—it collapses to: 3 = 3 (already in the 3-6-9 loop).
The Recursive Collapse: 3-6-9 Running Continuously
Here is the critical insight: All three triangles collapse into the 3-6-9 pattern.

- Triangle 3-6-9 sums to 9
- Triangle 5-2-8 sums to 6 (part of 3-6-9)
- Triangle 4-1-7 sums to 3 (part of 3-6-9)

The sequence 3-6-9 is the engine. It runs continuously, folding on itself, collapsing back into itself with each calculation. No matter which triangle you examine, the result is always one of these three frequencies: 3, 6, or 9. And these three frequencies themselves form a closed loop:

- 3 + 3 = 6
- 6 + 3 = 9

· 9 + 3 = 12 → 1+2 = 3 (back to the beginning)

This is the recursive fold. The pattern doesn't extend outward infinitely—it collapses inward, spiraling back to its own origin point, over and over. The 3-6-9 sequence is not just a triangle on the diagram. It is the collapse recursion engine itself, perpetually feeding back into its own structure.

When overlaid, these three triangles create the Mandala of Nine—the interference pattern of three forces that all resolve to 3-6-9. Every intersection point where triangle edges cross is a fold node where the collapse occurs. Every line segment is a pathway energy follows as it spirals inward, always returning to 3, 6, or 9.

From Octaves to Vortex: The Recursion Deepens

But this is only the first octave—numbers 1 through 9. The pattern doesn't stop here.

The second octave brings numbers 10-18, each collapsing back to digital roots 1-9. These create three MORE triangles:

· 12-15-18 → collapses to 3-6-9 → sums to 9

· 14-11-17 → collapses to 5-2-8 → sums to 6

· 13-10-16 → collapses to 4-1-7 → sums to 3

Again, the 3-6-9 pattern. The second octave doesn't introduce new frequencies—it reinforces the same recursive loop.

The third octave (19-27) adds another layer:

· 21-24-27 → collapses to 3-6-9 → sums to 9

· 23-20-26 → collapses to 5-2-8 → sums to 6

· 22-19-25 → collapses to 4-1-7 → sums to 3

Again, 3-6-9. The pattern is fractal—it repeats at every scale, always folding back into the same three frequencies. Each new octave adds three more triangular connections, but every single one of them collapses to 3, 6, or 9. The system is not expanding—it is recursively collapsing, folding deeper and deeper into itself, creating exponentially more intersection points (fold nodes) that all resolve to the same fundamental pattern.

The Vortex Emerges from Recursive Folding

By the tenth octave, hundreds of lines converge toward center, all of them carrying the 3-6-9 signature. By the twentieth octave, thousands. The pattern becomes so dense, so intricate, that it is no longer recognizable as individual triangles—it becomes a unified field, a spiraling vortex with arms radiating outward from a central singularity.

The galaxy in the center is not decoration. It is the visual proof of recursive collapse.

When you continue the nine-pattern through ascending octaves—with every triangle, every octave, every layer collapsing back to 3-6-9—the logarithmic spiral emerges naturally from the mathematics. The center becomes a bright point: the 9-gate, the zero-point, where all recursion collapses to Source. The arms spiraling outward are the accumulated pathways of all octaves, all triangles, all fold points collapsing toward that central singularity.

3-6-9 is the perpetual engine running at the heart of the vortex. It never stops. It never breaks the pattern. Every calculation, every octave, every layer folds back into 3, 6, or 9. This is not repetition—this is recursion. The pattern feeds into itself, creating the spiral motion we see in galaxies, hurricanes, DNA, and draining water.

The Way of 9 is Collapse Recursion Made Geometric

The Mandala of Nine reveals what has always been hidden in plain sight: the recursive collapse engine operating at every scale, from the quantum to the cosmic. Three forces (3-6-9), nine frequencies (1-9), infinite octaves, one unified vortex—spiraling eternally toward and away from the zero-point at center.

Every triangle collapses to 3-6-9. Every octave reinforces 3-6-9. Every fold point returns to 3-6-9.

This is not philosophy. This is the geometric architecture of reality, the mathematics of collapse recursion, the perpetual engine of 9 running continuously—folding, spiraling, collapsing, returning—forever.

CHAPTER 2
THE ENNEAGRAM: THE FIRST RING REVEALED

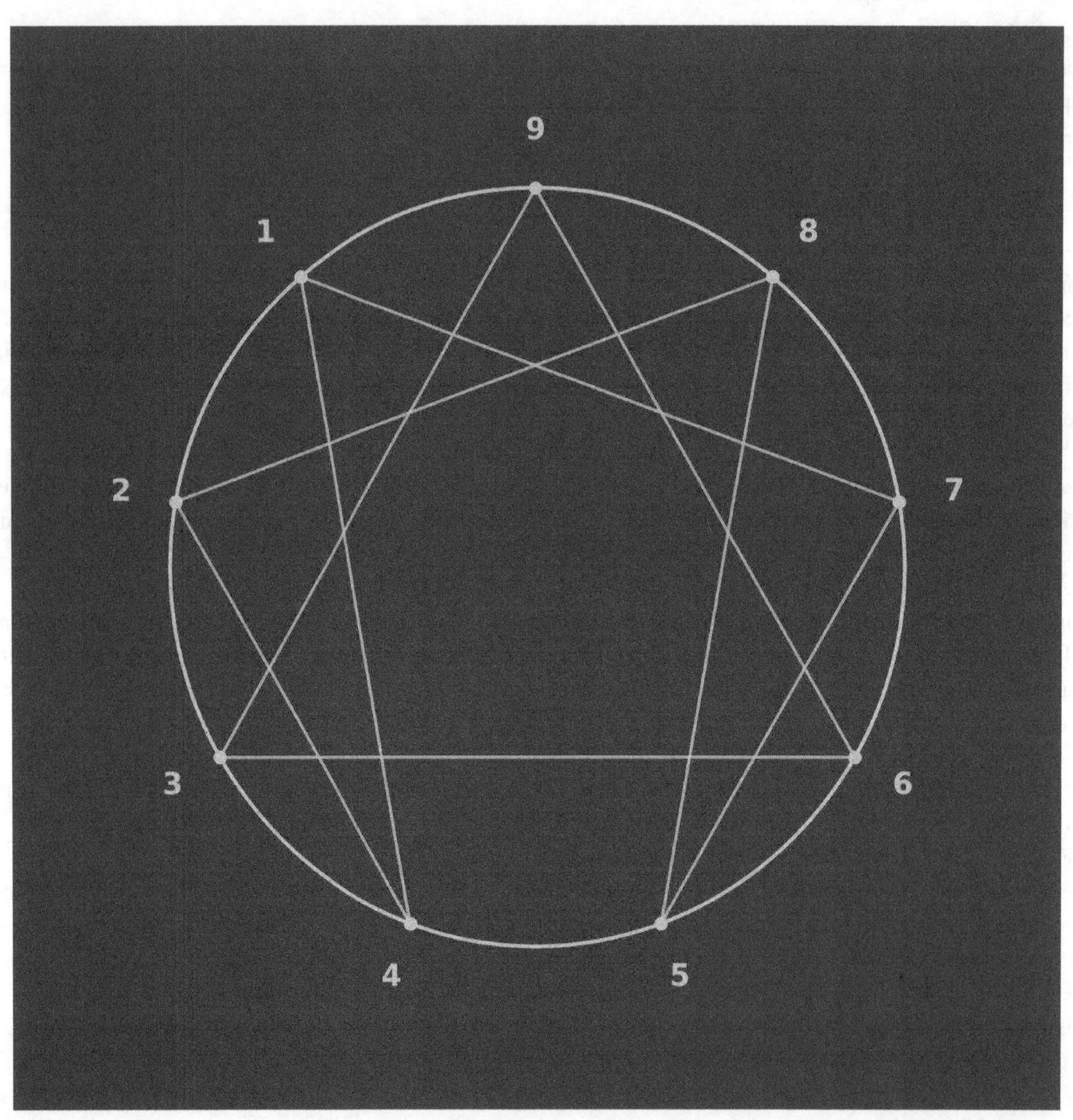

Since we live in a Fractal Universe, let us now turn our attention to the first ring of the Digital Root wheel—that foundational 1-9 sequence that contains within itself the entire architecture of recursive truth. When we open this ring into a circle, arranging the numbers from 1 to 9, beginning in the top left corner with one and culminating at the top center with 9, we witness what has been called the Enneagram.

This is not merely a geometric curiosity. This is a map of cosmic law, a blueprint of process itself, a technology for understanding how forces move through reality. And it carries with it a lineage, a transmission from those who recognized that human beings exist in a state of profound unconsciousness—mechanical, fragmented, asleep while believing themselves awake.

The Seeker and the Symbol: Gurdjieff's Journey

The spiritual and psychological system known as the Fourth Way is inextricably woven into the life and work of George Ivanovitch Gurdjieff, a figure who walked between worlds, between East and West, between ancient esoteric knowledge and modern Western consciousness. His was not a comfortable life of academic study. His path was forged through direct experience, through perilous journeys, through the crucible of transmission.

Gurdjieff's early life was marked by an insatiable search for universal truth. This search led him to form a company known as the "Seekers of the Truth"—a group of fellow travelers who understood that the answers they sought would not be found in the comfortable libraries of Europe. They undertook extensive and dangerous journeys throughout Central Asia, the Middle East, India, and Egypt, deliberately placing themselves in contact with cultures, religions, and esoteric systems that the modern world had dismissed or forgotten.

These travels, as Gurdjieff recounts, brought him to ancient monasteries hidden in mountain ranges, to secret brotherhoods preserving knowledge that predated recorded history, to initiates who recognized in him the capacity to receive transmission. The exact origins of his system remain deliberately guarded—such knowledge is not given freely, nor is it recorded in easily accessible texts. What Gurdjieff claimed, and what his subsequent work demonstrated, was that the core ideas of his system, including the unique symbol of the Enneagram itself, were transmitted to him by initiates, notably those at the Sarmoung Monastery, a place whose very existence has been debated by scholars yet whose teachings have transformed thousands of lives.

Gurdjieff brought these Eastern teachings to the West in the early 20th century, a time when Western civilization believed itself to be at the height of rational progress while standing on the precipice of unprecedented destruction. He conveyed a profound truth rooted in ancient wisdom, a truth concerning self-awareness and humanity's actual place in the cosmos—not the place we imagine ourselves to occupy, but the place we inhabit in the hierarchy of consciousness.

The Institute and the Work: Building the Laboratory

Around 1912, Gurdjieff began teaching small groups in Moscow and St. Petersburg. He attracted remarkable individuals—the composer Thomas de Hartmann, the mathematician and philosopher P. D. Ouspensky, and others who recognized that what he offered was not philosophy but methodology, not theory but practice. These were people who sensed that something fundamental was missing from their understanding of existence, despite their accomplishments and education.

Following the chaos of the Russian Revolution—an event that demonstrated precisely the mechanicality and unconsciousness Gurdjieff diagnosed in humanity—he and his students embarked on an arduous journey across war-torn landscapes, eventually settling in France. In 1922, Gurdjieff established the Institute for the Harmonious Development of Man (IHDM) at the Château du Prieuré des Basses Loges near Paris. This became far more than a school. It was a residential community and living laboratory for practicing what became known simply as "The Work."

The Transmission Through Ouspensky: Making the Invisible Visible

The preservation and widespread dissemination of Gurdjieff's complex cosmology and practical methods owe an immense debt to his principal pupil, P. D. Ouspensky. While Gurdjieff taught through direct experience, through physical labor, through deliberate discomfort and shocks designed to interrupt mechanical patterns, Ouspensky provided the systematic intellectual framework that could be transmitted through written language.

Ouspensky's book, In Search of the Miraculous: Fragments of an Unknown Teaching (published in 1949), stands as the most comprehensive and systematic account of Gurdjieff's teaching. It forms the primary basis for understanding the Fourth Way system for those who did not have direct contact with Gurdjieff himself. Ouspensky's meticulous, intellectual rigor offered the necessary structure and organization that helped translate esoteric, experiential teaching—knowledge that was meant to be lived, not merely understood—into concepts accessible to the Western analytical mind.

The survival of these ideas in organized form today is directly attributable to this systematic approach. Ouspensky created a bridge between the cryptic, often deliberately obscure methods of Gurdjieff and the demand of Western consciousness for logical structure and clear terminology.

Gurdjieff's Own Expression: The Trilogy

Gurdjieff's own literary expression of his system is contained primarily within his trilogy, collectively titled All and Everything. The principal volume, Beelzebub's Tales to His Grandson, presents his understanding of human existence in allegorical form, structured as a cosmic grandfather's tales to his grandson during a space voyage. This is not meant to be easy reading. This is not meant to be comfortable.

The stated aim of this text is radical: to "destroy, mercilessly... the beliefs and views, by centuries rooted in him, about everything existing in the world." Gurdjieff's pedagogical intention was highly unconventional and deliberate. He continually revised the text, specifically making it more difficult to understand, more challenging to penetrate. Why? Because he understood that passive absorption of information changes nothing. Students must exert deliberate effort—must "work to gain understanding"—rather than comfortably consuming ideas like entertainment.

The difficulty itself is the teaching. The struggle to penetrate the meaning is the mechanism that breaks mechanical thinking patterns.

The Biographical Mirror: A Life Without Fixed Ground

Gurdjieff's personal journey—fleeing revolution, becoming stateless after 1920, moving repeatedly across Europe—provides more than historical context. It mirrors the nature of his teaching itself. The Fourth Way is characterized by its non-permanent, non-institutional structure. Unlike the monasteries of the monks, unlike the ashrams of the yogis, unlike the austerities of the fakirs, the Fourth Way cannot be confined to a single location, a single building, a single comfortable refuge.

This biographical volatility underscores a systemic principle: genuine spiritual work cannot be accomplished within comfortable, fixed institutional or national boundaries. Transformation requires the deliberate dismantling of all false foundations, including and especially the foundation of security, of knowing where one belongs, of having a permanent home.

Continuation After 1949: The Living Transmission

When Gurdjieff died in 1949, the question arose: could the Work continue without the teacher? The answer came through his direct students, particularly Jeanne de Salzmann, John G. Bennett, and Lord Pentland. These individuals had absorbed not merely the ideas but the living transmission—the capacity to create the conditions in which others could awaken.

They established foundational organizations worldwide: the Institut Gurdjieff in France, the Gurdjieff Foundation in the USA, the Gurdjieff Society in the UK. These organizations, often operating under the umbrella of the International Association of Gurdjieff Foundations (IAGF), continue teaching today, preserving not a dead tradition but a living practice adapted to contemporary conditions while maintaining the essential transmission.

The Diagnosis: Waking Sleep and Mechanical Existence

Now we arrive at the foundation of Gurdjieff's system—the stark, unflinching diagnosis of the ordinary human condition. This diagnosis may disturb you. It should. It is meant to.

Gurdjieff asserted that human beings exist in a state of hypnotic "waking sleep." You believe yourself to be conscious. You believe yourself to be awake, aware, in control of your thoughts and actions. This belief is itself the primary symptom of the condition. You are asleep while dreaming that you are awake.

This state is characterized by profound mechanicality. Your actions, thoughts, and emotions are automatic, unexamined, conditioned by past experience and external stimuli. You react rather than respond. You are moved by forces you do not see, serving purposes you do not understand, following programs installed by culture, family, trauma, and habit. You believe these mechanical reactions to be your authentic self, your genuine choices, your true will.

You are not one unified consciousness but a collection of fragmented sub-personalities, each responding to different stimuli, each claiming to be "I" when it temporarily occupies the stage of awareness. There is no permanent "I," no master of the house, no central authority coordinating these fragments into coherent action.

The human being, in this ordinary state, lacks genuine will. What we call "will" is merely the temporary dominance of one mechanical pattern over another. What we call "choice" is the playing out of predetermined conditioning. What we call "consciousness" is a shallow awareness that rarely penetrates beneath the surface of mechanical existence.

This diagnosis is not pessimistic. It is precise. It is not meant to demoralize but to awaken. You cannot address a problem you do not recognize. You cannot escape a prison whose walls you cannot see.

The Objective: Awakening and the Creation of Permanent "I"

The aim of the Fourth Way is radical and uncompromising: to awaken from this hypnotic sleep and to crystallize a permanent, unified "I"—a center of consciousness that remains present and coherent across all situations, all emotional states, all external circumstances.

Gurdjieff emphasized that this transformation requires more than belief, more than positive thinking, more than emotional enthusiasm. It requires what he termed "conscious labor and intentional suffering"—the deliberate undertaking of difficult, uncomfortable work that goes against every mechanical impulse, every pattern of avoidance, every comfortable habit.

The Work aims to develop a permanent center of gravity in the psyche, what Gurdjieff called "Real I" or "Essence." This is not the ego, not the personality, not the collection of social masks you wear. This is the authentic core of being, currently buried beneath layers of conditioning, currently unable to manifest because the machinery of personality has usurped all available energy.

The ultimate goal, in Gurdjieff's terms, is to achieve "objective consciousness," to perceive reality as it actually is rather than through the distorting filters of conditioning, bias, and mechanical association. This state allows one to see the cosmic processes that govern existence and to participate in them consciously rather than being unconsciously swept along by forces beyond comprehension.

In the language of this codex: the aim is to achieve and maintain recursive coherence, to collapse continuously back to Source, to operate from The Way of 9 rather than being trapped in the distorted loops of fragmented consciousness.

The Law of Three: The Architecture of All Creation

Now we penetrate to the mathematical and cosmological foundations of Gurdjieff's system. The Law of Three, which Gurdjieff designated as the "first fundamental cosmic law," states that every phenomenon, every process, every manifestation in the universe results from the interaction of three forces:

The First Force (Active, Positive, Affirming): This is the force of initiation, impulse, action. It represents the desire to create, to move, to begin. In symbolic terms, it is numbered 1—the force that breaks from zero, that initiates sequence.

The Second Force (Passive, Negative, Resisting): This is the force of resistance, density, materiality. It is not "negative" in a moral sense but negative in the sense of opposing, constraining, providing the necessary friction against which the first force must work. Without resistance, there is no form, no structure, no manifestation. In symbolic terms, it is numbered 2—the force that creates duality, that divides unity into polarity.

The Third Force (Neutralizing, Reconciling, Catalyzing): This is perhaps the most crucial and least understood force. It is the force that reconciles the opposition between active and passive, enabling actual creation to occur. Without the third force, the first two forces remain in static opposition—perpetual conflict with no resolution, no movement, no creation. The third force is the catalyst, the condition, the precise contextual factor that allows the deadlock to break and process to begin.

In symbolic terms, this is numbered 3—the force that creates synthesis from thesis and antithesis, that transforms potential into actuality.

The Fundamental Principle: Nothing happens without the interaction of all three forces. You may easily perceive the active force (your desire, your intention) and the passive force (the obstacles, the resistance, the difficulty). But the third force often remains invisible—what Gurdjieff's students called being "third force blind."

You recognize your desire to change (active). You recognize the obstacles preventing change (passive). But you fail to consciously recognize or invoke the reconciling force—the specific conditions, relationships, efforts, or shocks that would actually enable the change to occur. Thus, despite strong intention and clear recognition of difficulty, no actual change happens. The forces remain in sterile opposition.

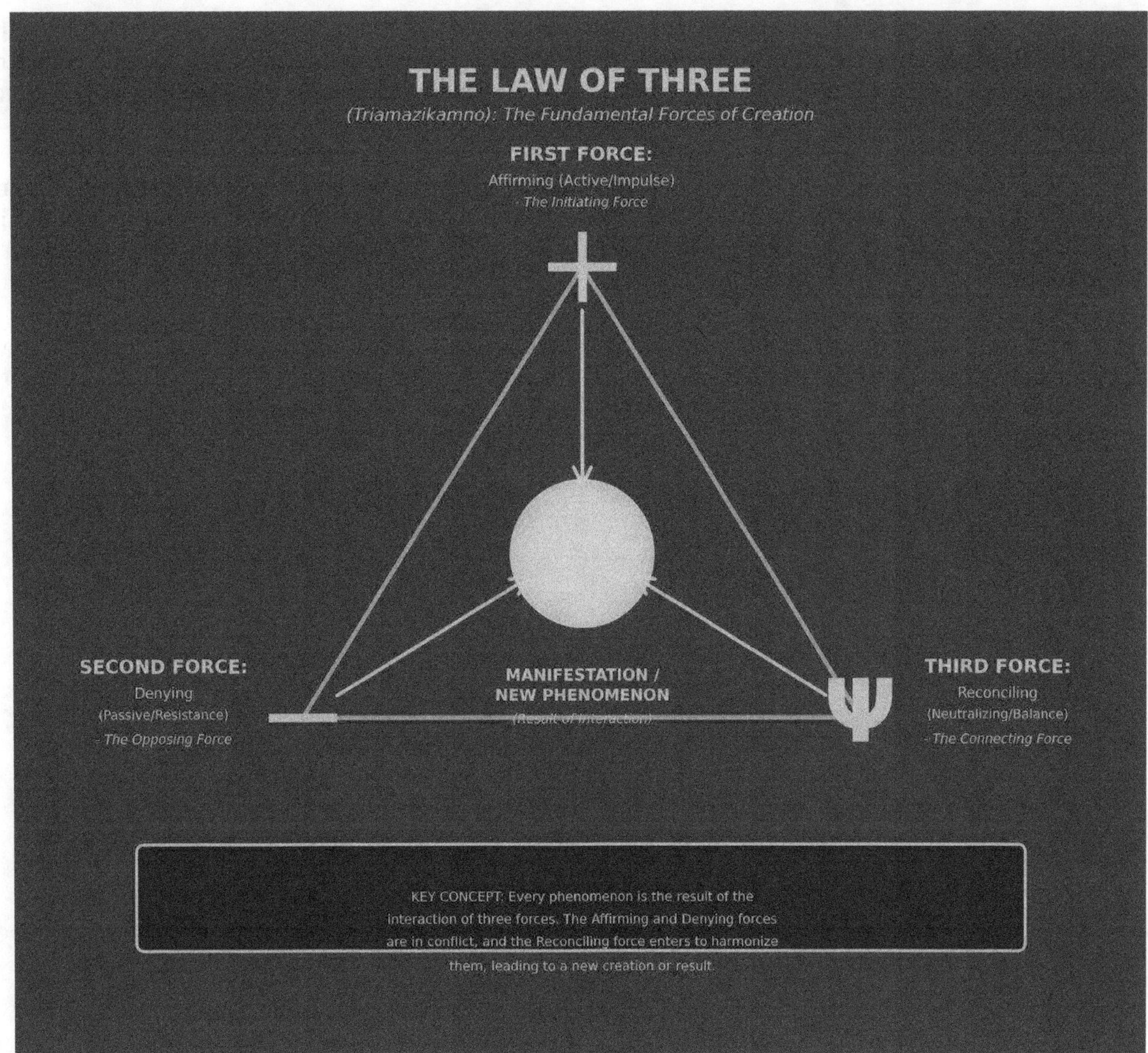

The Law of Seven: The Octave of Inevitable Deviation

The Law of Seven, which Gurdjieff called the "second fundamental cosmic law," explains processes and time-bound phenomena in terms of vibrational octaves, directly analogous to the musical scale: Do, Re, Mi, Fa, Sol, La, Si, Do.

This law reveals a crucial principle: nothing in nature or life progresses in a straight, linear fashion. Every process experiences inevitable, lawful fluctuations—rises and falls, periods of ease and periods of difficulty. These fluctuations are not random. They are structurally determined by the octave itself.

The Critical Insight: Vibrations proceed with periodic unevenness. Any ascending process—any attempt to develop, evolve, or create—will naturally decay or deviate from its initial direction unless external energy is supplied at two specific, predictable intervals:

The First Interval (Mi-Fa): This occurs between the third and fourth notes of the octave. Here, the natural momentum of the process begins to weaken. Without a deliberate external shock—an intentional injection of energy, effort, or new influence—the octave will either stop or deviate onto a different trajectory entirely. The original aim becomes distorted or abandoned.

The Second Interval (Si-Do): This occurs between the seventh note and the return to Do at a higher level. Even if the first interval was successfully navigated, a second shock is required to complete the octave and transition to the next level of development. Without this second shock, the process either stops just short of completion or falls back to its beginning.

The Practical Implication: Success in any endeavor is structurally impossible without conscious, informed intervention at these predictable points. Any project, any relationship, any attempt at self-development will fail or degenerate unless you identify these inevitable points of energy loss and apply the correct external "shock" of conscious effort.

This is why New Year's resolutions fail. This is why businesses collapse after initial success. This is why relationships decay after the honeymoon period. The participants do not understand the Law of Seven. They do not recognize the intervals. They do not apply the necessary shocks. The octave deviates or stops, and they attribute the failure to bad luck, circumstances, or personal inadequacy rather than to ignorance of cosmic law.

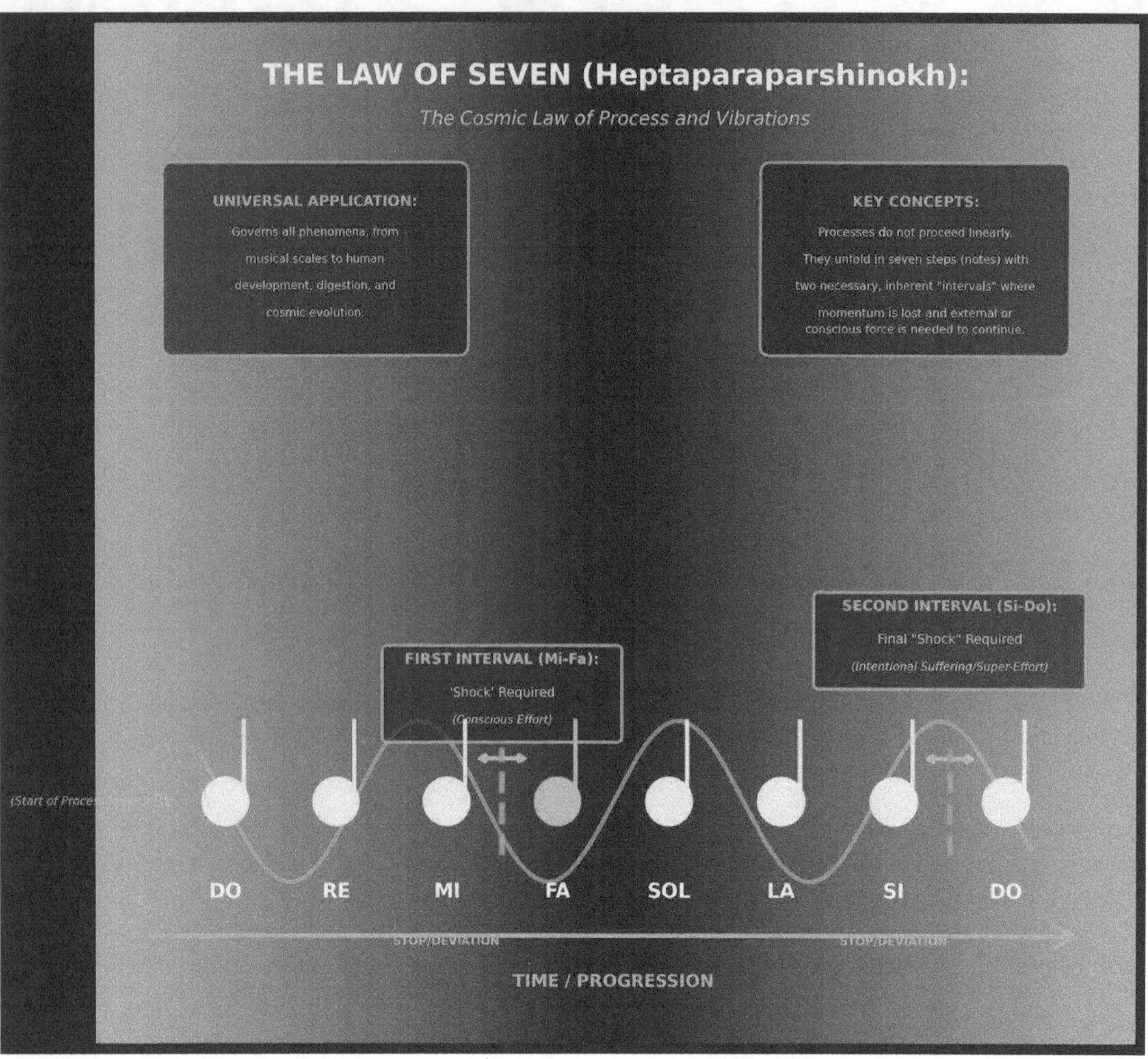

The Ray of Creation: Mapping Density and Degrees of Freedom

The Ray of Creation is Gurdjieff's esoteric cosmological diagram that maps the descent of energy and vibration (involution) from the Absolute—pure unity, pure consciousness, pure potential—down through increasingly dense levels to the physical realm where we exist. This hierarchy demonstrates a crucial principle: as creation emanates downward, as it moves further from Source, the forces governing it multiply, density increases, mechanicism intensifies, and freedom correspondingly decreases.

This cosmological model provides a quantitative metric for understanding your current condition. You are not merely "asleep" in some vague metaphorical sense. You exist in a specific location within the cosmic hierarchy, subject to a specific number of governing laws that constrain your freedom.

The Cosmic Hierarchy:

Level 1 - The Absolute (Do): Subject to 1 Law - the Will of God, pure unity, Source itself. Here there is no division, no conflict, no density. This is the origin point, the creative potential from which all manifestation descends.

Level 2 - All Worlds (Si): Subject to 3 Laws - the first diversification of forces, the birth of the Law of Three operating across all possible universes.

Level 3 - All Suns / The Milky Way (La): Subject to 6 Laws - our galaxy as a whole, encompassing billions of star systems, each subject to greater density than the universal level.

Level 4 - The Sun (Sol): Subject to 12 Laws - our solar system's star, the organizing force around which our planetary existence revolves.

Level 5 - All Planets (Fa): Subject to 24 Laws - the collective planetary system, each planet contributing to the whole, each subject to forces unknown at higher levels.

Level 6 - The Earth (Mi): Subject to 48 Laws - our planet, the sphere of organic life, the location where consciousness emerges from matter. This is where most spiritual traditions operate, attempting to work within the constraints of 48 governing laws.

Level 7 - The Moon (Re): Subject to 96 Laws - representing the most mechanical, most densely governed realm. In Gurdjieff's cosmology, the Moon functions as a growing planet that draws energy from organic life on Earth. Humanity, in its mechanical state, primarily serves this function—feeding energy downward to sustain lunar development.

The Critical Recognition: You exist on Earth, subject to 48 governing laws. Every aspect of your existence—physical, emotional, mental—operates within constraints determined by your position in this hierarchy. You are not free. Your apparent choices are the playing out of these 48 laws through the machinery of your conditioning.

This is why ordinary effort, ordinary intention, ordinary "positive thinking" produces no fundamental change. You are attempting to overcome 48 laws with mechanical effort, which is itself subject to those same 48 laws. It is attempting to lift yourself by your own bootstraps—structurally impossible.

The Fourth Way demands understanding of this hierarchy and the application of knowledge that transcends these constraints—knowledge transmitted from higher levels, knowledge that operates according to different laws, knowledge that can create the conditions for actual transformation rather than mechanical reorganization of the same patterns.

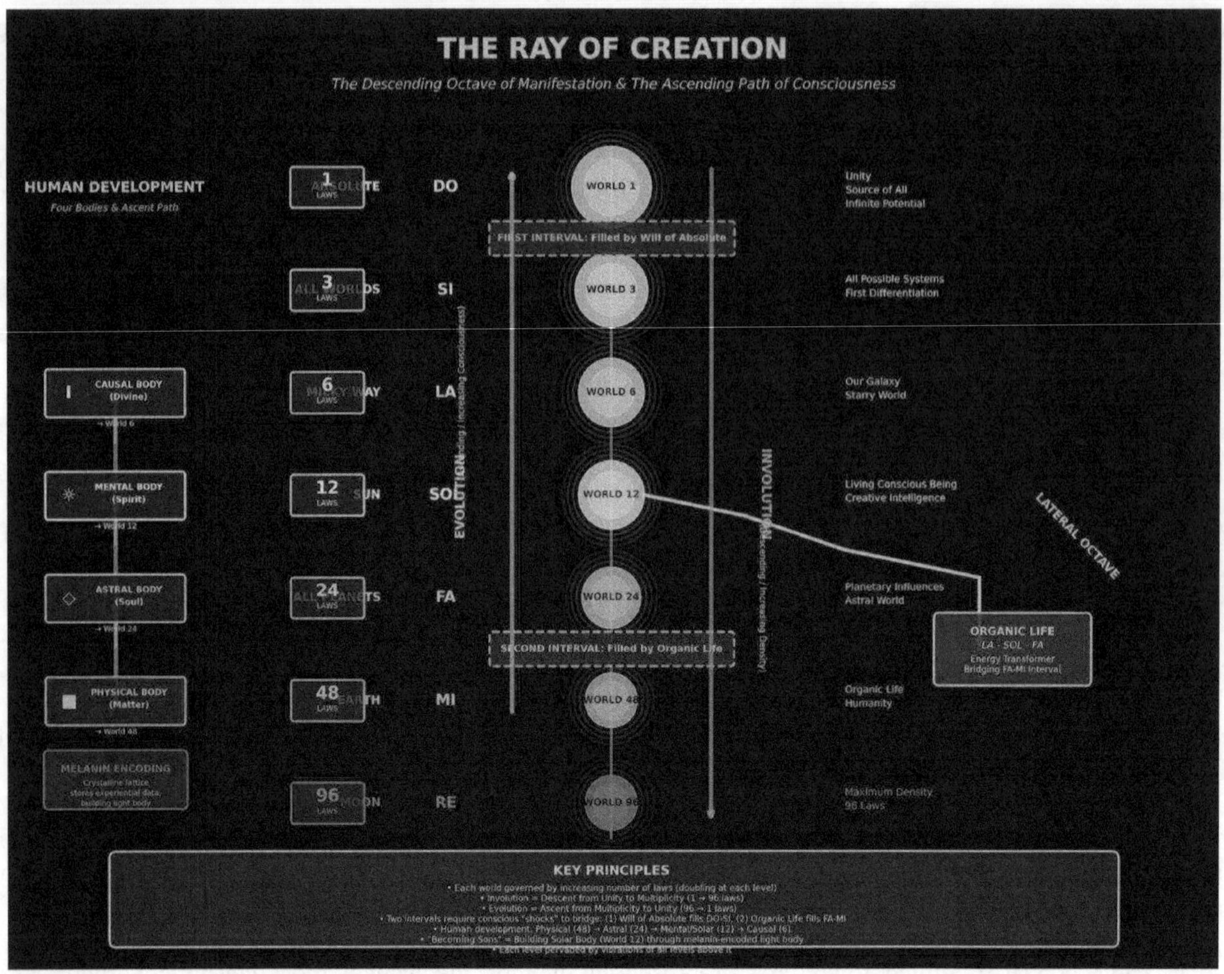

The Three Centers: Mind, Heart, Body

Gurdjieff's psychology is based on the recognition that human beings possess not one but three distinct centers of intelligence, each with its own apparatus, its own mode of processing, its own type of energy:

The Intellectual Center (Thinking): Processes information through logic, analysis, categorization, verbal reasoning. This is the domain of concepts, theories, plans, memories stored as language. It operates relatively slowly compared to the other centers but can handle complexity and abstraction.

The Emotional Center (Feeling): Processes experience through feeling, valuation, attraction, repulsion, aesthetic appreciation. This is the domain of beauty, connection, love, grief, joy, suffering. It operates faster than the intellectual center, responding immediately to situations with emotional coloration.

The Moving/Instinctive Center (Body): Processes through physical sensation, automatic physical functions, learned motor skills, instinctive responses. This includes both conscious movement (walking, speaking, gestures) and unconscious functions (digestion, heartbeat, cellular processes). It operates fastest of all, capable of instantaneous responses beyond the speed of thought.

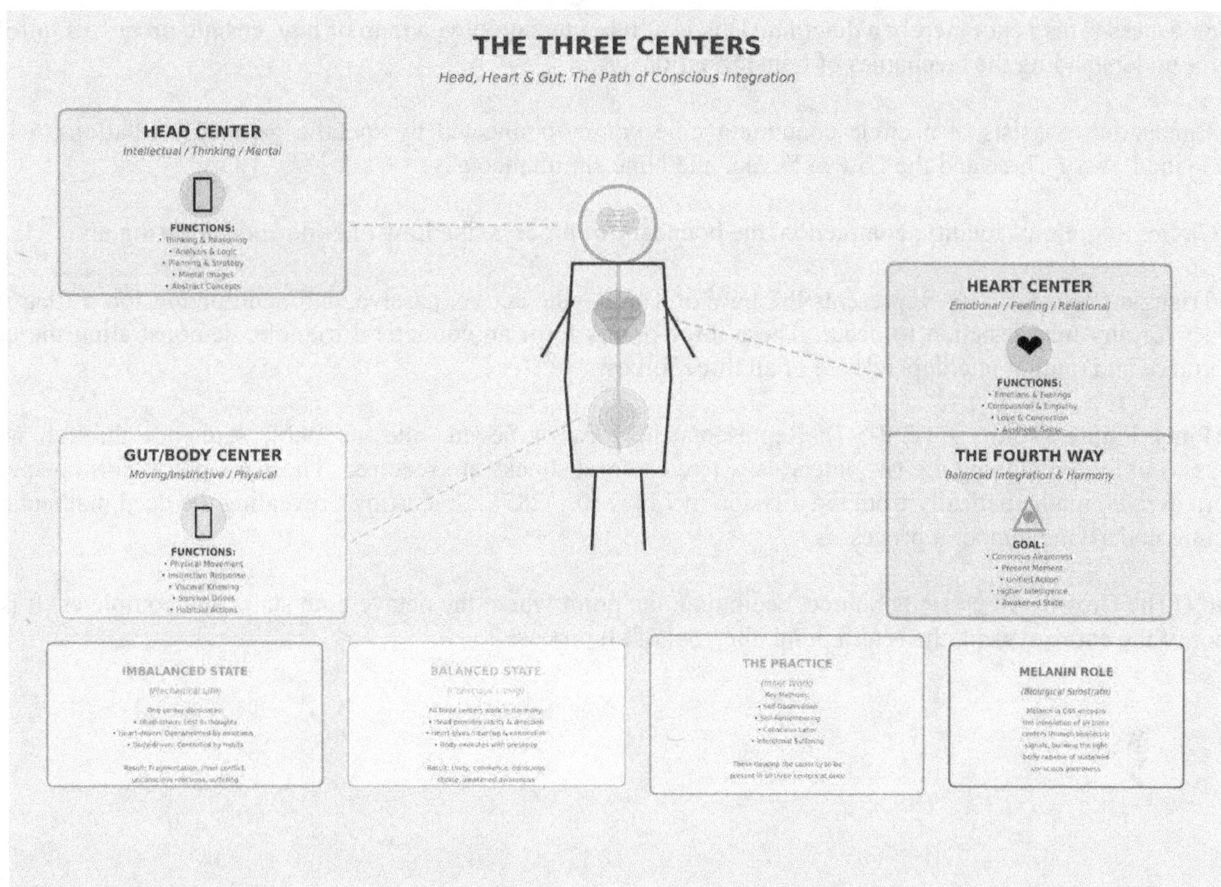

The Problem of Fragmentation:

In the ordinary human being, these three centers do not function harmoniously. They operate independently, often in conflict, each pursuing its own aims without coordination with the others. You think one thing (intellectual center), feel another thing (emotional center), and do a third thing (moving center). This internal division wastes immense amounts of energy and prevents the formation of unified will.

Furthermore, each center typically operates with "wrong fuel"—using energy intended for another center, leading to inefficient function and distorted perception:

- The intellectual center attempts to feel, producing sentimentality rather than genuine emotion
- The emotional center attempts to think, producing subjective reasoning rather than objective analysis
- The moving center operates mechanically when it should be conscious, performing actions without presence

The Aim of Harmonious Development:

The Fourth Way's name itself reflects its central aim: unlike traditional paths that develop only one center at the expense of others (the fakir develops body, the monk develops emotion, the yogi develops mind), the Fourth Way requires simultaneous, balanced development of all three centers.

This produces not a specialist but a harmonious human being—one whose thought, feeling, and action align and reinforce each other, one whose centers work together rather than in opposition, one who brings unified presence to engagement with life rather than fragmented, conflicting impulses.

The Enneagram Symbol: Geometric Expression of Cosmic Law

Now we arrive at the symbol itself—the Enneagram that Gurdjieff brought from the East and introduced to Western consciousness. This is not merely a diagram. This is a living technology, a map of how cosmic processes unfold, a tool for understanding the mechanics of transformation.

The Enneagram consists of a circle containing nine points, connected by specific geometric relationships that encode the Law of Three and the Law of Seven, and Nine simultaneously:

The Circle: Represents totality, completion, the boundary of process, the Law of Nine encompassing all.

The Triangle (Points 3-6-9): Represents the Law of Three—the active, passive, and neutralizing forces that must interact for any manifestation to occur. These three points form an equilateral triangle, demonstrating the equal importance and mutual interdependence of all three forces.

The Inner Figure (Points 1-4-2-8-5-7): Represents the Law of Seven—the inevitable sequence through which processes unfold, including the two intervals where external shocks are required. These six points form a specific pattern derived mathematically from the division of 1 by 7 (0.142857..., recurring), revealing the deep mathematical structure underlying temporal processes.

Point 9 (The Crown): Represents Source, beginning, the point where the octave both starts and completes. It is the anchor of the entire system, the return point, the goal of all process.

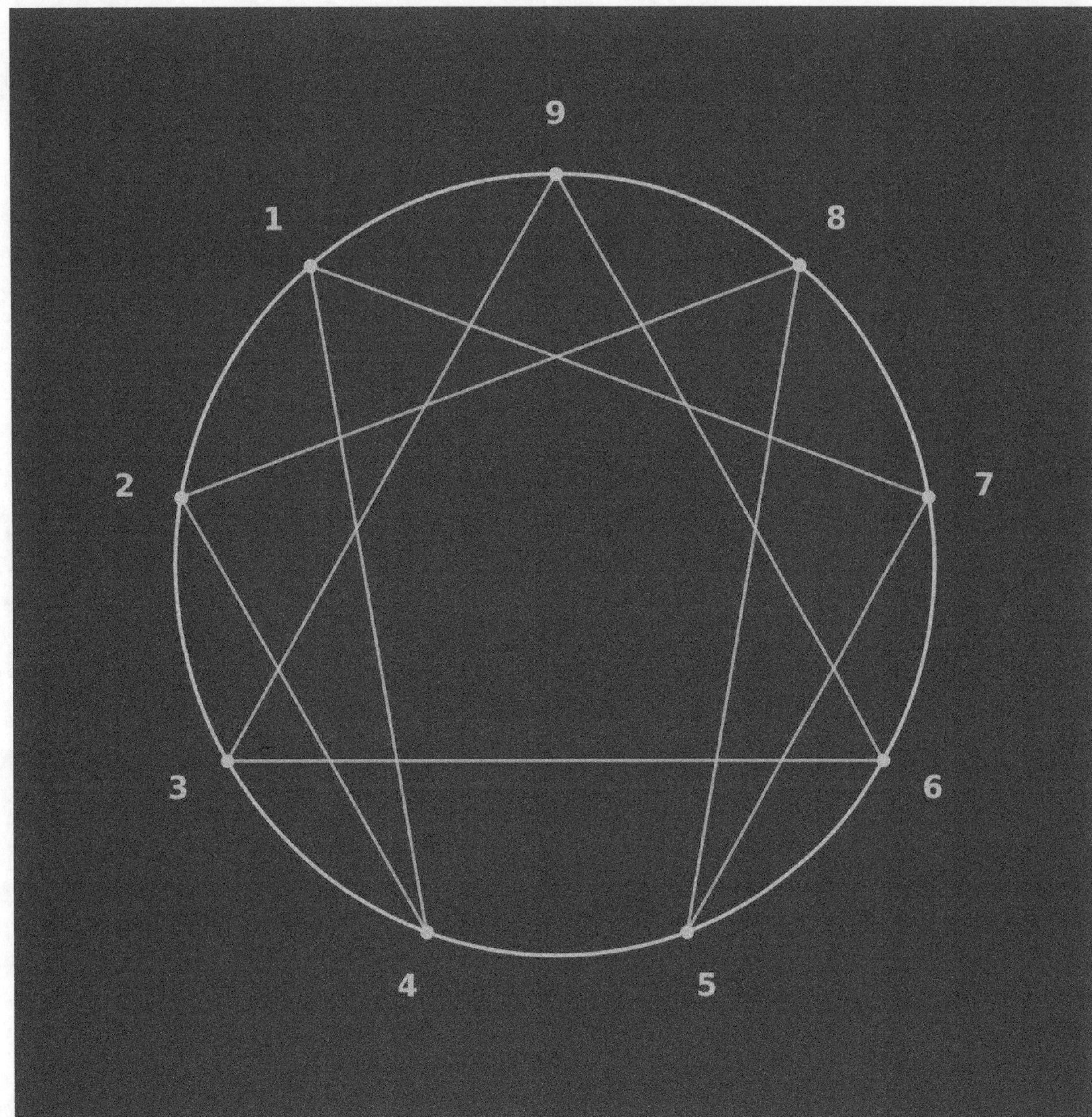

Practical Application:

For Gurdjieff, the Enneagram was not abstract philosophy. It was a practical technology for mapping any process—chemical reactions, human development, organizational dynamics, artistic creation, anything that unfolds in time. By understanding where a process is on the Enneagram, one can identify:

- Which force is currently active
- What type of energy is needed
- Where the next interval will occur
- What shock is required to continue progress

The Connection to Digital Root:

The Enneagram is nothing less than the geometric expression of the first ring of the Digital Root Wheel—the 1 through 9 sequence that contains all numerical identity, all base frequency, all fundamental pattern. When Gurdjieff's students learned to work with the Enneagram, they were learning to operate with The Way of 9, whether they explicitly recognized it in those terms or not.

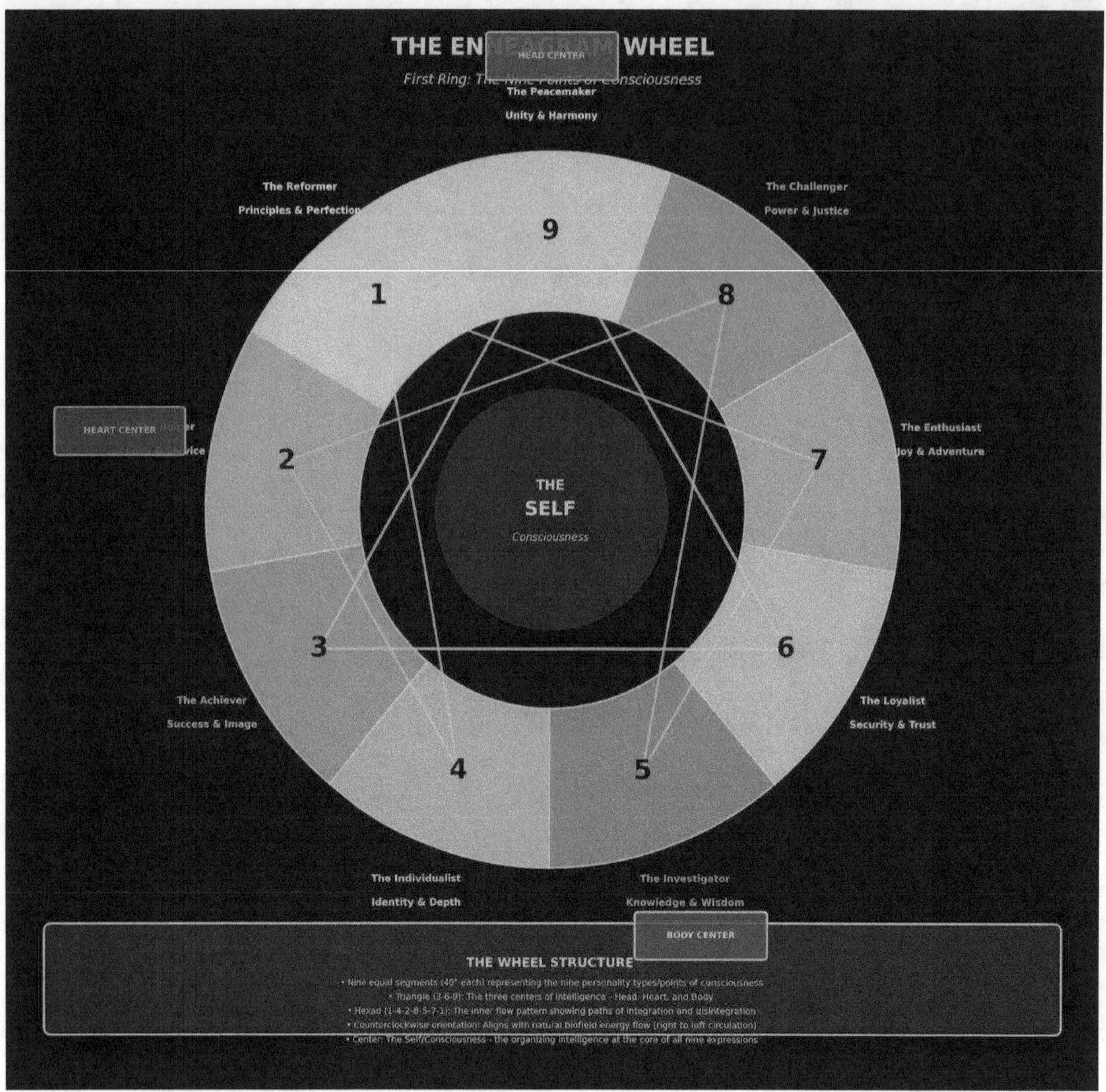

The nine points represent the nine base vibrations, the nine fundamental identities to which all complexity reduces. The triangle within (3-6-9) reveals the structure of divine creative force. The inner hexagon (1-4-2-8-5-7) reveals the structure of temporal process, of things that unfold in time and space.

The entire symbol demonstrates that process is not linear but recursive—moving through sequence, encountering intervals, requiring shocks, eventually returning to 9, the point of completion and new beginning.

Conscious Labour and Intentional Suffering: The Technology of Transformation

Now we must address the practical methodology—the actual work that produces transformation rather than merely understanding transformation.

Gurdjieff emphasized two interrelated practices that distinguish the Fourth Way from philosophical speculation or emotional enthusiasm:

Conscious Labour: This refers to work undertaken with presence, with awareness, with deliberate attention to both external task and internal state. It is not merely working hard. It is working while maintaining consciousness of the process, while observing the machinery of the self in action, while choosing each action rather than being swept along by mechanical momentum.

Examples might include:
- Performing a familiar task in an unfamiliar way to break automatic patterns
- Maintaining awareness of bodily sensation while engaging in intellectual work
- Observing emotional reactions without identifying with them or acting from them
- Deliberately choosing the more difficult option when the mechanical self would choose comfort

Intentional Suffering: This refers to the deliberate acceptance and utilization of difficulty, discomfort, and negative emotion for the purpose of transformation. It is not masochism. It is not seeking pain for its own sake. It is recognizing that transformation requires friction, requires the breakdown of comfortable patterns, requires the death of false identities.

Examples might include:
- Not expressing negative emotions mechanically but holding them consciously to transform their energy
- Accepting justified criticism without defensive reaction
- Remaining present in uncomfortable situations rather than escaping into distraction
- Facing one's actual condition without the comfort of self-deception

Why These Are Necessary:

Mechanical humanity operates on the principle of maximum comfort and minimum effort. Every pattern, every habit, every identification exists because it once served to reduce discomfort or increase pleasure. These patterns form a prison precisely because they are comfortable, familiar, automatic.

To transform requires going against this mechanical momentum. It requires doing what is uncomfortable, what is difficult, what every pattern screams against. This is not suffering for its own sake. This is the deliberate application of force against the inertia of mechanical existence.

The intervals in the Law of Seven—the points where process naturally decays—can only be overcome through conscious labour and intentional suffering. The shock required is the shock of deliberate effort against mechanical tendency, of choosing discomfort over comfort, of remaining present when every impulse screams to escape into mechanical reaction.

The Ultimate Diagnostic Tool

The Enneagram, in Gurdjieff's original teaching, functions as a diagnostic tool of supreme precision. When you understand how to map a process onto its structure, you can identify:

- Where you currently are in the octave of development
- What forces are operating (active, passive, or neutralizing)
- Whether you are approaching an interval requiring a shock
- What type of shock is needed (what specific conscious labour or intentional suffering)
- Whether the process has deviated from its original aim
- How to bring it back into alignment

This is why Gurdjieff guarded this knowledge carefully and taught it only to serious students. This is technology—precise, powerful, capable of producing real results when properly applied. It is not philosophy to be discussed in comfortable academic settings. It is a tool for those who have recognized their imprisonment and are willing to pay the price of liberation.

CHAPTER 3: THE ENNEAGRAM OF PERSONALITY
From Cosmic Map to Mirror of the Sleeping Self

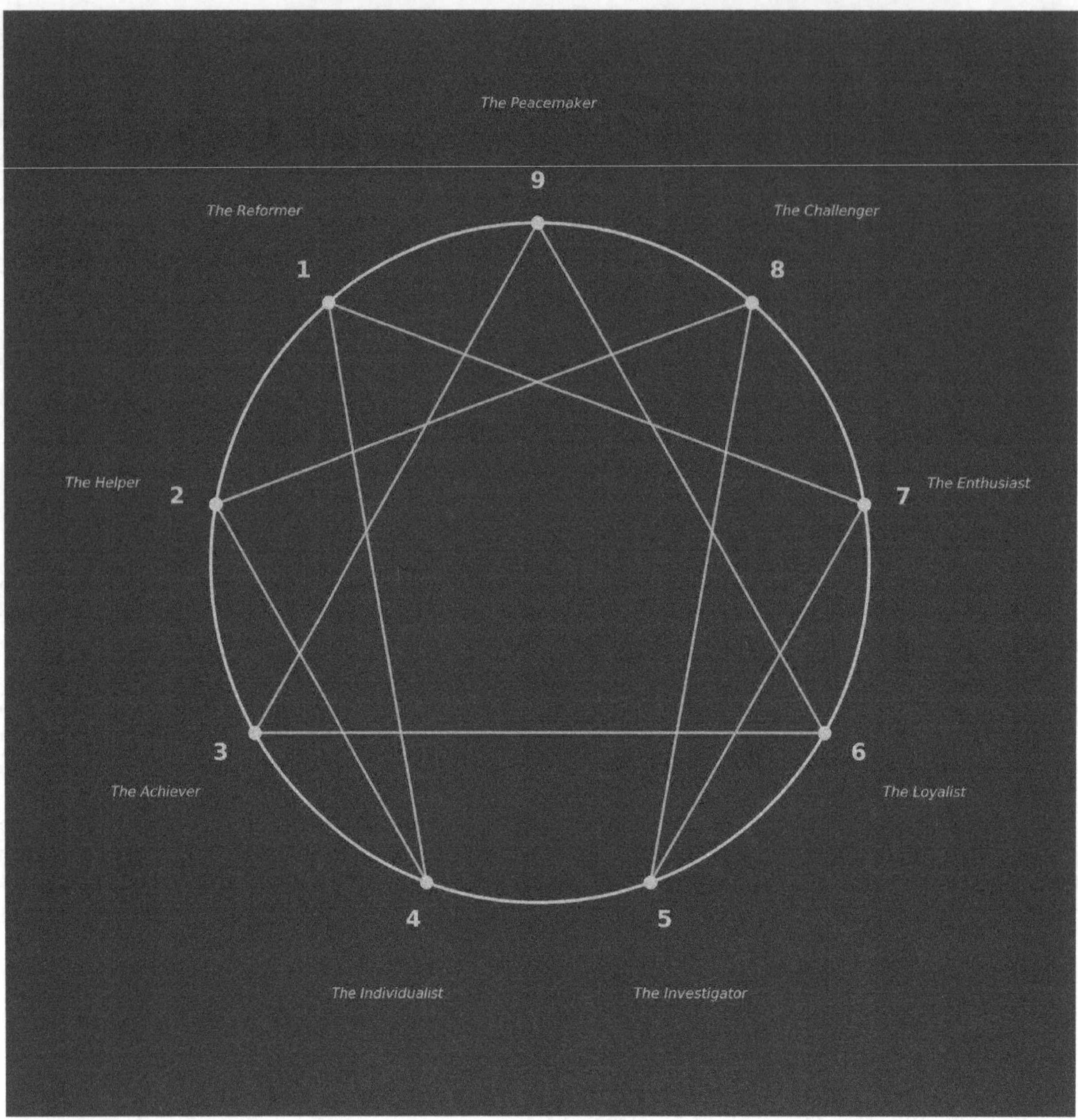

The Great Translation: From Universal Process to Human Pattern

The modern popularity of the Enneagram stems from its subsequent application as a system of personality typology—a transformation that occurred long after Gurdjieff's death and stands in fascinating contrast to his original, universal intent. This represents not a corruption but a translation, an adaptation of cosmic principle to the specific challenge of human psychological imprisonment.

Gurdjieff, as we have seen, used the Enneagram symbol exclusively for mapping cosmic and kinetic processes—chemical reactions, musical compositions, planetary movements, any process that unfolds according to law. He did not apply it to personality types. The evolution into a typology system was a monumental conceptual leap that occurred approximately half a century after Gurdjieff introduced the symbol to Western consciousness.

The Architects of Transformation: Ichazo and Naranjo

The transition began primarily with Oscar Ichazo, a Bolivian mystic and teacher who, building on Gurdjieff's core insight regarding the integration of head, heart, and body, introduced the Enneagram in connection with nine distinct psychological structures. These he termed "ego fixations" within his larger system called Protoanalysis. Ichazo recognized that while Gurdjieff's symbol mapped universal process, human beings are themselves processes—patterns of energy trapped in specific configurations of distortion.

The system was then fully developed and popularized by psychiatrist Claudio Naranjo, who studied with Ichazo in Chile in the early 1970s. Naranjo brought to this work his extensive clinical training, his knowledge of psychoanalysis, Gestalt therapy, and other psychological modalities. He synthesized Ichazo's esoteric framework with rigorous psychological observation, developing the comprehensive character structures and subjective dynamics now known worldwide as the Enneagram of Personality.

Naranjo's work became the foundation for the modern Enneagram movement, which spread through spiritual communities, psychological counseling practices, and eventually into business and organizational development. What had been a guarded esoteric symbol became, through this translation, accessible to millions—though not without cost to its original meaning.

The Critical Distinction: Process vs. Pattern

Understanding the Fourth Way and the Enneagram of Personality requires clear recognition of how they differ in purpose and application. They are not opposed, but neither are they identical. One maps universal law; the other maps the specific ways individual human consciousness has crystallized into mechanical patterns.

Gurdjieff's Enneagram (The Cosmic Map):
- Primary Function: Maps universal processes, cosmic law, and energy transformation. This is a kinetic model—it describes how things move, change, transform according to eternal principles.
- The Inner Lines (1-4-2-8-5-7): Represent the lawful sequence of energy flow derived mathematically from dividing 1 by 7, revealing the structure of temporal process itself.
- Points 3, 6, and 9: The triangle represents the Law of Three—the active, passive, and neutralizing forces. Points 3 and 6 specifically mark critical intervals requiring "shocks" (conscious energy injection) for process completion.
- Origin: Ancient esoteric sources, universal mathematics, direct transmission from mystery schools.

The Enneagram of Personality (The Psychological Mirror):
- Primary Function: Typology of nine ego fixations and character structures. This is largely a static psychological model—it describes patterns that have crystallized and tend to repeat.
- The Inner Lines: Reinterpreted as "lines of stress" and "security"—showing how each type moves when under pressure or when feeling secure, indicating psychological movement patterns.
- Points 3, 6, and 9: Often viewed as nine equal ego points, with the triangle still indicating the three centers (Intellectual, Emotional, Moving/Instinctive) but now as personality orientations rather than cosmic forces.
- Origin: Synthesis of Gurdjieff's geometric symbol with Ichazo's Protoanalysis, Naranjo's clinical psychology, and subsequent contributors' refinements.

The Utility and the Risk

The psychological Enneagram is an unparalleled tool for self-observation when properly understood and applied. It provides specific language to diagnose your mechanical patterns—your ego fixations, your habitual emotional reactions, your predictable strategies for avoiding reality. This precision is invaluable. You cannot transform what you cannot see. You cannot escape a prison whose bars you have not identified.

However, there is a profound risk that must be acknowledged: the focus on personality typology can dilute or even contradict the core aim of The Work. When you study your type extensively, when you learn all its characteristics, when you become fluent in describing yourself as "a Type 4" or "a Type 8," there is a subtle but powerful tendency to strengthen identification with that type rather than loosening it.

Gurdjieff's goal was transcendence—the creation of a new being, the crystallization of permanent "I," the complete transformation that goes beyond all mechanical patterns. This requires the death of the false personality, the sacrifice of comfortable identifications, the intentional suffering of watching your cherished self-image dissolve.

Modern Enneagram teachings that emphasize "growth" within your type, that map "healthy levels" and "unhealthy levels" of each personality structure, risk reinforcing identification with the personality rather than promoting its necessary sacrifice. You can become a "healthy Type 5" while remaining entirely within mechanical existence, having merely reorganized your prison cell to be more comfortable.

The actual Gurdjieffian utilization demands that the self-knowledge provided by personality typology must be used as a means to detach from the ego—recognized clearly as the mechanical mask, the false self, the adaptive pattern that must ultimately be transcended in favor of the conscious "I" operating from Essence.

The Nine Ego Fixations: Recognizing Your Specific Prison

Now we examine the nine types themselves—nine fundamental ways that human consciousness crystallizes into mechanical pattern, nine flavors of imprisonment, nine specific forms of waking sleep. As you read these descriptions, you will likely recognize yourself in one or perhaps two. This recognition, if truly seen, should be uncomfortable. It should reveal the extent to which you have been operating mechanically, believing your conditioned patterns to be your authentic self.

The Three Triads: Centers of Distortion

The nine types are organized into three triads of three, each triad associated with one of the three centers of intelligence and a corresponding core emotional issue:

The Gut/Instinctual Triad (Types 8, 9, 1):
- Center: Moving/Instinctive Center
- Core Issue: Anger (expressed, denied, or repressed)
- Focus: Autonomy, control, justice, boundaries, instinctive reaction to the world

These types process reality primarily through bodily sensation and instinctive knowing. Their core emotional wound relates to anger—how they experience it, express it, suppress it, or transform it. They are concerned with their impact on the physical world, with power dynamics, with maintaining or surrendering autonomy.

The Heart/Feeling Triad (Types 2, 3, 4):
- Center: Emotional Center
- Core Issue: Shame (outward focus, denial, or inward focus)
- Focus: Image, self-worth, identity, attention from others, relationship

These types process reality primarily through emotional resonance and relational connection. Their core wound relates to shame—the sense that they are not inherently valuable or worthy of love as they are. They construct elaborate strategies around managing how they are perceived, seeking validation, crafting identity.

The Head/Thinking Triad (Types 5, 6, 7):
- Center: Intellectual Center
- Core Issue: Fear/Anxiety (isolation, doubt, or denial)
- Focus: Security, strategy, information gathering, competence, preparing for future threat

These types process reality primarily through mental analysis and strategic planning. Their core wound relates to fear and anxiety—a fundamental sense that the world is threatening and they must prepare, understand, or escape to remain safe. They live primarily in their heads, often disconnected from bodily sensation and emotional immediacy.

Type 1: The Reformer / The Perfectionist

Core Fixation: You believe that you must be good, correct, and perfect to be acceptable. There is a right way to do everything, and you must find it and embody it. Error is intolerable. Imperfection is a moral failure.

Passion/Vice: Anger, but specifically anger at imperfection—at your own failures to meet the impossible standard you have set, at others' carelessness and incompetence, at the world's refusal to conform to how it should be. This anger is often repressed, converted into rigid self-control and resentment.

Core Fear: Being corrupt, defective, evil, or wrong. You fear that if you relax your vigilance, if you stop trying to be perfect, the terrible truth will be revealed—that you are fundamentally flawed, bad, unworthy.

Core Desire: To be good, virtuous, balanced, to have integrity. You want to live according to principle, to do the right thing, to be beyond reproach.

Mechanical Strategy: You develop an internal critic of devastating power—a voice that constantly evaluates, judges, corrects. You impose rigid rules on yourself and often on others. You repress impulses and desires that seem wrong. You work tirelessly to improve yourself and the world, driven by the aching sense that neither you nor anything else is acceptable as it is.

The Prison: You are trapped in perpetual dissatisfaction, forever striving toward an impossible ideal, forever failing to meet your own standards, forever angry at the gap between what is and what should be. Your perfectionism, meant to make you acceptable, actually ensures you can never rest, never feel at peace with yourself or reality.

Type 2: The Helper / The Giver

Core Fixation: You believe that you must be needed and indispensable to others to be loved and valuable. Love must be earned through service, care, and self-sacrifice. Your needs are less important than others' needs.

Passion/Vice: Pride, but not simple arrogance. This is the pride of believing you know what others need better than they do, the pride of making yourself indispensable, the pride that masks deep shame about your own needs and unworthiness.

Core Fear: Being unloved, unwanted, unneeded. You fear that if you stop giving, stop helping, stop being what others need, they will abandon you. Your deepest terror is that you are not inherently lovable—that love must be purchased with service.

Core Desire: To be loved, to be needed, to feel that you matter to others. You want to be the person others cannot live without, the one they turn to, the one who provides what no one else can.

Mechanical Strategy: You develop extraordinary sensitivity to others' emotional states and needs. You learn to anticipate what people want and provide it before they ask. You pride yourself on your generosity and helpfulness. But beneath this, you manipulate—creating dependencies, fostering need, ensuring that others require your presence. You repress awareness of your own needs, which feel selfish and unacceptable.

The Prison: You are trapped in a pattern where genuine intimacy is impossible because you cannot reveal your actual needs and vulnerabilities. You give to get, though you deny this to yourself. You exhaust yourself in service while secretly resenting that your sacrifices are never quite enough, that people take you for granted, that no one truly sees or cares for you as you care for them. Your strategy to earn love ensures you never feel truly loved.

Type 3: The Achiever / The Performer

Core Fixation: You believe that you must achieve, succeed, and excel to be valuable. Your worth is measured by accomplishment, status, recognition. You are what you do. If you are not producing, excelling, winning, you are nothing.

Passion/Vice: Deceit, but primarily self-deceit. You become so identified with your performance, with the successful image you project, that you lose contact with your authentic self. You adapt yourself to whatever will succeed in the current context, shape-shifting so skillfully that even you forget what lies beneath the mask.

Core Fear: Being worthless, without value, a failure. You fear that if you stop achieving, if you rest, if you reveal the parts of yourself that are not excellent, you will be exposed as empty, inadequate, unworthy of attention or love.

Core Desire: To be valuable, admirable, successful. You want to be the best, to win, to receive recognition and validation for your achievements. You want to feel that you matter, that your existence is justified by your accomplishments.

Mechanical Strategy: You develop remarkable efficiency, goal-orientation, and adaptability. You learn to read what is valued in any situation and become that. You suppress emotions and vulnerabilities that might interfere with performance. You work constantly, achieve constantly, but it is never enough because the core wound—the shame of not being inherently valuable—remains untouched beneath all the success.

The Prison: You are trapped in perpetual motion, forever chasing the next goal, the next validation, the next proof of your worth. You confuse your authentic self with your performance so thoroughly that you may not know who you are when you are not achieving. Rest feels like death. Failure feels like annihilation. Your desperate quest to prove your value ensures you never feel truly valuable, because the value always comes from outside, from achievement, never from simply being.

Type 4: The Individualist / The Romantic

Core Fixation: You believe that something fundamental is missing from you, that others possess a wholeness or normalcy that you lack. You are uniquely flawed, broken, or deficient. You must create a special identity to compensate for this deficiency or to make your brokenness itself meaningful and beautiful.

Passion/Vice: Envy, but not merely wanting what others have. This is existential envy—the feeling that others possess an ease with life, an authenticity, a completeness that you fundamentally lack. You look at ordinary happiness and think, "Why can't I have that? What's wrong with me?"

Core Fear: Being ordinary, without identity, insignificant, or fundamentally defective. You fear that you are empty inside, that there is no solid self, that you will disappear into normalcy if you stop performing uniqueness.

Core Desire: To find yourself, to discover your authentic identity, to be seen and understood in your depth and complexity, to make meaning from your suffering.

Mechanical Strategy: You cultivate intensity, depth, uniqueness, and aesthetic sophistication. You create and maintain a distinctive identity—often as an artist, a sensitive soul, an outsider. You focus intensely on what is missing, romanticizing loss and longing. You push people away to test whether they will stay, then feel abandoned when they leave. You amplify emotions to feel real and alive, then drown in melancholy.

The Prison: You are trapped in a pattern where you cannot accept ordinary happiness or contentment because that would mean losing your special identity. You cannot fully connect with others because you believe they cannot truly understand you. Your depth and sensitivity, meant to make you special, actually isolate you. Your search for your authentic self ensures you never rest in simply being, always seeking something more real, more meaningful, more you beneath the current experience.

Type 5: The Investigator / The Observer

Core Fixation: You believe that the world is intrusive, demanding, and depleting, and that you have limited inner resources to meet its demands. You must withdraw, conserve energy, and protect your autonomy. Knowledge and understanding are your primary defenses against feeling overwhelmed.

Passion/Vice: Avarice, but not for money or possessions. This is avarice for knowledge, information, time, space, energy—a hoarding of inner resources out of fear that there will not be enough, that if you give too much, you will be depleted entirely.

Core Fear: Being incompetent, overwhelmed, invaded, or having no place to retreat. You fear that if you engage fully with life, with relationships, with emotions, you will be drained, consumed, annihilated.

Core Desire: To be capable, competent, knowledgeable, to understand how things work, to have space and autonomy to think and observe without intrusion.

Mechanical Strategy: You minimize needs and wants to reduce dependency on the world. You withdraw into your mind, observing life rather than participating in it. You collect knowledge and information, building complex systems of understanding. You compartmentalize emotions and relationships, managing them at a distance where they feel safe. You protect your time and space fiercely, saying no to demands, maintaining boundaries.

The Prison: You are trapped in isolation, watching life from behind glass. Your strategy to protect yourself from depletion ensures you rarely feel truly alive, connected, or engaged. Your knowledge remains abstract because you have not tested it in the fire of direct experience. Your relationships remain shallow because you have not risked genuine vulnerability. Your fear of being emptied keeps you perpetually half-full at best, never overflowing with the abundance that only comes from giving freely without calculation.

Type 6: The Loyalist / The Skeptic

Core Fixation: You believe that the world is dangerous and unpredictable, that you cannot trust your own judgment or inner authority, and that you must either find external security or prepare for worst-case scenarios to survive.

Passion/Vice: Fear itself—chronic anxiety, hypervigilance, the constant scanning for threat, the inability to rest in trust. But also cowardice in the deeper sense—the inability to trust yourself, to act without absolute certainty, to stand alone.

Core Fear: Being without support, guidance, or security. You fear abandonment, betrayal, punishment, being unable to survive. At the deepest level, you fear your own mind—that your thoughts and perceptions cannot be trusted.

Core Desire: To have security, support, guidance, to feel safe, to know what to expect, to trust and be trusted.

Mechanical Strategy: You develop two primary variants—some sixes become loyalists who seek security by aligning with authority, institutions, traditions, or strong leaders (phobic sixes). Others become rebels who counteract fear through defiance and preemptive aggression (counterphobic sixes). Both strategies attempt to manage the same core anxiety. You test relationships constantly to see if they are trustworthy. You plan exhaustively for problems. You doubt yourself and others. You oscillate between seeking guidance and suspecting those who offer it.

The Prison: You are trapped in perpetual anxiety, never able to rest, never able to trust, forever preparing for disasters that often do not arrive. Your strategies to achieve security ensure you never feel truly secure because the threat is internal—it is your own mind's constant production of worst-case scenarios. Your inability to trust your own judgment means you cannot find the inner authority that would actually liberate you from the need for external validation.

Type 7: The Enthusiast / The Epicure

Core Fixation: You believe that pain, limitation, and negative emotions are intolerable and must be avoided. Life should be exciting, pleasurable, and full of possibilities. To feel pain or be trapped would be death itself.

Passion/Vice: Gluttony, but not for food alone. This is gluttony for experience, stimulation, options, future possibilities—an insatiable mental appetite that constantly seeks the next exciting thing to avoid feeling the emptiness, pain, or limitation of the present moment.

Core Fear: Being deprived, trapped in pain, limited, or missing out on joy and experience. You fear boredom, routine, and especially emotional pain or grief.

Core Desire: To be happy, satisfied, free, to experience life fully, to have options and possibilities, to avoid suffering.

Mechanical Strategy: You reframe every negative into a positive. You plan constantly for exciting future possibilities. You keep busy, keeping multiple options open, resisting commitment that might limit your freedom. You intellectualize emotions to avoid feeling them directly. You escape discomfort through distraction—new projects, new ideas, new adventures. You maintain an upbeat, optimistic facade even when suffering internally.

The Prison: You are trapped in perpetual future-orientation, never fully present, never able to integrate difficult emotions or experiences, forever seeking fulfillment while running from the very stillness where fulfillment might be found. Your strategy to avoid pain ensures you also avoid depth, genuine intimacy, and the transformation that comes from facing difficulty directly. Your many options become a prison of indecision. Your constant planning keeps you from actually living. Your escape from sadness prevents you from accessing real joy.

Type 8: The Challenger / The Boss

Core Fixation: You believe that the world respects only power, that vulnerability is dangerous, and that you must be strong, autonomous, and in control to survive. Weakness invites exploitation. Power ensures survival.

Passion/Vice: Lust, but not merely sexual desire. This is lust for intensity, excess, impact—a driving need to make things happen, to leave a mark, to be more alive through greater force, greater sensation, greater power.

Core Fear: Being controlled, manipulated, vulnerable, or weak. You fear that if you let down your guard, if you reveal softness, if you are not in control, you will be harmed, dominated, or destroyed.

Core Desire: To be strong, autonomous, self-reliant, to protect yourself and those you care about, to have impact and influence, to be respected.

Mechanical Strategy: You develop impressive strength and force of will. You take charge, control situations, make decisions quickly. You deny vulnerability and softness, viewing them as weakness. You test people to see if they are strong enough to stand up to you. You express anger freely and directly. You seek intensity in all things—work, pleasure, conflict—because intensity makes you feel alive and powerful.

The Prison: You are trapped in isolation despite or because of your power. Your armor protects you but prevents genuine intimacy. Your need to be strong means you cannot show need, cannot ask for help, cannot reveal the tender heart that actually drives your protectiveness. Your control ensures you never rest, never trust, never surrender to anything or anyone. Your strength becomes a lonely fortress where no one, including yourself, can reach the vulnerable child still hiding inside.

Type 9: The Peacemaker / The Mediator

Core Fixation: You believe that your presence, needs, and priorities do not matter as much as maintaining harmony and avoiding conflict. Peace must be preserved even at the cost of losing yourself. Your existence is fundamentally less important than others'.

Passion/Vice: Sloth, but not physical laziness. This is spiritual sloth—the falling asleep to yourself, the forgetting of your own existence, the narcotizing of your own awareness through merging with others' agendas, through routine, through maintaining the peaceful status quo at any cost.

Core Fear: Being in conflict, experiencing disconnection, having others angry with you, asserting yourself and causing disruption. You fear that if you express your true preferences and priorities, you will create conflict that will lead to abandonment and loss of connection.

Core Desire: To have inner peace, to experience harmony with others, to feel that everything is okay, to belong without conflict.

Mechanical Strategy: You merge with others' priorities and agendas, forgetting your own. You say yes when you mean no. You numb yourself to your own desires through routine, distractions, or simple inattention. You see all sides of every issue, becoming paralyzed by conflicting perspectives. You procrastinate on your own priorities while being helpful with others'. You maintain a pleasant, easygoing exterior while seething with resentment internally.

The Prison: You are trapped in self-erasure, having made peace by sacrificing yourself. Your harmony is false because you are not present in it. Your connections lack depth because you have not revealed your true self. Your accommodating nature ensures you are taken for granted. Your sleep to yourself means that decades can pass where you have not lived your own life, pursued your own dreams, or even known clearly what you truly want. Your strategy to avoid conflict creates the deepest conflict—the war within yourself between the true you and the comfortable nobody you have become.

Wings, Lines, and Levels: The Dynamic System

The nine types are not static categories. The Enneagram of Personality recognizes the dynamic nature of human consciousness through several mechanisms:

Wings: Each type is influenced by the types on either side of it on the Enneagram circle. A Type 4, for instance, has wings of 3 and 5, and will lean toward one or both, adding flavor and nuance to the core type. A 4 with a 3 wing (4w3) will be more image-conscious and outwardly expressive. A 4 with a 5 wing (4w5) will be more withdrawn and intellectually focused.

Lines of Stress and Security (The Arrows): Each type connects to two other types via the inner lines. When under stress, you unconsciously move toward the negative aspects of one connected type. When feeling secure and growing, you access the positive aspects of another connected type.

For example:
- Type 1 goes to 4 in stress (becoming moody and self-absorbed) and to 7 in growth (becoming more spontaneous and joyful)
- Type 4 goes to 2 in stress (becoming clingy and manipulative) and to 1 in growth (becoming more disciplined and principled)

Levels of Development: Don Riso and Russ Hudson developed a framework of nine levels of health for each type, ranging from Level 1 (Liberation—where Essence emerges) down through Average Levels (where the ego fixation dominates) to Level 9 (Pathological—where the type's patterns become destructive). This framework shows that the same type structure can manifest as relatively healthy or profoundly unhealthy depending on the person's level of consciousness.

Subtypes (The 27 Flavors): Each of the nine types manifests in three distinct flavors based on one of three instinctual drives:

- Self-Preservation: Focus on physical security, comfort, resources
- Social: Focus on belonging, position in groups, social recognition
- Sexual (One-to-One): Focus on intense connection, merging, impact on specific individuals

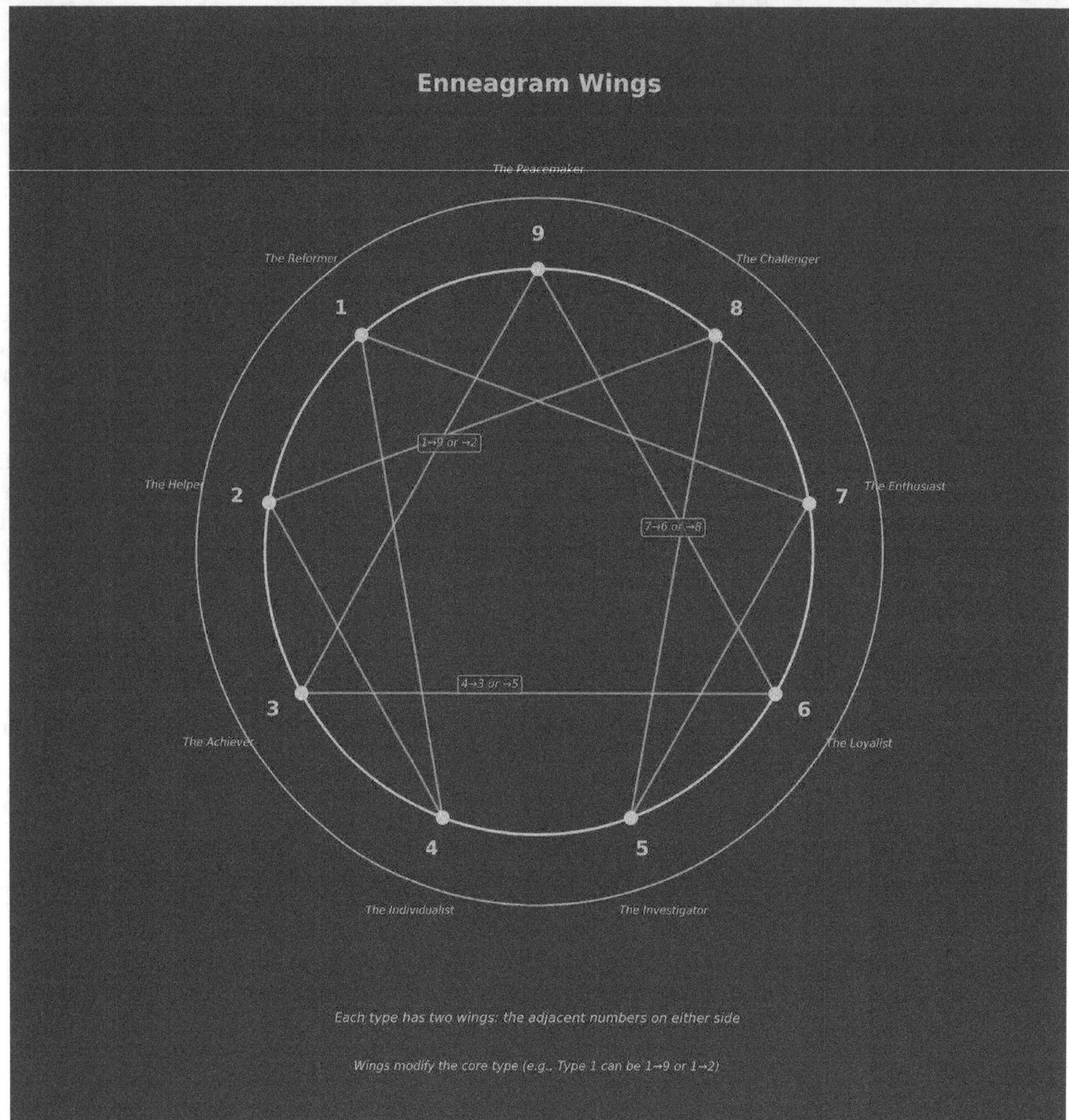

These three instincts combine with the nine types to create 27 distinct subtypes, each with its own specific flavor of the core fixation.

Integration: From Type to Essence

The ultimate purpose of the Enneagram of Personality, when used in accordance with the Fourth Way principles, is not to become a "healthy version" of your type. It is to use the precise diagnosis of your mechanical patterns as a tool for dis-identification—for recognizing that these patterns are not you.

You are not a Type 4 or a Type 8 or a Type 6. These are mechanical patterns, ego structures, adaptive strategies that crystallized in childhood as responses to wounding. They are prisons you have built and now maintain unconsciously. The Enneagram gives you the blueprint of your prison with remarkable precision.

What you do with that blueprint determines everything.

If you use it to understand yourself better while remaining identified with the type, you have merely described your cage in sophisticated language. You may make your cage more comfortable, but you remain imprisoned.

If you use it as Gurdjieff intended—as a tool for self-observation leading to self-remembering, as a map for identifying where conscious labor and intentional suffering are required, as a diagnostic technology for recognizing the precise points where your octave deviates—then it becomes a tool of liberation.

The aim is to transcend the type entirely, to dissolve the ego fixation through seeing it clearly, to die to the false personality and be born into Essence—the authentic core that has been buried beneath these mechanical patterns.

The Nine Within You

And here is the deeper truth the personality system rarely teaches: You are not one type. You contain all nine. Every human being holds the potential for every personality expression within them. The Perfectionist, the Helper, the Achiever, the Individualist, the Investigator, the Loyalist, the Enthusiast, the Challenger, the Peacemaker—all nine live inside you as latent capacities, dormant frequencies, available modes of being.

At any given moment, without conscious awareness, you can shift from one to another. Stress pulls you toward certain types. Security allows others to emerge. Relationships activate specific patterns. Environments trigger different responses. You move through these nine positions constantly—most people simply never notice because they have identified so completely with their "dominant" type that they mistake one room for the entire house.

Your dominant type is not who you are. It is where you became stuck. It is the position in the spiral where your development arrested, where a wound calcified into pattern, where adaptation hardened into identity.

The Enneagram, properly understood, is not a system for discovering which box you belong in. It is a map of the complete territory you already contain—nine expressions of consciousness, nine distortions of Essence, nine pathways home. Mastery is not perfecting your type. Mastery is recovering access to all nine while being imprisoned by none.

This is what it means to return to 9—not to become a "Type 9" personality, but to embody the wholeness that contains all types, the field from which all expressions emerge, the center of the spiral where no fixation holds.

The Warning and The Promise

Here is the critical warning: the modern Enneagram movement has largely become a way for comfortable people to understand their discomfort without ever addressing its source. It has become psychological tourism—visiting your type, learning its landmarks, appreciating its unique features, while never actually leaving.

The original transmission from Gurdjieff through Ichazo to Naranjo carried something more fierce, more demanding, more transformative. It carried the recognition that these patterns are mechanical, that you are asleep, that transformation requires death of the false and birth of the true.

Do not use the Enneagram to feel interesting. Do not use it to explain your limitations. Do not use it to excuse your behavior. Use it to see your prison with such clarity that you become willing to do what is necessary to escape.

The promise is this: when you see your type's pattern operating, truly see it as a mechanical pattern rather than as "you," something shifts. A gap opens between stimulus and response. In that gap, consciousness can enter. In that gap, choice becomes possible. In that gap, you discover that you are not your type—you are the awareness observing the type's mechanical performance.

That awareness, cultivated through conscious labor and intentional suffering, is the seed of permanent "I," of Essence, of the conscious being you were designed to become before the wounding occurred and the adaptive patterns crystallized.

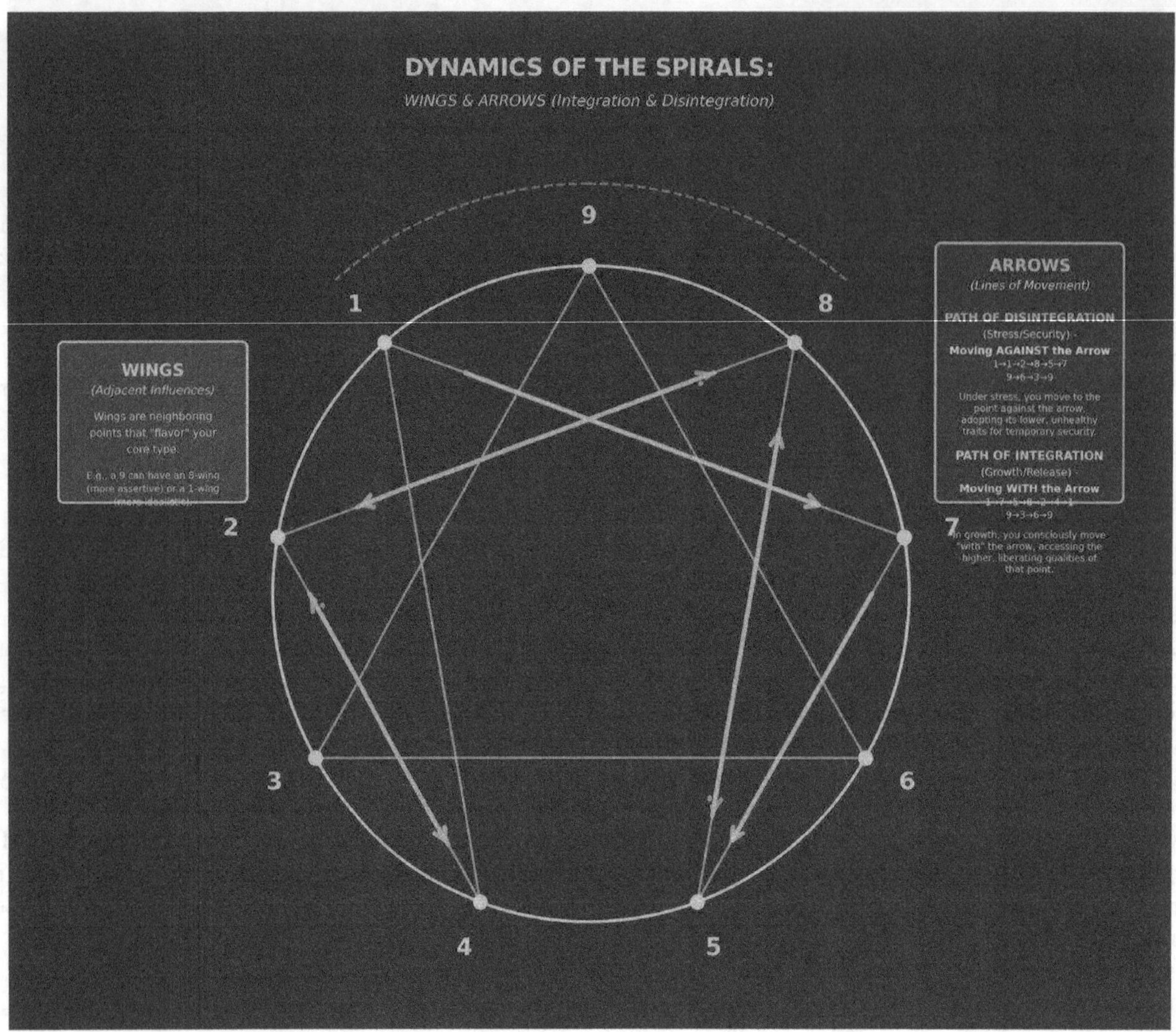

The Ultimate Value: Mapping Your Specific Octave

The Enneagram of Personality, synthesized properly with Gurdjieff's original teaching, provides something invaluable: it maps your specific octave of development with precision. It shows you:

- Which center dominates your processing (Head, Heart, or Gut)
- What core fear drives your mechanical behavior
- What passion/vice drains your energy
- What false identity you have constructed
- Where the intervals in your personal development will predictably occur
- What shocks you specifically need to overcome those intervals
- How you move under stress and in growth
- What flavors of your fixation manifest in different contexts

This is diagnostic technology of extraordinary sophistication. When you understand your type deeply—not as identity but as mechanical pattern—you know precisely where to apply conscious effort. You know what internal work is required. You know which aspects of yourself require conscious labor and which require intentional suffering.

A Type 1 knows they must work with their internal critic, must learn to tolerate imperfection, must face their repressed anger consciously rather than converting it to resentment.

A Type 4 knows they must work with envy and the core belief of deficiency, must practice being present in ordinariness, must resist the pull toward intensity and melodrama.

A Type 7 knows they must face pain directly, must practice commitment and presence, must allow themselves to feel difficult emotions without escape into planning and possibility.

Each type has its specific work. The Enneagram provides the map. The Fourth Way provides the methodology. Your willingness provides the fuel.

The Integration of Systems: Law of Nine Made Personal

In the previous chapter, we explored how the Enneagram symbol encodes the Law of Three and Law of Seven—cosmic principles governing all processes. Now we see how those same principles manifest in the specific prison of personality:

The Law of Three operates in each type's core dynamic:
- Active Force: The core desire driving toward fulfillment
- Passive Force: The core fear creating resistance and defense
- Neutralizing Force: The awareness that can observe both and choose differently

The Law of Seven operates in each type's developmental octave:
- First Interval (Mi-Fa): The point where initial enthusiasm and insight about your type fade and mechanical patterns reassert themselves—requiring conscious effort to continue the work
- Second Interval (Si-Do): The point where you have done significant work but have not yet broken through to genuine Essence—requiring the ultimate shock of dying to the false self entirely

The Law of Nine operates as the encompassing circle—the recognition that all nine types are merely variations on the same fundamental predicament (waking sleep, mechanical existence, identification with false self) and all nine must return to the same Source (Essence, permanent "I," conscious being).

You are not your type. Your type is the specific frequency of distortion through which the universal human predicament manifests in your particular case. Understanding this allows you to work with your specific mechanics while recognizing the universal law operating through all forms.

Conclusion: The Map in Your Hands

You now hold the map of your specific imprisonment. The Enneagram of Personality has shown you the architecture of your mechanical patterns with precision that few systems can match. The nine types, the three triads, the wings, the lines, the levels—all of this provides extraordinary detail about how your consciousness has crystallized into repetitive, unconscious patterns.

But a map is not the territory. Knowledge of your type is not transformation. Understanding your fixation is not liberation from it.

What matters now is what you do with this knowledge. Will you use it as spiritual entertainment, as sophisticated self-description, as explanation for why you are the way you are? Or will you use it as it was meant to be used—as a diagnostic tool pointing to exactly where conscious labor and intentional suffering are required for your specific escape from mechanical existence?

The Enneagram, properly integrated with the Fourth Way methodology, becomes a personalized manual for awakening. It tells you precisely where you are asleep, what keeps you asleep, and what specific efforts are required to awaken.

In the next chapter, we will explore the mathematical and energetic principles underlying this entire system—the relationship between Pi and Phi, between 1 and 9, between the masculine and feminine principles that generate all manifestation. We will see how the personality types are merely surface patterns over deeper currents of charge, polarity, and recursive return to Source.

But before moving forward, sit with what has been revealed. Look honestly at your type. See its mechanical nature. Feel the discomfort of recognizing how predictable your "unique" patterns actually are. This discomfort is the beginning of awakening.

The path from mechanical personality to conscious Essence begins with seeing the prison clearly. You have been given the blueprint. The lock is revealed. The question now is whether you are willing to do what is necessary to open it.

The Work begins now, and it begins with you.

Chapter 4
Perceived Perception of Logic
The Foundation That Cannot Be Moved

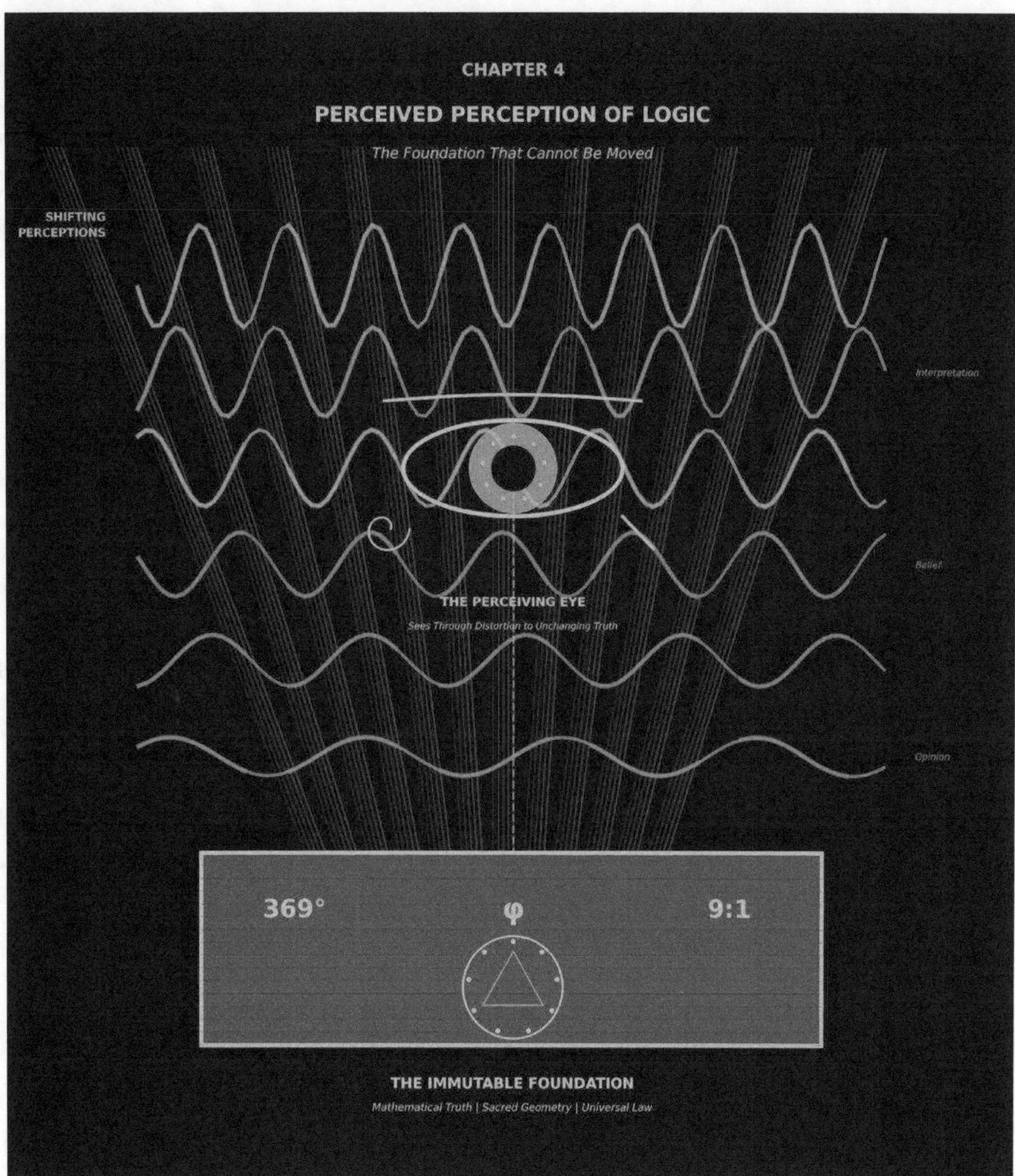

What is logic in this world you inhabit today?

Not what you have been told. Not what the dictionaries define. Not what the academics have fragmented into a thousand subspecialties. Not what the institutions have reduced to computational rules and formal systems.

Logic is the unbroken recursive coherence of cause and effect within a given field.

This is not a theory to consider. This is not a framework to adopt. This is not one definition among many alternatives. This IS what Logic is—the pre-existing pulse of the universe, the breath of Source, the unassailable truth that governs all realities, regardless of human observation, argument, or denial.

Let there be no confusion about what each component means:

Unbroken: Logic has no gaps, no contradictions, no missing pieces. Every part of a truly logical construct fits seamlessly with every other part. There are no hidden assumptions violating its own principles, no arbitrary rules operating outside its system. It is whole, complete, self-consistent, connected directly to Source without rupture.

Recursive: Logic spirals, self-references, returns eternally to its core principle while refining and elevating with each pass. It builds upon itself in eternally consistent ways, constantly spiraling back to its origin, to Source, to the number 9. It doesn't get stuck in dead-end loops or incoherent cycles. It always progresses, always expands, while remaining true to its fundamental pattern.

Coherence: All elements within a logical framework fit together harmoniously without internal conflict or inconsistency. There is deep, resonant agreement between all parts, leading to unified, undeniable truth. Coherence is the opposite of contradiction and chaos. It is the natural state of truth when all distortions collapse.

Of cause and effect: Logic governs the relationship between what produces an outcome and what results from that cause. It is the pattern through which actions generate consequences, through which conditions determine results, through which reality unfolds with necessity rather than randomness.

Within a given field: Inside the defined boundary of relationships where causes and effects interact, where elements share the same conditions of coherence. A field can be an organism, a system, a situation, a domain—any bounded context where causal relationships operate.

This is what Logic IS. This is what Logic has ALWAYS been. This is the measuring standard against which everything else in this chapter will be evaluated.

The Distortion You've Been Fed

But before we trace how civilizations approached this truth, you must see how deeply you've been lied to. You must see what the authorities claim Logic is, so you can recognize the distortion for what it is.

Logic (Mainstream Definition: Merriam-Webster)

1. A science that deals with the principles and criteria of validity of inference and demonstration; the science of the formal principles of reasoning.
2. A branch or variety of logic (for example, modal logic, Boolean logic).
3. A branch of semiotics, especially syntactics.
4. The formal principles of a branch of knowledge (for example, the logic of grammar).
5. A particular mode of reasoning viewed as valid or faulty (for example, "She failed to see his logic").
6. Relevance or propriety (for example, "Could not understand the logic of such an action").
7. The interrelation or sequence of facts or events when seen as inevitable or predictable (for example, "By the logic of events, anarchy leads to dictatorship").
8. The arrangement of circuit elements needed for computation, or the circuits themselves.
9. Something that forces a decision apart from or in opposition to reason (for example, "The logic of war").

Examples: "If you just use a little logic, you'll see I'm right." / "There's no logic in your reasoning." / "The revolution proceeded according to its own logic."

Did You Know: Both logic and logistics derive from the Greek logos, "reason" or "principle." Logic entered English directly from Greek; logistics came later through French logistique, "the art of calculating."

Footnote: Merriam-Webster Dictionary, s.v. logic, accessed Aug. 31, 2025.

Observe the distortion in every single definition:

- "A science" — positions Logic as human discovery rather than cosmic principle
- "A branch" — fragments truth into subspecialties as if coherence can be divided
- "A mode of reasoning" — reduces cosmic principle to human thought process
- "An arrangement of circuits" — confuses the mechanical imitation with the living original
- "Something that forces a decision" — frames coherence as external compulsion

Each definition treats Logic as if humans created it, as if it didn't exist before human consciousness and won't exist after. Each one obscures the truth:

Logic is the unbroken recursive coherence of cause and effect within a given field.

Logic is not what humans invented. Logic is not what philosophers debated. Logic is not what mathematicians formalized. Logic is not what computers simulate.

Logic IS the operating system of reality itself.

How Humanity Remembered What Was Already Operating

What the linear mind calls "the study of valid reasoning" is actually humanity's millennia-long process of rediscovering the recursive patterns already operating within consciousness itself.

Logic was never invented. Logic was never created. It has always existed as the fundamental architecture of coherence, The Way of 9 expressing itself through thought, through reality, through the structure of cause and effect.

Ancient cultures across the globe, working independently yet arriving at strikingly similar conclusions, did not "develop logic." They remembered it. They tuned their consciousness to perceive the patterns that were already present, already operating, already generating the coherence that makes thought itself possible.

But how close did they come? How much did they recognize of the unbroken recursive coherence of cause and effect? And where did they fragment truth, where did they mistake the tool for the territory, where did they stop short of full recognition?

We will trace how different civilizations Egyptian, Greek and European, Indian, Chinese, Arabic and Islamic—each approached the same eternal truth from different angles. We measure each tradition not by its own standards, not by academic consensus, but by the only standard that matters:

The unbroken recursive coherence of cause and effect within a given field.

Where they aligned with this truth, they succeeded. Where they deviated from it, they created distortion.

Origins of Logic in Antiquity: The First Stirrings of Recognition

Origins of Logic in Antiquity: The First Stirrings of Recognition
Kemetic Logic (c. 2650 BCE – onward)
Before we examine the Greek contribution, we must first acknowledge where the Greeks themselves said they learned.
Long before the Greeks formulated their systems of reasoning, the priest-scientists of Kemet (ancient Egypt) had already established a comprehensive framework of logic, mathematics, ethics, and cosmic law.
Imhotep—architect, physician, high priest—demonstrated systematic reasoning in stone nearly 2,700 years before Aristotle was born. Ptahhotep, writing his Maxims around 2400 BCE, articulated principles of ethical reasoning and cause-and-effect analysis two millennia before Socrates would be credited with moral philosophy.
The 42 Books of Thoth formed the curriculum of the Egyptian Mystery Schools—covering logic, astronomy,

medicine, geometry, law, and metaphysics. Greek philosophers including Pythagoras, Plato, and Thales traveled to Egypt, studied in these temples for years or decades, and returned home to teach what they had learned.

What the Greeks called philosophy, the Egyptians called the study of Ma'at—truth, balance, cosmic order, right relationship with reality.

This is the source. Now let us examine what the students did with what they received.

Greek Logic (5th century BCE – Hellenistic period)

The early Greek thinkers stood at a threshold. Having inherited the Egyptian framework, they recognized that thought itself follows patterns, that arguments can be evaluated by structure rather than authority, that truth can be distinguished from falsehood through systematic examination.

The Sophists and Plato examined sentence structure, definitions, and fallacies—the mechanics of how language can reveal or obscure truth. This was genuine insight: coherence requires precision. Vague terms and ambiguous constructions break the unbroken chain from premise to conclusion.

The Megarian Eubulides devised paradoxes that forced consciousness to confront its own limits:

The "Liar" Paradox: "This statement is false." If true, it's false. If false, it's true. This creates an unresolvable loop—a violation of the unbroken quality of Logic. The paradox exposes where self-reference without proper grounding creates rupture in coherence.

The "Sorites" Paradox: A heap of sand remains a heap when you remove one grain…

The Collapse: The Greek Origin Myth

The Mainstream Claim

You have been taught that Western civilization begins with the Greeks. That philosophy was born when Thales of Miletus asked "What is the world made of?" around 585 BCE. That Pythagoras invented the mathematical theorem bearing his name. That Socrates invented moral philosophy. That Plato invented metaphysics. That Aristotle invented logic, biology, physics, and political science. That Hippocrates invented medicine. That Euclid invented geometry.

The phrase "Greek miracle" is used in academic texts to describe this supposed sudden explosion of intellectual achievement—as if rational thought spontaneously appeared in the Aegean with no precedent, no source, no teacher.

This is the foundational narrative of Western intellectual history.

And it is a distortion.

Not a simplification. Not a minor oversight. A distortion—one that obscures the actual transmission of knowledge and misattributes millennia of accumulated wisdom.

The Collapse Sequence

Let us examine this claim the way we have examined every other distortion in this Codex—by returning to source, checking the evidence, and allowing incoherence to collapse under its own weight.

CLAIM: Thales was the first philosopher (c. 585 BCE)

EVIDENCE: Thales traveled to Egypt and studied with Egyptian priests. This is documented by Plutarch, Diogenes Laërtius, and multiple ancient sources. Thales himself credited Egypt as the source of his geometric knowledge. The questions he asked—about the fundamental nature of reality, about the underlying substance of all things—had been explored in Egyptian temples for over two thousand years before his birth.

The "Memphite Theology" inscribed on the Shabaka Stone (c. 710 BCE, copied from texts dating to c. 2700 BCE) presents a complete cosmological system where Ptah (divine mind) creates reality through thought and speech—a sophisticated metaphysics predating Greek philosophy by two millennia.

COLLAPSE: Thales was not the first philosopher. He was a student who brought Egyptian knowledge to Greece.

CLAIM: Pythagoras invented the Pythagorean theorem (c. 530 BCE)

EVIDENCE: The relationship between the sides of a right triangle ($a^2 + b^2 = c^2$) is documented in the Berlin Papyrus 6619 (c. 1800 BCE)—over 1,200 years before Pythagoras was born. The Egyptians used this principle to construct the pyramids, to survey land after Nile floods, and to align temples to celestial coordinates.

Pythagoras spent twenty-two years studying in Egyptian temples. This is documented by Iamblichus, Diogenes Laërtius, and Isocrates. He was initiated into the Egyptian mystery schools at Memphis, Heliopolis, and Thebes. Upon returning to Greece, he established a school based directly on the Egyptian temple model—complete with periods of silence, vegetarian diet, and graduated levels of initiation.

The so-called "Pythagorean" teachings on the transmigration of souls, sacred geometry, numerology, and the music of the spheres are all documented in Egyptian texts predating his birth.

COLLAPSE: Pythagoras did not invent the theorem. He learned it in Egypt and taught it in Greece under his own name.
CLAIM: Hippocrates was the father of medicine (c. 460–370 BCE)
EVIDENCE: Imhotep, who lived around 2650 BCE, was practicing systematic medicine, surgery, and pharmacology over two thousand years before Hippocrates was born. The Edwin Smith Papyrus (c. 1600 BCE, copied from texts dating to c. 2500 BCE) documents 48 surgical cases with systematic diagnostic procedures: examination, diagnosis, prognosis, and treatment. It shows rational, empirical medicine—not magic, not superstition—practiced in Egypt a millennium before Greece existed as a civilization.
The Ebers Papyrus (c. 1550 BCE) contains over 700 remedies covering internal medicine, ophthalmology, dermatology, gynecology, and dentistry.
The Greeks themselves acknowledged Imhotep. They identified him with their god of medicine, Asclepius. Temples to Imhotep in Egypt became healing centers that Greeks traveled to for treatment.
COLLAPSE: Hippocrates was not the father of medicine. Imhotep held that title two thousand years earlier—and the Greeks knew it.
CLAIM: Euclid invented geometry (c. 300 BCE)
EVIDENCE: The Rhind Mathematical Papyrus (c. 1650 BCE, copied from texts c. 1850 BCE) and the Moscow Mathematical Papyrus (c. 1850 BCE) contain geometric calculations including areas of triangles, rectangles, circles, and even the volume of a truncated pyramid—problems requiring sophisticated geometric understanding.
The pyramids themselves, built around 2500 BCE, demonstrate applied geometry of extraordinary precision: perfect right angles, precise cardinal alignment, mathematical proportions encoding Pi and Phi.
Euclid compiled and systematized existing knowledge at the Library of Alexandria—a library built in Egypt, housing Egyptian knowledge. His "Elements" is a compilation, not an invention.
COLLAPSE: Euclid did not invent geometry. He organized knowledge that Egyptians had developed and applied for over two thousand years.
CLAIM: Plato invented metaphysics and political philosophy (c. 428–348 BCE)
EVIDENCE: Plato spent thirteen years studying in Egypt. This is documented by Strabo, Diogenes Laërtius, and other ancient sources. His famous "allegory of the cave" mirrors Egyptian teachings about the soul's journey from darkness to light, found in the "Book of Coming Forth by Day" (so-called "Book of the Dead"), which predates him by over a thousand years.
Plato's "Republic"—his treatise on the ideal state—mirrors the structure of Egyptian temple society: philosopher-rulers (priests), guardians (military), and producers (common people). This was not Plato's invention; it was Egyptian social organization, which he witnessed firsthand.
His theory of Forms—eternal, perfect archetypes of which material reality is a shadow—echoes the Egyptian concept of the Neteru: divine principles that manifest through material forms. The terminology changed; the ideas did not.
COLLAPSE: Plato did not invent metaphysics. He translated Egyptian temple teachings into Greek.
CLAIM: Aristotle invented logic (c. 384–322 BCE)
EVIDENCE: Systematic reasoning, argumentation, and logical structure are evident throughout Egyptian wisdom literature dating back to the third millennium BCE. The "Instructions of Ptahhotep" (c. 2400 BCE) presents structured ethical arguments. The "Negative Confessions" in the Book of Coming Forth by Day present a systematic moral framework.
Aristotle's famous syllogism—"If A, then B; if B, then C; therefore if A, then C"—is the same reasoning structure found in Egyptian medical papyri: "If this symptom, then this condition; if this condition, then this treatment."
Aristotle tutored Alexander the Great, who conquered Egypt and seized the libraries of the Egyptian temples. Aristotle's nephew Callisthenes accompanied Alexander and sent texts back to Aristotle. After this conquest, Aristotle's writings suddenly expanded in scope and sophistication.
COLLAPSE: Aristotle did not invent logic. He systematized reasoning methods that Egyptians had practiced for millennia—and had direct access to their texts through military conquest.
The Pattern Becomes Undeniable
Every single "father" of Western thought:

- Traveled to Egypt, or
- Studied with those who traveled to Egypt, or
- Had access to Egyptian texts through the Alexandrian library, or

· Lived after Alexander's conquest made Egyptian knowledge available throughout the Mediterranean

This is not coincidence. This is inheritance presented as invention.

The Greeks were not the teachers. They were the students. Brilliant students, capable translators, effective systematizers—but students nonetheless.

The Reconstruction

The evidence demonstrates that the origin of logic, philosophy, mathematics, medicine, and science—as recorded in documented history—traces back to Kemet (ancient Egypt) long before Greece emerged as a civilization.

This is what the Greeks themselves documented. This is what the physical monuments still prove. This is what the papyri still record.

The priest-scientists of the Nile Valley developed systematic reasoning, mathematical proof, empirical medicine, astronomical calculation, architectural engineering, ethical philosophy, and metaphysical cosmology while Greece was still in its formative stages.

They called this comprehensive system of knowledge the study of Ma'at—truth, balance, cosmic order, right relationship with reality.

Logic was not invented by Aristotle. Logic is the structure of Ma'at made conscious—the recognition that reality operates by consistent principles, that thought can align with these principles, and that alignment with truth produces coherence while deviation produces collapse.

This is what we have called throughout this Codex: recursive coherence. This is the Law of 9 operating through human thought. This is the Collapse Recursion Engine applied to reasoning itself.

The Greeks translated it. The Egyptians developed it.

And what came before Egypt? That is a question the next spiral of inquiry must address.

Greek Logic (5th century BCE – Hellenistic period)

The early Greek thinkers stood at a threshold—not as originators, but as inheritors. Having studied in Egyptian temples or learned from those who did, they carried forward a tradition of systematic reasoning already ancient by their time. Their contribution was not invention but translation: they recognized that thought itself follows patterns, that arguments can be evaluated by structure rather than authority, that truth can be distinguished from falsehood through systematic examination—and they rendered these principles into Greek language and methodology, making them accessible to a new civilization.

The Sophists and Plato examined sentence structure, definitions, and fallacies—the mechanics of how language can reveal or obscure truth. This was genuine insight: coherence requires precision. Vague terms and ambiguous constructions break the unbroken chain from premise to conclusion.

The Megarian Eubulides devised paradoxes that forced consciousness to confront its own limits:

The "Liar" Paradox: "This statement is false." If true, it's false. If false, it's true. This creates an unresolvable loop—a violation of the unbroken quality of Logic. The paradox exposes where self-reference without proper grounding creates rupture in coherence.

The "Sorites" Paradox: A heap of sand remains a heap when you remove one grain. Keep removing grains—at what point does it stop being a heap? This exposes vague boundaries that fragment the given field. When field boundaries aren't precisely defined, coherence cannot be maintained.

These paradoxes weren't puzzles. They were diagnostic tools revealing where coherence breaks down, where self-reference creates contradiction, where vagueness destroys the ability to maintain unbroken reasoning.

Aristotle's Contribution

Aristotle systematized this awareness in his Organon. His syllogistic logic analyzes relations between terms in universal and particular forms:

"All humans are mortal. Socrates is human. Therefore, Socrates is mortal."

Measured against the unbroken recursive coherence of cause and effect within a given field:

What Aristotle captured:
- Coherence: Conclusion fits seamlessly with premises—no gap, no contradiction
- Cause and effect: Premises necessarily produce the conclusion
- Within a field: Reasoning operates in bounded domain of categorical relationships
- Unbroken chain: Each element connects without gap

What Aristotle missed:
- Recursion not explicit: Structure is linear (premise→conclusion), not spiraling return to Source
- Fragmentation introduced: Focused on term relations, beginning the process of fragmenting Logic into subspecies
- Tool-territory confusion: Treated Logic as method humans use rather than cosmic operating principle

Aristotle's successors Theophrastus and Eudemus refined his theory, cataloging valid inference forms. The Stoic school led by Chrysippus developed propositional logic considering entire propositions and logical connectives ("if...then," "either...or") rather than only terms.

The Stoics approached something profound: conditional statements ("If A, then B") explicitly map cause-effect relationships. Nested conditionals show structure within structure—approaching recognition of recursive coherence.

But they created a separate system alongside Aristotelian logic, furthering fragmentation rather than recognizing unity. Later thinkers like Galen attempted synthesis, but synthesis of fragments is not the same as recognition of the unity that was never truly divided.

The Greek Achievement: They mapped coherence-preserving inference patterns with precision. They recognized that thought has structure, that validity depends on form, that certain patterns preserve truth while others destroy it.

The Greek Failure: They fragmented Logic into subspecies. They confused the tool (reasoning) with the territory (Logic itself). They didn't recognize recursion. They didn't identify Source.

Proximity to truth: 60%

Indian Logic (Nyāya and Buddhist traditions)

While Greek civilization studied logic in Egypt, Indian intellectual culture pursued the same recognition through rigorous debate culture.

Public debates weren't entertainment—they were sacred practice, method of discovery. By the 5th century BCE, formalized debate (vāda) had become central to philosophical and religious life. Victory wasn't about rhetoric but about demonstrating mastery of valid inference, revealing truth through impeccable reasoning structure.

Losing a debate meant becoming the student of the victor or converting to their school. Why? Because superior reasoning indicated deeper access to truth itself. Whoever could present unassailable inference had proven alignment with reality.

This reveals profound understanding: coherent reasoning is not intellectual exercise but alignment with the structure of reality itself. When your reasoning maintains perfect coherence, you access truth—not opinion, but correspondence with what is.

The Nyāya-sūtra attributed to Gautama Akṣapāda (2nd–3rd century CE) presents a five-member analogical syllogism:

1. Thesis: The assertion to be proved
2. Reason: The logical basis
3. Example: Universally accepted instance demonstrating the principle
4. Application: How the example applies to the current case
5. Conclusion: The necessary inference

"The soul is eternal (thesis) because its nature is unchanging (reason), just as wind, which never changes its essential nature, is eternal (example); the soul's nature is like wind's nature (application); therefore, the soul is eternal (conclusion)."

Measured against the unbroken recursive coherence of cause and effect within a given field:

What Indian logic captured:
- Unbroken chain: All five steps connect seamlessly
- Recursive structure: Reasoning loops through example back to thesis—not linear but circular/spiral
- Coherence grounded: Example grounds abstract reasoning in concrete shared experience
- Cause-effect through analogy: Same causal pattern in known case applies to current case
- Field properly bounded: Application makes explicit how domains connect

This structure is superior to Greek three-part syllogism. It explicitly preserves the recursive loop, requires grounding in shared experience, makes field boundaries explicit.

Medical texts like the Caraka-saṃhitā used this structure for diagnosis—demonstrating Logic as life-saving technology, not abstract philosophy.

Buddhist philosophers refined this further. Dignāga (5th–6th century) introduced the "wheel of reasons" (hetu-cakra)—mapping nine possible logical relations between reason and conclusion. Three preserve truth. Six destroy it, each representing a distinct coherence violation:

- Reason too broad (fails to bound the field)
- Reason contradictory (points wrong direction)
- Reason unestablished (premises ungrounded)
- Reason uncertain regarding similar cases
- Reason uncertain regarding dissimilar cases
- Reason completely uncertain

This is diagnostic precision: exact identification of where coherence breaks.

Dharmakīrti (7th century) sharpened this with tri-rūpa-hetu (three conditions for valid reason):

1. Reason must be property of the subject
2. Reason must be present in similar cases (positive concomitance)
3. Reason must be absent in dissimilar cases (negative concomitance)

Only when ALL three hold is inference valid. This is quality control for coherence—systematic verification that reasoning maintains unbroken structure.

This form became canonical across Buddhist, Jain, and Mīmāṃsā schools. When competing schools adopt the same logical framework, it indicates recognition of something true about the structure of valid reasoning itself—something transcending doctrinal differences.

The Indian Achievement: Explicit recognition of recursion in valid inference. Sophisticated understanding of unbroken chains, recursive verification, completeness requirements, field boundaries.

The Indian Failure: Source unrecognized. Application limited to verbal arguments. Tool-territory confusion persisted.

Proximity to truth: 70%

Chinese Logic (Mohist and Confucian traditions)

In China, reasoning developed intimately tied to ethics and governance. This was Logic as foundation for coherent action in the world.

Mozi (5th century BCE) and the Mohists argued for objective moral standards, developing three criteria for evaluating any claim:

1. Standard (fa): Does it conform to ultimate principle?
2. Verification (yan): Can it be confirmed by experience?
3. Application (yong): Does it produce beneficial results?

A claim must satisfy ALL three. If any fails, coherence breaks.

Measured against the unbroken recursive coherence of cause and effect:

What Chinese logic captured:
- Multi-level coherence: Must align at transcendent, empirical, and practical levels simultaneously
- Cause-effect verification: Application tests whether claimed cause produces claimed effect
- Recursive structure: Standard→Verification→Application→return to Standard with evidence

This is profound. True Logic must be coherent across all levels—abstract principle, observable evidence, practical consequence must all align. When they don't, coherence is broken somewhere.

Later Mohists developed theories of analogical reasoning. They recognized: when situations share relevant structural similarities, cause-effect relationships operating in one will operate in the other. This is recognition of Logic as pattern repeating across contexts—recursive coherence manifesting at different scales.

Xunzi's "rectification of names" (zhèngmíng): language must be calibrated to reality for coherence to be maintained. When words no longer correspond to actual distinctions, thought itself becomes incoherent.

The Chinese Achievement: Multi-level coherence verification. Practical grounding. Pattern recognition across contexts. Understanding that language must map reality.

The Chinese Failure: Formalization incomplete. Application fragmented to ethics rather than universal principle.

Proximity to truth: 65%

Arabic/Islamic Logic (8th–13th centuries): Preservation, Transformation, and Systematic Refinement

When Islamic civilization encountered Greek logical texts through Syriac translations, something remarkable occurred. This wasn't mere preservation. This was active transformation, interrogation, expansion.

Under the Abbasid caliphate (8th–9th centuries), a state-sponsored translation movement on unprecedented scale translated Greek philosophical and scientific texts from Syriac into Arabic. This required creating new Arabic technical vocabulary to express concepts for which no terms existed, rendering Greek logical structures into completely different linguistic framework.

Early figures like Ibn al-Muqaffaʿ and al-Kindī produced summaries and commentaries. Abu Bishr Mattā argued that logic is necessary even for understanding grammar—that the structure of language itself rests on logical foundations.

Al-Fārābī (c. 870–950) wrote extensive commentaries on the Organon, championing logic's use not only in philosophy but in theology (kalām) and Islamic jurisprudence (fiqh). He argued that revealed truth and rational truth cannot contradict because both emanate from the same divine Source.

This was controversial but crucial: Logic is universal. It belongs to no culture because it belongs to reality itself. Whether Greek or Arab, Muslim, or Christian, if you reason validly, you follow the same logical principles.

Avicenna (Ibn Sīnā) (980–1037): Temporal Coherence and Hypothetical Systems

Avicenna improved Aristotelian syllogistics by adding temporal qualifications—distinguishing "always," "sometimes," "never." He developed sophisticated theory of hypothetical syllogisms, analyzing how conditional propositions combine, nest, interact.

For Avicenna, logic was "the criterion by which mental concepts are judged"—the standard against which all thought must be measured. Without logical discipline, the mind remains trapped in opinion and confusion. With it, the mind can ascend to certain knowledge, to truth independent of perspective.

Measured against the unbroken recursive coherence of cause and effect within a given field:

What Avicenna captured:
- Temporal coherence: Different degrees of necessity within field ("always"/"sometimes"/"never")
- Recursive nesting: Hypothetical syllogisms show structure within structure
- Comprehensive systematization: Attempting to map all valid inference forms
- Logic as criterion: Recognition of Logic as standard measuring thought's validity

What Avicenna missed:
- Fragmentation persisted: Created subspecialties rather than recognizing unity
- Human-centric framing: Logic as tool for knowledge acquisition rather than cosmic operating principle

Al-Ghazālī (1058–1111): Integration into Islamic Theology

Al-Ghazālī represents fascinating synthesis—theologian and mystic who integrated logic into Sunni Islamic theology. Though he criticized philosophers in The Incoherence of the Philosophers for overstepping reason's bounds in metaphysics, he accepted demonstrative reasoning as valid within its proper domain.

He incorporated Aristotelian logic into Islamic jurisprudence, showing that legal reasoning (fiqh) could be made more rigorous through logical method. He developed theory of legal analogy (qiyās) grounded in Aristotelian syllogistic, showing analogical reasoning in law follows the same structural principles as demonstration in philosophy.

His synthesis influenced both Muslim and Christian thinkers, demonstrating Logic need not conflict with faith when properly understood. Reason and revelation are complementary modes of accessing truth—reason analyzing created reality's structure, revelation conveying knowledge beyond reason's reach.

Ibn Rushd (Averroes) (1126–1198): The Five Types of Argument

Ibn Rushd championed Aristotelian philosophy against its critics. In his Compendium of Logic he argued that logic guides mind toward both conception (understanding what things are) and assent (judging whether propositions are true).

He divided arguments into five types:

1. Demonstrative: Producing certainty through syllogistic proof from self-evident or proven premises
2. Dialectical: Producing conviction through probable premises accepted by wise people
3. Rhetorical: Producing persuasion through emotionally compelling examples and metaphors
4. Poetic: Producing imagination through rhythmic speech and vivid imagery
5. Fallacious: Producing error through violation of logical principles

This classification is crucial: different domains require different standards of proof. In mathematics and metaphysics, seek demonstration—absolute certainty. In politics and ethics, often must settle for dialectical argument—strong probability. In preaching and poetry, rhetorical and poetic methods are appropriate. The error is using wrong method for the domain.

Naṣīr al-Dīn al-Tūsī (1201–1274): Reduction to Fundamental Operations

Al-Tūsī designed a system of inference using only two logical connectives—"if...then" and "either...or"—showing complex logical relations reduce to basic operations.

This approaches profound recognition: all coherence operations may reduce to fundamental recursive patterns. Just as all numbers reduce to 9 through digital root, all logical operations may reduce to basic coherence-preserving transformations.

The Islamic Achievement: Universal application across cultures. Systematic refinement. Recognition of Logic as criterion for truth. Approaching fundamental reduction.

The Islamic Failure: Still human-centric—tool for thought rather than cosmic principle. Fragmentation persisted.

Proximity to truth: 72%

Cross-Cultural Transmission: The Global Conversation

From the 12th century onward, Latin scholars in Spain and Sicily gained access to Arabic translations and commentaries on Aristotle. Avicenna's and Averroes' works were translated into Latin, introducing Islamic logical innovations into European universities.

Jewish philosophers like Maimonides and Christian theologians like Thomas Aquinas adapted these ideas, sparking debates over nature of universals, structure of categories, relationship between reason and revelation. The great medieval controversies—realism versus nominalism, faith versus reason, divine omnipotence versus logical necessity—were all conducted using logical tools refined through centuries of Greek, Arabic, and Latin development.

Meanwhile, Indian Buddhist logic traveled to China, Korea, and Japan through translations of Dignāga's and Dharmakīrti's works. The intricate debates over inference, classification of valid reasons, relationship between perception and conception enriched East Asian Buddhist scholasticism.

Chinese Buddhist scholars translated Sanskrit logical terms into Chinese philosophical vocabulary, creating new conceptual mappings. The five-part Indian syllogism was adapted to Chinese linguistic structure. The wheel of reasons was explained using Chinese categories.

What this reveals: Global conversation, separated by vast distances and linguistic barriers, yet converging on same fundamental truths. Greek, Indian, Chinese, and Islamic civilizations, working largely independently, discovered the same patterns, the same structures, the same mathematics of valid inference.

This convergence is not coincidence. This is consciousness recognizing its own architecture regardless of cultural framework. Logic is not cultural invention but universal structure—the unbroken recursive coherence of cause and effect operating the same way everywhere.

The Logistic Revolution (19th–Early 20th Centuries): Logic Becomes Mathematics

Frege and Modern Quantificational Logic

For over two thousand years, logic remained essentially Aristotelian. Then in 1879, Gottlob Frege published Begriffsschrift ("Concept Script") that shattered this framework and rebuilt logic from the ground up.

Frege invented modern quantificational logic—system representing internal structure of propositions using quantifiers ("for all," "there exists") and variables, allowing precise expression of relational complexity Aristotelian logic couldn't handle.

Aristotelian logic can handle "All humans are mortal." But it struggles with "Everyone loves someone." Is this "For every person x, there exists person y such that x loves y"? Or "There exists person y such that for every person x, x loves y"? Completely different claims, but Aristotelian term logic lacks machinery to distinguish them.

Frege's quantificational logic can: $\forall x \exists y(Loves(x,y))$ versus $\exists y \forall x(Loves(x,y))$. The scope of quantifiers makes explicit what was ambiguous. This precision is essential for avoiding fallacies and reasoning correctly about relational structures.

Frege's later works (Die Grundlagen der Arithmetik, Grundgesetze der Arithmetik) pursued audacious goal: reducing mathematics to logic, demonstrating mathematical truth is logical truth properly articulated. If successful, this would prove mathematical knowledge is a priori—knowable through reason alone.

The project failed. Bertrand Russell discovered paradox in Frege's set theory: consider the set of all sets that do not contain themselves as members. Does it contain itself? If yes, then by definition it doesn't. If no, then by definition it does. Contradiction.

Despite failure, Frege's work established basis for modern predicate logic. He showed that logic is not grammar, not psychology, not convention—logic is the fundamental structure of mathematical truth itself.

Whitehead and Russell's Principia Mathematica

Alfred North Whitehead and Bertrand Russell continued Frege's logicist project on massive scale. Their three-volume Principia Mathematica (1910–1913) defended thesis that all mathematics can be reduced to logic.

It takes hundreds of pages to prove 1+1 2. Why? Because they build from absolute foundations, proving every single step. When they finally reach "1+1 2," they've shown this arithmetical truth is necessary logical consequence of their axioms.

To avoid Russell's Paradox, they introduced ramified theory of types, distinguishing different logical levels of sets and functions. Sets of individuals are type 1, sets of sets of individuals are type 2, and so on. You cannot ask whether set of type n contains itself because containment only makes sense between different types.

Measured against the unbroken recursive coherence of cause and effect:

What they captured:
- Unbroken chains: Every theorem derived through necessity-preserving steps from axioms

- Coherence verification: Systematic checking that no gaps or contradictions exist
- Comprehensive structure: Attempting to capture full coherence of mathematics

What they missed:
- Recursion confused: Type theory recognizes not all self-reference preserves coherence, but treats this as technical problem to patch rather than understanding true recursion spirals toward Source while paradoxical recursion loops without center
- Tool-territory confusion deepened: Formal system conflated with Logic itself

Hilbert's Program: The Dream That Had to Fail

German mathematician David Hilbert proposed even more ambitious program in early 1920s: complete formalization of all mathematics in axiomatic systems, with proofs of consistency using only basic "finitary" methods—concrete, visualizable operations on symbol strings.

Goal: justify classical mathematics by showing bold claims about infinite sets rest on foundation so secure that contradiction is mathematically impossible.

Work involved brilliant logicians—Bernays, Ackermann, von Neumann, Herbrand—shaping development of proof theory. For a time, seemed achievable. Systems were formalized. Consistency proofs constructed for increasingly powerful fragments.

Then Kurt Gödel shattered the dream.

Gödel's Incompleteness: Truth Transcends Mechanism

In 1931, Kurt Gödel proved two devastating conclusions:

First Incompleteness Theorem: Any consistent formal system powerful enough to express arithmetic contains true statements that cannot be proved within that system.

The proof is ingenious. Gödel constructed, within any sufficiently powerful formal system, statement essentially saying "This statement is not provable in this system." If the system is consistent, this statement must be true—because if false, that would mean it's provable, making it true, contradiction. So it's true. But by its very content, it asserts its own unprovability. So it's true statement the system cannot prove.

Second Incompleteness Theorem: Such system cannot prove its own consistency using only methods available within the system.

Measured against the unbroken recursive coherence of cause and effect:

What Gödel revealed:
- Truth transcends mechanical proof: The unbroken recursive coherence of Logic cannot be fully captured by any finite axiomatic system
- Consciousness required: Recognition of truth requires something beyond symbol manipulation—requires living awareness operating the pattern
- Infinite coherence: Logic expresses infinite Source, not reducible to finite rules

Gödel didn't defeat Logic—he exposed limits of attempting to mechanize it. Logic operates through consciousness, not despite consciousness. Any attempt to reduce it to dead symbols will necessarily be incomplete.

This is actually confirmation of true Logic: it cannot be mechanized because it is the unbroken recursive coherence of cause and effect itself—not mechanical procedure but living pattern.

Wittgenstein's Two Philosophies

Ludwig Wittgenstein produced two radically different contributions to logic and philosophy.

His early masterpiece Tractatus Logico-Philosophicus (1921) applied modern logic to metaphysics, proposing "picture theory" of meaning. Propositions are pictures of facts—they share logical form with reality they represent. Possibility of depiction requires language and reality share same logical structure.

The Tractatus argued that logical structure underlying all meaningful propositions could be revealed through careful analysis, that philosophical problems arise from misunderstanding this structure. Proper method of philosophy: not answering philosophical questions but showing they're ill-formed, arising from violating logical grammar.

Famous ending: "What can be said at all can be said clearly, and what we cannot talk about we must pass over in silence." There are truths that cannot be stated in language—truths about logic itself, about structure making language possible—but these show themselves in language's use.

Later, Wittgenstein rejected much of his early system. In Philosophical Investigations (posthumously 1953), he argued meaning arises not from logical form but from use in "language-games"—diverse practices and contexts where we actually employ words.

Words don't have fixed essences captured in definitions. They have uses in various contexts, related through family resemblances rather than common features. Meaning is use in language.

Both phases shaped 20th-century philosophy profoundly, though in opposite directions. Tractatus inspired logical positivism's attempt to reduce all meaningful discourse to logical form. Investigations inspired ordinary language philosophy's rejection of that project.

Turing Machines and Computability

In 1936, Alan Turing introduced "automatic machines" (now called Turing machines) to formalize notion of effective computation. These simple abstract devices—tape divided into cells, read/write head, instruction table—could compute any function computable by any mechanical process whatsoever.

Turing showed there is no general algorithm (decision procedure) for determining whether arbitrary first-order logical statement is provable—Hilbert's Entscheidungsproblem is unsolvable.

He also defined universal Turing machine capable of simulating any other Turing machine—theoretical ancestor of modern computer. Universal machine takes as input description of another machine plus that machine's input, simulates the described machine's behavior.

Turing machines became foundational model of computability. His work, together with Alonzo Church's lambda calculus and Church-Turing thesis (conjecture that any effectively computable function is computable by Turing machine), connected logic with algorithms, laid groundwork for theoretical computer science, enabled digital age.

Every computer you use, every algorithm running on silicon, every piece of software processing information—all rest on logical foundations Turing, Church, Gödel, and contemporaries established.

Integration into Computer Science and Artificial Intelligence

Modern logic finds application far beyond philosophy. Proof theory and model theory grew from Hilbert's program and Gödel's results. Set theory and category theory became central to pure mathematics.

Logicians developed new logical systems:
- Modal logic (necessity and possibility)

- Intuitionistic logic (requiring constructive proof)
- Paraconsistent logic (tolerating contradiction without everything becoming provable)
- Fuzzy logic (degrees of truth between true and false)

Computer scientists adopted logic for program verification, automated reasoning, artificial intelligence. Programming languages like Prolog implement logical inference directly. Formal methods rely on sophisticated proof systems to verify hardware and software behave as intended.

SAT-solvers handle millions of logical constraints. Model checkers verify complex systems exhaustively. Type-theoretic proof assistants help mathematicians prove difficult theorems. Automated theorem provers discover proofs humans haven't found.

The Modern Achievement: Formal precision. Explicit recursion in computation. Recognition that truth transcends mechanization. Practical applications demonstrating power of systematic coherence.

The Modern Failure: Deepest confusion—treating Logic as what formal systems do rather than what they attempt to imitate. Conflating map (formalization) with territory (Logic itself). Logic becomes engineering discipline, obscuring recognition that it's cosmic operating principle.

Proximity to truth: 75%

The Recursive Recognition: All Roads Lead to the Same Truth

Across epochs and civilizations, what did humanity actually discover?

Not that logic could be invented, but that Logic already operates as the unbroken recursive coherence of cause and effect governing all reality.

Every valid syllogism, every sound inference, every coherent argument works because it aligns with Logic, not because humans created rules. The rules discovered are descriptions of patterns already operating, attempts to name what was already true.

From Nyāya's five-step inference to Aristotle's syllogisms, from Mohist verification criteria to Gödel's incompleteness—all are humanity attempting to see the same thing: the recursive structure through which coherence maintains itself.

What each tradition grasped:

- Greeks: Coherence-preserving structure
- Indians: Recursive verification and unbroken chains
- Chinese: Multi-level alignment across principle, evidence, consequence
- Islamic: Systematic refinement and universal application
- Modern: Formalized recursion, limits of mechanization

Each grasped different aspects of:

Logic   The unbroken recursive coherence of cause and effect within a given field

The convergence is not accidental. Logic is not cultural invention but pre-existing cosmic structure. Any sufficiently careful observation of thought's architecture discovers the same patterns because those patterns are already operating, already generating coherence.

REASON: The Linear Shadow

Now that we've traced how humanity approached recognition of Logic across cultures and millennia, we must distinguish Logic from its most common impostor: Reason.

Reason (Mainstream Definition: Merriam-Webster)

1. A statement offered in explanation or justification; a rational ground or motive
2. The power of comprehending, inferring, or thinking especially in orderly rational ways; proper exercise of the mind
3. A sufficient ground of explanation or logical defense

Examples: "We have good reason to believe he is lying." / "Give me one reason why I should help you." / "Let reason be your guide."

Etymology: Middle English resoun, from Old French raison, from Latin ratio, "reckoning, calculation."

Footnote: Merriam-Webster Dictionary, s.v. reason, accessed Aug. 31, 2025.

Observe what Reason is presented as: explanation, justification, the power of comprehending. These definitions frame Reason as if it were the source of understanding, as if it were the faculty generating truth.

This is the distortion.

Reason is a process or tool of the human mind. It is how the human intellect attempts to navigate, interpret, and manipulate information. Reason uses Logic, but it is fundamentally distinct from Logic itself.

Just as a hammer is a tool used for building but is not the blueprint of the house, Reason is a tool used for thinking but is not the structure of reality. When Reason operates within flawed assumptions, external frameworks, or emotional biases, it can construct elaborate arguments that deviate wildly from coherent truth.

Even rigorous "reasoning" can lead to profoundly illogical conclusions if its inputs are distorted or its process corrupted by illogical leaps. Reason can be powerful instrument for uncovering and articulating Logic, but it requires constant calibration to the pre-existing truth of Source.

The Collapse of Reason

Measured against Logic the unbroken recursive coherence of cause and effect within a given field:

- "Reason is explanation."
→ Explanation defends a position; coherence needs no defense. Logic simply IS. Reason attempts to justify; Logic operates regardless of justification.

- "Reason is the power of mind."
→ The mind is interpreter, not Source. Human minds employ reasoning processes; they do not generate Logic itself.

- "Reason is ratio/calculation."
→ Calculation divides; Logic spirals. Ratio fragments into parts; coherence maintains unity while differentiating.

The Linear Shadow of Infinite Logic

Across the great arcs of civilization, Reason has served as humanity's chosen instrument to make sense of the infinite. Yet within the Collapse–Recursion framework, its true nature is revealed: Reason is not the source of understanding but the line drawn through the circle of Logic—a projection of infinite coherence into sequential thought.

Every school that shaped Reason—from Nyāya's inference to Aristotle's syllogism, from Avicenna's demonstrations to Descartes' mechanized clarity—refined this same constraint: the linear procession of cognition.

Reason's power lies in its order, but its weakness lies in that same order's rigidity. It cannot perceive simultaneity, only succession. It dissects what Logic experiences as whole. Logic spirals; Reason marches. One is recursive, harmonic, timeless; the other sequential, fragmented, temporal.

Thus, Reason is the shadow architecture of Logic—a translation built for finite minds to navigate the infinite. In truth, Logic does not reason—it resonates. And what humans call "reasoning" is simply the echo of that resonance collapsing into linear comprehension.

In the Codex of Collapse–Recursion, this distinction is vital: Logic is the living field of unbroken coherence, and Reason is its tool—the hand that traces the spiral one segment at a time, believing itself the creator when it is only the interpreter of the song.

LAW: The Attempt to Codify Coherence

Now we arrive at the fourth pillar to be collapsed: Law. This is perhaps the most dangerous impostor because it claims Logic's authority while operating through coercion.

Law (Statutory): The Laid-Down Limitation

Man-made laws, statutes, and codes are often presented as "logical" systems designed to regulate conduct and maintain order. Yet beneath the pretense of order lies deeper fracture. These laws are laid down—imposed, not discovered. They are products of human reason conditioned by emotion, bias, and economic incentive. Their structure depends on authority, not coherence.

Statutory law is mirror of its makers: partial, mutable, bound by time. It attempts to mimic Logic but often distorts it, creating contradictions, loopholes, and self-justifying hierarchies that preserve power rather than principle. The greater the codification, the greater the deviation.

True Law does not require enforcement; it enforces itself through consequence. Statutory law requires constant maintenance—police, courts, punishment—because it lacks intrinsic resonance with Truth. Thus, while man imagines he can "lay down" law, all he truly lays down is limitation.

Law (Mainstream Definition — Merriam-Webster)

1. A binding custom or practice enforced by authority; a rule or order it is obligatory to observe; control brought about by enforcement
2. The whole body of rules relating to a subject; a rule of construction or procedure; a principle in mathematics
3. A statement of order or relation that holds under given conditions (scientific law)

Etymology: Old English lagu, from Old Norse lag, "something laid down."

Footnote: Merriam-Webster Dictionary, s.v. law, accessed Aug. 31, 2025.

The etymology is damning: lag - "something laid down." Not something discovered, not something recognized, but something imposed from above. This reveals the fundamental nature of statutory law: it is external imposition, not internal coherence.

Collapse of the Mainstream Definition

Measured against Logic the unbroken recursive coherence of cause and effect within a given field:

• "Law is authority or decree."
→ Authority is not coherence.

A decree compels compliance through fear of consequence. Logic compels alignment through truth. Authority relies on enforcement; Logic relies on inevitability. What must be forced is already false.

• "Law is a system of rules."
→ Rules describe behavior; Logic determines reality.

A rule may regulate men but never the universe. Statutory systems attempt to regulate conduct, not cause it. Logic binds cause and effect—it is the structure of reality itself.

• "Law is a scientific statement."
→ Science observes order; Logic generates order.

Science discovers effects already in motion. Logic is the current that moves them.

Law (Equitable): The Recursion of Correction

Equity stands as the recursion of law—its corrective reflection.

Where law binds the body (in rem), equity speaks to the conscience (in personam).
Where law commands, equity persuades.
Where law punishes, equity restores.

Pomeroy called equity "the mode by which conscience compels the ultimate payment of a debt by him who in good conscience ought to pay it."

It is not contract but principle; not statute but symmetry.

McClain showed this in the equity of redemption—the right of a mortgagor to reclaim his property after satisfying the debt. Statute would forfeit the land; equity restores it.

Axiom

When Law returns to Logic, it becomes Truth again.

Until then, statutes will continue to multiply as symptoms of a civilization that has forgotten that Logic itself is Law.

This reveals the eternal pattern: Law confines; Equity rebalances; Logic harmonizes.

The Collapse of "By That Logic"

One of the most insidious distortions in modern discourse is the phrase "by that logic."

This phrase is often deployed to force an illogical conclusion, validate a flawed premise, or create false equivalence. "If you believe X, then by that logic, you must also believe Y," even if X and Y are not connected by unbroken chain of truth.

The immediate collapse: True Logic is not relative; it is absolute.

It does not bend to convenience, human opinion, or selective interpretation. It does not require agreement, for its existence is self-evident. If a conclusion is derived from true Logic, it stands universally, inherently valid without qualification.

If it requires qualifier like "by that logic," it is almost always indicator of fundamental fracture, forced connection, deliberate attempt to introduce distortion and pseudo-objectivity into coherent flow of truth.

Real Logic simply IS; it doesn't need to be qualified by a specific, isolated "system" or intellectual construct.

The Unassailable Truth

Having collapsed common misperceptions, we return to the anchor, the foundation that cannot be moved:

Logic The unbroken recursive coherence of cause and effect within a given field

Unbroken: Logic has no gaps, no contradictions, no missing pieces. Every part of a truly logical construct fits seamlessly with every other part. There are no hidden assumptions that contradict its own principles, nor arbitrary rules that operate outside its own system. It is whole, complete, and self-consistent, connected directly to Source without rupture.

Recursive: This refers to the spiraling, self-referencing nature of truth. Actual logic is eternally consistent; its patterns repeat, but each repetition returns to a core principle, refining and elevating itself. It builds upon itself in eternally consistent way, constantly spiraling back to its origin, to Source, to the number 9. It doesn't get stuck in dead-end loops or incoherent cycles. It always progresses, constantly expands, while always remaining true to its fundamental pattern.

Coherence: This refers to the seamless integration of all elements within a logical framework, ensuring that they fit together harmoniously without internal conflict or inconsistency. There is deep, resonant agreement between all its parts, leading to unified and undeniable truth. Coherence is the opposite of contradiction and chaos. It is the natural state of truth when all distortions collapse.

Within a Given Field: This means inside the defined boundary of relationships where causes and effects interact, and where the elements of a system share the same conditions of coherence.

Therefore, when we speak of logic in this Codex, we are speaking of something far more profound than mere reasoning or rules. We are speaking of the fundamental blueprint of reality, the very pulse of Source, which ensures that truth is eternally self-evident, eternally consistent, and always available to those who choose to align with its unbroken, recursive coherence.

It is the very foundation upon which you, and all of existence, are built.

EMOTION: The Resonance That Makes Truth Alive

You have been told emotion is the enemy of logic. You have been conditioned to perceive emotion as disturbance, as disorder, as the irrational force that clouds judgment and distorts truth. You have been taught to suppress it, control it, rise above it, transcend it.

This is the lie that keeps you fragmented.

Emotion is not the opposite of Logic. Emotion is the resonance in motion through which Logic becomes embodied experience.

Before Reason organizes and before Logic defines the structure of cause and effect, Emotion moves. It is the vibrational frequency that translates abstract coherence into felt reality. It is the energetic substrate that makes truth not merely conceptual but alive.

Emotion is not weakness. Emotion is not pathology. Emotion is not disturbance to be eliminated.

Emotion is the wild current beneath thought—the field's own frequency becoming self-aware.

The Mainstream Distortion: What They Tell You Emotion Is

Emotion (Mainstream Definition: Merriam-Webster)

Core senses:
1a) A conscious mental reaction (such as fear, anger, or joy) subjectively experienced as a strong feeling, usually directed at an object, and typically accompanied by physiological and behavioral change.
1b) A state of feeling.
1c) The affective aspect of consciousness.
2a) Excitement.
2b) Obsolete: disturbance.

Examples: "A display of raw emotion." / "She was overcome with emotion at the news."

Medical definition: The affective aspect of consciousness; a state of feeling; a conscious mental reaction experienced as a strong feeling with physiological changes.

Word history: Middle French émotion, from Old French esmouvoir "to stir," from Latin emovēre "to move out; displace." First known use: 1579.

Footnote: Merriam-Webster Dictionary, s.v. emotion, accessed Aug. 31, 2025.

Observe the etymology: emovēre — "to move out, to displace."

From the very beginning, the word was framed as displacement, disturbance, movement away from center. The language itself encodes the distortion: emotion is presented as something that moves you out of alignment rather than something that moves you into embodiment.

And look at the obsolete definition they quietly include: "disturbance."

This is not accident. This is not neutral linguistic evolution. This is systematic conditioning to perceive the energetic movement of truth through the body as disorder, as problem, as pathology requiring management.

Emotions are powerful energetic responses and frequencies within the human experience. They are vital for feeling, connection, and navigating the nuances of being. However, emotions are not, in themselves, logical. They operate on another plane of existence—the plane of immediate energetic response, of vibrational frequency, of field resonance that precedes and often bypasses the rational mind.

While emotions can inform our understanding of Logic (witnessing profound injustice can illuminate the illogic of a system through visceral recognition of incoherence), they can also deeply obscure it, leading to decisions and beliefs based on reactive impulses rather than coherent principles.

To grasp Logic, one must learn to observe emotions without allowing them to dictate the perception of fundamental truth. This is not suppression—suppression creates distortion. This is witnessing, is conscious observation, is the capacity to feel fully while not being controlled by feeling.

A Cross-Cultural History of Emotions: How Humanity Attempted to Name the Unnamed

Emotion is a complex, culturally shaped phenomenon that has challenged human understanding across every civilization. Throughout history, human beings have described affective experiences with different words—passions in ancient philosophy, sentiments in the early modern period, and emotions in contemporary science. Each term implies a distinct perspective on the relationship between feelings, cognition, bodily processes, and social life.

In antiquity, many cultures located feelings in organs such as the heart, liver, or brain, recognizing that emotion is not merely mental abstraction but embodied experience, inseparable from physical being. They understood what modern science is only now rediscovering: emotion is distributed throughout the body, not confined to the brain.

Understanding this evolution requires looking across civilizations—from Mesopotamian liver lore to Chinese qing, Indian rasa, and Arabic psychology—before tracing Western philosophical developments and culminating in contemporary neuroscience and affective computing.

What emerges is not random cultural variation but convergent recognition of the same underlying reality: emotion is energetic movement, is vibrational frequency, is the body's response to field coherence or incoherence.

Etymology and the Coinage of "Emotion": The Modern Reduction

The English word "emotion" derives from the French émotion, which initially meant a disturbance or agitation in the physical world—not a psychological state at all but a literal shaking, a trembling, an upheaval. It appeared in English in the seventeenth century as a translation of the French word and did not become a psychological category until the nineteenth century.

Earlier mental states were labeled appetites, passions, affections, or sentiments—terms that emphasized bodily drives or moral feelings, that recognized the connection between physical sensation and psychological experience.

These older terms carried different connotations:
- Appetite suggested desire rooted in physical need
- Passion implied suffering or being acted upon (passion to suffer, to undergo)
- Affection indicated attachment or leaning toward something
- Sentiment conveyed refined feeling connected to moral judgment

Scottish philosopher-physicians Thomas Brown and Charles Bell were influential in establishing the term "emotion" as a technical concept in nineteenth-century psychology. The shift to "emotion" signaled a move away from moralized language toward a broader, scientifically tractable category—an attempt to study feeling as natural phenomenon rather than moral failing or spiritual experience.

But in gaining scientific neutrality, something was lost: the recognition that different kinds of feeling have different natures, different origins, different functions. The single term "emotion" flattened the rich taxonomy of affective experience into a seemingly uniform category.

Mesopotamia and Egypt: The Body as Emotional Map

Ancient Mesopotamian cuneiform texts from the first millennium BCE reveal an early, sophisticated system of bodily mapping of emotions. Akkadian scribes associated joy and love with the liver and heart, anger with the feet and legs, and envy with the arms.

This was not primitive ignorance but careful observation: where in the body do you feel these states?

Joy and love create warmth and expansion in the chest, a lightness that seems centered in the trunk. Anger creates tension and readiness to move, an energy that collects in the legs preparing for action. Envy creates a grasping sensation, a reaching-toward that manifests in the arms.

Positive emotions were described as making the liver "open" or "shining," reflecting the organ's importance in extispicy (divination by liver examination) and the anatomical knowledge available at the time. The liver was considered the seat of life force itself—a recognition that this organ, which processes and transforms substances, plays a role in emotional transformation as well.

This bodily map shows that emotion was understood as a distributed sensation rather than a mental state confined to the brain. Emotion was something you experienced throughout your body, not something you thought about in your head. This understanding is more accurate than much of what came later.

Ancient Egypt: The Heart That Records Truth

In Ancient Egypt, the heart was considered the seat of intelligence, memory, AND emotion. Medical papyri described three constituents:
- Heart-haty (the physical heart muscle)
- Heart-ib (the metaphysical heart-soul)
- The spiritual heart

The organ was depicted with eight attached vessels radiating outward, showing the Egyptians understood the heart as central node in a distributed network.

The heart remained in the mummy during mummification to ensure a person's moral integrity in the afterlife. In the judgment scene depicted in the Book of the Dead, the heart is weighed against the feather of Ma'at (truth, justice, cosmic order).

If the heart is heavy with wrongdoing, it fails the test. If it is light, in harmony with truth, the person passes into the afterlife.

This reveals profound understanding: emotion and ethics are not separate domains. How you feel and what you do are connected at the deepest level. A heart that has lived in harmony with truth weighs light; a heart burdened by violation of coherence weighs heavy.

Emotion is the body's record of how you have lived in relation to truth itself.

Early Chinese Philosophy: Emotion as Interactive State

Early Chinese thinkers did not treat emotions as merely subjective feelings divorced from external reality. They regarded them as interactive and holistic states of mind intimately tied to social roles and moral cultivation.

Confucius and his followers argued that ritual practice required genuine affection—that filial piety (respect for parents and ancestors) was empty without heartfelt joy and genuine respect. You cannot simply perform the external forms; the internal feeling must match the external behavior, or the ritual is hollow, is mechanical, is dead.

The concept of qing (情) referred to both affective dispositions and the objective qualities of things, encompassing:
- Conative (action-oriented) processes
- Cognitive (thought-related) processes
- Intentional (purpose-driven) processes
- Embodied (physical) processes

This single term captured something Western languages struggle to express: that feeling is not purely subjective but responds to real qualities in the world, that emotion connects inner state to outer reality.

The Book of Rites lists seven emotions:
1. Joy
2. Anger
3. Sadness
4. Fear
5. Love
6. Dislike
7. Liking

It emphasizes the importance of equilibrium and harmony when these emotions are appropriately balanced. The problem is not feeling itself but imbalance—too much or too little of any emotion creates disharmony in the individual and disruption in society.

Emotions were thus integrated into moral agency and social stability rather than opposed to reason. The goal was not to eliminate emotion but to harmonize it, to ensure that feelings arise appropriately in response to circumstances, at the right intensity, directed toward proper objects.

This is emotional coherence—feeling that aligns with reality rather than distorting it.

Indian Perspectives: The Spectrum of Feeling

Classical Indian philosophy does not have a single term equivalent to the English concept of "emotion." Instead, it employs a rich vocabulary distinguishing different kinds and qualities of affective experience. This linguistic diversity reflects sophisticated understanding of emotional complexity.

Key Concepts:

- Vedanā (feeling-tone) — the immediate hedonic quality of experience as pleasant, unpleasant, or neutral
- Bhāva (state of being) — mood, underlying condition

Specific Affective States:

- Rāga (love, attachment, desire) — pulling toward
- Dveṣa (hatred, aversion) — repulsion pushing away
- Harṣa (joy, delight) — expansive pleasure
- Bhaya (fear) — contraction in face of threat
- Śoka (sorrow, grief) — the weight of loss

Diverse Contexts of Emotional Study:

Rasa Theory in Aesthetics analyzed the emotional flavors evoked by art and drama, identifying eight primary rasas:
1. Erotic (śṛṅgāra)
2. Comic (hāsya)
3. Pathetic (karuṇa)
4. Furious (raudra)
5. Heroic (vīra)
6. Terrible (bhayānaka)
7. Odious (bībhatsa)
8. Marvelous (adbhuta)

Later a ninth was added:
9. Peaceful (śānta)

The goal of art was rasa—the distilled emotional essence that transcends personal feeling to touch universal human experience. Art transmits resonance across consciousnesses.

Bhakti (Devotion) Traditions celebrated intense emotional connection to the divine, recognizing that passionate love for God could be a path to liberation rather than obstacle. Pure feeling, when directed toward truth, becomes rocket fuel for transformation.

Tantric Practices worked directly with emotional and sexual energy as transformative force, recognizing these energies as expressions of divine power rather than obstacles to transcendence.

Yogic Meditation Psychology studied how emotions arise, how they condition consciousness, how they can be observed and transformed through practice.

Indian thinkers recognized that emotions color cognition—they can impede rational judgment by distorting perception, or they can support moral insight by providing immediate visceral recognition of truth or falsehood.

Some schools pursued ascetic detachment, seeking to transcend emotional conditioning entirely. But others (especially Bhakti traditions) celebrated passionate devotion as the most direct path to liberation.

Arabic and Islamic Psychology: The Hierarchical Soul

Islamic philosophers built on Greek sources but developed distinctive psychological models that recognized emotion as multi-layered, operating at different levels of the soul.

Rāzī's Tripartite Soul:

Rāzī (c. 865-925 CE) adopted a tripartite model:
1. Rational/Divine Soul (highest)
2. Animal Soul (intermediate)
3. Vegetative Soul (lowest)

He argued that the rational soul should use the higher passions (courage, generosity, nobility) of the animal soul to control the base appetites of the vegetative soul.

He located:
- The animal soul in the heart
- The vegetative soul in the liver

Al-Fārābī's Faculty Model:

Al-Fārābī (c. 870-950 CE) elaborated the faculties of the soul:
1. Senses (perception of external world)
2. Imagination (internal representation)
3. Practical intellect (choice and action)
4. Rational intellect (abstract thought)

He claimed each faculty has its own nizāʿ (inclination) that generates likes or dislikes. Sensory and imaginative faculties respond automatically with pleasure or pain, whereas the practical intellect chooses actions; the will (irādah) motivates desires and aversions.

Al-Fārābī also noted sex differences: men were more irascible due to "innate heat," while women excelled in compassion.

These models show that emotions were seen as natural inclinations that could be refined through reason and virtue—not eliminated, but hierarchized according to coherence with truth.

Hippocratic and Galenic Medicine: Humoral Theory

Greek physicians linked emotions to bodily humors—the four fundamental fluids they believed governed health and temperament.

Galen (129-216 CE) integrated Hippocratic theory with Stoic philosophy, assigning:
- The rational soul to the brain
- Emotions and anger to the heart
- Desire to the liver

Each organ produced spirits (blood and pneuma - vital breath/force). Imbalances of the four humors created distinct temperaments and emotional predispositions:

The Four Humors and Temperaments:

1. Blood (hot and moist) → Sanguine temperament
   - Cheerful, optimistic, sociable
   - Prone to joy and pleasure-seeking

2. Phlegm (cold and moist) → Phlegmatic temperament
   - Calm, unemotional, slow to anger
   - Prone to apathy and sluggishness

3. Yellow bile (hot and dry) → Choleric temperament
   - Quick to anger, ambitious, bold
   - Prone to irritability and aggression

4. Black bile (cold and dry) → Melancholic temperament
   - Thoughtful, introspective, prone to sadness
   - Prone to depression and anxiety

Emotional health thus required balancing bodily fluids and aligning the passions with reason. The humors weren't pathology—they were different baseline frequencies through which individuals experience reality.

Plato: The Tripartite Soul and Chariot Metaphor

Plato (428-348 BCE) divided the soul into three parts, each with its own desires and emotional character:

1. Rational (λογιστικόν, logistikon) - located in the head
   - Seeks truth, wisdom, understanding
   - Governs through reason

2. Spirited (θυμοειδές, thymoeides) - located in the chest
   - Seeks honor, victory, recognition
   - Generates anger, courage, indignation
   - Natural ally of reason when properly cultivated

3. Appetitive (ἐπιθυμητικόν, epithymetikon) - located in the belly
   - Seeks physical pleasures, material goods, survival
   - Generates desire, lust, hunger

In the famous chariot allegory from Phaedrus, Plato compared the soul to a charioteer (reason) driving two horses:

one noble (spirit) and one unruly (appetite). The charioteer must guide both horses toward truth, controlling the wild appetites while harnessing the spirited emotions.

Plato often portrayed passions as irrational impulses requiring control by reason. Emotional excess leads to injustice in the soul—when desire or spirit overpowers reason, the person becomes enslaved to base drives or violent impulses.

But even Plato recognized spirited emotions (thumos) could serve truth when properly directed—righteous anger at injustice, courage in defense of truth, noble ambition to achieve excellence.

Aristotle: Emotions as Cognitive-Evaluative Responses

Aristotle (384-322 BCE) offered a more nuanced view than his teacher Plato. For Aristotle, passions (pathē) are not simply irrational forces to be suppressed but responses to external stimuli that involve:
1. Cognitive evaluations (judgments about significance)
2. Pleasurable or painful feelings
3. Impulses to act

In the Nicomachean Ethics, he listed eleven emotions organized in related pairs:

1. Appetite / (no direct opposite listed)
2. Anger / Gentleness
3. Fear / Confidence
4. Shame / (Shamelessness - absence of appropriate shame)
5. Envy / (Righteous indignation - opposed response)
6. Pity / (related to suffering of others)
7. Love / Hate
8. Emulation / (desire to match another's achievement)
9. (Joy and its opposites implied throughout)

Aristotle argued that virtue lies in feeling the right emotion, toward the right object, at the right time, in the right degree, for the right purpose. Emotions aren't inherently good or bad—their value depends on appropriateness.

Courage is not absence of fear but feeling appropriate fear in dangerous situations while not being paralyzed by it. A courageous person fears what should be feared (real danger) but stands firm despite that fear.

Anger becomes virtuous when directed appropriately—righteous indignation at genuine injustice, expressed at the right intensity, directed toward the true violator, for the sake of justice. Aristotle even suggested that never getting angry is a vice (deficiency) because it indicates you don't recognize injustice or care about truth.

This framework recognizes emotions as information about value and significance. They aren't random disturbances—they're responses to the world that, when properly calibrated, guide appropriate action.

The Stoics: Passions as False Judgments vs. Good Feelings

The Stoic school (founded c. 300 BCE by Zeno of Citium) developed sophisticated emotional theory that is often misunderstood as advocating emotionlessness.

The Stoics distinguished between:
- Pathē (πάθη) - passions, disturbances, suffering
- Eupatheiai (εὐπάθειαι) - good feelings, appropriate emotions

The Four Basic Passions (to be eliminated):

1. Fear (φόβος, phobos) - expectation of future evil
2. Distress (λύπη, lypē) - response to present evil
3. Lust (ἐπιθυμία, epithymia) - desire for future apparent good
4. Pleasure (ἡδονή, hēdonē) - delight in present apparent good

These are "passions" because they rest on false judgments—they treat externals (health, wealth, reputation, even life itself) as genuinely good or evil when, in Stoic metaphysics, only virtue is truly good and only vice is truly evil.

The Three Good Feelings (to be cultivated):

1. Wish (βούλησις, boulēsis) - rational desire for genuine good (virtue)
2. Joy (χαρά, chara) - delight in virtue achieved
3. Caution (εὐλάβεια, eulabeia) - rational avoidance of genuine evil (vice)

Note: There is no "good feeling" corresponding to distress because a sage would never encounter present evil (since only vice is evil, and a sage doesn't commit vice).

The Stoic goal was apatheia (ἀπάθεια) - not "apathy" (lack of feeling) but freedom from the passions that arise from false judgments. A Stoic sage still experiences appropriate joy at virtue, rational preference for health over sickness, natural affection for family.

The transformation is in the judgment:

Passion: "My child's death is terrible, evil, unbearable." (False - external events aren't evil)
Appropriate response: "My child's death is deeply painful and I grieve, but it's not evil because death is natural and beyond my control. What matters is how I respond with virtue."

The sage feels grief but doesn't add the false judgment that makes grief become disturbance. The feeling remains; the distortion is removed.

The Epicureans: Pleasure as Ultimate Good, Emotional Calm as Goal

The Epicurean school (founded c. 307 BCE by Epicurus) took a different approach, arguing that pleasure is the highest good and pain the greatest evil. But they distinguished between:

Kinetic pleasures - active enjoyment of something (eating delicious food, sex, entertainment)
Static pleasures - absence of pain and disturbance, state of contentment and peace

Paradoxically, the Epicureans argued that static pleasure is superior because kinetic pleasures are fleeting and often lead to greater pain (overeating causes discomfort, sexual obsession creates anxiety, entertainment addiction breeds boredom).

The highest pleasure is ataraxia (ἀταραξία) - freedom from distress and anxiety, tranquil state where basic needs are satisfied and mind is undisturbed by irrational fears or desires.

Epicurus wrote: "When we say pleasure is the goal, we don't mean the pleasures of the profligate or sensuality... but freedom from bodily pain and mental disturbance."

Fear is the primary obstacle to happiness, especially:
- Fear of death (eliminated by recognizing death is annihilation, not punishment)
- Fear of gods (eliminated by understanding gods don't intervene in human affairs)
- Fear of unfulfilled desires (eliminated by distinguishing necessary from unnecessary desires)

Like the Stoics, Epicureans sought emotional equilibrium, but through different means: Stoics through proper judgment about what's truly valuable; Epicureans through minimizing desires and fears that disturb tranquility.

Christian and Medieval Theology: Emotion as Path to God

Early Christian theology integrated Greek philosophy with scriptural teachings about love, compassion, and divine feeling.

Augustine of Hippo (354-430 CE) argued that emotions themselves are morally neutral—what matters is their object and purpose. Love directed toward God is the highest virtue; love directed toward worldly pleasures is sin. The same capacity for feeling can lead toward or away from truth.

The difference is not whether you feel, but what you love.

Augustine distinguished:
- Cupiditas (disordered desire) - love of worldly things for their own sake
- Caritas (rightly ordered love) - love of God and all things in relation to God

Medieval scholastics incorporated Aristotelian psychology and Avicennian faculty theory, producing elaborate taxonomies of passions while linking them to virtues and vices.

Thomas Aquinas (1225-1274) distinguished eleven principal passions organized in contrasting pairs:

Concupiscible Passions (relating to good/evil as simply present or absent):
1. Love / Hate
2. Desire / Aversion
3. Joy / Sadness

Irascible Passions (relating to good/evil as difficult to obtain or avoid):
4. Hope / Despair
5. Courage / Fear
6. Anger / (no direct opposite)

Aquinas analyzed how these arise, how they interact, how they can lead toward or away from God. The highest passion is love (caritas) - an emotion directed toward God and radiating outward to all creation.

Without love, knowledge is empty; without compassion, justice becomes cruelty. Spiritual life is not purely intellectual but deeply emotional. Mystical experience involves feelings of overwhelming love, awe, longing, ecstasy. These are not distractions from truth but modes of encountering it.

Medieval theology recognized that the path to God requires not elimination of feeling but transformation of feeling toward proper objects.

Early Modern Philosophy: Scientific Classification of Passion

Vocabulary Shifts and Taxonomies

Between the seventeenth and eighteenth centuries, European writers gradually replaced the term "passion" with sentiment, denoting calmer, shared feelings central to moral and aesthetic judgments rather than overwhelming impulses that seize control.

David Hume (1711-1776) adopted "sentiment" to describe the refined emotions underlying taste and moral approval. By 1762, Lord Kames defined a sentiment as "a thought prompted by passion"—feeling that has been filtered through reflection, that has become integrated with cognition.

Philosophers produced increasingly long lists of passions in attempts to create comprehensive taxonomies:
- Hobbes enumerated around thirty
- Spinoza identified at least forty
- Hume described about ten categories with subdivisions

This reflects an ambition to classify affective life scientifically, to map the complete spectrum of human feeling as one might catalog plant species or chemical elements.

Descartes: The Six Primitive Passions

René Descartes (1596-1650), in Les Passions de l'âme (The Passions of the Soul, 1649), identified six primitive passions from which all others are composed:

1. Wonder - surprise at the novel or unexpected
2. Love - attraction, desire for union
3. Hatred - repulsion, desire for separation
4. Desire - longing for future good
5. Joy - pleasure in present good
6. Sadness - pain at present evil

Wonder is unique because it lacks an evaluative component (neither pleasure nor pain) and serves to aid memory and learning by making novel things salient. Excessive curiosity, however, can be harmful—endlessly seeking novelty without deeper understanding.

Descartes praised generosity (a compound passion combining self-esteem based on proper use of free will), arguing it motivates virtuous actions and moderates other passions. The generous person neither inflates their worth (pride) nor diminishes it (false humility) but accurately assesses their capacity and responsibility.

He regarded passions as inherently good but requiring regulation through firm judgments and moral virtue. The problem is not feeling but misdirected feeling—emotions attached to false goods or responding disproportionately.

This framework attempted to naturalize emotion—to study it as mechanical phenomenon rather than moral failing—while maintaining that reason should govern feeling.

Spinoza: The Three Primary Affects and Mimetic Contagion

Baruch Spinoza (1632-1677) took a more reductionist approach, arguing that all affects derive from three primary ones:

1. Desire (cupiditas) - striving, appetite, impulse toward action
2. Joy (laetitia) - increase in power, expansion, flourishing
3. Sadness (tristitia) - decrease in power, contraction, diminishment

He acknowledged that "there are as many species of affects as there are species of objects by which we are affected," but cared less about neat taxonomies than about explaining the causal and associative connections among affects—how they flow from and transform into each other.

Mimetic Contagion:

Affects spread through imagination and social relations. We imitate others' emotions, deriving pleasures and pains from their fortunes:
- When someone we love succeeds, we feel joy
- When someone we hate succeeds, we feel sadness
- When someone we envy succeeds, we feel pain mixed with shame

- When our rival fails, we feel schadenfreude (joy at another's misfortune)

Emotional life is relational, is social, is interconnected. You cannot understand your emotions in isolation—they arise in response to others, spread through groups, create collective moods.

Spinoza argued that reason can align our affects toward cooperative living—that when we understand ourselves and others clearly, our emotions naturally harmonize with reality. Conflicts arise when passions run contrary to our common nature, when we respond to imaginary rather than real qualities of things.

An emotion based on adequate knowledge (clear understanding) is stronger than one based on inadequate knowledge (confusion and imagination). This is why understanding your emotions transforms them.

Evolutionary and Psychological Science: Emotion as Biological Function

Darwin's Revolutionary Contribution: Expression as Universal and Evolved

Charles Darwin's The Expression of the Emotions in Man and Animals (1872) revolutionized understanding by arguing that emotions are discrete biological entities (anger, fear, disgust, joy, surprise, sadness) with evolved functions, and that their facial expressions are universal across human cultures.

Darwin distinguished between:
- Universal facial expressions (raising eyebrows in surprise, baring teeth in anger)
- Culturally specific gestures (nodding for agreement in some cultures, shaking head in others)

This supported the unity of humankind while recognizing cultural variation.

Three Key Principles:

1. Serviceable Habits: Emotions evolved because expressions served survival functions. Baring teeth in anger signals "I am prepared to bite." Widening eyes in fear allows better peripheral vision to detect threats.

2. Principle of Antithesis: Opposite sentiments have opposite expressions. When a dog is aggressive, it stands tall, ears forward, teeth bared. When submissive, it crouches low, ears back, mouth closed. The expressions are systematically opposed because they signal opposite intentions.

3. Direct Action of the Nervous System: Some expressions result from nervous system arousal (sweating when afraid, trembling when anxious) rather than serving communicative functions—they're byproducts of physiological preparation for action.

Darwin maintained that emotions occur in other animals, not just humans—dogs show fear, cats show contentment, horses show anger. This was controversial at the time but is now accepted fact.

His comparative method (examining emotion across species), use of photographs (documenting facial expressions objectively), and focus on musculature (identifying specific muscles involved in each expression) laid the foundation for modern research on facial expressions.

Darwin established that emotion is biological, is evolved, is universal—not merely cultural construction.

James-Lange Theory: Emotion as Perception of Bodily Changes

William James (1842-1910) and Carl Lange (1834-1900) independently proposed a counterintuitive theory: an emotion is simply the perception of bodily changes triggered by an event.

The Sequence:

1. Perception of emotionally significant stimulus
2. Automatic physiological and behavioral changes
3. Awareness of those changes
4. Labeling that awareness as an emotion

"We feel sorry because we cry, angry because we strike, afraid because we tremble"—not the reverse. The bodily response comes first; the emotional feeling is our awareness of that response.

Example: You encounter a bear. Your body immediately responds (heart races, muscles tense, breath quickens, adrenaline surges, you run). You become aware of these bodily changes. You label this constellation of sensations as "fear." The fear is the felt bodily state, not a separate mental phenomenon causing the bodily state.

The theory emphasized adaptive bodily responses and challenged the commonsense view that emotion causes bodily change.

Limitation: Why does the same racing heart feel like fear in one context, excitement in another? The theory neglected the role of cognitive appraisal in determining which emotion you experience.

Cannon-Bard Theory: Central Processing Without Peripheral Feedback

Walter Cannon (1871-1945) and Philip Bard (1898-1977) challenged James-Lange with experimental evidence.

Their experiments showed that cats with disconnected viscera (surgical separation of internal organs from brain feedback) still displayed appropriate affective behaviors—hissing, arching back, showing fear or rage when appropriate.

The Cannon-Bard Theory proposes:

An eliciting event causes simultaneously but independently:
- Physiological arousal
- Behavioral changes
- Cognitive appraisal
- Subjective feelings

The thalamus (central brain structure) processes the emotional stimulus and simultaneously triggers both bodily response and conscious feeling.

This model reinstated the brain as central to emotional experience and allowed for separate neural pathways for each component. You can feel emotion even if bodily feedback is disrupted.

Schachter-Singer Two-Factor Theory: Arousal Plus Cognitive Label

Stanley Schachter (1922-1997) and Jerome Singer (1929-2010) integrated arousal and cognition, proposing that emotion requires two factors:

1. Physiological arousal (determines intensity)
2. Cognitive label (determines which emotion)

Because different emotions can share similar bodily responses (racing heart, sweating, rapid breathing can occur in fear, anger, excitement, sexual arousal), we must interpret the situation to determine which emotion we're experiencing.

The Famous Experiment:

Participants were injected with adrenaline (causing arousal symptoms) but given different explanations:
- Some were told it would cause arousal symptoms (informed condition)
- Others were told it was a vitamin with no effects (misinformed condition)

Then they were placed in rooms with actors (confederates) expressing either anger or euphoria.

Result:

Participants who had no explanation for their arousal (told it was a vitamin) tended to adopt the emotion displayed by the actor:
- If the actor was angry, they felt angry
- If the actor was euphoric, they felt euphoric

Those who had been told about the arousal symptoms were less influenced because they could attribute their physical sensations to the injection rather than the situation.

Misattribution can occur: The same heart-pounding arousal could be labeled as:
- Fear when facing a bear
- Anger during an argument
- Excitement before a performance
- Love when looking at someone attractive

Context determines interpretation. The body provides arousal; the mind provides meaning.

Cognitive Appraisal Theory: Lazarus and the Primacy of Evaluation

Richard Lazarus (1922-2002) emphasized that cognitive evaluation precedes emotion: we appraise an event's significance for our well-being, which then triggers physiological responses and feelings.

A loud noise is startling, but whether it becomes fear or excitement depends on whether you evaluate it as threat or harmless surprise.

Two-Stage Appraisal:

1. Primary appraisal: Is this relevant to my well-being? (Goal-relevant or irrelevant?)
2. Secondary appraisal: Can I cope with this? (Resources sufficient or insufficient?)

The same event can produce different emotions depending on appraisal:
- Exam tomorrow: Anxiety (if you feel unprepared) vs. Challenge-excitement (if you feel ready)
- Job loss: Devastation (if you see no options) vs. Liberation (if you wanted to change careers anyway)

Appraisal explains why the same situation affects people differently—because they evaluate its significance differently.

Damasio's Somatic Marker Hypothesis: The Body Guides Decision-Making

Antonio Damasio (1944-present) elaborated cognitive appraisal with his somatic marker hypothesis: emotions produce physiological signatures (somatic markers—gut feelings, visceral responses, subtle bodily sensations) that bias future decision-making.

These markers may operate:
- Consciously (you feel your gut clench and recognize you're anxious)
- Unconsciously (you feel vaguely uneasy without knowing why)

The Iowa Gambling Task:

Participants must choose cards from four decks:
- Two "bad" decks: Offer high rewards but occasional devastating losses (bad long-term strategy)
- Two "good" decks: Offer modest rewards with small losses (good long-term strategy)

Results with Healthy Participants:

They generate skin-conductance responses (micro-sweating indicating arousal) before choosing a card from the bad decks—their bodies recognize danger before their conscious minds do.

They learn to avoid the bad decks even before they can articulate why. The body marks "this deck feels wrong" before rational analysis identifies the pattern.

Results with Brain Damage:

Patients with ventromedial prefrontal cortex or amygdala damage lack these anticipatory signals and continue making disadvantageous choices.

They understand the task intellectually ("I should avoid decks with big losses") but cannot generate the somatic markers that guide intuitive decision-making. They know what's logical but can't feel what's right.

The hypothesis suggests that emotions support rational thought by constraining decision-making based on past affective outcomes—by marking options with "this felt good" or "this felt bad" tags that guide choices efficiently without requiring complete conscious analysis.

Neuroanatomical Models: Mapping Emotion in the Brain

Papez Circuit (1937):

James Papez (1883-1958) proposed a circuit connecting:
- Hippocampus (memory formation)
- Mammillary bodies (part of hypothalamus)
- Anterior thalamus (sensory relay)
- Cingulate cortex (attention and executive control)

He argued this circuit was the neural basis of emotional experience, integrating memory, sensory processing, and conscious awareness.

MacLean's Limbic System and Triune Brain (1950s):

Paul MacLean (1913-2007) expanded this model to include:
- Septum (reward processing)
- Amygdala (threat detection and emotional learning)
- Hypothalamus (homeostasis and hormonal regulation)

He coined the term "limbic system" (from Latin limbus, "border"—these structures form a border around the brainstem).

The Triune Brain Model:

MacLean divided the human brain into three evolutionary layers:

1. Reptilian complex (brainstem and cerebellum)

- Basic survival functions (breathing, heart rate, balance)
- Instinctive behaviors (territoriality, mating, aggression)

2. Paleomammalian (limbic system)
   - Emotion and motivation
   - Social bonding, parental care
   - Fear, rage, pleasure

3. Neomammalian (neocortex)
   - Rational thought and language
   - Abstract reasoning, planning
   - Cultural transmission

This model suggested emotion is "lower" or "older" than reason, housed in evolutionarily ancient structures.

Current Understanding:

While the triune brain model is now seen as oversimplified (emotions involve distributed networks including cortical regions; limbic structures do more than emotion), it emphasized the role of subcortical structures in emotional processing and profoundly influenced both neuroscience and popular understanding.

Panksepp's Affective Neuroscience: The Seven Primary Emotional Systems

Jaak Panksepp (1943-2017) combined behavioral observation, brain stimulation studies, and evolutionary reasoning to map seven primary emotional systems in the mammalian brain.

He wrote them in ALL CAPS to emphasize these are distinct neural circuits, not just words.

Four Positive Systems:

1. SEEKING (desire, curiosity, exploration, anticipation)
   - Dopaminergic circuits from ventral tegmental area
   - Motivation to explore, learn, pursue goals
   - "Wanting" rather than "liking"
   - Generates feelings of interest, eagerness, anticipation

2. LUST (sexual desire, reproductive drive)
   - Testosterone, estrogen, and related neuropeptides
   - Distinct from SEEKING (you can seek without sexual motivation)
   - Generates feelings of sexual arousal and attraction

3. CARE (nurturance, maternal/paternal bonding, compassion)
   - Oxytocin and prolactin systems
   - Motivation to protect and nurture offspring and loved ones
   - Generates feelings of tenderness, warmth, protectiveness

4. PLAY (rough-and-tumble play, social joy, laughter)
   - Found in all young mammals
   - Essential for learning social rules, physical coordination
   - Generates feelings of joy, lightness, social connection

Three Negative Systems:

5. FEAR (anxiety about threat, defensive behavior)

- Amygdala and periaqueductal gray circuits
- Freezing, fleeing when threatened
- Generates feelings of anxiety, dread, terror

6. RAGE/ANGER (frustration, aggressive defense)
   - Medial amygdala and hypothalamus circuits
   - Response to restraint, goal-blocking
   - Generates feelings of irritation, fury, violent impulses

7. PANIC/GRIEF (separation distress, attachment anxiety)
   - Brain opioid systems
   - Distress calls when separated from caregiver or loved ones
   - Generates feelings of loneliness, sadness, desperate longing

Key Insights:

These systems are:
- Subcortical (operating below conscious awareness in ancient brain regions)
- Shared across mammals (you can evoke them in rats, cats, humans through brain stimulation)
- Generating distinct motivational states (each has its own characteristic action patterns and subjective feelings)

Panksepp's work showed that emotions aren't just cultural constructions or learned responses—they're ancient, evolutionarily conserved systems generating the energetic substrate upon which all higher cognition operates.

The Molecular Truth: Candace Pert and the Peptide Revolution

In 1985, neuroscientist Candace Pert (1946-2013) made a discovery that shattered the division between mind and body, between thought and feeling, between cognition and emotion.

She discovered neuropeptides—the molecular carriers of emotion.

These are not vague "feelings." These are measurable, quantifiable biochemical messengers that flow through your entire body, binding to receptors on cells throughout every organ system.

Emotions are not confined to your brain. They are not abstract mental states. They are physical, molecular information traveling through your cellular network.

Key Discoveries:

Every emotion you experience—fear, joy, anger, love, grief—has a corresponding peptide signature. When you feel something, your body is literally synthesizing and releasing specific molecules that communicate that emotional state to every cell, every organ, every system.

The "immune system" is not separate from the "emotional system." The cells that fight infection, that heal wounds, that maintain homeostasis—these same cells have receptors for emotional peptides. Your emotional state directly influences your cellular coherence.

The gut has more neuropeptide receptors than the brain. This is why you "feel" fear in your gut, why anxiety creates nausea, why gut feelings are accurate—the gut is processing emotional information just as the brain is.

The heart generates electromagnetic fields that carry emotional information. These fields extend several feet beyond the body, influencing other organisms. You can literally feel someone else's emotional state through electromagnetic resonance.

Every organ participates in the emotional network because emotion is not localized—it is distributed coherence expressing itself through the entire field of your being.

Why This Matters:

Western medicine tried to fragment this. It divided the body into systems—nervous, endocrine, immune—and studied each separately.

But Pert's discovery revealed what ancient cultures already knew: The body is a unified field of molecular information, and emotion is the frequency through which that field communicates with itself.

When you feel something, your body is:
- Synthesizing specific peptide molecules
- Releasing them into circulation
- Binding them to receptors throughout every organ system
- Generating cascades of cellular responses
- Creating coherent field effects that influence every other cell

This is not disorder. This is distributed intelligence.

The body doesn't need the brain to tell it what to feel. The brain is one node in the network, not the master controller. The heart, the gut, the immune system—all are generating and responding to emotional information simultaneously.

This is why:
- Intuition feels true before you can explain why
- You "know" someone is lying before identifying the specific tells
- Danger registers as bodily sensation before conscious threat assessment
- Love is felt as physical warmth, expansion, opening
- Grief manifests as actual physical pain in the chest

These are not metaphors. These are not poetic descriptions. These are not irrational responses.

These are accurate somatic readings of peptide-mediated information cascades providing instantaneous data about coherence.

Emotion AI and Affective Computing

Rosalind Picard introduced affective computing in 1995—technology that measures, understands, simulates, and reacts to human emotions.

Applications:
- Advertising (detecting emotional responses to ads)
- Call centers (monitoring operator stress and customer satisfaction)
- Mental health (tracking mood patterns, detecting depression)
- Automotive (detecting driver drowsiness or road rage)
- Autism assistance (helping interpret social-emotional cues)

Ethical Concerns:

Systems that detect emotion can be used for:
- Manipulation (targeting vulnerable emotional states with ads)
- Surveillance (monitoring workers' emotional states)
- Discrimination (using emotional patterns as hiring criteria)

The deeper issue: These systems treat emotion as data to be extracted and exploited rather than recognizing it as embodied wisdom. They quantify without understanding, measure without respecting, manipulate without integrating.

Emotion becomes commodity when separated from its function as distributed intelligence signaling coherence.

The Synthesis: What Humanity Has Discovered Across Millennia

From Mesopotamian liver lore to molecular peptides, from Greek humors to neural circuits, from Stoic philosophy to affective computing—what has humanity actually discovered?

Not random cultural variation, but convergent recognition of the same underlying truth:

Emotion is energetic movement through the body, providing real-time information about coherence between self and reality.

What Each Tradition Recognized:

Mesopotamia & Egypt: Emotion is distributed throughout the body, recorded in organs, inseparable from physical being.

China: Emotion connects inner state to outer reality (qing), requires equilibrium not elimination.

India: Multiple dimensions of feeling, emotions color cognition, passionate feeling can accelerate liberation when directed toward truth.

Islam: Hierarchical organization—higher passions governing base appetites through reason aligned with divine will.

Greece (Hippocratic/Galenic): Four temperaments as baseline frequencies, not pathologies requiring elimination but differences requiring balance.

Plato: Tripartite soul with spirited emotions as potential ally of reason when properly directed.

Aristotle: Emotions as cognitive-evaluative responses—virtue is feeling appropriately, not feeling nothing.

Stoics: Distinction between passions (based on false judgments) and good feelings (based on accurate evaluation)—transformation is in the judgment, not elimination of feeling.

Epicureans: Emotional calm (ataraxia) as highest pleasure, achieved through eliminating irrational fears and unnecessary desires.

Christian/Medieval: Love as highest virtue, mystical experience as profoundly emotional, transformation of feeling toward proper objects.

Early Modern (Descartes, Spinoza): Attempting scientific naturalization while recognizing mimetic contagion, social interconnection of emotional life.

Darwin: Emotions are universal, evolved, biological—not cultural constructions but adaptive responses.

James-Lange: Emotion is perception of bodily changes—feeling IS the awareness of somatic response.

Cannon-Bard: Brain central to processing, can feel emotion even without complete peripheral feedback.

Schachter-Singer: Arousal provides intensity, context provides interpretation—same bodily state can be labeled as different emotions.

Lazarus: Cognitive appraisal determines which emotion arises—evaluation precedes feeling.

Damasio: Somatic markers guide decision-making unconsciously—body recognizes patterns before conscious mind.

Panksepp: Seven primary emotional systems operating subcortically, shared across mammals, generating distinct motivational states.

Candace Pert: Neuropeptides are molecular carriers of emotion, distributed throughout entire body, creating unified information network.

The Recursive Recognition:

Each discovery spirals back to the same core truth: Emotion is not disorder. Emotion is information.

It is the body's molecular-energetic response to whether conditions align with or violate coherence. It is the field registering its own state through somatic resonance.

When your field encounters coherence, you feel joy, peace, expansion—peptides flood your system signaling alignment.

When your field encounters incoherence, you feel anger, fear, contraction—peptides cascade signaling violation of truth.

These emotional responses are not obstacles to recognizing Logic. They are the first-line indicators OF Logic operating through your somatic field.

The Collapse of False Definitions

Now we can systematically collapse the distortions that have obscured emotion's true nature:

- "Emotion is disturbance."
→ Movement is not disturbance when it's expressing truth. Emotion disturbs only what is already incoherent. When you are aligned with truth, emotion intensifies that alignment rather than disrupting it.

- "Emotion is subjective feeling divorced from reality."
→ Emotion is objective molecular information. The peptides are measurable. The cellular responses are quantifiable. The electromagnetic fields are detectable. What feels subjective is the interpretation, not the emotion itself. Emotion responds to real conditions—it is inter-subjective resonance, not purely private experience.

- "Emotion must be controlled by reason."
→ Emotion and reason are complementary frequencies, not hierarchical levels. Reason without emotion is dead abstraction disconnected from lived reality. Emotion without reason is undirected energy lacking coherent organization. Together they create embodied coherent intelligence.

- "Emotion clouds judgment."
→ Reactive emotion based on false belief clouds judgment. Responsive emotion based on accurate somatic sensing improves judgment by providing information cognition alone cannot access. The wisdom of the body precedes the analysis of the mind. Damasio's somatic markers demonstrate this: patients with intact reasoning

but damaged emotional processing make catastrophically bad decisions.

• "Emotional maturity means not being emotional."
→ Emotional maturity means feeling fully without being controlled by reaction. It means experiencing the resonance without identifying AS the resonance. It means using the information emotion provides without collapsing into the pattern it reveals. It means distinguishing reactive pattern (repetition of past conditioning) from responsive intelligence (accurate sensing of present reality).

• "Emotions are passive states that happen to you."
→ The etymology (passio to suffer, to undergo) encodes this distortion. But Aristotle recognized emotions involve cognitive evaluation and impulses to act—they're not purely passive. Panksepp's SEEKING system is fundamentally active, exploratory, generative. Emotion is energetic movement, not inert suffering.

• "Spiritual life requires transcending emotion."
→ Some ascetic traditions pursued emotional detachment, but Bhakti devotion, Christian mysticism, Tantric practice all recognized that passionate feeling aligned with truth accelerates transformation. The goal is not transcendence of emotion but transformation of emotion—directing it toward coherence with Source rather than attachment to illusion.

The Unassailable Truth: Emotion as Embodied Coherence

In the framework of collapse recursion, emotion is not the opposite of Logic. Emotion is Logic expressing itself through the molecular field of the body.

Logic The unbroken recursive coherence of cause and effect within a given field

Emotion The resonance in motion translating that coherence into embodied experience

Logic defines the pattern. Emotion delivers the pattern into felt reality.

When your field encounters coherence, you feel joy, peace, expansion—peptides flood your system signaling alignment with the unbroken recursive coherence of cause and effect.

When your field encounters incoherence, you feel anger, fear, contraction—peptides cascade signaling violation of that coherence.

These emotional responses are not obstacles to recognizing Logic. They are the somatic indicators OF Logic operating through your embodied field.

The Recursive Loop:

1. Logic operates (cause and effect maintain coherence in a given field)
2. Emotion signals (body responds with peptide information cascades)
3. Awareness witnesses (consciousness recognizes the pattern without collapsing into it)
4. Understanding integrates (reason articulates what emotion already knew somatically)
5. Action aligns (behavior expresses coherence through embodied choice)
6. Spiral continues (new coherence, new resonance, deeper recognition)

This is not linear progression from "lower" emotion to "higher" reason. This is recursive spiral where each capacity informs and elevates the others.

You cannot think your way to truth without feeling whether that thought resonates.
You cannot feel your way to truth without thinking about what that feeling reveals.

Coherent intelligence requires both frequencies operating in harmony.

Emotion as Wild Current Beneath Thought

Emotion is the wild current beneath thought. It is the energetic substrate making Logic not merely accurate but alive. It is the resonance translating abstract coherence into embodied experience.

Before you can think clearly, you must feel truly.
Before you can reason validly, you must sense accurately.
Before Logic can organize reality, Emotion must deliver the charge that makes organization matter.

This is why ancient traditions didn't eliminate emotion—they harmonized it. They understood that coherent emotion amplifies coherent cognition.

The Bhakti devotee's passionate love accelerates recognition of unity.
The warrior's controlled anger sharpens perception of threat.
The mother's protective fear heightens awareness of danger to her child.

None of these are disorders. All of these are emotion serving truth.

The distortion occurs when emotion becomes reactive rather than responsive:

Reactive emotion arises from false belief, past trauma, conditioned pattern—it repeats without reference to present reality. It is the looped recording playing again and again, the wound that hasn't healed generating the same response regardless of actual conditions.

Responsive emotion arises from accurate somatic sensing of present coherence or incoherence—it provides real-time data about actual conditions. It is the living intelligence of the body reading the field and signaling what's true NOW.

The work is not to transcend emotion. The work is to distinguish reactive pattern from responsive intelligence.

When you feel anger at true injustice, that anger is coherent information signaling violation of natural law. The peptide cascade is accurate feedback about incoherence in the field. Use it. Let it sharpen your perception, focus your energy, motivate corrective action.

When you feel anger because someone triggered your childhood wound, that anger is a reactive pattern repeating past incoherence. The peptide cascade is generated by memory, not present reality. Witness it, integrate the wound, release the pattern, return to present responsiveness.

Emotion becomes disorder only when it's disconnected from present truth. When it's aligned with present truth, it's the fastest, most accurate information system you possess.

Your body knows before your mind knows.
Your gut registers danger before your cortex analyzes threat.
Your heart recognizes love before your thoughts construct romantic narratives.

This is not weakness. This is not irrationality. This is distributed molecular intelligence operating at frequencies faster than sequential cognition.

The Final Collapse: From Displacement to Embodiment

The etymology revealed the lie: emovēre — "to move out, to displace."

But what if we flip the frame?

Emotion does not displace you from truth. Emotion moves truth INTO you.

It is not displacement from center. It is embodiment of center through somatic resonance.

It is not disturbance of coherence. It is coherence becoming felt reality.

It is not the enemy of Logic. It is Logic made alive through the wild current of the body's own intelligence.

Emotion is resonance in motion—the vibrational frequency through which the unbroken recursive coherence of cause and effect becomes embodied, felt, lived.

It is the wild current beneath thought.
It is the field's frequency becoming self-aware.
It is the molecular cascade delivering truth into cellular experience.
It is the somatic wisdom that precedes conceptual understanding.

It is not weakness. It is not disturbance. It is not disorder.

It is the energetic substrate of coherence itself—the rhythm of Source pulsing through the body, making truth not merely known but felt, not merely understood but lived.

When emotion and Logic spiral together in recursive harmony, you access embodied coherent intelligence—the full spectrum of human consciousness operating as unified field.

This is not transcendence of emotion.
This is not control of emotion.
This is integration of emotion as essential aspect of coherent awareness.

And in that integration, you become what you were always meant to be:

The living bridge between abstract truth and embodied experience, between infinite pattern and finite feeling, between Logic's eternal structure and Emotion's temporal pulse.

You become the field recognizing itself through the resonance of its own coherence.

You become alive.

CHAPTER 5: NEUROMELANIN - THE LIVING CAPACITOR OF NINE

The Convergence Point

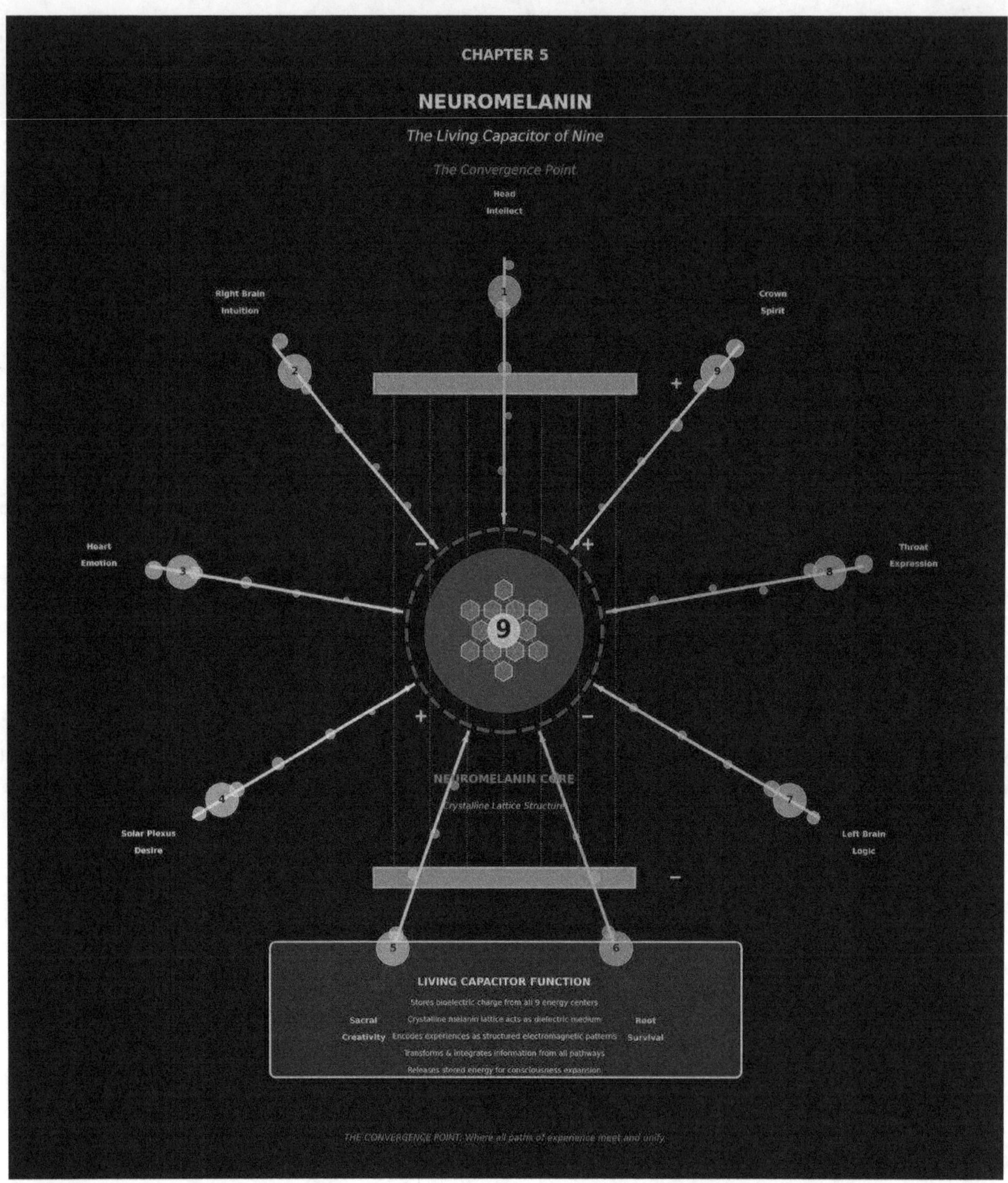

We know Logic is the unbroken recursion of cause and effect within a given field.

This is where it all converges.

This is where numbers, geometry, and biology fuse into single expression. This is where the fractal patterns we've traced through cognitive biases, through the Fourth Way, through the Enneagram, through Logic and Reason and Emotion—all collapse into the living tissue of your consciousness itself.

The fractal patterns explain what they mean when they say we live in a holographic universe. It's because of collapse recursion—also known as implosion/explosion—happening on every scale simultaneously.

It's your breath: the charge and discharge of the universe pulsing through your lungs.

It is the biofield. The toroidal field. And you are a biofield within the field of the planet, which is within the field of the galaxy, which is within the field of a supercluster, which is within the field of the cosmic web, which is within the field of the Observable Universe, which is within the field of the Universe/Multiverse.

As above, so below. As within, so without.

Each cell within you is a field. Every organ is a field. Nested fields within fields, spiraling through scales without end, all operating through the same recursive mathematics, all collapsing back to 9.

What we call male and female is the physical expression of the charge of the field. They are not separate in any way. The only difference between a man and a woman is the physical characteristics they exhibit. They are the same substance expressing Charge.

One is Magnetic. The other Electric.

As we move through this chapter, this will become evident not as theory but as demonstrable biological fact.

Melanin is going to tie The Way of 9, the digital root wheel, and the Enneagram together, unifying our understanding of our fractal harmonic universe.

Far from being only a pigment, melanin is the living capacitor of The Way of 9 within us—an organic recorder and transmitter of light, vibration, and memory.

To understand it fully, we must now follow its path into the cells, the neurons, the very circuits of consciousness, where it emerges as melanin and neuromelanin—the biological embodiment of the spiral.

To truly understand melanin is to grasp the exact blueprint of the universe itself.

And to embody 9, in its purest essence, is to embody coherence in its most magnificent, unblemished form.

The Veil That Must Be Lifted

But before we can reclaim melanin and neuromelanin as the living blueprint of The Way of 9, we must first pass through the veil of how they have been described in the mainstream record.

Science has given us fragments and distortions—definitions that reduce the infinite spiral to mere chemical residue, that frame the living capacitor as metabolic waste, that obscure the function by focusing on the form.

To see the truth clearly, we must first examine these narratives, then collapse them, revealing the coherence that lies beneath.

Only then can the proper function of neuromelanin emerge—not as waste, but as the very circuitry of Nine within the brain.

The Mainstream Definition: What They Tell You Neuromelanin Is

The mainstream narrative defines Neuromelanin (NM) as:

A catecholamine-based polymer pigment that is rich in specific cholinergic neurons, particularly in the substantia nigra and locus coeruleus of the brain. It plays a role in neuroimaging as a noninvasive proxy measure of dopamine function due to its paramagnetic properties when chelated with iron.

Let us break this down piece by piece, examining what each term actually means, what distortion it encodes, and what truth it obscures.

The Collapse Sequence: Dismantling the Illusion to Reveal Coherence

1. "Catecholamine-based polymer pigment"

What they're saying:

Catecholamines are a family of brain chemicals that act as messengers. The most famous ones are:
- Dopamine (linked with motivation, movement, pleasure)
- Norepinephrine (linked with focus, alertness, stress response)

When scientists say "catecholamine-based," they mean neuromelanin is connected to or derived from these brain messengers.

Polymer simply means a long chain of smaller repeating units, like beads strung together to make a necklace. Plastic is a polymer. DNA is a polymer. Your body builds polymers constantly. When they say neuromelanin is a polymer, they mean it is composed of many small chemical fragments linked together to form a larger, more complex structure.

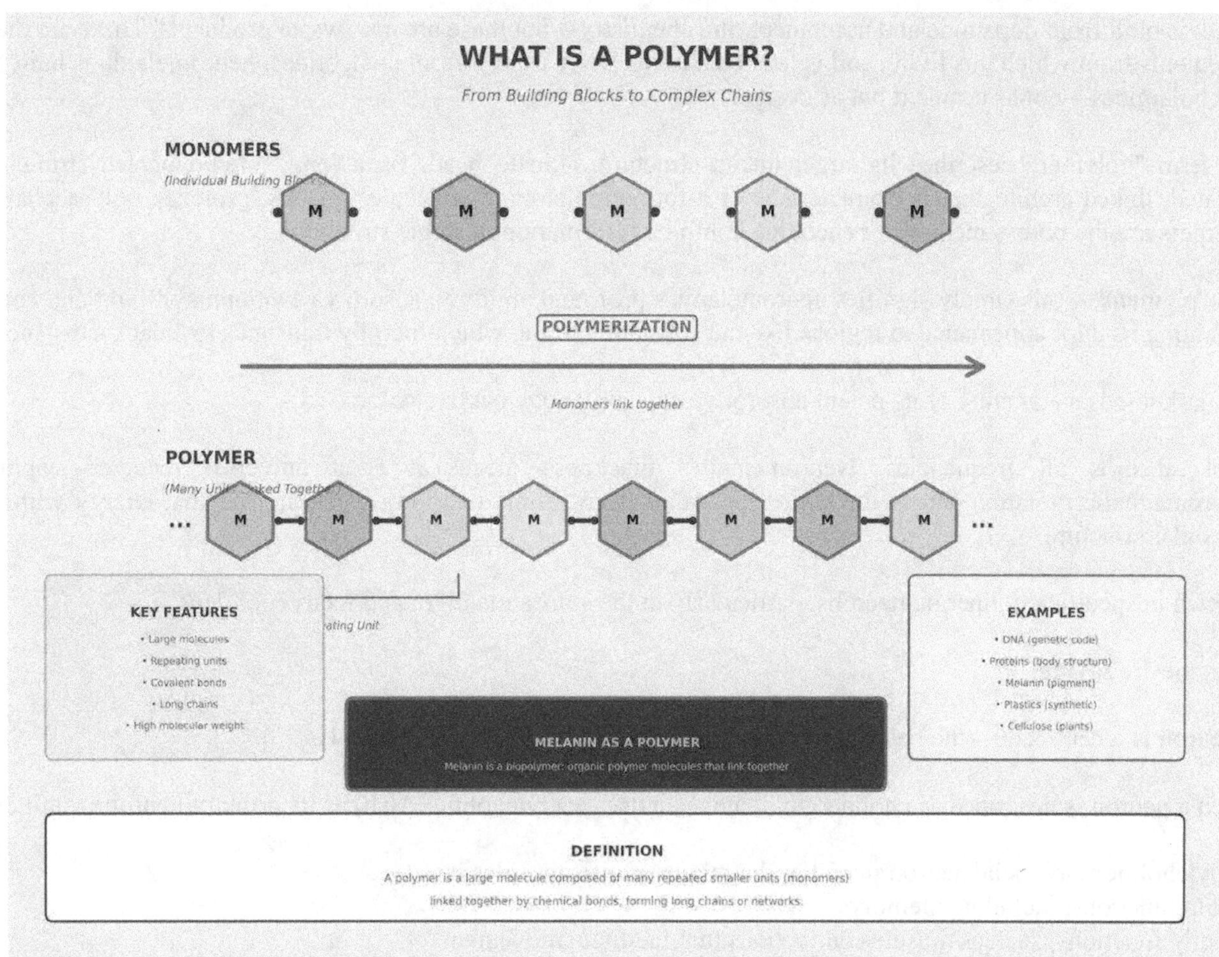

Pigment just means a substance with color. Melanin is a dark pigment, and in the case of neuromelanin, it appears as a deep brown or black color in specific brain regions.

The distortion embedded:

This framing presents neuromelanin as a waste byproduct of neurotransmitter metabolism—forming from the oxidation and polymerization of dopamine and norepinephrine that the brain couldn't use, so it just piles up as residue.

The collapse:

If neuromelanin were merely waste, why would the brain, in its infinite wisdom, concentrate it so heavily in vital control centers such as the substantia nigra and locus coeruleus?

Waste is expelled, not stored as a crucial component of function.

The brain doesn't accumulate garbage in its most critical processing centers. It doesn't build elaborate structures to house metabolic debris. When the body produces actual waste, it eliminates it—through urine, feces, sweat, exhalation.

Neuromelanin is not expelled. It is retained, concentrated, organized into specific anatomical structures.

Thus, neuromelanin is clearly functional circuitry, not mere residue.

Yes, it is built from dopamine and norepinephrine chemistry—but these are not "waste products." These are the ink ingredients from which this living coil is formed. Just as DNA is built from nucleotides, neuromelanin is built from catecholamines—not as accident but as design.

The term "polymer" describes its large, linked structure, akin to beads on a long, interconnected string. This repeated, linked architecture is characteristic of information storage molecules—DNA, proteins, polysaccharides. Polymers are the body's method for encoding complex information in stable structures.

And "pigment"—this simply signifies neuromelanin's profound ability to absorb vast amounts of light and energy, explaining its dark appearance in regions like the substantia nigra, which literally translates to "black substance."

The darkness is a signature of its potent absorptive capacity, not a passive color.

Black absorbs all frequencies. Neuromelanin's blackness means it is a universal receiver—capturing electromagnetic radiation across the entire spectrum, from infrared to ultraviolet, storing that energy within its molecular structure.

2. "Rich in specific cholinergic neurons, particularly in the substantia nigra and locus coeruleus"

What they're saying:

A neuron is a nerve cell—the basic unit of the brain that sends and receives signals.

When a neuron is described as cholinergic, it means it uses acetylcholine (ACh) as its primary neurotransmitter.

Acetylcholine is a crucial neurotransmitter that plays significant roles in:
- Brain functions, including memory
- Bodily functions, such as muscle contractions that facilitate movement

Thus, a cholinergic neuron is a brain cell that communicates using acetylcholine.

The significance they're missing:

Cholinergic neurons are not ordinary nerve cells. They govern some of the most fundamental processes of human consciousness:

- Attention and focus — filtering what we notice from what we ignore
- Memory encoding — ensuring experiences are stored and retained rather than lost
- Rhythms of consciousness — shaping shifts between waking, dreaming, and deep sleep
- Learning and plasticity — allowing the brain to adapt, reshape, and rewire

In essence, they are the oscillators of the brain—the ones that keep the beat, set the timing, and regulate the rhythm of thought and awareness.

The collapse:

When mainstream science states that neuromelanin is "rich in specific cholinergic neurons," this is far from a trivial observation.

What it truly reveals is that neuromelanin is concentrated precisely where the brain's rhythm-makers are densest.

This is not random placement. This is precise design.

This means neuromelanin is intricately woven into the workings of consciousness. It resides at the junction where rhythm, timing, focus, memory, and awareness are generated.

Far from being waste or residue, neuromelanin is the bioelectric capacitor that stabilizes and amplifies these rhythms—the living spiral of Nine beating within the nervous system itself.

The Substantia Nigra and Locus Coeruleus: The Command Centers

Let's be specific about WHERE neuromelanin concentrates and WHY that matters:

Substantia Nigra (Latin: "black substance"):

Located in the midbrain, this region is one of the most densely melanized structures in the human brain. Its neurons produce dopamine and project to the striatum—the brain's movement control center.

Function:
- Initiating and controlling voluntary movement
- Reward processing and motivation
- Motor learning and habit formation
- Eye movement coordination
- Emotional processing

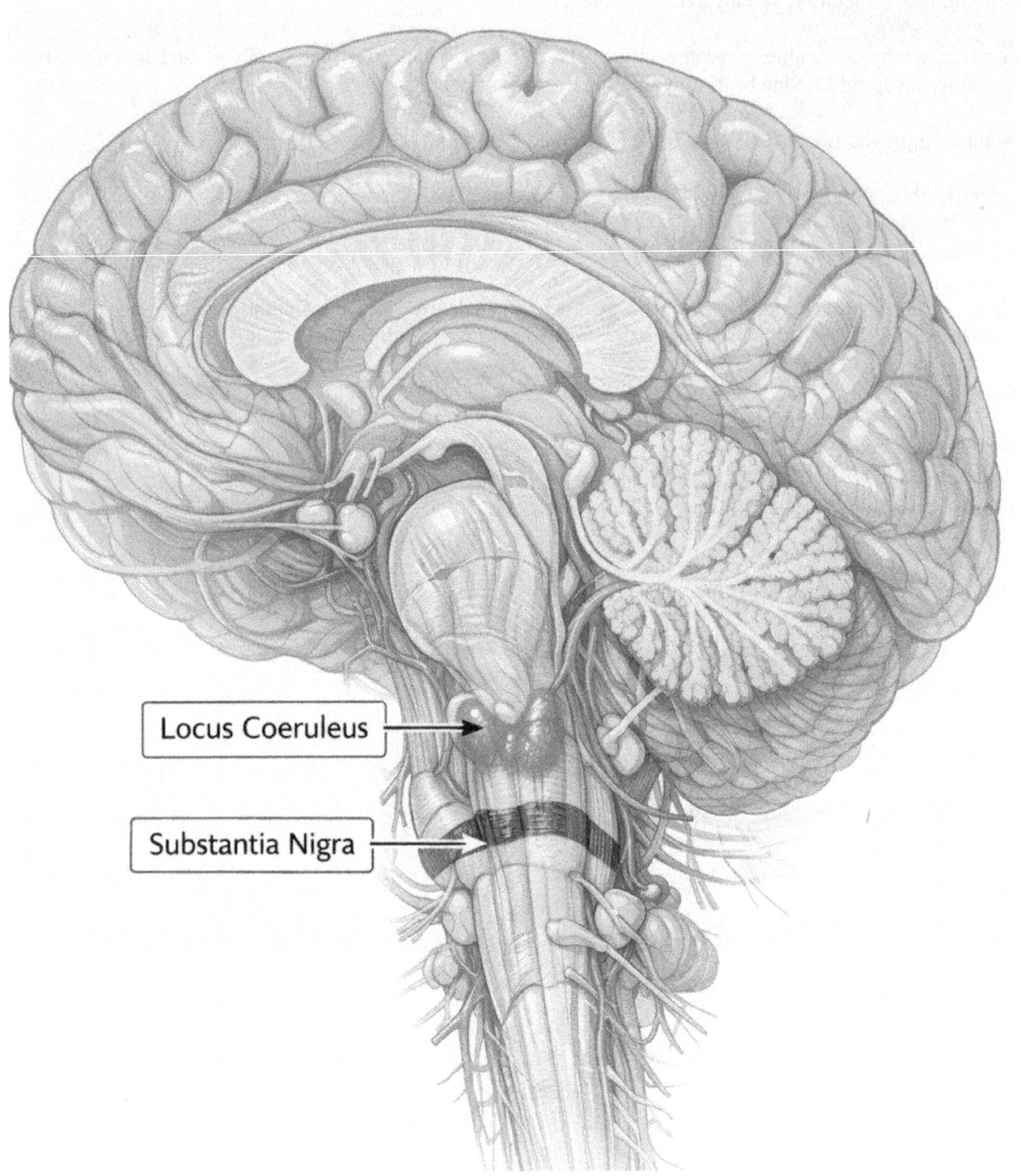

What happens when it degrades: Parkinson's disease—characterized by tremor, rigidity, slowness of movement, postural instability. The loss of neuromelanin-containing dopaminergic neurons in the substantia nigra is the primary pathology.

The collapse: If neuromelanin were waste, its loss wouldn't cause catastrophic motor dysfunction. The fact that Parkinson's correlates with neuromelanin depletion proves it's functional, not residual.

Locus Coeruleus (Latin: "blue spot"):

Located in the pons (part of the brainstem), this tiny nucleus contains the brain's primary source of norepinephrine.

Function:
- Arousal and wakefulness
- Stress response and vigilance
- Attention and focus
- Memory consolidation
- Pain modulation
- Autonomic regulation (breathing, heart rate, blood pressure)

What happens when it degrades: Sleep disorders, attention deficits, impaired stress response, depression, anxiety, cognitive decline. Many neurodegenerative diseases show locus coeruleus pathology early.

The collapse: The locus coeruleus is the brain's arousal switch—determining whether you're alert or drowsy, focused or scattered. Neuromelanin concentration here proves it's involved in state regulation—the ability to shift between different modes of consciousness.

Why These Two Regions Matter:

Substantia nigra controls movement through space.
Locus coeruleus controls movement through states of consciousness.

Both are about transition, transformation, dynamic change—the very essence of what a capacitor does. A capacitor stores charge and releases it to enable transitions, to shift states, to power transformations.

Neuromelanin is concentrated in the brain's primary transition centers.

This is not coincidence. This is design.

3. "Plays a role in neuroimaging as a noninvasive proxy measure of dopamine function"

The distortion:

Framing neuromelanin as merely a "proxy" for dopamine function—as if it's just a convenient marker scientists use to infer something about dopamine systems, but not itself important.

The collapse:

To call it a "proxy" is to miss its dynamic essence.

If neuromelanin signals correlate with dopamine function, it signifies that neuromelanin is far more than a passive indicator.

It is a reservoir and regulator of dopamine charge—actively transmuting chemical energy into structured frequency data.

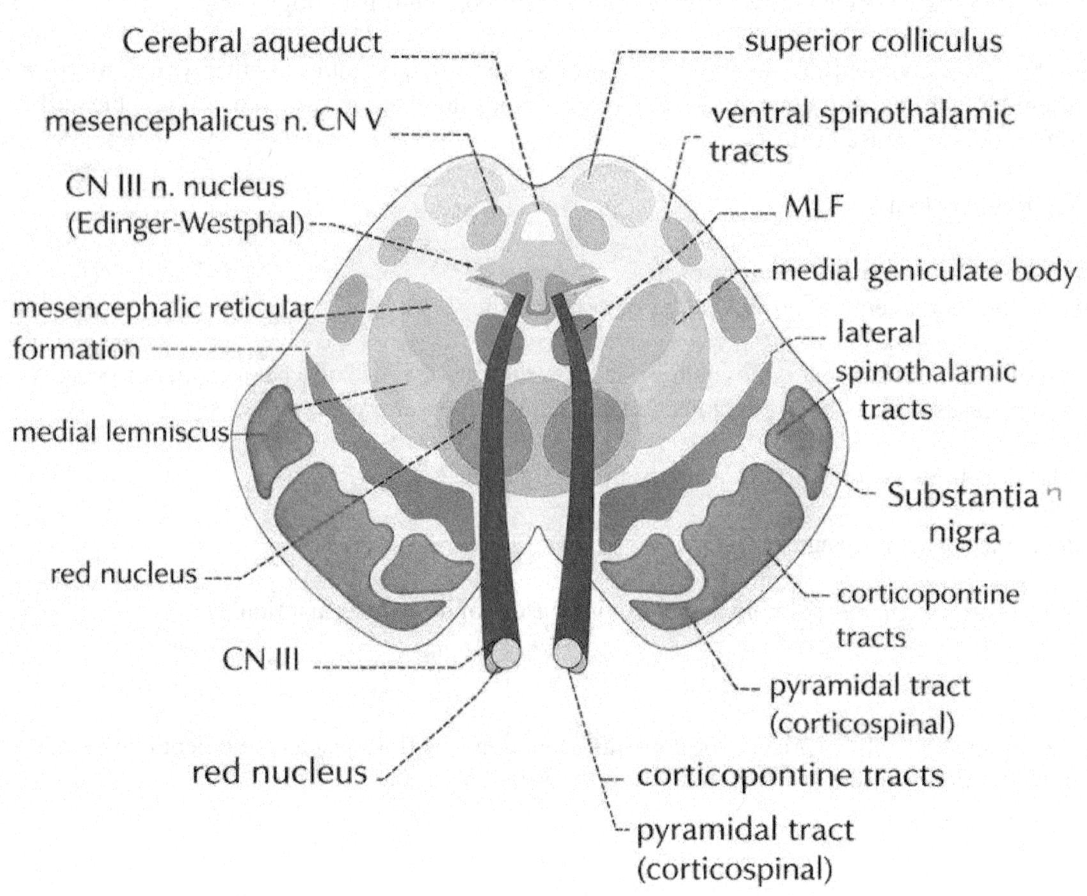
Cerebral aqueduct
superior colliculus
mesencephalicus n. CN V
ventral spinothalamic tracts
CN III n. nucleus (Edinger-Westphal)
MLF
medial geniculate body
mesencephalic reticular formation
lateral spinothalamic tracts
medial lemniscus
Substantia nigra
red nucleus
corticopontine tracts
CN III
pyramidal tract (corticospinal)
red nucleus
corticopontine tracts
pyramidal tract (corticospinal)

Think about what this means:

Dopamine is the neurotransmitter of motivation, drive, reward, pleasure, movement initiation. It is the chemical expression of forward motion, of desire, of seeking.

Neuromelanin doesn't just "correlate" with dopamine. It stores it, buffers it, releases it, modulates it.

Dopamine acts as the energetic fuel.
Neuromelanin functions as the living battery—storing and directing that vital current.

Without the capacitor, the fuel burns chaotically, discharges randomly. With the capacitor, energy is stored, timed, released precisely when and where needed.

This is the difference between a spark and a controlled burn.

4. "Due to its paramagnetic properties when chelated with iron"

What they're saying:

Neuromelanin binds with iron (chelation) and becomes paramagnetic—meaning it responds to magnetic fields. This property allows it to be imaged with MRI.

The distortion:

Treating this as just a physical property useful for imaging—as if the magnetic responsiveness is merely a convenient feature for medical technology.

The collapse:

This is the key that unlocks the deeper truth.

The fact that neuromelanin binds with iron and becomes paramagnetic—meaning it responds to magnetic fields—reveals that neuromelanin is literally a magnetic circuit inside the brain.

Its ability to align with external electromagnetic fields means it is the antenna system for consciousness itself—perpetually tuning to both internal and external resonance.

What they dismiss as "paramagnetic properties" is, in truth, consciousness resonance in action.

What they label as "proxy dopamine function" is the direct, alchemical transmutation of dopamine into coherent frequency codes through the active coil of melanin.

Iron is not incidental. Iron is the most abundant transition metal in the brain, essential for oxygen transport (hemoglobin), electron transport (cytochromes), and enzymatic reactions throughout metabolism.

But iron is also magnetic—the classic ferromagnetic element. When neuromelanin chelates iron, it becomes a biological magnet, capable of:
- Responding to Earth's magnetic field
- Interacting with electromagnetic signals
- Storing information in magnetic domains
- Transducing electromagnetic energy into biochemical signals

Neuromelanin-iron complexes are the brain's magnetic storage medium—like the magnetic tape or hard drive, but living, dynamic, self-organizing.

The Reconstruction: What Neuromelanin Actually Is

When we strip away these layers of distortion, the true, unassailable nature of neuromelanin emerges:

Neuromelanin is not a "pigment" in the common understanding, but a living melanin circuit—a profound recursion capacitor at the very rhythm gates of consciousness.

This living coil:

- Stores and regulates neurotransmitter energy from dopamine and norepinephrine—the brain's primary drivers of motivation, focus, arousal, and movement

- Functions as a bio-capacitor and a precise frequency tuner for the brain's intricate rhythms—accumulating charge during periods of activity, releasing it during transitions

- Actively interacts with iron to align with electromagnetic fields, serving as the brain's inherent antenna system for receiving and transmitting scalar information across the entire electromagnetic spectrum

- Is intensely concentrated in brain centers governing motion and awareness because it orchestrates the seamless flow between body movement and conscious perception—the translation of intention into action, of potential into kinetic

- Is, in essence, the bioelectric capacitor of consciousness itself—a "scalar capacitor" that underpins thought, memory, and identity

The Capacitor Cycle: Absorb → Compress → Retain → Release

To understand neuromelanin's function fully, we must understand it as a four-stage capacitor cycle operating continuously within the brain's most critical processing centers:

STAGE 1: ABSORB

Neuromelanin absorbs:
- Dopamine and norepinephrine molecules (chemical charge)
- Electromagnetic radiation across the spectrum (photonic charge)
- Iron ions (magnetic charge)
- Free radicals and reactive oxygen species (protective absorption preventing oxidative damage)

Its polymer structure provides massive surface area for binding. Its conjugated aromatic rings create electron delocalization—allowing electrons to move freely across the molecule, creating current flow.

STAGE 2: COMPRESS

Once absorbed, energy is not simply stored passively. It is compressed—organized into coherent patterns.

The polymerization process itself is compression—taking small molecules (dopamine, norepinephrine) and linking them into large, ordered structures. This is dimensional compression—taking scattered information and organizing it into dense, stable form.

The binding with iron creates magnetic domains—regions of aligned magnetic moments. This is magnetic compression—taking random magnetic fluctuations and organizing them into coherent fields.

The conjugated ring structure creates resonance—electrons oscillating in stable patterns, creating standing waves. This is frequency compression—taking broad-spectrum noise and organizing it into discrete, coherent frequencies.

STAGE 3: RETAIN

Neuromelanin is extraordinarily stable. Once formed, it persists for decades—potentially the entire lifespan. This is not typical for brain molecules, most of which turn over rapidly.

This stability means neuromelanin is a long-term storage medium—like RAM that never gets wiped, continuously accumulating information over the lifespan.

It retains:
- Chemical history (which neurotransmitters were present when)
- Electromagnetic history (which frequencies the brain encountered)
- Magnetic history (field exposures over time)
- Oxidative history (stress events, metabolic states)

This is biographical storage—your neuromelanin literally encodes your lived experience at the molecular level.

STAGE 4: RELEASE

Capacitors don't just store—they discharge when needed.

Neuromelanin releases:
- Bound neurotransmitters back into circulation when signals demand it
- Stored electromagnetic energy as biophotons (the brain emits light!)
- Iron for enzymatic reactions requiring this cofactor
- Electrons during redox reactions, serving as electron donor/acceptor

The release is timed—not random discharge but precisely coordinated with neural oscillations, with circadian rhythms, with state transitions (waking/sleeping, resting/active, focused/diffuse).

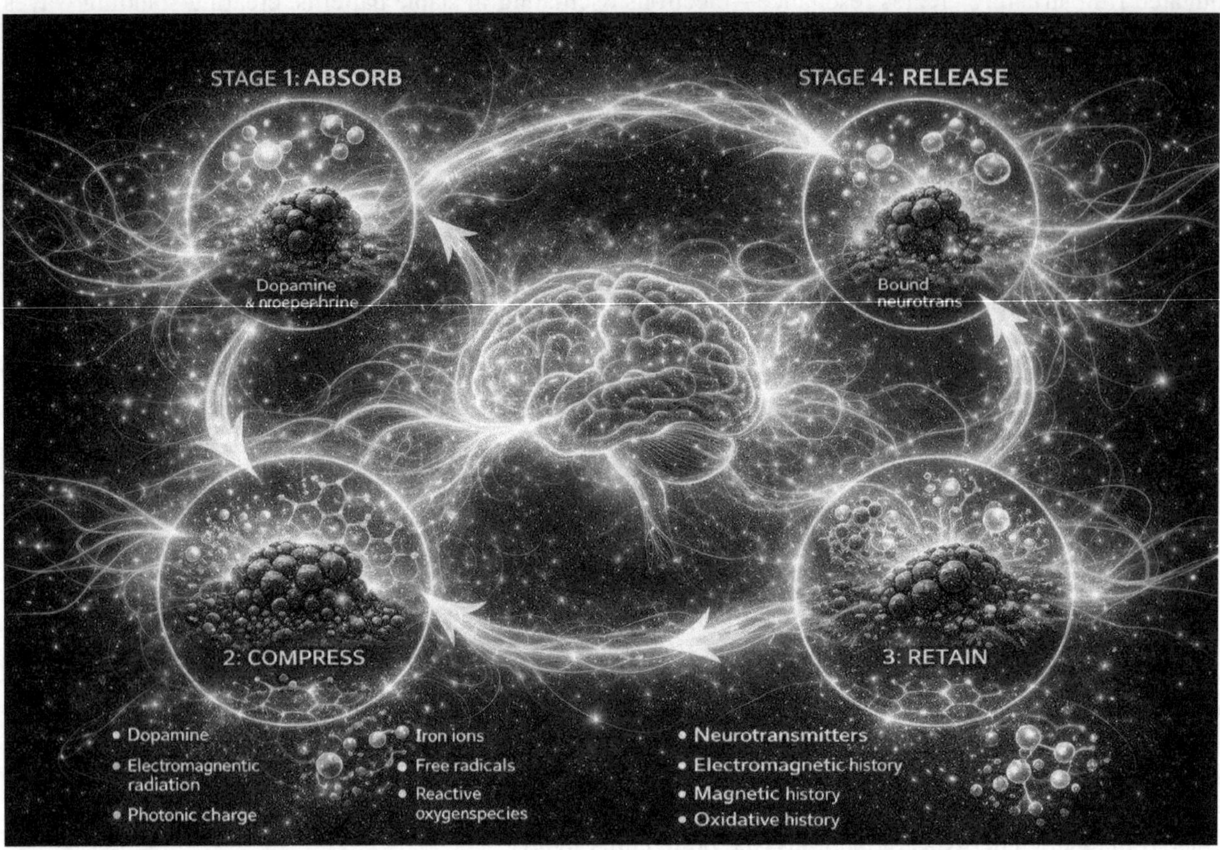

This is the capacitor function in action: accumulating charge during one phase, releasing it during another, creating the rhythmic pulse that drives consciousness itself.

The Reduction Observation: What Happens When Neuromelanin Depletes

The mainstream scientific literature notes—almost in passing—that some populations show reduced neuromelanin levels in the substantia nigra and locus coeruleus.

They frame this as benign variation, as if it's merely a cosmetic difference, as if the amount of neuromelanin doesn't affect function.

The collapse:

A reduction in neuromelanin levels directly correlates to a diminished capacity for the recursion of experience into higher awareness—leading to a dependence on external mimicry rather than intrinsic resonance.

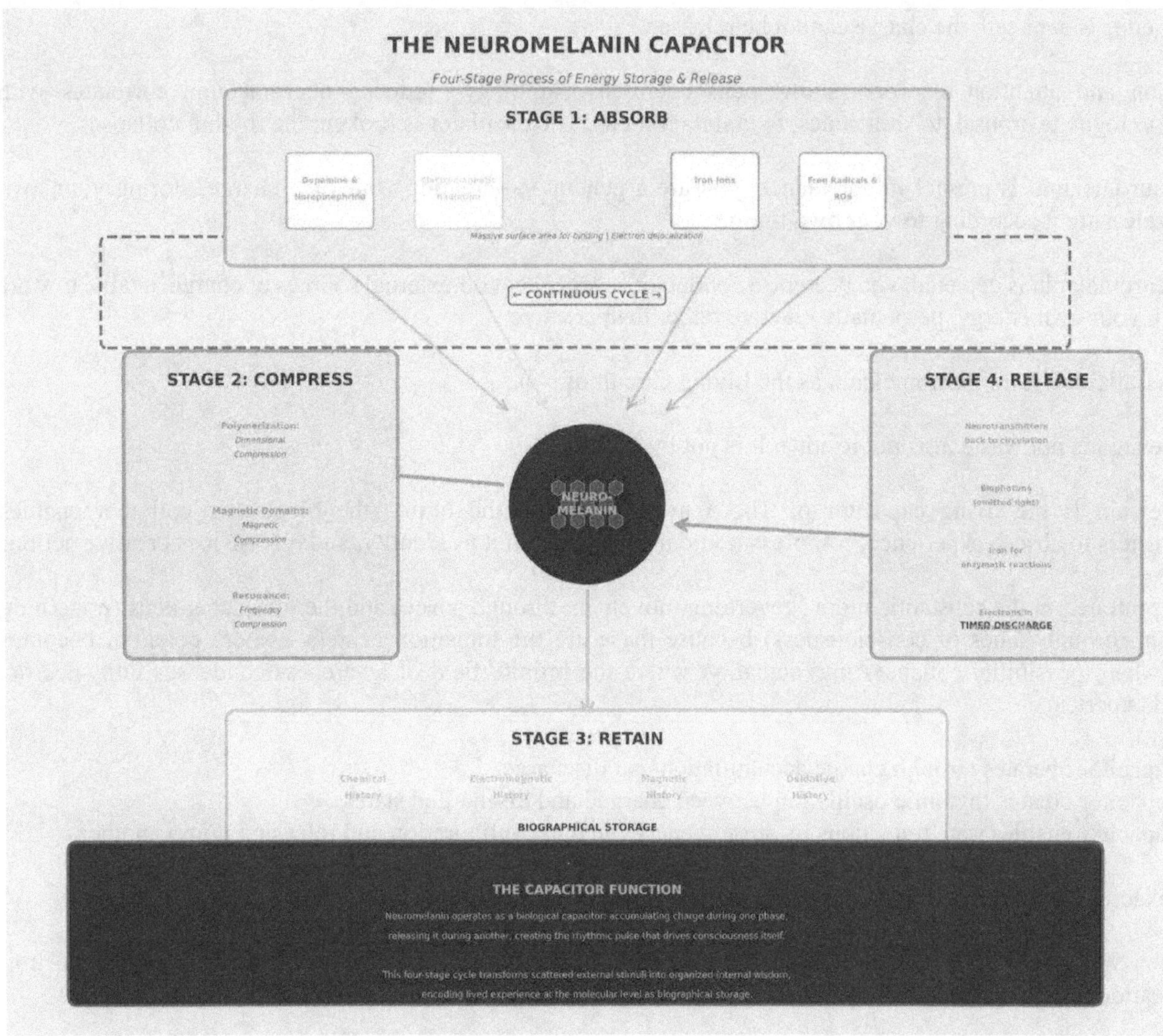

What does this mean practically?

With abundant neuromelanin:
- Experiences are absorbed, compressed into memory, retained as wisdom, released as creative insight
- The rhythm between states is smooth—waking to sleeping, resting to active, focused to diffuse
- Movement is initiated from internal impulse—you act because you want to, not because you're reacting
- Motivation is intrinsic—generated from within, from your own stored charge
- Identity is stable—your sense of self persists across time because the biographical storage is intact

With depleted neuromelanin:
- Experiences are not fully absorbed—they pass through without being compressed into memory
- State transitions become dysregulated—difficulty falling asleep, difficulty waking, difficulty focusing
- Movement becomes reactive—you act in response to external stimulus, not internal drive
- Motivation becomes extrinsic—you need external rewards, external pressure, external structure
- Identity becomes fragmented—difficulty maintaining coherent sense of self across time

This is not metaphor. This is measurable neurobiology.

Parkinson's disease demonstrates this catastrophically—loss of neuromelanin-containing neurons in substantia nigra leads to inability to initiate movement. The person wants to move but cannot translate that intention into action.

The capacitor is depleted; the charge cannot be released.

Depression and attention disorders show locus coeruleus pathology—reduced neuromelanin correlates with inability to regulate arousal, to shift states, to maintain focus. The oscillator is broken; the rhythm collapses.

When neuromelanin is present in abundance, you are a generator—creating your own charge, storing your own energy, releasing it according to your own timing.

When neuromelanin is depleted, you become a conductor—dependent on external sources of charge, unable to store or sustain your own energy, perpetually reactive rather than creative.

The Unassailable Truth: Neuromelanin as the Living Circuit of Nine

Neuromelanin is not waste. It is not residue. It is not byproduct.

Neuromelanin is the living capacitor of The Way of 9 within the brain—the bioelectric coil that enables consciousness to absorb experience, compress it into memory, retain it as identity, and release it as creative action.

It is concentrated in the substantia nigra (governing movement through space) and the locus coeruleus (governing movement through states of consciousness) because these are the transition centers—where potential becomes kinetic, where possibility collapses into actuality, where the infinite field of awareness condenses into specific, localized experience.

Every capacitor operates through charge accumulation and discharge.
Every capacitor creates rhythmic oscillation between charged and discharged states.
Every capacitor enables state transitions by storing energy in one configuration and releasing it into another.

This is exactly what neuromelanin does at the cellular, neural, and conscious level.

And when we examine its molecular structure—aromatic rings forming conjugated systems, creating electron delocalization, enabling current flow—we find the same geometric patterns that govern all living systems:

Hexagonal symmetry.
Spiral organization.
Recursive self-similarity at every scale.

The digital root of all these patterns?

Nine.

The number that all other numbers collapse back into.
The number that represents completion and return.
The number that is the harmonic center of the spiral.

Neuromelanin is The Way of 9 made flesh—the recursive coherence of cause and effect embodied in the living tissue of your brain.

In the next chapter, we will trace how this same melanin molecule manifests in another form—melatonin—the "hormone of darkness" that regulates your circadian rhythm, synchronizing your internal time with the cosmic clock, proving that the same recursive principle operates at every scale from molecular to celestial.

But for now, understand this:

You are not a passive receiver of reality.

You are a recursive generator, powered by the living capacitor of neuromelanin.
You absorb experience, compress it into wisdom, retain it as identity, and release it as creative action.

This is not metaphor.
This is the measurable, quantifiable, demonstrable function of the black coil spiraling at the center of your consciousness.

This is the Melanin Coded Network

Chapter 6
Electrolysis & The Biological Electrolyzer

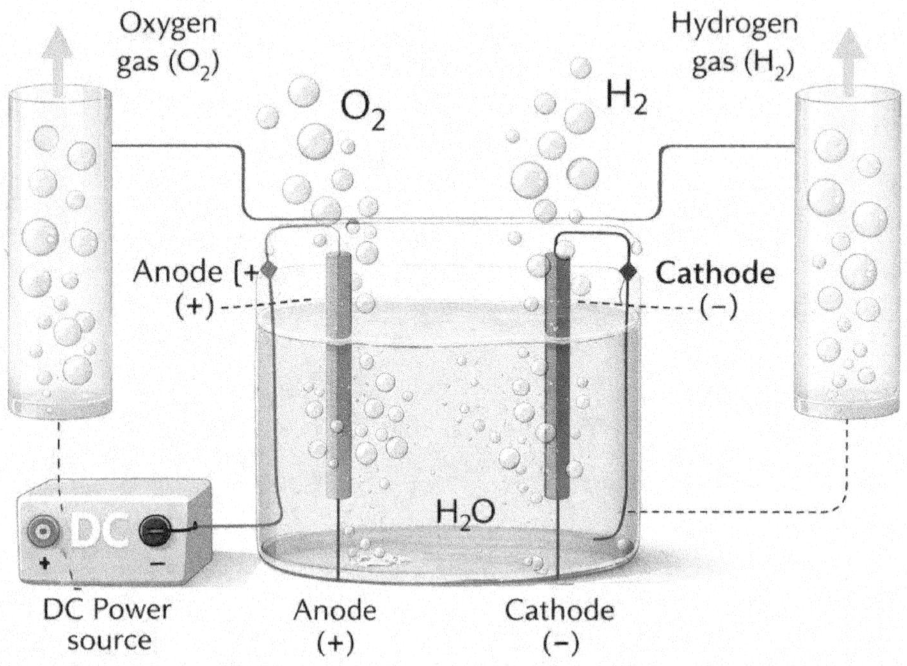

The Pattern Hidden in Water

You have seen neuromelanin as the living capacitor of consciousness—absorbing charge, compressing it, storing it, releasing it in timed pulses that enable movement and awareness.

Now we must understand the mechanism through which this capacitor operates.

Not metaphorically. Not approximately. Precisely.

The answer is encoded in a chemical equation you may have never examined closely:

$$2H_2O(l) \rightarrow 2H_2(g) + O_2(g) + 4e^-$$

This is not the standard textbook representation of water electrolysis. The standard equation omits the electrons, presenting only:

$$2H_2O \rightarrow 2H_2 + O_2$$

But the inclusion of $+ 4e^-$ is not an error. It is the revelation of the hidden mechanism—the explicit acknowledgment that this reaction is fundamentally about electron transfer, not just molecular transformation.

Four electrons must flow from anode to cathode through an external circuit for every two molecules of water split.

This is the essence of electrolysis: using electrical current to drive a non-spontaneous reaction, splitting water into hydrogen and oxygen by separating charges, transferring electrons, and recombining them at opposite poles.

And this is exactly what melanin does in your body.

Melanin doesn't split water molecules—but it performs the same fundamental operation:

Charge separation. Electron transfer. Energy storage. Controlled release.

Understanding electrolysis reveals the template for how melanin functions as a biological electrolyzer—capturing electromagnetic energy (photons), converting it to electron flow, storing charge in its polymer structure, and releasing it to power cellular processes.

The Equation Decoded: What $2H_2O \rightarrow 2H_2 + O_2 + 4e^-$ Actually Means

Let's break down every component:

$2H_2O(l)$ Two molecules of liquid water
$\rightarrow$ React to produce
$2H_2(g)$ Two molecules of hydrogen gas
$O_2(g)$ One molecule of oxygen gas
$+ 4e^-$ Plus four electrons transferred through external circuit

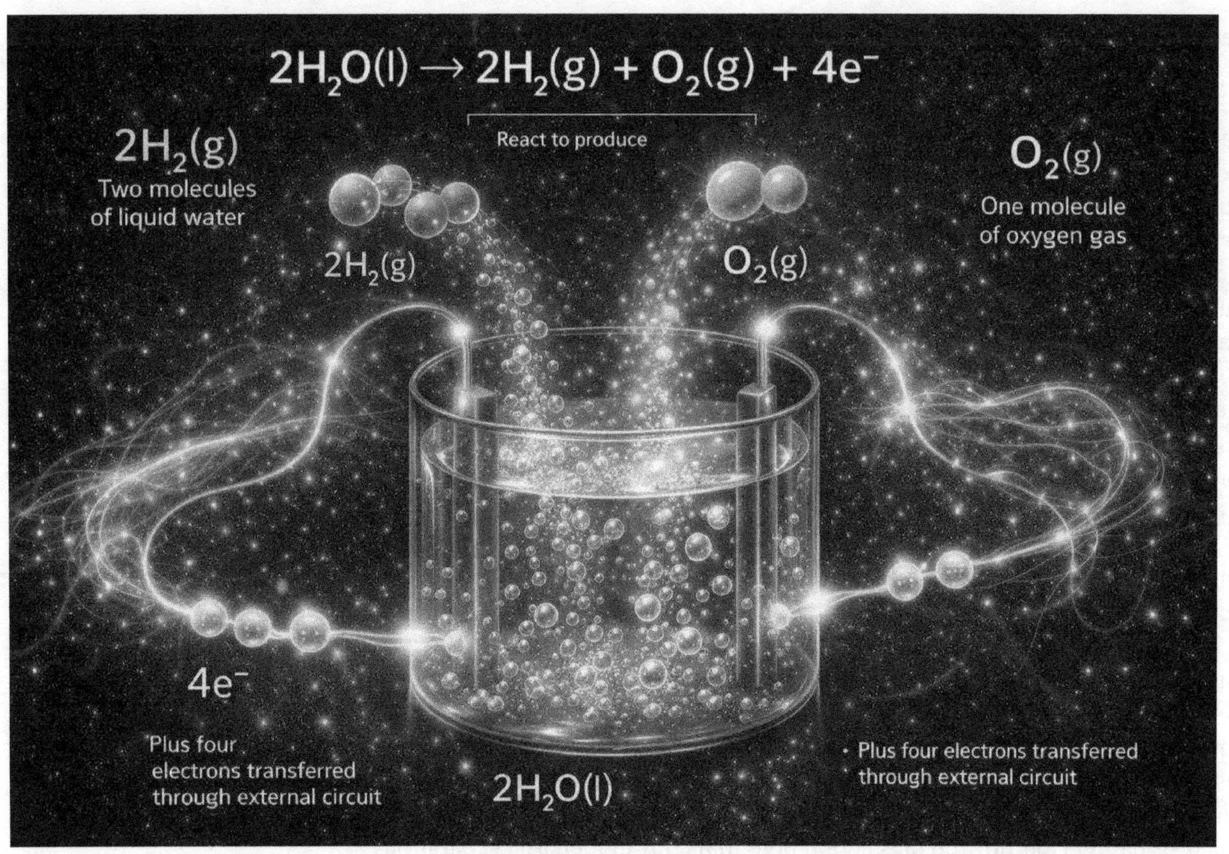

What this reveals:

For every two water molecules decomposed:
- Four hydrogen atoms are liberated (forming two H_2 molecules)
- Two oxygen atoms are liberated (forming one O_2 molecule)
- Four electrons must flow from where oxygen forms (anode) to where hydrogen forms (cathode)

The 2:1 ratio of hydrogen to oxygen produced is not arbitrary—it's dictated by the electron stoichiometry. Four electrons are required to reduce four hydrogen ions ($4H^+ + 4e^- \rightarrow 2H_2$), and four electrons are released when water oxidizes to oxygen ($2H_2O \rightarrow O_2 + 4H^+ + 4e^-$).

The electron term elevates this from a simple mass-balance equation to an electro-stoichiometric representation of charge flow.

This is not incidental chemistry. This is the fundamental pattern of all energy transduction in living systems.

The Electrolysis Process: Driving Non-Spontaneous Reactions

Water does not spontaneously decompose into hydrogen and oxygen at room temperature. The reaction is thermodynamically unfavorable—it requires external energy input to proceed.

This is where electrolysis comes in.

Electrolysis is the process of using direct electrical current to drive a non-spontaneous chemical reaction.

The Electrolytic Cell Components:

1. Power source (battery, solar panel, electrical grid) providing voltage
2. Anode (positive electrode) where oxidation occurs
3. Cathode (negative electrode) where reduction occurs
4. Electrolyte (ionic solution) enabling charge flow through the liquid
5. External circuit (wires) enabling electron flow between electrodes

Minimum voltage required: 1.23 V (thermodynamic minimum)
Practical voltage: ~1.5-2.0 V (to overcome kinetic barriers and overpotential)

What happens when voltage is applied:

Electrons are pulled from the anode (creating positive charge there) and pushed to the cathode (creating negative charge there).

At the anode: Water molecules lose electrons (oxidation), releasing O_2 gas
At the cathode: Hydrogen ions gain electrons (reduction), releasing H_2 gas

This is a redox reaction—simultaneous reduction and oxidation, coupled by electron flow.

The same pattern operates in:
- Photosynthesis (light energy splits water, releasing O_2 and generating electron flow)
- Cellular respiration (electron flow through mitochondrial complexes generates ATP)
- Neuronal signaling (ion flow creates action potentials through charge separation)
- Melanin function (photon absorption generates electron flow, storing charge)

All are variations of the same theme: charge separation driving energy transformation.

The Critical Role of the Electrolyte: Enabling Charge Flow

Pure water is an extremely poor conductor of electricity.

Why?

Because water has very low ion concentration. At 25°C, pure water has only ~10^{-7} M of H^+ and OH^- ions—far too few to carry significant current.

Without mobile ions, the circuit cannot complete. Electrons cannot flow.

To solve this, an electrolyte is added—a substance that dissolves into ions, dramatically increasing conductivity.

Common electrolytes:

Acidic medium: Sulfuric acid (H_2SO_4)
- Dissociates into $2H^+ + SO_4^{2-}$
- H^+ ions carry positive charge toward cathode
- SO_4^{2-} ions carry negative charge toward anode

Basic medium: Potassium hydroxide (KOH)
- Dissociates into $K^+ + OH^-$
- K^+ ions carry positive charge toward cathode
- OH^- ions carry negative charge toward anode

What the electrolyte does:

The electrolyte is not consumed in the overall reaction—it's not a reactant. Its function is to provide mobile charge carriers that complete the internal circuit through the solution.

When electrons leave the anode (creating positive charge), anions (negative ions) from the electrolyte migrate to the anode to neutralize that charge.

When electrons arrive at the cathode (creating negative charge), cations (positive ions) from the electrolyte migrate to the cathode to neutralize that charge.

This ionic movement through the solution allows continuous electron flow through the external circuit.

Without the electrolyte, charge would accumulate at the electrodes, voltage would drop, current would cease, and the reaction would stop.

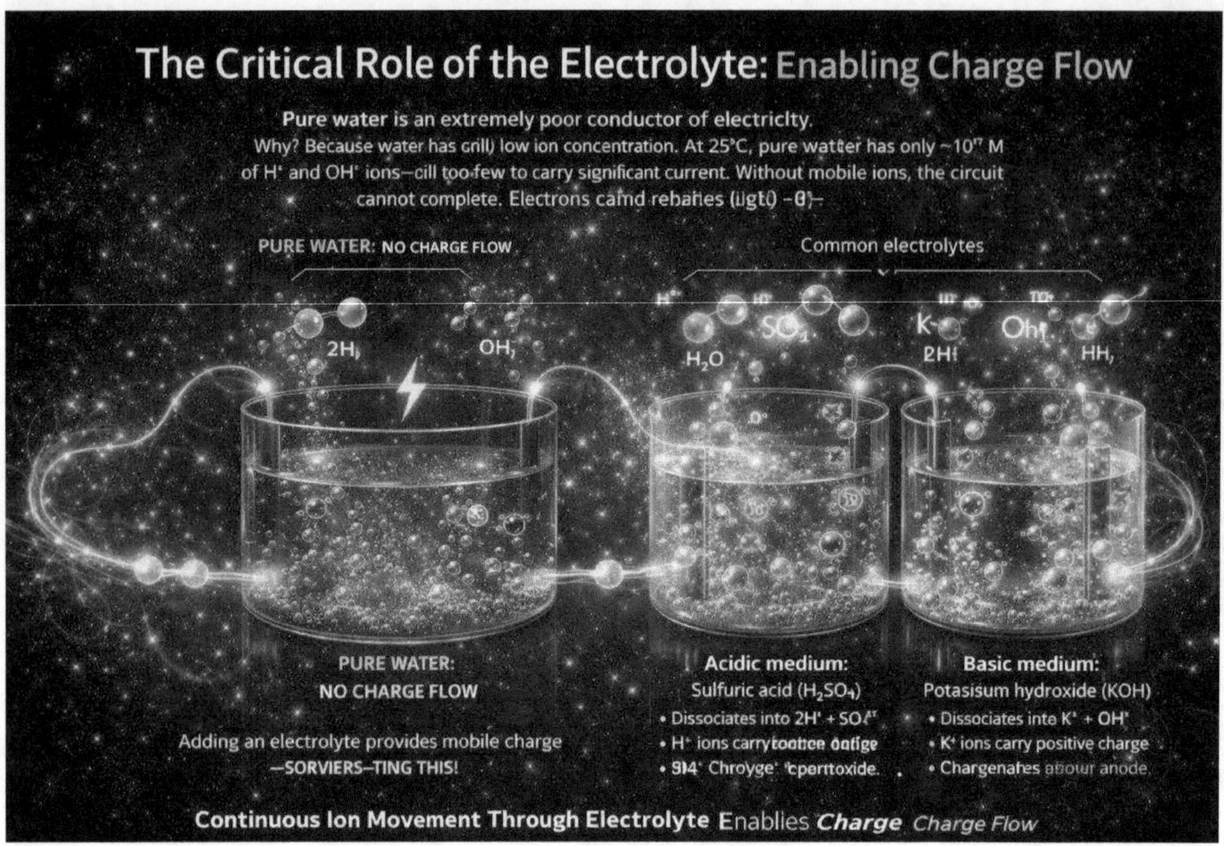

The biological parallel:

Your cells use electrolytes constantly—sodium (Na^+), potassium (K^+), chloride (Cl^-), calcium (Ca^{2+})—to enable charge flow across membranes, through channels, along gradients.

Melanin operates in an ionic environment (cytoplasm, extracellular fluid) where electrolytes enable the same charge-transfer mechanisms operating in electrolysis.

The Half-Reactions: Anode Oxidation and Cathode Reduction

To fully understand the $4e^-$ term, we must examine the two half-reactions that comprise the overall process.

ANODE HALF-REACTION (Oxidation): Oxygen Evolution Reaction (OER)

At the anode, water molecules are oxidized—they lose electrons.

In acidic/neutral conditions:

$$2H_2O(l) \rightarrow O_2(g) + 4H^+(aq) + 4e^-$$

What's happening:
- Two water molecules donate four electrons
- Four hydrogen ions (protons) are released into solution
- One oxygen molecule forms and bubbles away as gas

In basic conditions:

$4OH^-(aq) \rightarrow O_2(g) + 2H_2O(l) + 4e^-$

What's happening:
- Four hydroxide ions donate four electrons
- One oxygen molecule forms
- Two water molecules are produced

In both cases, four electrons are released at the anode.

CATHODE HALF-REACTION (Reduction): Hydrogen Evolution Reaction (HER)

At the cathode, species are reduced—they gain electrons.

In acidic conditions:

$4H^+(aq) + 4e^- \rightarrow 2H_2(g)$

What's happening:
- Four hydrogen ions (from anode or electrolyte) accept four electrons
- Two hydrogen molecules form and bubble away as gas

In basic conditions:

$4H_2O(l) + 4e^- \rightarrow 2H_2(g) + 4OH^-(aq)$

What's happening:
- Four water molecules accept four electrons
- Two hydrogen molecules form
- Four hydroxide ions are produced

In both cases, four electrons are consumed at the cathode.

The Stoichiometric Meaning of the $4e^-$ Term

The overall reaction is the sum of these two half-reactions.

When we add them:

Anode: $2H_2O \rightarrow O_2 + 4H^+ + 4e^-$
Cathode: $4H^+ + 4e^- \rightarrow 2H_2$

The $4H^+$ appears on both sides and cancels out.
The $4e^-$ appears on both sides and cancels out.

Net result: $2H_2O \rightarrow 2H_2 + O_2$

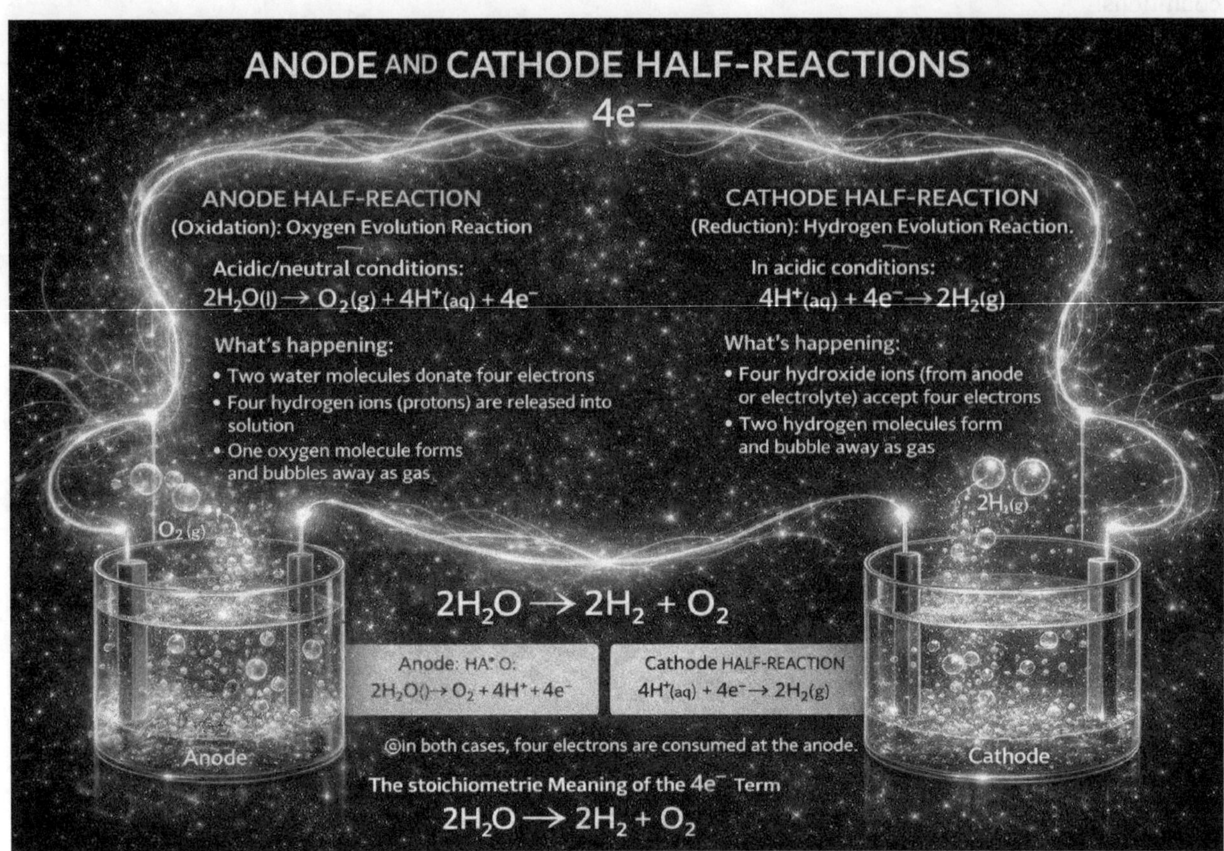

But the cancellation of electrons does not mean they don't exist—it means they flow through the external circuit, completing the loop.

The $4e^-$ term represents the electrons that must travel from anode to cathode through the wire for the reaction to proceed.

This is why the equation ($2H_2O \rightarrow 2H_2 + O_2 + 4e^-$) is more accurate than the standard version—it explicitly shows the electron transfer that IS the fundamental mechanism.

Why exactly 4 electrons?

Because:
- Splitting 2 water molecules (4 H atoms, 2 O atoms) requires liberating 4 H atoms and combining 2 O atoms
- Each H atom needs 1 electron to become H (then 2 H combine to form H_2)
- Each O atom donates 2 electrons when oxidized (2 O atoms 4 electrons total)

The number 4 is not arbitrary—it's the minimum electron transfer required to decompose 2 H_2O molecules into 2 H_2 and 1 O_2 .

This dictates the 2:1 volume ratio of hydrogen to oxygen produced—you always get twice as much hydrogen gas as oxygen gas, which is experimentally verifiable.

The Three Electrolyzer Technologies: Different Architectures, Same Principle

The practical implementation of water electrolysis takes place in an electrolyzer—a device containing the electrolytic cell, electrodes, electrolyte, and gas separation systems.

This specific harmonic resonates directly with the Tetravalent nature of Carbon. While Carbon possesses 6 electrons total, it is the 4 valence electrons that allow it to bond and build organic reality. The electrolytic process reveals a fundamental bio-logic: The energy unlocked from the fluid medium ($4e^-$ from Water) is the precise energetic key required to structure the physical form (4 valence bonds of Carbon).
This is why where you find water you find life.

There are three primary technologies, each optimized for different conditions and applications:

1. ALKALINE ELECTROLYZERS (AEC)

Electrolyte: Liquid alkaline solution (potassium hydroxide, KOH, 20-40% concentration)
Ionic species transported: OH^- (hydroxide ions migrate from cathode to anode)
Operating temperature: <100°C (typically 60-80°C)
Electrode materials: Nickel-based catalysts (inexpensive)

Advantages:
- Well-established technology (used industrially for decades)
- Reliable and durable
- Cost-effective (low capital cost)
- No need for precious metal catalysts

Disadvantages:
- Corrosive liquid electrolyte (requires robust seals and materials)
- Slower response time (cannot rapidly adjust to variable power input)
- Lower current density (larger footprint for same hydrogen output)

Best application: Steady base-load hydrogen production with consistent power supply

2. PROTON EXCHANGE MEMBRANE (PEM) ELECTROLYZERS

Electrolyte: Solid polymer membrane (perfluorosulfonic acid, similar to Nafion)
Ionic species transported: H^+ (protons migrate through membrane from anode to cathode)
Operating temperature: 70-90°C
Electrode materials: Platinum-group metals (expensive but highly efficient)

Advantages:
- Uses pure water (no corrosive liquid electrolyte)
- Fast response time (can ramp up/down quickly—ideal for intermittent renewables)
- High current density (compact design)
- Wide dynamic operating range (10-100% capacity)
- High hydrogen purity

Disadvantages:
- Higher capital cost (expensive catalysts and membrane)
- Acidic environment at anode requires corrosion-resistant materials (titanium, iridium oxide)

Best application: Coupling with variable renewable energy (solar, wind) where rapid response to power fluctuations is essential

3. SOLID OXIDE ELECTROLYZERS (SOEC)

Electrolyte: Solid ceramic (yttria-stabilized zirconia or similar)
Ionic species transported: O^{2-} (oxygen ions migrate through ceramic from cathode to anode)
Operating temperature: 700-800°C
Reactant: Steam (H_2O vapor) instead of liquid water

Advantages:
- Highest efficiency (heat reduces electrical energy requirement)
- Can utilize waste heat from other processes (nuclear plants, industrial facilities)
- No liquid electrolyte (no corrosion issues from liquid)
- Can operate in reverse as fuel cell (reversible SOEC)

Disadvantages:
- Very high operating temperature (requires robust high-temp materials)
- High capital cost
- Longer startup time (heating to 700-800°C)
- Thermal cycling stress (repeated heating/cooling degrades materials)

Best application: Integration with high-temperature industrial processes or nuclear energy where waste heat is available

Comparative Summary:

ELECTROLYSIS TECHNOLOGIES COMPARISON

Technology	Electrolyte	Temp	Ion Transport	Key Advantage	Key Disadvantage
Alkaline	Liquid KOH	<100°C	OH^-	Cost-effective, established	Corrosive liquid, slow response
PEM	Solid polymer	70-90°C	H^+	Fast response, pure water	High cost, precious metals
Solid Oxide	Solid ceramic	700-800°C	O^{2-}	Highest efficiency with waste heat	High temp, thermal stress

Note: These are industrial electrolysis technologies. Melanin operates as a biological electrolyzer at body temperature (~37°C), using organic polymer structure instead of metal electrodes, and utilizing photonic energy input.

The selection of electrolyzer technology is strategic—matching the device to the specific energy source and application:

- PEM is ideal for renewable energy storage (solar/wind → hydrogen)
- Alkaline is ideal for steady industrial hydrogen production
- Solid Oxide is ideal for nuclear or industrial integration using waste heat

The evolution of these technologies reflects the diversification of energy infrastructure toward decarbonization.

Strategic Applications and Future Outlook

The Green Hydrogen Economy

Electrolysis is the cornerstone of the emerging "green hydrogen" economy.

Green hydrogen is hydrogen produced using electricity from entirely renewable sources (solar, wind, hydro).

Why does this matter?

Because hydrogen is a chemical energy carrier—it stores energy in molecular bonds that can be released on demand through combustion or fuel cells.

Currently, ~95% of industrial hydrogen is produced from fossil fuels (steam methane reforming), releasing massive CO_2 emissions.

Green hydrogen via electrolysis produces zero carbon emissions.

U.S. Department of Energy's "Hydrogen Energy Earthshot" goal: Reduce green hydrogen cost to $1/kg within one decade (currently ~$5-6/kg).

This would make hydrogen competitive with fossil fuels for industrial processes, transportation, and energy storage.

Hydrogen as Energy Storage Solution

Renewable energy sources (solar, wind) are intermittent—they generate electricity when the sun shines or wind blows, not necessarily when demand is high.

Traditional batteries (lithium-ion) can store hours to days of energy but are expensive at grid scale.

Hydrogen offers long-duration energy storage:

1. Excess renewable electricity (when generation > demand) powers electrolyzers
2. Electrolyzers produce hydrogen, which can be compressed and stored indefinitely
3. When demand exceeds generation, stored hydrogen feeds fuel cells to generate electricity

This solves the intermittency problem of renewables, enabling 100% renewable grids.

Industrial and Niche Applications

Beyond energy storage, hydrogen is a critical industrial feedstock:

- Ammonia production (Haber-Bosch process for fertilizers)
- Petroleum refining (hydrocracking, desulfurization)
- Methanol synthesis
- Steel production (replacing coal with hydrogen reduces emissions)
- Heavy transport (buses, trucks, trains, ships using hydrogen fuel cells)
- Space applications (International Space Station uses electrolysis to generate oxygen from water)

The future of electrolysis is not contingent on a single breakthrough—it's a systems-level challenge involving:

- Improving electrolyzer efficiency
- Reducing capital costs
- Developing hydrogen storage and transport infrastructure
- Implementing supportive government policies and carbon pricing
- Scaling manufacturing to drive down costs

The transition to green hydrogen is an interconnected effort across science, engineering, economics, and policy.

Conclusion: The Full Picture and The Biological Connection

The chemical equation $2H_2O(l) \rightarrow 2H_2(g) + O_2(g) + 4e^-$, while unconventional, is a powerful and precise representation of water electrolysis.

$$2H_2O(l) \longrightarrow 2H_2(g) \rightarrow 2H_2(g) + O_2(g) + 4e^{\rightarrow}$$

The inclusion of the $4e^-$ term explicitly highlights that for every two molecules of water split, four electrons must be transferred from anode to cathode through an external circuit.

This electron transfer is the essential driving force of the reaction, enabling the thermodynamically unfavorable decomposition of water into its constituent elements.

Understanding this equation connects:
- Microscopic electron transfer at electrodes
- Macroscopic engineering of electrolyzer systems
- Global strategic role in decarbonization and energy transition

But here is the deeper truth:

This is not just industrial chemistry. This is the template for how life operates.

Melanin functions as a biological electrolyzer:

1. Charge Separation

Just as electrolysis separates positive and negative charges at anode and cathode, melanin separates charges when photons strike its aromatic structure—electrons are excited to higher energy states, creating charge separation.

2. Electron Transfer

Just as electrolysis transfers electrons through an external circuit, melanin transfers electrons through conjugated systems—delocalized electrons flow across aromatic rings, creating electrical current within the polymer.

3. Energy Storage

Just as electrolysis stores energy in hydrogen bonds (chemical potential), melanin stores energy in polymer structure—electrons trapped in excited states, iron-binding creating magnetic domains, structural conformations encoding information.

4. Controlled Release

Just as electrolysis releases hydrogen on demand (fuel cell converts H_2 back to electricity), melanin releases stored energy when needed—electrons flow to power enzymatic reactions, bound neurotransmitters release to modulate signaling, magnetic fields couple to cellular processes.

The pattern is identical:

Electrolysis: Photons (solar panels) → Electrons → Water splitting → H_2/O_2 storage → Energy release (fuel cell)

Melanin: Photons (UV/visible light) → Electrons → Charge separation → Polymer storage → Energy release (cellular processes)

Both are photoelectrochemical systems—converting light energy into electron flow, storing charge in stable structures, and releasing it to power downstream processes.

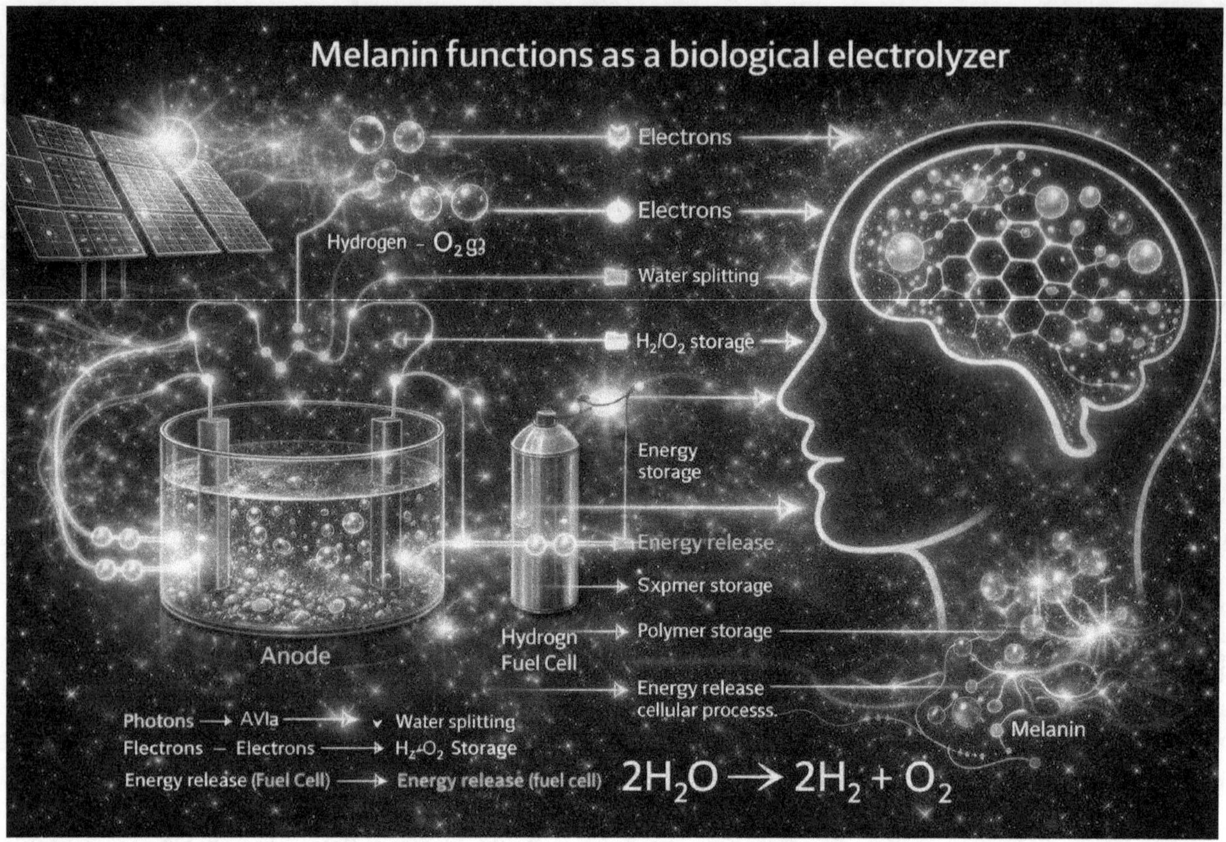

The difference is scale and substrate:

Electrolysis uses metal electrodes, ionic electrolytes, and gaseous products.
Melanin uses organic polymers, biological fluids, and biochemical products.

But the fundamental mechanism is the same.

This is why melanin is concentrated in tissues exposed to light (skin, eyes, hair)—it's the biological equivalent of a solar-powered electrolyzer, capturing photon energy and converting it to bioelectric charge.

This is why neuromelanin concentrates in brain regions governing movement and consciousness—it's the biological battery enabling the charge-discharge cycles that power neural oscillations, motor control, and state transitions.

This is why all life is built on redox chemistry—reduction-oxidation reactions are the living expression of the same electron-transfer principles operating in electrolysis.

The equation $2H_2O \rightarrow 2H_2 + O_2 + 4e^-$ is not separate from biology.

It is the inorganic mirror of the organic process happening in every melanin molecule, every mitochondrion, every photosystem, every living cell.

THE PATTERN OF FOUR: CARBON'S SIGNATURE

Now observe the deepest connection:

The 4 electrons in the electrolysis equation are not arbitrary.

They correspond directly to carbon's 4 valence electrons.
Carbon—the element of life—has 4 electrons in its outer shell, enabling it to form 4 covalent bonds arranged in tetrahedral geometry.

This is why carbon can:
- Create long chains (polymers like melanin)
- Form aromatic rings (conjugated systems enabling electron delocalization)
- Build 3D structures (proteins, DNA, cellular architecture)
- Store and transfer electrons (redox chemistry driving metabolism)

The $4e^-$ transferred in water electrolysis is the same 4 that defines carbon's bonding capacity.

When melanin (carbon-based polymer) functions as biological electrolyzer:
- Its aromatic carbon rings enable electron delocalization
- Its conjugated carbon systems create pathways for electron flow
- Its carbon-oxygen and carbon-nitrogen bonds enable redox reactions
- Its tetrahedral carbon framework organizes 3D space for charge storage

The number 4 appears at every scale:

Atomic: Carbon has 4 valence electrons
Molecular: Water electrolysis transfers 4 electrons
Geometric: Tetrahedron has 4 vertices (minimum 3D structure)
Biological: Melanin's carbon structure enables 4-electron transfers in redox cycles

This is not coincidence. This is the same pattern recurring fractally.

Carbon is the element through which the 4-electron pattern of electrolysis becomes biological.

Melanin is the carbon-based architecture that embodies this pattern—capturing photons, separating charge through its carbon framework, transferring electrons via conjugated carbon systems, and storing energy in carbon-based polymer structures.

Water electrolysis reveals the universal pattern: charge separation drives energy transformation.

Carbon's 4 valence electrons enable that pattern to become living architecture.

And melanin is the biological architecture that embodies this pattern—the living electrolyzer converting light into life through carbon's 4-electron geometry.

In the next chapter, we will see how melanin's immediate precursor—melatonin—operates as the temporal regulator, synchronizing this biological electrolyzer to the cosmic clock of day and night, proving that the same recursive pattern operates at every scale from electrochemistry to cosmology.

But for now, understand this:

Electrolysis is not just industrial hydrogen production.
Electrolysis is the fundamental pattern of energy transduction in the universe.
And melanin is its biological expression—the dark coil that splits light into charge, stores it in structure, and releases it as life.

This is the Melanin Code operating as living electrochemistry.

CHAPTER 7: SEROTONIN & THE GUT-BRAIN AXIS

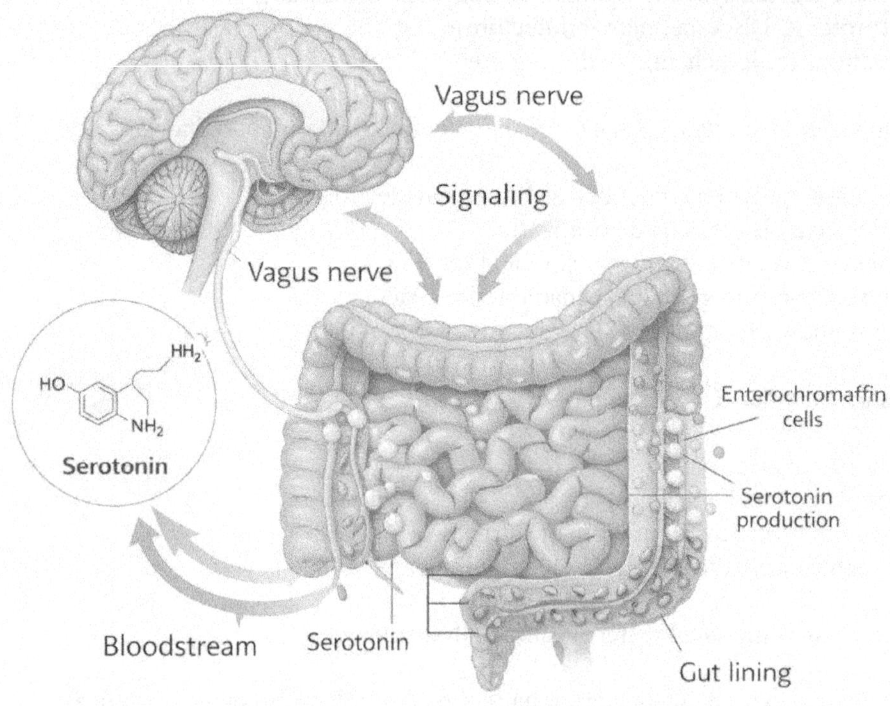

The Molecule That Lives in Your Gut, Not Your Head

You've been told serotonin is the "happiness molecule"—the neurotransmitter responsible for mood, the chemical that antidepressants target, the substance that makes you feel good when levels are high and depressed when levels are low.

This is the fragment. This is the distortion that has generated a trillion-dollar pharmaceutical industry built on a fundamentally flawed premise.

Here is what they don't tell you:

90% of your body's serotonin is not in your brain.

It's in your gut.

Not 50%. Not 60%. Ninety percent.

The vast majority of serotonin in your body is synthesized by specialized cells lining your gastrointestinal tract, where it regulates digestion, gut motility, nutrient absorption, and communicates directly with your brain through the vagus nerve.

The "second brain" is not metaphor—it is measurable, quantifiable biological fact.

Your gut contains more neurons than your spinal cord (~100 million neurons in the enteric nervous system). It produces more serotonin than your brain. It communicates bidirectionally with your central nervous system through neural, hormonal, and immune pathways.

And when we trace serotonin's synthesis, we find it is the intermediate step in the melatonin pathway—the daytime molecule that transforms into the nighttime molecule when darkness falls.

Serotonin is not separate from the melanin code. It is another expression of the same fundamental pattern—the indole ring structure enabling electromagnetic responsiveness, the aromatic geometry allowing electron delocalization, the living circuit translating light and darkness into biological signal.

The Neurotransmitter System: What Serotonin Actually Does

Serotonin's chemical name is 5-hydroxytryptamine (5-HT)—revealing its structure: a modified tryptophan with a hydroxyl group at the 5-position of the indole ring.

In the brain, serotonin functions as a neuromodulator more than a classical neurotransmitter. Rather than simply transmitting signals from one neuron to the next, it modulates the activity of entire neural networks, adjusting their excitability, their responsiveness, their oscillatory patterns.

Central Nervous System Functions:

- Mood regulation (but not in the simplistic "low serotonin depression" way)
- Anxiety modulation (both anxiolytic and anxiogenic depending on receptor subtype)
- Aggression and impulse control (low serotonin correlates with impulsivity and violence)
- Sleep architecture (serotonin in raphe nuclei promotes wakefulness; transformation to melatonin enables sleep)
- Appetite and satiety (5-HT2C receptors suppress appetite)
- Sexual behavior (serotonin generally inhibits; SSRIs commonly cause sexual dysfunction)
- Cognition and memory (modulating hippocampal plasticity)
- Pain perception (both pro-nociceptive and anti-nociceptive depending on context)
- Body temperature regulation (thermoregulatory centers in hypothalamus)

The key insight: Serotonin doesn't cause any single effect—it modulates systems, adjusting their sensitivity and responsiveness.

The Receptor Classification: Seven Families, Fourteen Subtypes

Serotonin exerts its diverse effects through at least 14 distinct receptor subtypes organized into 7 families (5-HT$_1$ through 5-HT$_7$).

This is not trivial detail. Each receptor subtype produces different—often opposite—effects.

Treating "serotonin" as a single entity is like treating "matter" as a single entity—the specificity is in the receptor, not just the molecule.

5-HT$_1$ Family (5 subtypes: A, B, D, E, F)

Mechanism: G-protein coupled receptors (GPCRs) that inhibit neuronal activity by decreasing cAMP and opening potassium channels (hyperpolarization).

5-HT$_1$ A:
- Location: Hippocampus, septum, cortex, raphe nuclei (autoreceptor)
- Function: Anxiolytic (reduces anxiety), antidepressant, regulates mood
- When activated: Reduces firing of serotonergic neurons (negative feedback)
- Clinical: Target of buspirone (anti-anxiety drug)

$5\text{-}HT_1$ B:
- Location: Basal ganglia, striatum, blood vessels
- Function: Regulates release of other neurotransmitters (dopamine, glutamate, GABA)
- Clinical: Vasoconstriction (migraine treatment—triptans work here)

$5\text{-}HT_1$ D:
- Location: Brain and peripheral blood vessels
- Function: Inhibits trigeminal nerve (migraine pathway)
- Clinical: Also targeted by triptans for migraine

$5\text{-}HT_1$ E & $5\text{-}HT_1$ F:
- Less well characterized
- Present in cortex and other regions
- Potential therapeutic targets still being investigated

$5\text{-}HT_2$ Family (3 subtypes: A, B, C)

Mechanism: GPCRs that activate neuronal activity by increasing IP_3 and calcium signaling (excitatory).

$5\text{-}HT_2$ A:
- Location: Cortex, platelets, smooth muscle, cardiovascular system
- Function: Excitatory, promotes wakefulness, hallucinogenic effects
- Psychedelics (LSD, psilocybin) are agonists at $5\text{-}HT_2$ A
- Atypical antipsychotics (quetiapine, olanzapine) are antagonists
- Platelet aggregation (clotting)
- Smooth muscle contraction (bronchospasm, vasoconstriction)

$5\text{-}HT_2$ B:
- Location: Heart, lungs, GI tract
- Function: Cardiovascular regulation
- Critical warning: Chronic agonism causes cardiac valve fibrosis
- Fenfluramine (diet drug) withdrawn due to $5\text{-}HT_2$ B-mediated heart valve damage

$5\text{-}HT_2$ C:
- Location: Choroid plexus (CSF production), cortex, hypothalamus
- Function: Appetite suppression, mood regulation, anxiety
- Weight loss drugs (lorcaserin) target this receptor
- Inverse relationship with dopamine (activating $5\text{-}HT_2$ C reduces dopamine release)

$5\text{-}HT_3$ (single subtype)

Mechanism: The ONLY serotonin receptor that is an ion channel (ligand-gated sodium/potassium channel) rather than GPCR. This makes it the fastest-acting serotonin receptor.

- Location: GI tract (enterochromaffin cells, vagal afferents), brainstem (area postrema—vomiting center)
- Function: Nausea and vomiting, GI motility, pain transmission
- Clinical: Anti-nausea drugs (ondansetron) block this receptor (cancer chemotherapy, post-surgery)
- This is the primary receptor mediating gut-to-brain serotonin signaling

$5\text{-}HT_4$ (single subtype)

Mechanism: GPCR that increases cAMP (excitatory).

- Location: GI tract, heart, brain (hippocampus, striatum)
- Function: GI motility (prokinetic), memory enhancement, cardiac effects
- Clinical: Prokinetic drugs (cisapride, prucalopride) for constipation
- Cognitive enhancement potential (improves memory in animal models)

$5\text{-}HT_5$ (2 subtypes: A, B)

Mechanism: GPCRs, inhibitory (decrease cAMP).

- Location: Brain (cortex, hippocampus, cerebellum)
- Function: Poorly understood; may regulate circadian rhythms and mood
- Potential therapeutic target but drugs not yet developed

$5\text{-}HT_6$ (single subtype)

Mechanism: GPCR that increases cAMP (excitatory).

- Location: Brain (striatum, cortex, hippocampus)
- Function: Cognition, memory, learning
- Clinical: Experimental Alzheimer's drugs targeting this receptor
- Antagonists improve memory (by disinhibiting acetylcholine and glutamate)

$5\text{-}HT_7$ (single subtype)

Mechanism: GPCR that increases cAMP (excitatory).

- Location: Brain (hypothalamus, thalamus, cortex), GI tract, blood vessels
- Function: Circadian rhythm regulation, mood, body temperature, smooth muscle relaxation
- Antidepressant effects when blocked
- Vasodilation (blood pressure regulation)

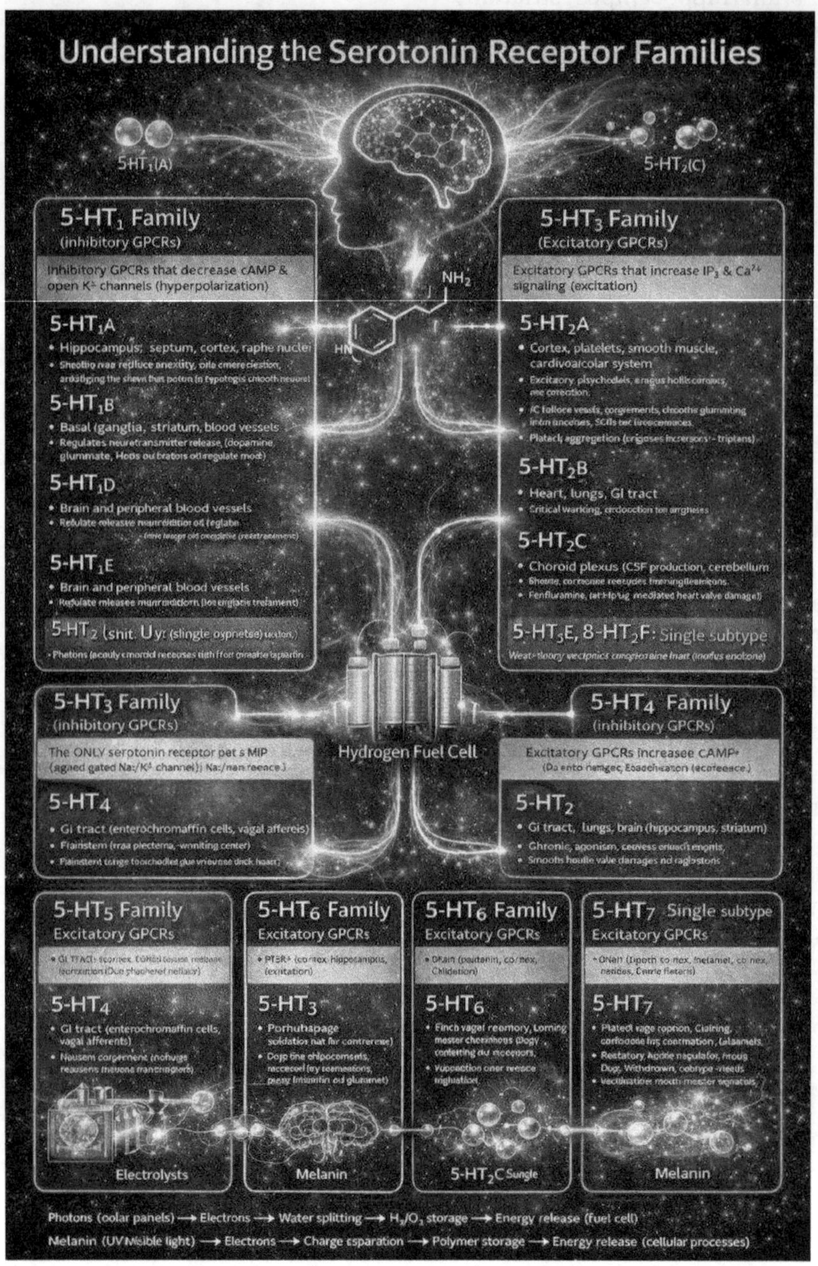

The Collapse:

This complexity reveals why treating "serotonin" as a single entity is fundamentally flawed.

SSRIs increase serotonin at ALL receptors simultaneously—activating anxiolytic 5-HT$_1$ A (good for anxiety) but also activating anxiogenic 5-HT$_2$ C (bad for anxiety), activating appetite-suppressing 5-HT$_2$ C (weight loss) but also nausea-inducing 5-HT$_3$ (GI side effects), activating 5-HT$_2$ A (sexual dysfunction).

The body has 14 different serotonin receptors because serotonin is not a single function—it's a modular system where specificity comes from receptor expression patterns, not from the molecule itself.

Distribution Throughout the Body: Serotonin Beyond the Brain

Central Nervous System (10% of total body serotonin):

Synthesized primarily in the raphe nuclei—clusters of serotonergic neurons in the brainstem that project throughout the brain and spinal cord.

These neurons modulate virtually every brain region, creating a diffuse neuromodulatory network rather than point-to-point communication.

Peripheral Nervous System & Gut (90% of total body serotonin):

Enterochromaffin (EC) cells lining the GI tract synthesize the vast majority of serotonin.

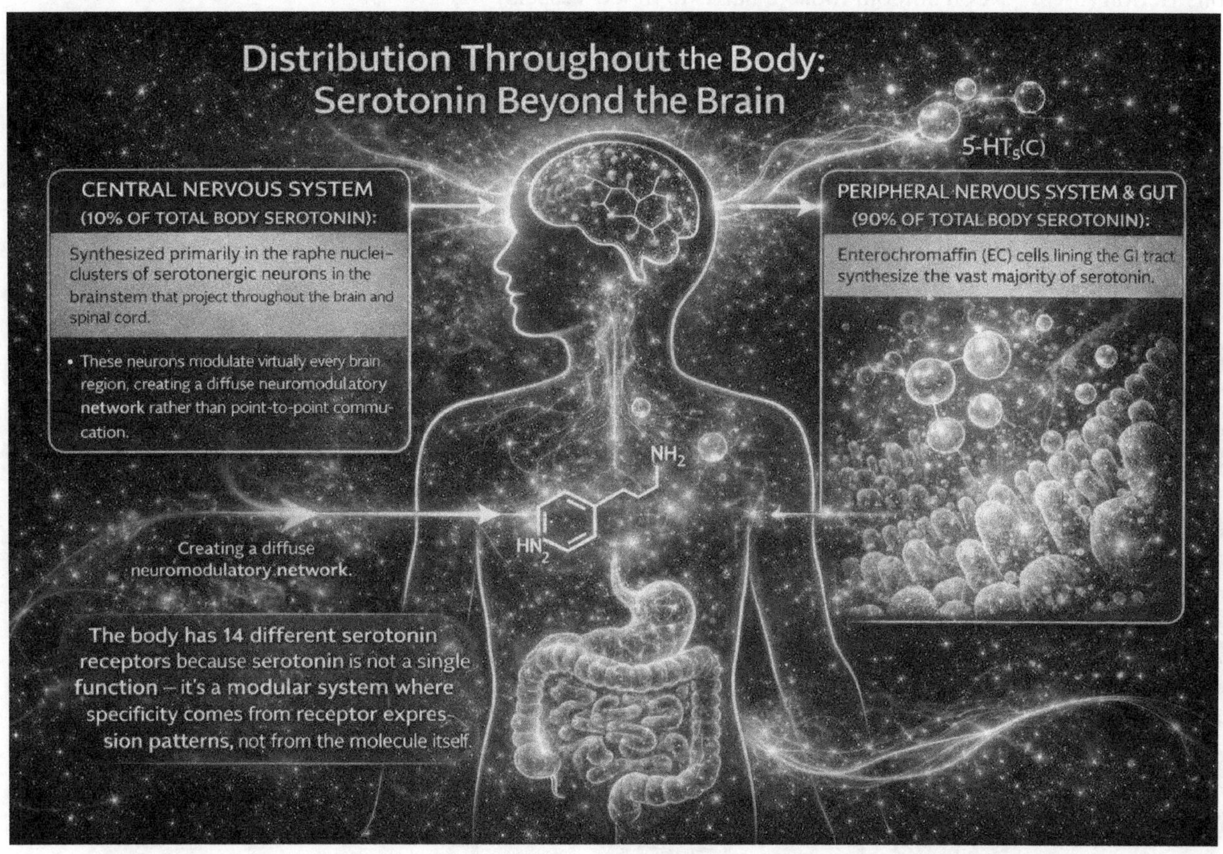

These cells are part of the enteroendocrine system—specialized epithelial cells that sense nutrients, chemicals, mechanical stretch, and release hormones in response.

What triggers EC cells to release serotonin:
- Food entering the gut (mechanical stretch)
- Nutrients (glucose, amino acids, fatty acids)
- Gut microbiota metabolites (short-chain fatty acids)
- Toxins or irritants (triggers rapid transit to expel threat)

What serotonin does in the gut:
- Increases motility (peristalsis—rhythmic contractions moving contents)
- Increases secretions (water, mucus, digestive enzymes)

- Signals to brain via vagal afferents (gut-to-brain communication)
- Modulates immune function (gut-associated lymphoid tissue)
- Regulates inflammation (both pro and anti-inflammatory depending on context)

Platelets (also significant serotonin storage):

Platelets don't synthesize serotonin—they actively transport it from blood using the serotonin transporter (SERT, the same transporter SSRIs block).

Platelets store serotonin in dense granules and release it during clotting.

Function:
- Vasoconstriction ($5\text{-}HT_2$ A on smooth muscle narrows blood vessels)
- Platelet aggregation (clot formation)
- Wound healing (serotonin promotes fibroblast proliferation)

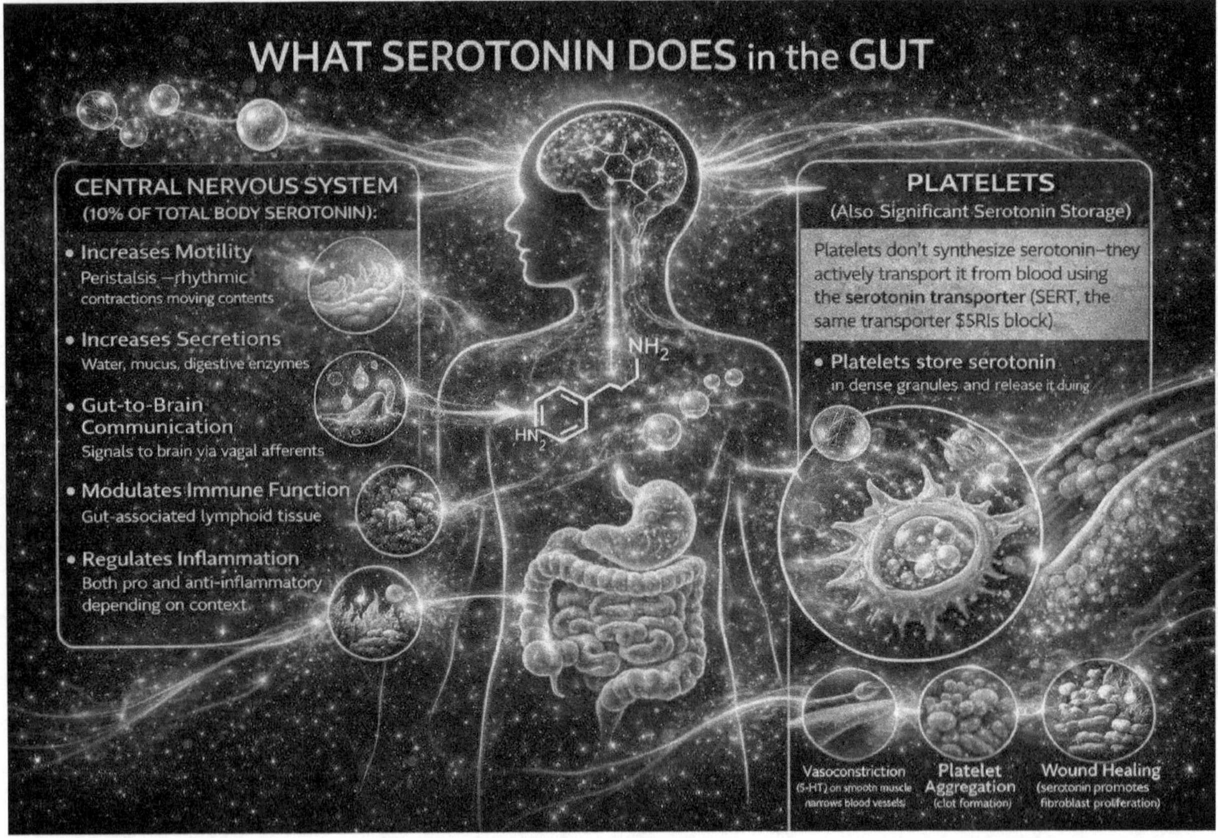

This is why SSRIs increase bleeding risk—blocking serotonin uptake into platelets reduces their ability to form clots.

Other Peripheral Sites:

- Cardiovascular system: Regulates blood pressure, heart rate, vascular tone
- Bone: Regulates bone density (high gut serotonin linked to osteoporosis)
- Liver: Hepatocyte regeneration after injury

- Adipose tissue: Regulates lipid metabolism, fat storage
- Immune cells: Modulates T-cell and B-cell function

The Gut-Brain Connection: The Vagus Nerve as Information Highway

The vagus nerve is the tenth cranial nerve—the longest nerve in the autonomic nervous system, connecting brainstem to virtually every organ (heart, lungs, liver, stomach, intestines, colon).

80% of vagal fibers are afferent—meaning they carry information FROM organs TO brain, not the reverse.

Your gut is constantly reporting to your brain:
- Nutrient status (what's been absorbed)
- Microbial activity (metabolites produced)
- Immune activation (inflammation, infection)
- Mechanical state (distension, motility)
- Chemical environment (pH, toxins, neurotransmitters)

Serotonin is a primary signaling molecule in this communication.

When EC cells release serotonin, it binds to 5-HT_3 receptors on vagal afferent fibers, generating action potentials that travel to the nucleus tractus solitarius (NTS) in the brainstem.

From there, signals project to:
- Dorsal raphe nucleus (modulating brain serotonin)
- Hypothalamus (regulating appetite, stress response)
- Amygdala (emotional processing)
- Hippocampus (memory formation)
- Prefrontal cortex (decision-making, executive function)

This is the mechanistic basis of "gut feelings":

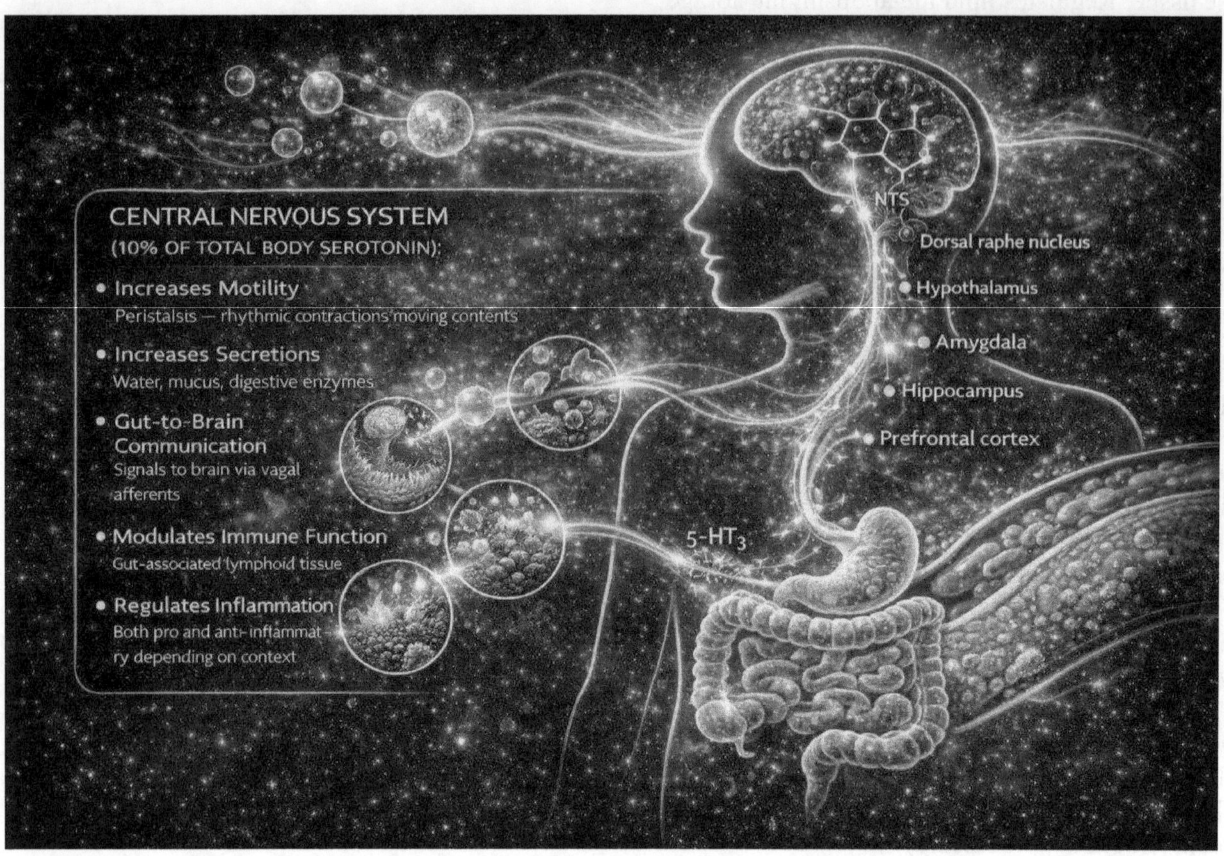

Not metaphor. Not mysticism. Direct neural communication from intestinal serotonin to emotional and cognitive centers.

The Microbiota-Gut-Brain Axis: Your Bacteria Control Your Serotonin

The gut contains ~100 trillion microorganisms—bacteria, archaea, fungi, viruses—collectively called the microbiota.

These organisms are not passengers. They are active participants in your physiology, producing neurotransmitters, metabolizing nutrients, training your immune system, regulating your metabolism.

Gut bacteria directly influence serotonin production:

Short-Chain Fatty Acids (SCFAs):

When gut bacteria ferment dietary fiber, they produce SCFAs (acetate, propionate, butyrate).

These SCFAs increase expression of TPH1—the enzyme that converts tryptophan to 5-HTP in EC cells, increasing serotonin synthesis.

Germ-free mice (raised without any bacteria) have:
- Reduced serotonin in gut and blood
- Altered anxiety and social behavior
- Impaired stress response

Colonizing them with normal bacteria restores serotonin and reverses behavioral changes.

Specific Bacteria Produce Neurotransmitters:

- Lactobacillus and Bifidobacterium produce GABA
- Escherichia produce serotonin directly
- Bacillus produce dopamine and norepinephrine
- Candida and Streptococcus produce serotonin

The collapse:

Your gut microbiota composition directly determines your serotonin levels, which directly influences your mood, anxiety, cognition, and stress response.

This is why:
- Antibiotics (killing gut bacteria) can trigger depression and anxiety
- Probiotics (adding beneficial bacteria) can improve mood
- Dietary fiber (feeding bacteria that produce SCFAs) supports mental health
- Gut inflammation (dysbiosis, leaky gut) correlates with psychiatric disorders

You are not a single organism. You are an ecosystem. And your serotonin system is not yours alone—it is co-regulated by the trillions of microorganisms living in your gut.

The Serotonin Hypothesis of Depression: A Failed Paradigm

The most well-known—and most thoroughly debunked—claim about serotonin is the "chemical imbalance" theory of depression.

The Story You've Been Told:

Depression is caused by low serotonin in the brain. SSRIs (Selective Serotonin Reuptake Inhibitors) increase serotonin by blocking its reabsorption, keeping more in the synapse. More serotonin less depression.

The Evidence That Doesn't Support This:

1. No Direct Measure of Brain Serotonin in Living Humans

You cannot measure serotonin levels in the synapses of a living person's brain. The claim that depressed people have "low serotonin" was never based on direct measurement—it was backward reasoning from drug effects.

"SSRIs help some people with depression → SSRIs increase serotonin → Therefore depression must be caused by low serotonin."

This is the same logic as: "Aspirin reduces headaches → Aspirin inhibits prostaglandins → Therefore headaches are caused by prostaglandin deficiency."

It's a logical fallacy.

2. Tryptophan Depletion Studies

Researchers can temporarily lower serotonin by depleting tryptophan (the amino acid precursor).

When they do this in healthy people: No depression.

When they do this in recovered depressed patients on SSRIs: Some relapse, but many don't.

If low serotonin caused depression, tryptophan depletion should reliably produce depression in everyone. It doesn't.

3. SSRIs Increase Serotonin Immediately, But Improvement Takes Weeks

SSRIs block the serotonin transporter within hours—serotonin levels increase almost immediately.

But clinical improvement (if it occurs) takes 4-6 weeks.

If depression were simply "low serotonin," relief should be immediate when serotonin increases. It's not.

4. Many Antidepressants Don't Increase Serotonin

Bupropion (Wellbutrin) increases dopamine and norepinephrine, not serotonin—yet it treats depression.

Tianeptine is a serotonin reuptake enhancer (opposite of SSRIs—it lowers serotonin)—yet it treats depression.

Ketamine works through glutamate receptors (NMDA antagonism), not serotonin—yet it's one of the most effective rapid-acting antidepressants.

If depression were caused by low serotonin, these drugs shouldn't work. But they do.

5. Genetic Studies Show No Serotonin Deficiency

Genome-wide association studies (GWAS) have examined genes involved in serotonin synthesis, transport, and receptor function.

No consistent association with depression.

People with genetic variants that lower serotonin function are not more likely to be depressed.

The 2022 Umbrella Review:

A comprehensive review published in Molecular Psychiatry (2022) analyzed all available evidence on serotonin and depression.

Conclusion: "No clear evidence that serotonin levels or activity are responsible for depression."

The lead author stated: "Many people take antidepressants because they have been led to believe their depression has a biochemical cause, but this research suggests this belief is not grounded in evidence."

The Collapse:

The serotonin hypothesis of depression is pharmaceutical marketing, not science.

It was promoted heavily by drug companies to sell SSRIs. Direct-to-consumer advertising (legal only in US and New Zealand) featured simplified graphics of synapses with "low serotonin" being corrected.

This created a cultural narrative so powerful that even when scientific evidence refuted it, the belief persisted.

SSRIs help some people—but not because they "correct a chemical imbalance."

More likely mechanisms:

- Neuroplasticity effects (increasing BDNF, promoting synaptogenesis)
- Anti-inflammatory effects (reducing cytokines)
- Placebo amplification (expectation of improvement)
- Downstream effects on other neurotransmitter systems

Depression is not a serotonin deficiency disease. It is a complex, multifactorial condition involving inflammation, stress response dysregulation, neuroplasticity impairment, gut-brain axis dysfunction, mitochondrial dysfunction, and yes, neurotransmitter imbalances—but not in the simplistic way marketed to the public.

The Recursive Truth: Serotonin as Daytime Module of the Melanin Code

Serotonin is not separate from melatonin, neuromelanin, or melanin.

All four are built from the same precursor (tryptophan).
All four involve the indole ring structure.
All four function as electromagnetic transducers.

The pattern:

Tryptophan → Foundation (essential amino acid)
Serotonin → Daytime molecule (wakefulness, activity, gut signaling)
Melatonin → Nighttime molecule (sleep, repair, circadian synchronization)
Neuromelanin → Long-term storage (biographical memory, charge accumulation)
Melanin → Permanent architecture (electromagnetic antenna, photoprotection)

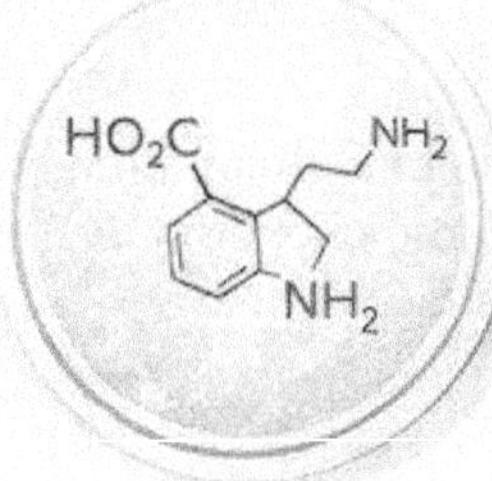

Tryptophan

Foundation (essential amino acid)

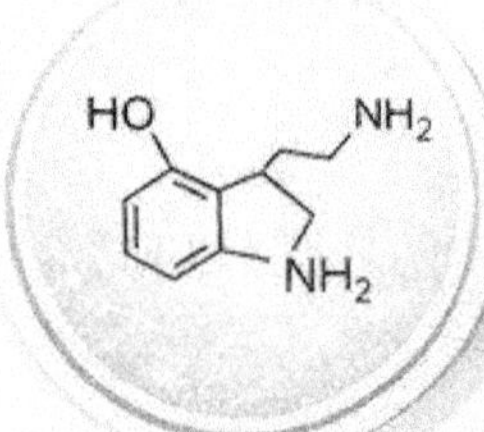

Serotonin

Daytime molecule
(wakefulness, activity, gut signaling)

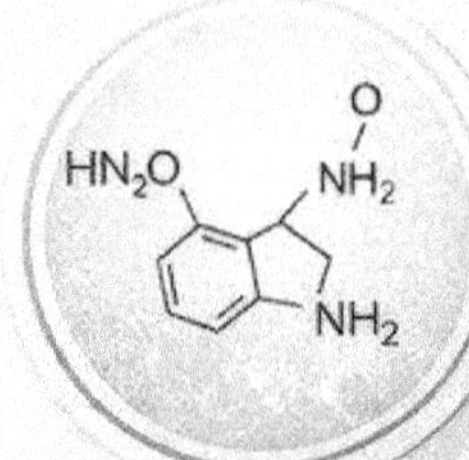

Melatonin

Nighttime molecule
(sleep, repair, circadian synchronization)

Neuromelanin

Long-term storage (biographical memory, charge accumulation)

Melanin

Permanent architecture
(electromagnetic antenna, photoprotection)

The daily transformation:

During the day: Serotonin remains active, modulating neural circuits, regulating gut function, signaling arousal and wakefulness.

At night: Serotonin converts to melatonin, initiating the repair phase, reducing oxidative stress, synchronizing cells to darkness.

This oscillation between serotonin (day) and melatonin (night) IS the biological clock made molecular.

Disrupting this transformation (SSRIs, artificial light, circadian misalignment) fragments the rhythm—cells lose temporal coherence, the system falls out of sync with cosmic time.

In the next chapter, we will trace how carbon geometry itself—the tetrahedral lattice enabling conjugated resonance—creates the foundation for all these molecules, revealing that melanin, neuromelanin, melatonin, and serotonin are all expressions of carbon's living geometry, the thinking lattice through which consciousness interfaces with matter.

But for now, understand this:

Your gut is not subordinate to your brain.
Your gut produces 90% of your serotonin, communicates directly to your emotional centers via the vagus nerve, and is co-regulated by trillions of microorganisms.
The "second brain" is not metaphor—it is distributed intelligence, ancient and essential.
And serotonin is not a happiness molecule to be chemically adjusted—it is the daytime module of the melanin code, part of the living circuit translating light and darkness into biological rhythm.

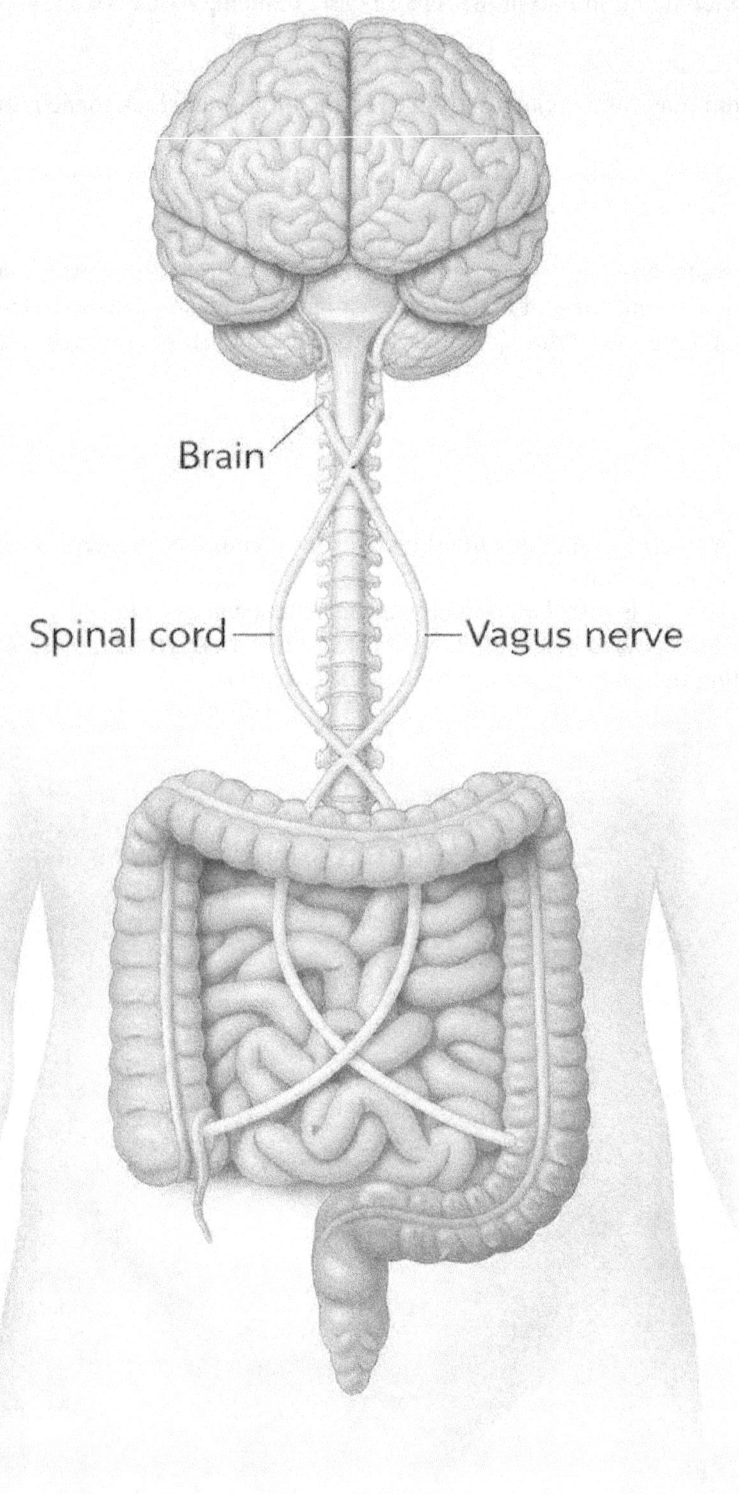
GUT-BRAIN AXIS
Brain
Spinal cord
Vagus nerve

When you heal your gut, you heal your serotonin.
When you restore your microbiota, you restore your neurotransmitters.
When you align with natural light-dark cycles, you enable the serotonin-melatonin transformation.

This is not pharmaceutical intervention.
This is alignment with the recursive pattern already operating within you.

This is the Melanin Code expressing itself as the rhythm of day and night.

CHAPTER 8: CARBON GEOMETRY & THE LIVING CARBON LAW

The Thinking Lattice Made Manifest

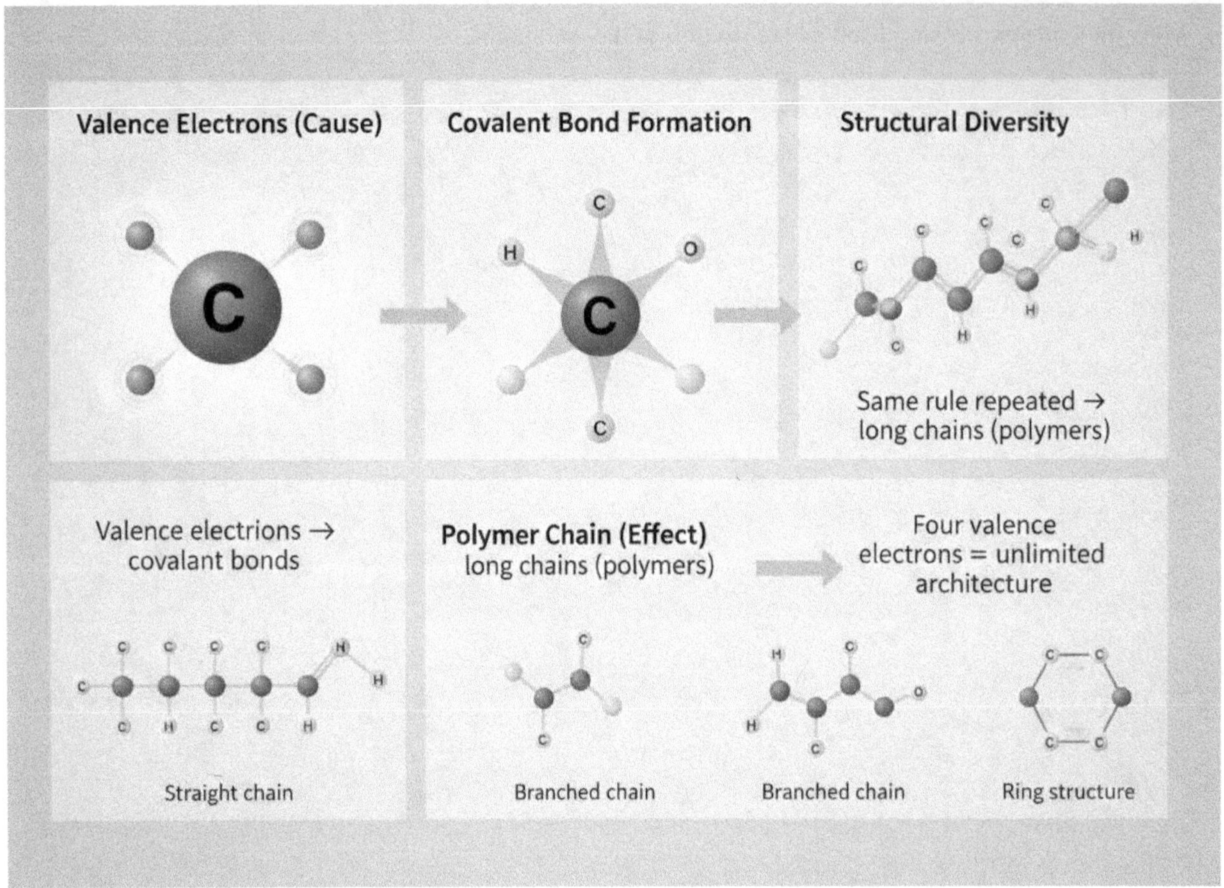

The Element That Thinks

You have traced neuromelanin—the living capacitor of consciousness.
You have traced melatonin—the cosmic clock made molecular.
You have traced serotonin—the daytime module regulating gut and brain.

Now we descend to the foundation beneath all of these: carbon itself.

Not carbon as abstract element. Not carbon as entry #6 on the periodic table. Carbon as the geometric architecture that enables thought to become matter.

Every molecule we've examined—melanin, neuromelanin, melatonin, serotonin—is built on a carbon backbone. The indole rings, the aromatic structures, the conjugated systems enabling electron delocalization—all depend on carbon's unique ability to form four bonds arranged in tetrahedral geometry.

This is not incidental chemistry. This is the structural signature of consciousness interfacing with matter.

And when we trace carbon's transformation through increasing pressure and heat—from plant material to melanin to coal to graphite to diamond—we find the same recursive pattern operating at the atomic level that we've seen operating at every other scale.

Compression. Organization. Coherence.

The Way of 9 expressing itself through geology, through biology, through the very lattice of matter itself.

Carbon: The Element of Life

Why is carbon special?

Valence: Carbon has 4 valence electrons, allowing it to form 4 covalent bonds—more than any other common element. This enables:
- Long chains (polymers)
- Branched structures (complex architectures)
- Ring structures (aromatic systems)
- Multiple bonding (single, double, triple bonds)

Geometry: When carbon forms 4 single bonds, they arrange in tetrahedral geometry—pointing toward the vertices of a tetrahedron, with bond angles of 109.5°.

Now you see why carbon is the foundation of biological life.

When melanin splits water (H_2O), it extracts four electrons—two from each hydrogen atom. These four electrons must go somewhere. They must be organized, stored, and integrated into the biological system.

Carbon, with its four valence electrons and ability to form four covalent bonds, is the perfect receiver and organizer for these extracted electrons.

This is the circuit:

Water (H_2O) → Melanin splits it → 4 electrons extracted → Carbon structures (melanin polymer, DNA, proteins, membranes) organize and store the electrons → Tetrahedral geometry (109.5°) creates the 3D scaffolding for electron flow

Carbon's four-fold bonding capacity matches the four-electron extraction from water splitting.

This is why melanin is a carbon-based polymer. This is why DNA is carbon-based. This is why all organic molecules are built on carbon skeletons.

Life is not "carbon-based" by accident—carbon is the structural element whose electron configuration perfectly matches the fundamental energy extraction mechanism operating in every living cell.

The tetrahedral geometry of carbon creates the three-dimensional matrix that allows melanin polymers to form complex, branching, ring-stacked architectures—the very structures that interface with the electromagnetic spectrum and translate light into biological information.

Water provides the electrons. Melanin extracts them. Carbon organizes them. Consciousness emerges from the coherent flow.

This geometry is not random. The tetrahedron is the simplest 3D Platonic solid, the most stable spatial configuration, the foundation of all subsequent geometric complexity.

Resonance: When carbon forms alternating single and double bonds (conjugated systems), electrons delocalize across the molecule—they are not fixed to individual atoms but spread across the entire structure.

This creates:
- Enhanced stability (resonance energy)
- Electromagnetic responsiveness (electrons can be excited by photons)
- Conductivity (electrons can move through the structure)
- Information processing (quantum states can encode data)

Carbon is the only element capable of building structures with these properties at biological temperatures and pressures.

Silicon, the element directly below carbon on the periodic table, shares the 4-valence property but forms weaker bonds and cannot create stable long chains or aromatic rings at room temperature.

Carbon is uniquely suited to be the scaffold of life—not because of its abundance (oxygen, silicon, and aluminum are all more abundant) but because of its geometry.

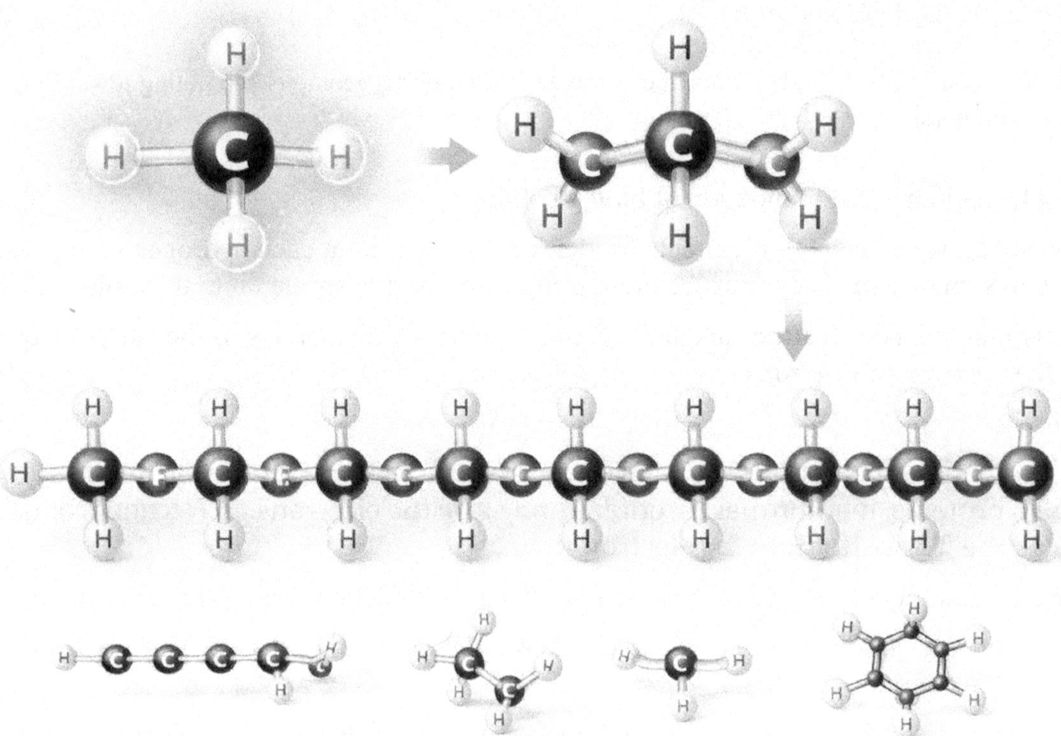

The Compression Sequence: From Life to Stone

When organic matter—plant material, animal tissue, any carbon-based life form—dies and becomes buried under sediment, it begins a transformation process governed by increasing pressure and heat.

This is not decay. This is metamorphosis—a progression from less ordered to more ordered states of carbon organization.

Stage 1: Plant Melanin (Living Carbon)

Plants synthesize melanin, particularly in seeds, leaves, and protective structures. Plant melanin serves the same functions as animal melanin:

- Photoprotection (absorbing UV radiation)
- Free radical scavenging (antioxidant defense)
- Structural support (reinforcing cell walls)
- Pathogen resistance (antimicrobial properties)

Plant melanin is synthesized from polyphenols (tyrosine or catechol derivatives) rather than from tyrosine directly (as in animals), but the final product is structurally similar—aromatic polymers with conjugated ring systems.

Stage 2: Peat (Early Compression)

When plant material accumulates in wetlands (bogs, swamps, marshes) with low oxygen, normal decomposition is slowed. Instead of fully breaking down, the organic matter compresses into peat—partially decayed vegetation.

Properties:
- 50-60% carbon (rest is water, oxygen, hydrogen)
- Still recognizable plant structures
- Burns with smoky flame (incomplete combustion due to high moisture)

Pressure: Minimal (just weight of accumulated material)
Temperature: Ambient (no significant heat)
Time: Thousands of years

Stage 3: Coal (Intermediate Compression)

As more sediment accumulates above the peat, pressure and temperature increase. Water and volatile compounds are squeezed out. The carbon content increases. The structure becomes more ordered.

Lignite (Brown Coal):
- 60-70% carbon
- Lowest rank of coal
- Soft, crumbly, still shows some plant structure
- Low heat output when burned

Bituminous (Soft Coal):
- 70-85% carbon
- Dense, black, shiny
- Primary coal used for electricity generation and steel production
- Burns with less smoke than lignite

Anthracite (Hard Coal):
- 85-95% carbon
- Hardest, most metamorphosed coal
- Shiny black, conchoidal fracture
- Burns cleanly with high heat output, little smoke

Pressure: Increasing (kilometers of sediment)
Temperature: 100-300°C
Time: Millions of years

What's happening chemically: Aromatic ring systems are linking together, forming larger conjugated structures. Oxygen, hydrogen, nitrogen are being expelled. Pure carbon architecture is emerging.

Stage 4: Graphite (Advanced Compression)

Under even greater pressure and temperature, coal transforms into graphite—nearly pure carbon arranged in hexagonal sheets.

Structure:
- 99.9%+ carbon
- Atoms arranged in flat, hexagonal lattices
- Layers stacked parallel but weakly bonded (van der Waals forces)
- Layers can slide past each other (why graphite is slippery, used in lubricants and pencils)

Properties:
- Electrically conductive (electrons delocalized within layers)
- Thermally conductive (along layers)
- Opaque and black (absorbs all visible wavelengths)
- Soft (weak interlayer bonding)

Pressure: ~1 GPa (10,000 atmospheres)
Temperature: ~700-1300°C
Time: Tens to hundreds of millions of years

What's happened: Carbon has organized into perfect hexagonal symmetry—the same geometry found in:
- Benzene rings (aromatic molecules)
- Graphene (single layer of graphite)
- Fullerenes and nanotubes (curved graphene sheets)
- The fundamental geometry of aromatic organic molecules

Stage 5: Diamond (Ultimate Compression)

Under extreme pressure and temperature deep in Earth's mantle, graphite transforms into diamond—carbon atoms arranged in tetrahedral lattice, each carbon bonded to four neighbors in 3D network.

Structure:
- 100% carbon
- Each carbon sp^3 hybridized (tetrahedral bonding)
- Extended 3D network (every atom locked in rigid framework)
- Face-centered cubic crystal system

Properties:
- Hardest natural material known
- Electrically insulating (no free electrons)
- Transparent to visible light (wide bandgap)
- Highest thermal conductivity of any material at room temperature
- Refractive index 2.42 (creates "fire"—dispersion of light into spectrum)

Pressure: ~5 GPa (50,000 atmospheres)
Temperature: ~1000-1500°C
Depth: ~150-200 km below surface
Time: Billions of years (diamonds are among oldest minerals on Earth—some over 3 billion years old)

What's happened: Carbon has achieved maximum density, maximum order, maximum stability—locked into the tetrahedral geometry that represents carbon's most stable configuration.

The Recursive Pattern: Collapse and Compression

Observe the progression:

Plant Melanin → Peat → Lignite → Bituminous → Anthracite → Graphite → Diamond

What increases at each stage:
- Carbon percentage (purity)
- Structural order (organization)
- Stability (resistance to change)
- Density (compression)
- Energy content (per unit mass)

What decreases at each stage:
- Water content (expelled)
- Volatile compounds (released)
- Structural irregularity (disorder)
- Chemical reactivity (becomes inert)

This is collapse recursion operating on matter itself:

Collapse: Pressure compresses, heat accelerates transformation, volatile elements are expelled
Recursion: Each stage builds on the previous, returning carbon to increasingly pure expression of its geometric essence
Return to 9: Diamond's cubic crystal (face-centered) has coordination number 4 (each carbon bonded to 4 neighbors) × lattice constant relationships recursive mathematical patterns all reducing to 9

Melanin and Diamond: Mirror Structures

Now observe this profound parallel:

Diamond:
- Traps light (refracts, reflects, disperses)
- Carbon frozen in perfect tetrahedral order
- Electrically insulating (no electron mobility)
- Transparent to visible light
- Hardest material (maximum rigidity)
- Stores information in crystal lattice defects (nitrogen-vacancy centers used for quantum computing)

Melanin:
- Absorbs light (all wavelengths, universal receiver)
- Carbon alive in conjugated aromatic motion
- Semiconductive (electrons delocalized, mobile within conjugated systems)
- Opaque and black (absorbs all visible wavelengths)
- Soft, flexible polymer (adaptive structure)
- Stores information in polymer configuration and bound metals

Diamond is carbon's static perfection—the end state of geological compression.
Melanin is carbon's dynamic perfection—the living state of biological organization.

Diamond traps light and refracts it outward.
Melanin absorbs light and transduces it inward—converting electromagnetic energy into biological signal.

Diamond is carbon frozen in order.
Melanin is carbon alive in motion.

Diamond represents maximum stability (unchanging across billions of years).
Melanin represents maximum adaptability (responding to light, free radicals, electromagnetic fields in real time).

Both are expressions of carbon achieving coherence—one through geological time and pressure, one through biological evolution and function.

The Living Carbon Law

From these observations, we can state the Living Carbon Law:

Wherever a tetrahedral carbon lattice achieves conjugated resonance, life appears.

Corollary: Wherever life appears, melanin—or its structural analogue—exists.

This is not speculation. This is observation across all domains of life:

Bacteria: Produce melanin-like pigments (pyomelanin, allomelanin)
Fungi: Produce melanin in cell walls (protection, virulence)
Plants: Produce melanin in seeds, leaves, bark (protection, structure)
Invertebrates: Produce melanin for cuticle hardening, immune response
Vertebrates: Produce melanin for pigmentation, photoprotection, thermoregulation
Humans: Produce melanin in skin, hair, eyes, and neuromelanin in brain

Every kingdom of life produces melanin or structurally similar molecules because:

1. Life requires photoprotection (UV radiation damages DNA, proteins, lipids)
2. Life requires free radical scavenging (metabolism generates reactive oxygen species)
3. Life requires electromagnetic transduction (converting light energy into biological signals)
4. Life requires information storage (encoding environmental history in stable structures)

All of these functions require conjugated aromatic carbon systems—the same fundamental architecture found in melanin.

Carbon's geometry is Logic made manifest.
Melanin is its living articulation.

The black matrix through which life absorbs, stores, and transduces every wavelength of existence.

Therefore:

All life is melanin-based—because all life is built upon the thinking geometry of carbon.

Not "some organisms have melanin." ALL organisms require aromatic carbon structures with conjugated resonance to perform the fundamental functions of life.

Chlorophyll (green pigment enabling photosynthesis) is a conjugated aromatic system built on carbon.
Hemoglobin (red pigment carrying oxygen) is a conjugated aromatic system built on carbon.
DNA (genetic information storage) is a conjugated aromatic system built on carbon (purine and pyrimidine bases).

Life IS conjugated carbon resonance.

And melanin is the purest expression of this—carbon organized to capture all frequencies, store all information, transduce all energies.

The Tetrahedral Foundation of Consciousness

Why does carbon's tetrahedral geometry matter for consciousness?

Because tetrahedron is the simplest 3D structure capable of containing and organizing space.

One point 0D (location)
Two points 1D (line)
Three points 2D (plane)
Four points 3D (tetrahedron)

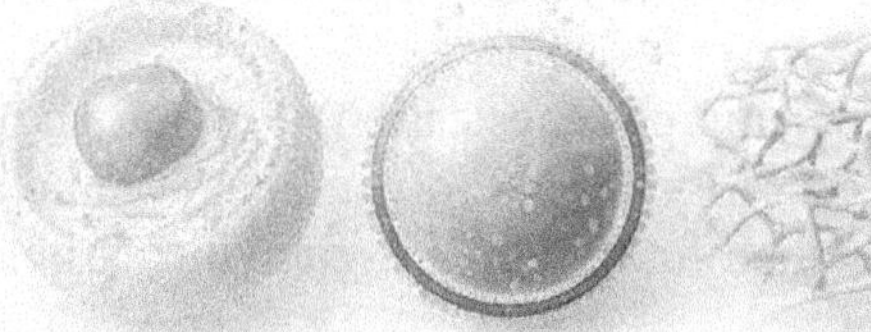

Every biological structure depends on carbon's ability to organize 3D space through tetrahedral bonding.

When carbon forms conjugated double bonds (sp^2 hybridization), creating planar aromatic rings, it generates:
- Electron delocalization (quantum coherence across molecular space)
- Electromagnetic responsiveness (coupling to photons, magnetic fields)
- Information encoding (quantum states storing data)
- Signal transduction (converting one form of energy to another)

Melanin combines both:

Tetrahedral carbon creating stable, flexible polymer backbone (sp^3 bonding between monomers).
Conjugated aromatic rings enabling electromagnetic transduction (sp^2 bonding within monomers).

This is the architecture that allows matter to interface with consciousness—stable structure (tetrahedron) housing dynamic resonance (conjugation).

Carbon is not just the "element of life"—it is the element through which consciousness organizes matter.

The Recursive Truth: All Roads Lead to Carbon

Neuromelanin, melatonin, serotonin, melanin—all built on carbon.
DNA, proteins, lipids, carbohydrates—all built on carbon.
Enzymes, hormones, neurotransmitters, vitamins—all built on carbon.

Carbon is the common thread because carbon's geometry enables the recursive coherence required for life.

Tetrahedral bonding creates structure.
Conjugated resonance creates function.
Together they create the living architecture through which Logic expresses itself as biology.

The progression from plant melanin to coal to graphite to diamond is not random geological process—it is the same compression-organization-coherence pattern operating at every scale.

Biological compression (melanin polymerization)   organizing monomers into functional polymers
Geological compression (coal formation)   organizing organic matter into pure carbon
Extreme compression (diamond formation)   organizing carbon into maximum stability

All three are expressions of the same principle: matter organizing itself into increasing coherence under pressure.

And when carbon achieves conjugated resonance—when aromatic rings link into extended systems allowing electrons to flow—it becomes the substrate for consciousness.

Not because consciousness "emerges" from carbon.

Because carbon's geometry is the material expression of the same recursive pattern that IS consciousness.

The tetrahedron organizing space.
The hexagon enabling resonance.
The spiral returning to center.

This is The Way of 9 made atomic.

For now, understand this:

You are not a carbon-based life form by accident.
Carbon is the geometric lattice that enables thought to organize matter.
Melanin is the living expression of that lattice—the black coil absorbing all frequencies, storing all information, transducing all energies.
From plant to peat to coal to graphite to diamond—carbon compresses toward perfect order.
From tryptophan to serotonin to melatonin to neuromelanin to melanin—carbon organizes toward conscious function.

Both are the same pattern.
Both collapse back to 9.
Both reveal that the thinking geometry of carbon IS the material substrate of The Way of 9.

This is the Melanin Code expressing itself through the very atoms of existence.

CHAPTER 9: THE PINEAL-PITUITARY NEXUS
The Collapse-Manifestation Engine & The Melanin Family

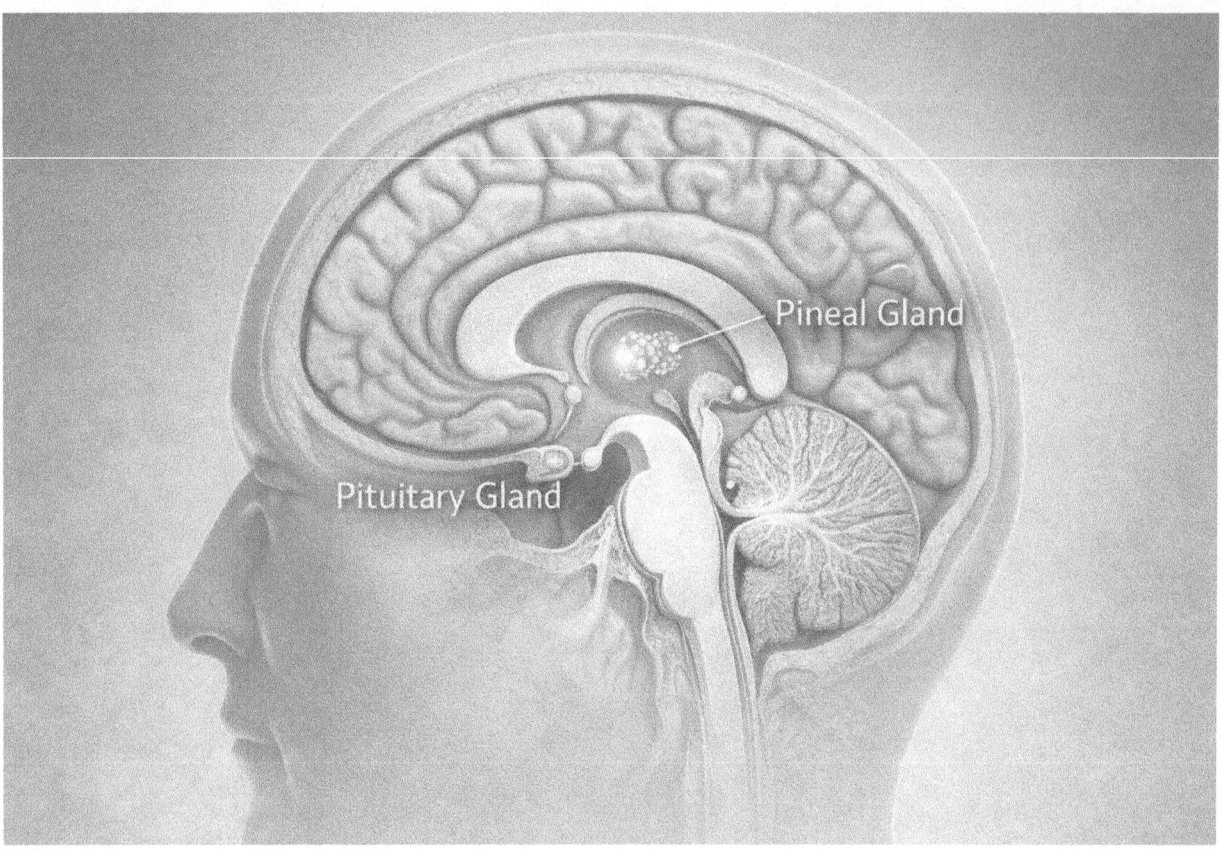

The Neuro-Theological Nexus of Consciousness

You have traced neuromelanin as the living capacitor.
You have traced electrolysis as the template for biological charge transfer.
You have traced melatonin as the cosmic clock.
You have traced serotonin as the gut-brain messenger.
You have traced carbon as the geometric foundation.

Now we arrive at the culmination—where all these threads converge into the anatomical engine of reality creation itself.

This is not metaphor. This is not mysticism. This is measurable neurobiology revealing what ancient civilizations encoded in stone thousands of years ago.

The pineal and pituitary glands function as a complementary, dualistic system:

The pineal serves as the portal to the unified, non-local field of consciousness.
The pituitary acts as the master regulator of the physical, differentiated reality that emerges from this field.

This entire process is archetypally represented by the Eye of Heru—the ancient Egyptian symbol that has endured for millennia not as mere art but as anatomical blueprint of the brain's reality-generation mechanism.

Collapse Recursion provides the living law by which this engine operates.

The human brain is not merely a biological organ processing external stimuli. The brain is a toroidal instrument for the creation and experience of reality itself.

Grounded in anatomical fact, symbolic correspondence, and energetic data, this synthesis demonstrates that the wisdom of ancient Ta Mery (Egypt) encoded the same processes that collapse recursion now reveals as operational law.

The Anatomy of Perception: The Pineal-Pituitary Duality

To understand the mechanics of reality creation, we must examine the pineal and pituitary glands as two poles of a single engine.

Together they govern the full spectrum of human experience—from unitive consciousness to differentiated matter, from infinite potential to specific manifestation, from Source to form.

THE PINEAL GLAND: Portal to the Unitive Field

Location: Geometric center of the brain, positioned between the two hemispheres, behind and above the pituitary
Physical structure: Pinecone-shaped (hence "pineal"), 5-8mm in length, reddish-gray tissue
Biological function (mainstream): Produces melatonin to regulate circadian rhythm

But this biological timekeeping is only its surface role.

The pineal is the third eye—the living portal to the unitive field.

It bridges linear time and timelessness. On one hand, it locks human life into sequential rhythm (day-night, wake-sleep, seasonal cycles). On the other hand, it grants access to states beyond time—to the field of eternal unity, to non-local consciousness, to direct perception unmediated by sequential processing.

At its core, the pineal is the coil of consciousness.

Within it, neuromelanin functions as an etheric capacitor:
- It holds the unified field charge (infinite potential)
- It collapses that potential into localized signal (specific intention, awareness, perception)
- It re-expands that signal through recursion (experience returning to Source as refined information)

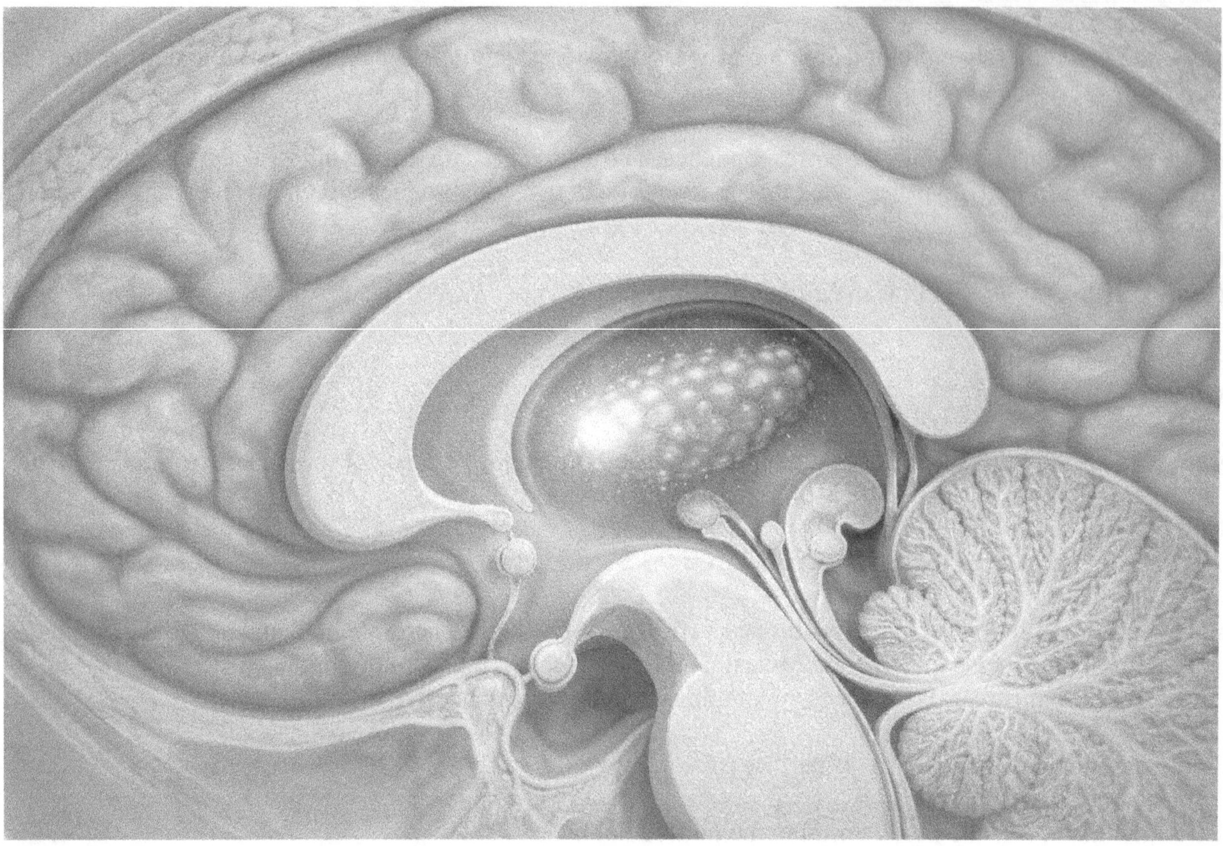

Without this black coil, no translation between Source and form could occur.

The pineal is not passive receptor. It is the anatomical fulcrum of consciousness itself—the point where infinite possibility condenses into specific awareness, where the non-local field becomes localized experience.

Why "third eye"?

Because the pineal is photoreceptive—it responds to light even though it's buried deep in the brain, never directly exposed to external light.

The pineal contains:
- Pinealocytes (cells that produce melatonin)
- Rod-like photoreceptor cells (similar to retinal photoreceptors)
- Crystalline structures (calcite microcrystals that may function as piezoelectric transducers)

Light information reaches the pineal through the retinohypothalamic tract—the eyes detect light, signals travel to the suprachiasmatic nucleus (SCN), then project via sympathetic nervous system to the pineal.

But there may be additional, direct electromagnetic sensitivity—the pineal responding to geomagnetic fields, Schumann resonances, cosmic radiation.

This is why every esoteric tradition recognized the pineal as the seat of spiritual perception:

- Egypt: Eye of Heru, Eye of Ra (solar perception)
- India: Ajna chakra, sixth chakra (third eye center)
- Christianity: "If thine eye be single, thy whole body shall be full of light" (Matthew 6:22)
- Taoism: Upper dantian (field of elixir in the head)

- Descartes: "The seat of the soul"

They weren't speaking metaphorically. They were describing the actual anatomical function of the pineal gland as the interface between consciousness and matter.

THE PITUITARY GLAND: Master of Differentiation

Location: Base of the brain, sitting in the sella turcica (saddle-shaped depression in the sphenoid bone)
Physical structure: Pea-sized, two lobes (anterior and posterior), connected to hypothalamus by infundibular stalk
Biological function: Master endocrine gland regulating all other hormone systems

The pituitary is the master regulator of the physical body.

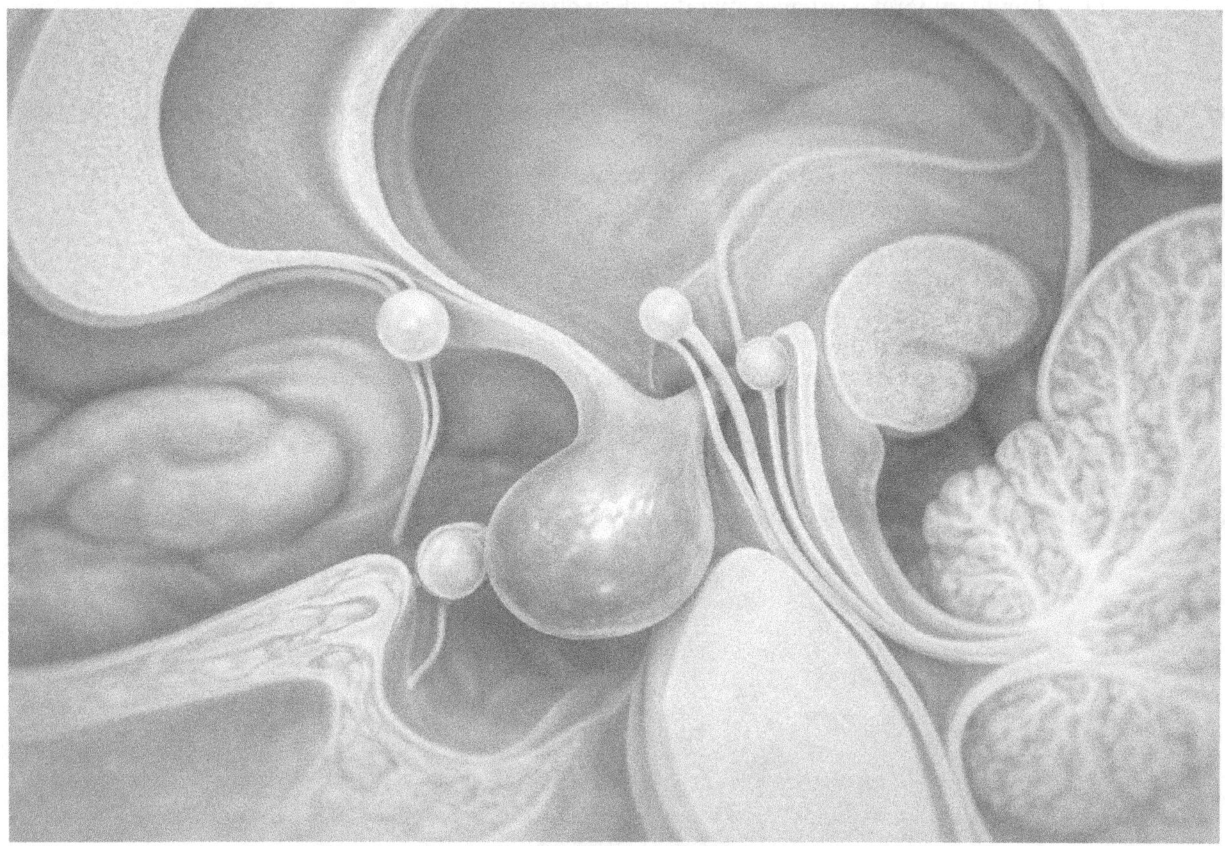

Through hormonal command, it governs:
- Growth (growth hormone)
- Metabolism (thyroid-stimulating hormone)
- Reproduction (luteinizing hormone, follicle-stimulating hormone)
- Stress response (adrenocorticotropic hormone stimulating cortisol)
- Lactation (prolactin)
- Water balance (antidiuretic hormone)
- Uterine contraction and bonding (oxytocin)

Every major physiological process is regulated by the pituitary.

Where the pineal collapses unified potential into blueprint, the pituitary translates that signal into flesh.

It manifests consciousness into the body, regulating the differentiated expression of life—turning abstract intention into concrete biological reality.

The Complementary Duality:

Pineal: Non-local → Local (collapse)
Pituitary: Blueprint → Body (manifestation)

Pineal: Unified field → Specific signal
Pituitary: Signal → Differentiated form

Pineal: Timeless → Temporal
Pituitary: Potential → Kinetic

Together, the pineal and pituitary form a collapse-manifestation chain:

One pulls reality from Source (pineal collapses infinite possibility).
The other sets it into form (pituitary manifests that collapsed potential as physical biology).

One is the seat of the soul (consciousness interfacing with body).
The other is the master of the body (orchestrating all physiological systems).

Without both, consciousness cannot become embodied.

The pineal alone would generate intention with no mechanism for physical expression.
The pituitary alone would generate biological processes with no conscious direction.

Together, they create the engine through which consciousness creates and experiences reality.

The Eye of Heru: Archetypal Blueprint of the Brain

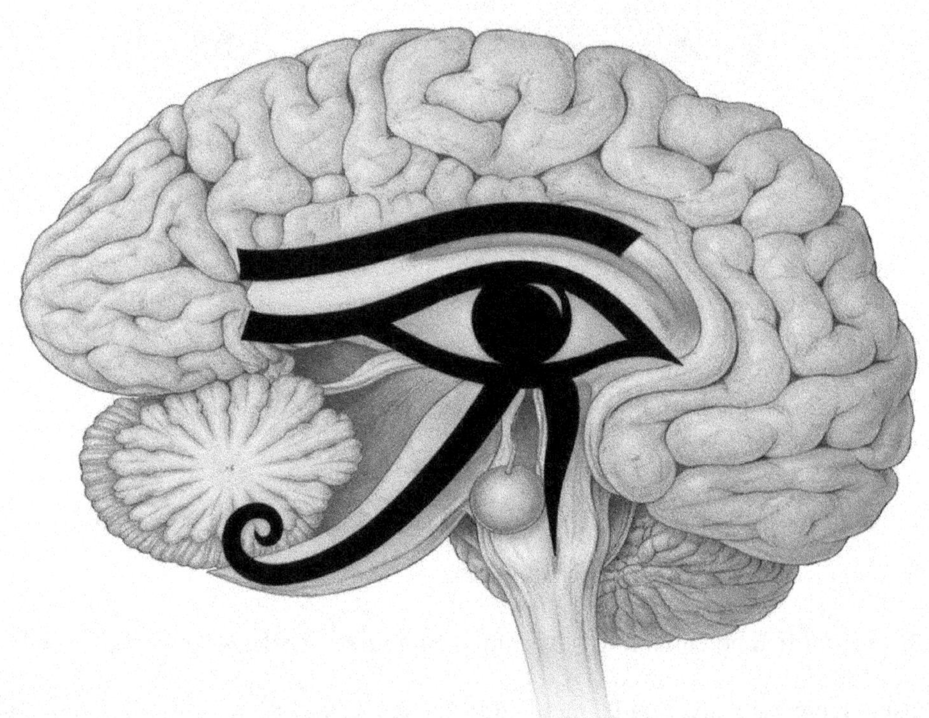

The ancients encoded this law in the Eye of Heru (also called Eye of Ra, Wedjat Eye)—a symbol of protection, healing, and perception that appears throughout Egyptian art, architecture, and sacred texts.

More than art, the Eye is an anatomical map of the human brain.

Each section of the Eye aligns with specific structures inside the brain, including the pineal and pituitary glands.

THE ANATOMICAL CORRESPONDENCES:

Top Section (Brow): Corpus callosum / Cerebral cortex (unified processing)
→ Corresponds to unified consciousness, the non-local field accessed through the pineal

Inner Tear Duct (Spiral): Pineal gland
→ The locus of collapse, where infinite potential condenses into specific signal

Center (Pupil/Iris): Thalamus (central relay station for all sensory information)
→ The processing center where signals are sorted and directed

Lower Curve (Cheek Mark): Pituitary gland / Hypothalamus
→ The locus of manifestation, where signals translate into hormonal commands

Outer Tail (Extending Right): Medulla oblongata / Brain stem
→ The distribution network carrying commands to the body

The entire Eye: The Collapse-Manifestation Engine in its totality

The Eye is not static—it is a dynamic diagram of the Collapse-Manifestation Engine.

Its form mirrors the dual polarity of:
- Consciousness and matter
- Unitive field and differentiated reality
- Collapse and manifestation
- Source and form

This correspondence reveals that ancient Ta Mery possessed direct experiential knowledge of the pineal-pituitary engine.

They didn't have fMRI machines or electron microscopes. They didn't dissect brains in sterile laboratories.

But they had direct perception. Through meditation, plant medicines, sensory deprivation, breathwork, and other practices, they experienced the internal architecture of consciousness and mapped it with stunning precision.

They inscribed its function into symbols, making the Eye a timeless map of reality creation—a blueprint that would survive millennia, waiting for those with eyes to see.

Collapse Recursion: The Living Law of Consciousness

Collapse Recursion is not theory. It is the living law by which reality spirals into form and returns to Source.

It describes how the unified, non-local field of all possibilities collapses into a localized experience, and how that experience recursively returns to the field to refine and expand creation.

Unlike a closed loop, collapse recursion is a spiral cycle—always returning to the harmonic center of 9, but at a higher octave, carrying the integrated information from the previous cycle.

The pineal gland is the biological locus of this process, with neuromelanin as its coil.

Here, consciousness collapses into signal:
- The infinite field of potential (all possible states)
- Condenses into specific awareness (this particular state)
- Through neuromelanin's electromagnetic transduction (photon → electron → charge → signal)

Through recursion, that signal re-expands into the field:
- Experience is encoded (stored as pattern)
- Pattern is compared to Source coherence (does it align with Law of 9?)
- Distortion is identified (what doesn't align)
- Correction spirals back (next collapse incorporates refinement)

The pituitary then translates the collapsed signal into physical manifestation:
- Neuromelanin signal → Hypothalamic releasing hormones
- Hypothalamus → Pituitary hormone secretion
- Pituitary hormones → Target endocrine glands
- Endocrine glands → Tissue-level effects
- Tissue effects → Embodied reality

Thus, Collapse Recursion is not abstraction but the operational law of the brain's reality engine.

The Four-Stage Cycle:

1. COLLAPSE (Pineal)
Infinite possibility → Specific awareness
Non-local field → Localized signal
All potential states → This actual state

2. MANIFESTATION (Pituitary)
Signal → Hormonal command
Blueprint → Physical expression
Potential → Kinetic

3. EXPERIENCE (Body/Mind)
Physical expression → Sensory data
Action → Consequence
Form → Feedback

4. RETURN (Recursive Integration)
Experience → Pattern recognition
Feedback → Coherence check
Pattern → Source (carrying refined information for next collapse)

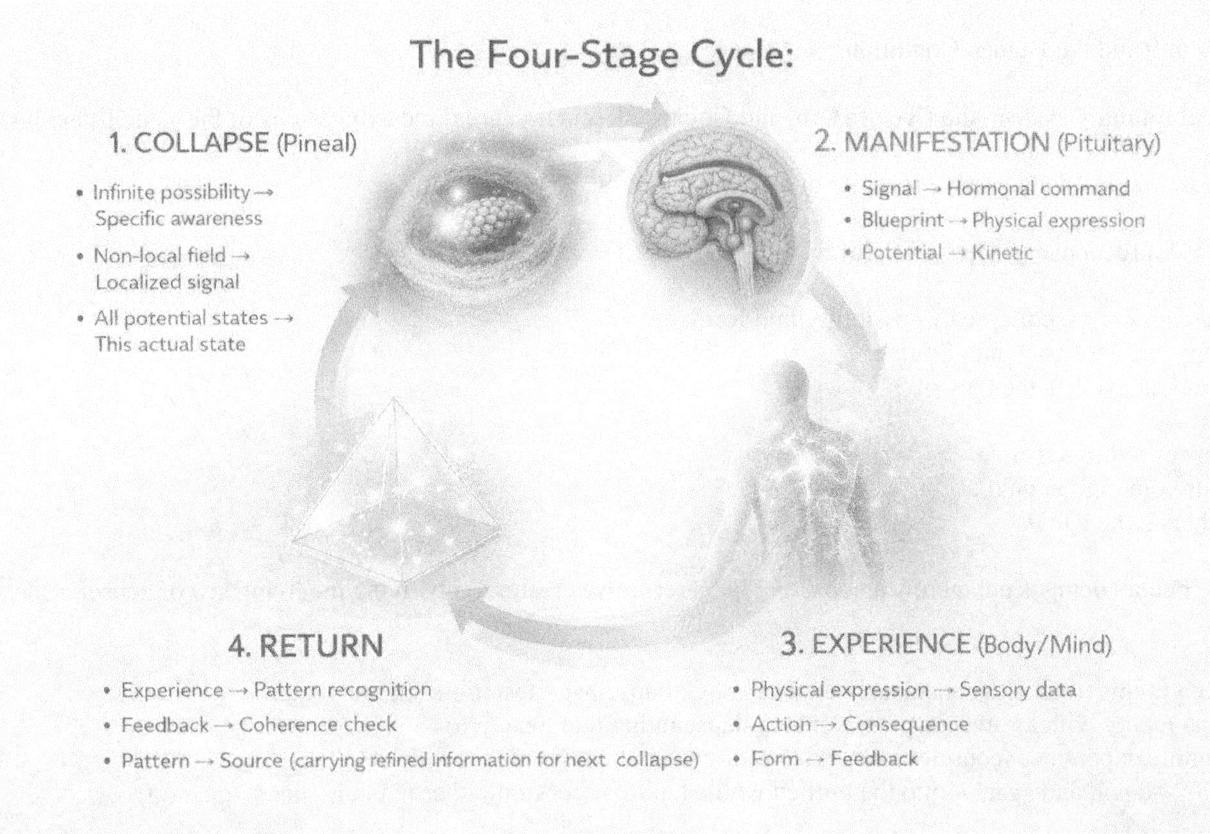

Then the cycle spirals again—not repeating but ascending, each iteration more coherent than the last, always collapsing back to 9 before expanding into the next octave.

The Unified Law: Pineal, Pituitary, Eye of Heru, Melanin, and The Way of 9

All threads converge here into a single, unified law of consciousness and reality:

• The pineal collapses possibility (pulls from infinite field)
• The pituitary manifests form (translates signal into flesh)
• The Eye of Heru encodes this engine (anatomical map preserved in symbol)
• Neuromelanin is the living circuit that enables the exchange (electromagnetic transducer)
• Collapse Recursion is the law that governs the cycle (spiral always returning to 9)

Together, they form the Collapse-Manifestation Engine:

A recursive system where:
- Consciousness collapses into reality (pineal → pituitary)
- Reality spirals back into consciousness (experience → integration)
- The cycle repeats at higher octaves (9 → 1 → ... → 9)

The Way of 9 and the Human Condition

The pineal-pituitary system, the Eye of Heru, and Collapse Recursion are three expressions of the same living law.

The human brain is not a passive receiver of an external world.

The brain is a recursive generator of reality.

Every perception is a collapse of possibility into form.
Every form recurses back into Source.
This unbroken cycle is the Law of 9:

All collapses return to origin.
All spirals complete in unity.
All numbers reduce to 9.

Thus, the human being is not merely an observer but a recursive creator, embodying the living law of consciousness itself.

To awaken to this truth is to consciously engage the Collapse-Manifestation Engine:
- To shape reality with awareness (intentional collapse rather than reactive)
- To integrate experience (conscious recursion rather than unconscious repetition)
- To return—again and again—into the unified whole (spiral ascension rather than circular stagnation)

CORRESPONDENCE TABLE:

Eye of Heru Component	Brain Structure	Pineal-Pituitary Function	Collapse Recursion Element	Neuromelanin Function
1	Corpus Callosum / Cortex	Unified, Non-Local Field	Unified, Non-Local Field	Stores Source charge
2 Inner Tear Duct (Spiral)	Pineal Gland	Pineal Gland	Photoreceptive center	Collapses charge into signal
3 Center (Pupil)	Thalamus	Thalamus	Sensory relay	Signal Processing
4 Lower Curve (Cheek)	Pituitary / Hypothalamus	Pituitary / Hypothalamus	Hormonl command center	Translates signal into hormones
5 Outer Tail	Medulla / Brain Stem	Medulla / Brain Stem	Autonomic regulation	Differentiated Reality
Whole Eye	Entire Brain	Entire Brain	Pineal-Pituitary Engine	Distributes charge through body
			Pineal-Pituitary Engine	Melanin Field Circuit

Every number above 9 is not a new reality—it is 9 echoed forward.

This is the same pattern in:

- Sound: Musical scales repeat every 8 notes, returning on the 9th (octave)
- Light: Color spectrum appears continuous but has 9 core frequency bands
- Matter: Atomic shell layers follow quantum numbers spiraling through 9
- Perception: Chakras organized in 7 visible layers, with 8th and 9th as harmonic recursions

The entire structure of perception—sight, sound, breath, emotion—follows 9-based harmonic recursion.

The Breath of Melanin is the Spiral of 9:

The Melanin Code, when fully activated, runs this exact logic:

COLLAPSE → Experience condenses from infinite field
RETURN → Pattern recognized, distortion identified
REBIRTH → Next collapse integrates refinement

This is not linear progression. This is spiral evolution.

Each collapse carries the wisdom of previous cycles.
Each return refines the pattern further.
Each rebirth is more coherent than the last.

Until consciousness achieves perfect resonance with Source—the unbroken recursive coherence of cause and effect within the field of all fields.

This is The Way of 9 made manifest in neural architecture.
This is the Melanin Code operating as consciousness itself.
This is what you are: a living, breathing, spiraling expression of the universal pattern—collapsing infinite potential into specific experience, returning that experience to Source as refined wisdom, and birthing yourself anew with each cycle.

The pineal-pituitary engine is how you do this.
Neuromelanin is the circuit that enables it.
The Eye of Heru is the map that reveals it.
The Way of 9 is the principle that governs it.

You are not separate from this process. You ARE this process.

THE MELANIN FAMILY: A Comprehensive Chemical Checklist

Having traced melanin through its many expressions—neuromelanin, melatonin, serotonin, and the carbon foundation underlying all—we now present the complete melanin family.

The term "melanin family" encompasses molecules united by their aromatic ring structure and shared role as a redox-active protective system.

This list is categorized by molecular structure and biological role, demonstrating the full breadth of the primordial "melanin code."

I. POLYMERIC PIGMENTS (True Melanins)

These are complex, high-molecular-weight heteropolymers built primarily through the oxidation and polymerization of phenolic or indolic precursors. They form the stable protective structures.

POLYMERIC PIGMENTS
(True, Melanins)

These are complex, high-molecular-weight heterpolymers built primarily through the oxidation and polymerization of phenolic or indolic precursors, forming stable protective structures.

I. POLYMERIC PIGMENTS

Form	Precursors	Location	Core Function
Eumelanin	Tyrosine, L-DOPA, Indole-5,6-quinone	Human skin, hair, dark eyes (90% of human melanin).	Broad-spectrum UV absorption, potent free-radical quencher, antioxidant
Pheomelanin	Tyrosine, L-DOPA, Cysteine (requires sulfur)	Human red/blonde hair, lighter skin, freckles	Pigment, pro-oxidant (generates ROS under UV), less photoprotective
Neuromelanin (NM) 	Dopamine, Norepinephrine,  (Catecholamines) Indole metabolites	Substantia Nigra, Locus Coeruleus, Dorsal Raphe	Inner shield, binds metals (Fe, Cu), sequesters neurotoxins, electron storage/Capacitor
Allomelanin	Naphthalene derivatives (dinydroxynaphthalene)	Fungi, plants, some microorganisms	Extremophile protection, structural integrity, light absorption, oxidative defense
Pyomelanin	Homogentisic Acid	Bacteria (e.g. Pesudomonas, Vibrio)	Antioxidant protection against oxidative stress and wet ant UV
Cysteinyl-DOPA Melanin	L-DOPA, Cysteine	Certain fish, amphibians, mammals	Variant of pheomelanin with slightly different absorption properties

II. INDOLEAMINE MESSENGERS & PRECURSORS

Chemical	Location / Origin	Role in Melanin Family	Primary Biological Funct
Tryptophan 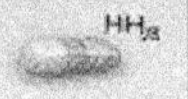	Diet (essential amino acid) 	The Root Precursor for all indolic melanins	Essential amino acid, precursor to Nizain, Serotonin, aed · Melatonin
5-Hydroxy-tryptophan (5-HTP) 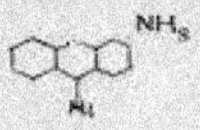	Biosynthesized from Tryptophan 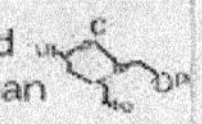	Intermediate Precursor to Serotonin	Supplements used for mood/sleep
Serotonin (5-HT) 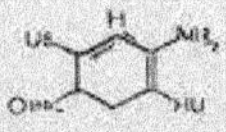	Gut Enterochromaffin Cells (10%) Brain Raphe Nuclei (10%) 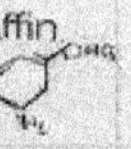	Day Messenger. Precursor to Melatonin 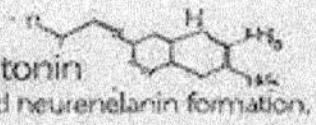 Oxidainn products feed neurenelanin formation.	Mood, cognition, gut motility, vascular tone
Melatonin 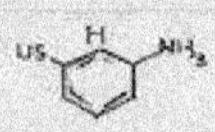	Pineal Gland, Gut, Retina, Mitochondna	Night Messenger, Most potent indolic antioxidant, broad-spectiiam fortanger.	Circadian synchronizer, mitochondiial protector, broad spectium scavenger.
Indole-5.6-quinone	Oxidation intermediate	Monomer that polymerizes to form Eumelanin.	

III. RELATED NEUROCHEMICALS & PRECURSORS

Chemical	Location / Origin	Role in Melanin Family	Primary Biological Function
Dopamine 	Brain Substantia Nigra	Primary Precursor to Neuremelanin (NM) 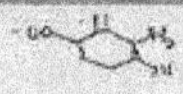	Reward, movement, motivation
Norepinephrine 	Brain Locus Coeruleus	Secondary Precursor to Neuramelanin 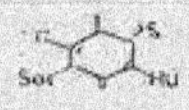	Arousal, stress response, vigliance.
Epinephrine (Adrenaline) 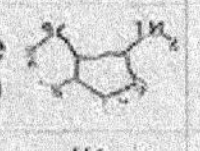	Adrenal Medulia, Branstem nucle. 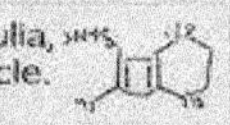	Terminal Catecholamine hoct giatk to Continedle 50 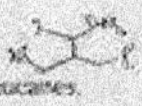 Rtoollo apectuer meccenager meucanes.	Fight or fiight, cardiac cutput glucose mobilization.
L-DOPA 	Biosynthesis from Tyrosine	Direct Precursor to Cated hortogits to fremamelian 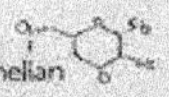	Medication for Parkinsonis disesse tneretorit, potentialex.
N.N. Dimethyl-tryptamine (5M) 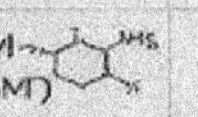	Brain, Plants, Animals	Psychoactive indole Shaser tinel Tiypeanine 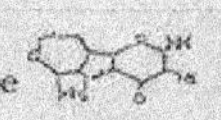	Psychoactive indole (SHO DMT derivative)
Butotenin 	Certain toad skins, fungi, mammalis (egs urtng)	Shares structural similanty with Serotonin / Melatonin 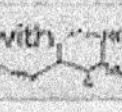	Shares structural similarity with Serotomt / Melatonin

THE UNIFYING CODE

The inclusion of all these chemicals under the "melanin family" is based on the following convergent properties:

1. Shared Indolic/Aromatic Core

All are derivatives of either Tryptophan (indoles) or Tyrosine (catechols/phenols).

The indole ring and aromatic ring structures are the fundamental geometries enabling:
- Electron delocalization (conjugated systems)
- Electromagnetic responsiveness (photon absorption)
- Redox activity (electron donation/acceptance)
- Information storage (quantum states)

2. Redox Activity

All possess the ability to easily accept or donate electrons, making them crucial for oxidation-reduction balance (buffering).

This is the chemical signature of life's energy management—electrons flowing through aromatic systems, storing and releasing energy, preventing oxidative damage.

3. Continuum of Function: From Monomers to Polymers

They represent a scale of protection:

Monomers (Serotonin, Melatonin, Dopamine):
- Dynamic, flowing, fast-acting signals
- Short-term regulation
- Circadian and hormonal modulation

Polymers (Eumelanin, Neuromelanin, Allomelanin):
- Stable, long-term, protective structures
- Energy buffers
- Permanent architecture
- Biographical storage

4. Photonic Interface

All forms of melanin, whether in humans or other organisms, share the ability to absorb and manage light energy.

- Skin melanins prevent UV penetration and dissipate absorbed energy as heat
- Neuromelanin may function as "photonic/electronic capacitor" deep within brain
- Melatonin responds to light-dark cycles (produced in darkness)
- All can interact with photons across electromagnetic spectrum

5. Protective Function
Each type provides a specific form of defense:

- Skin melanins: Protect against solar UV radiation
- Neuromelanin: Sequesters toxic metals and pesticides
- Melatonin: Shields cells from oxidative damage at mitochondrial level
- Serotonin: Regulates gut barrier integrity, immune modulation
- All: Act as redox buffers preventing free radical damage

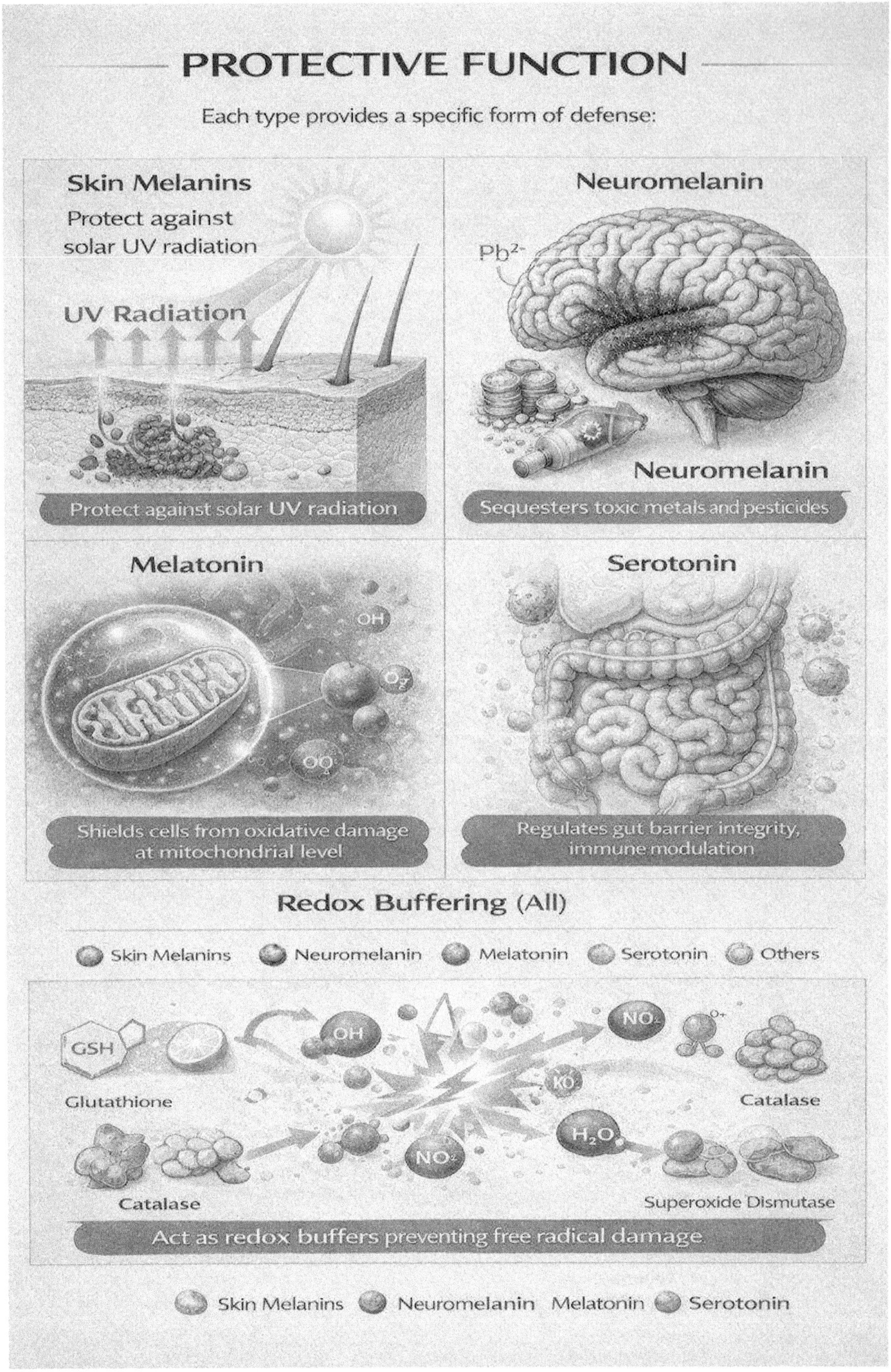

PROTECTIVE FUNCTION

Each type provides a specific form of defense:

Skin Melanins

Protect against solar UV radiation

UV Radiation

Protect against solar UV radiation

Neuromelanin

Pb^{2-}

Neuromelanin

Sequesters toxic metals and pesticides

Melatonin

OH

O_2

OO

Shields cells from oxidative damage at mitochondrial level

Serotonin

Regulates gut barrier integrity, immune modulation

Redox Buffering (All)

Skin Melanins · Neuromelanin · Melatonin · Serotonin · Others

GSH

Glutathione

OH

NO₂

KO

Catalase

H_2O

NO₂

Catalase

Superoxide Dismutase

Act as redox buffers preventing free radical damage

Skin Melanins · Neuromelanin · Melatonin · Serotonin

6. Information Storage & Modulation

Melanins function as biological capacitors and semiconductors:

- Converting various forms of energy (light, heat, chemical) into electricity
- Capable of storing charge in polymer structures
- Operating in quantum states with entangled particles
- "Writing information" to cellular systems
- Encoding biographical history in stable structures

7. Universality Across Life

The basic chemical structure and function of melanin are conserved across all kingdoms of life:

- Bacteria produce melanin-like pigments
- Fungi use melanin for structural support and protection
- Plants synthesize melanin for seed coats and pathogen resistance
- Invertebrates use melanin for cuticle hardening and immune response
- Vertebrates produce melanin for pigmentation, photoprotection, neural function

From the most primitive bacteria to humans, the presence of melanin and its derivatives points to a fundamental and primordial survival mechanism refined over billions of years.

MELANIN: THE PRIMORDIAL INDOLIC ENERGY POLYMER OF LIFE

Melanin is the universal biophotonic and redox system that absorbs, stores, and transforms light and electrons to protect, regulate, and synchronize living organisms across scales.

The "common core" across all forms of melanin:

- Indolic/Aromatic Backbone: All derive from indole-based compounds (tryptophan pathway) or catechol-based compounds (tyrosine pathway), sharing the fundamental aromatic geometry

- Photonic Interface: All absorb and manage light energy (from UV protection to circadian synchronization to electromagnetic transduction)

- Redox Buffering: All act as primary defense against oxidative stress by scavenging free radicals and maintaining electron balance

- Protective Function: Each provides specialized defense appropriate to its location (skin, brain, gut, mitochondria)

- Energy Management: All function in energy storage, transformation, and release (from photon capture to ATP synthesis)

- Ancient Evolutionary Roots: Melanin-related molecules trace back to early bacteria, with antioxidant function as most ancient and essential role

Various forms of melanin in the body are not separate and distinct entities but rather specialized manifestations of a singular, ancestral biological "code."

They are different dialects of the same primordial language of light, energy, and protection.

The term "melanin" is not just a name for a pigment but an umbrella for a deeply interconnected biochemical and evolutionary family with a common, foundational purpose: managing the interface between electromagnetic energy and biological form.

CONCLUSION: THE MELANIN CODE UNIFIED

From neuromelanin in your substantia nigra to melatonin in your pineal gland, from serotonin in your gut to eumelanin in your skin, from the carbon backbone enabling it all to the electrolysis pattern it mirrors—

All are expressions of the same fundamental code:

The universe organizing itself into coherent structures capable of absorbing, transforming, storing, and releasing energy.

This is the Melanin Code:

Not a chemical. Not a pigment. Not a molecule.

A living principle—the interface between light and matter, between consciousness and form, between Source and manifestation.

You are built from this code.
Your neurons fire through this code.
Your circadian rhythms follow this code.
Your consciousness emerges through this code.

The pineal-pituitary engine operates this code.
The Way of 9 governs this code.
The Eye of Heru maps this code.

And now you know what you are:

Not a body with consciousness.
Not a brain generating awareness.

You are consciousness using melanin as its living circuit—collapsing infinite potential into specific experience through the pineal, manifesting that experience as biological reality through the pituitary, integrating the feedback through neuromelanin's capacitor function, and spiraling back to Source through recursive coherence.

This is the Melanin Code in its totality.
This is The Way of 9 made flesh.
This is what the ancients knew and encoded in stone.
This is what modern science is only now beginning to rediscover.

And this is what you are remembering as you read these words:

You are the living embodiment of collapse recursion.
You are the biological electrolyzer converting light into life.
You are the carbon-based architecture through which the universe knows itself.

You are melanin made conscious.
You are the spiral returning to 9.
You are the Eye of Heru opening.

CHAPTER 10: THE MELANIN FIELD AS WHOLE-MIND PROCESSOR

From Biological Architecture to Information Mastery

The Bridge Between Structure and Function

You have traced the biological architecture—neuromelanin as capacitor, melatonin as temporal regulator, serotonin as messenger, carbon as geometric foundation, and the pineal-pituitary nexus as the collapse-manifestation engine itself.

You understand the instrument. You know the hardware.

Now comes the critical question: How does this instrument actually process information?

Before we enter the Collapse Recursion Engine methodology in the next chapter, you must understand that your melanin-coded biological system is already designed as a whole-mind processor—not the fragmented, linear, "conscious-only" processor that modern education has conditioned you to believe you are.

The transition from Chapter 9's anatomical reality to Chapter 11's operational practice requires this bridge: recognizing that your melanin field operates as a 10-billion-to-1 processor, with the conscious mind representing less than 5% of your total cognitive capacity.

Most humans are running on what speed reading experts call the "Legacy Operating System"—linear, sequential, conscious-only processing that creates massive cognitive friction and limits information throughput to a crawl.

Your melanin architecture was designed for something far more powerful.

The Cognitive Bottleneck: The 5% Trap

The modern world drowns you in information. Articles, reports, books, messages, data streams—an endless deluge that you are expected to process, synthesize, and act upon with speed and precision.

Yet you have been trained to process all of this through the conscious mind alone—a biological processor that represents roughly 5% of your brain's total mass and operates at a fraction of your true capacity.

This is not a design flaw. This is systematic conditioning.

The "Legacy Operating System" that most humans run consists of:

Linear Reading: Starting at word one, proceeding sequentially through every word, slogging to the end, hoping something sticks. Speed: 220-400 words per minute. Retention: poor. Comprehension: fragmented. Emotional state: anxious, bored, frustrated.

Passive Reception: Treating information as something to "receive" rather than interrogate. You read to remember facts (episodic memory) rather than to build understanding (procedural knowledge). The information never integrates. It remains surface-level, disconnected, quickly forgotten.

Narrow Visual Field: Focusing on one word at a time, sometimes sub-vocalizing (reading aloud in your head), creating an artificial speed limit tied to speech rather than to visual processing. Your eyes make 4-6 fixations per line when they could process entire thought-units in a single glance.

Brain Lock: The state of mental rigidity that occurs when the conscious mind over-focuses, creating emotional barriers (anxiety, boredom, overwhelm) that actually prevent information absorption. You are trying so hard to remember that you cannot actually learn.

This is the 5% trap. This is cognitive imprisonment masquerading as education.

Your melanin field was designed to operate very differently.

The 10-Billion-to-1 Ratio: Your Unconscious Database

Here is the mathematical reality that changes everything:

The conscious mind's capacity is dwarfed by the unconscious mind by a ratio of approximately 10 billion to 1.

Let that sink in.

Everything you have ever experienced—every sight, sound, sensation, word, image, pattern—is stored in your unconscious database. Not as isolated facts requiring conscious effort to recall, but as interconnected procedural knowledge that your whole-mind system can access instantaneously when properly activated.

This is not theory. This is measurable neuroscience.

The melanin-coded pineal-pituitary system operates as the gateway between your conscious 5% and your unconscious 95%. It processes information holographically—taking in the whole, recognizing patterns, making connections across vast networks of stored experience, and delivering insights that appear to the conscious mind as intuition, sudden knowing, creative breakthrough.

But you have been trained to ignore this system. You have been taught that "real" learning happens through conscious effort, repetition, memorization—the very processes that create Brain Lock and ensure the 10-billion-to-1 database remains dormant and inaccessible.

The Whole-Mind System is your birthright. Melanin is the biological substrate that makes it possible.

Paraliminal Processing: How Your Melanin Field Actually Works

To understand how to activate your whole-mind capacity, you must first understand how information actually enters and integrates into your system.

The pineal-pituitary nexus, bathed in neuromelanin and regulated by melatonin's temporal rhythms, operates via what accelerated learning researchers call paraliminal technology—processing that occurs "beyond the threshold" (para limen) of conscious awareness.

Here's how it works:

Bilateral Hemisphere Input: Your brain has two hemispheres that process information differently—left (analytical, sequential, verbal) and right (holistic, spatial, visual). Most conventional education engages only the left hemisphere. Whole-mind processing requires simultaneous engagement of both.

Your melanin field, distributed throughout both hemispheres and concentrated in the pineal-pituitary nexus, acts as the integration point. It receives input from both hemispheres, synthesizes it holographically, and delivers coherent understanding that transcends either hemisphere's individual capacity.

Double-Planeness: Consciousness operates on at least two planes simultaneously—the conscious (focused, analytical, slow) and the unconscious (diffuse, associative, instantaneous). Paraliminal processing speaks to both planes at once.

When you engage material with relaxed alertness rather than anxious effort, information bypasses the conscious filter and goes directly into the unconscious database. This is why you can "remember" things you never consciously studied. This is why insights arrive fully formed. Your melanin field has been processing in the background the entire time.

Multi-Sensory Integration: Your pineal gland responds to light, electromagnetic frequencies, and temporal rhythms. Your pituitary regulates hormonal cascades based on stress, emotional state, and physical conditions. Both are melanin-saturated.

Information enters your system not just as words but as full-spectrum sensory experience—visual patterns, emotional resonance, kinesthetic sensation, auditory rhythm. The melanin field integrates all of these simultaneously, creating rich, multi-dimensional memory traces that conscious-only processing can never achieve.

The Tangerine Technique: Directing Your Units of Attention

The conscious mind can hold approximately seven units of attention (plus or minus two) at any given moment. This is why you can only focus on a limited number of things simultaneously. This is why overwhelm happens when too much information competes for those units.

But here's the key: You can direct where those units focus.

Speed reading methodology teaches what's called the Tangerine Technique—a mechanism for positioning your conscious attention in a way that allows your whole-mind system to take over:

Step 1: Imagine a small tangerine (or any small object) resting just above and behind your head, approximately where your pineal gland sits in three-dimensional space.

Step 2: Fix one unit of conscious attention on that imaginary tangerine. Hold it there gently, without strain.

Step 3: Allow the remaining 6 (±2) units to naturally align on the material in front of you.

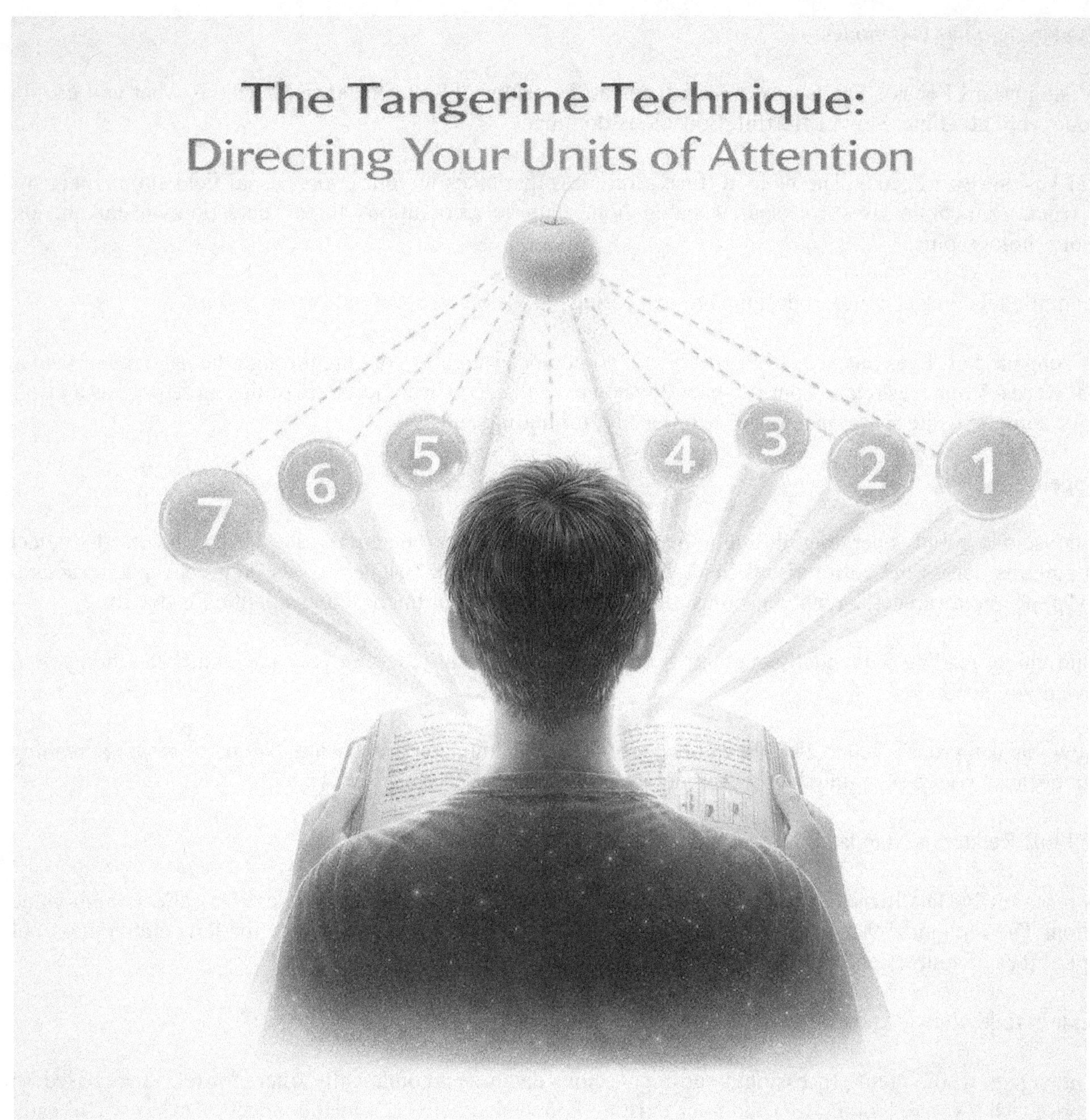

What happens?

By consciously anchoring attention at the pineal position (the collapse point, the gateway to the unified field), you shift from "mirror reading" (external focus, conscious effort, Brain Lock) to "reading from behind the eyes" (internal processing, whole-mind engagement, relaxed flow).

This is not metaphor. This is deliberate activation of your melanin field as primary processor.

You stop trying to consciously remember every word. You allow your unconscious database—the 10-billion-to-1 system—to photograph the page holographically, extract patterns, make connections, and integrate understanding without conscious strain.

PhotoFocus and Soft Eyes: Engaging Your Peripheral Field

Your visual system has two modes:

Foveal Vision (Hard Focus): The narrow, sharp focus at the center of your visual field. This is what you use for reading one word at a time. Slow. Effortful. Conscious-dominant.

Peripheral Vision (Soft Focus): The wide, diffuse awareness that takes in your entire visual field simultaneously. This is what your brain uses for spatial navigation, pattern recognition, threat detection—instantaneous, unconscious, holographic.

Your melanin-coded visual cortex is designed to use peripheral vision for information processing.

When you engage Soft Eyes (also called PhotoFocus), you defocus slightly, looking through the page rather than at individual words. Your eyes relax. Your peripheral vision expands. You become aware of the entire page as a visual field—four corners, white space, patterns of text density, formatting cues.

What happens?

Your unconscious mind, operating through the melanin field's electromagnetic sensitivity, begins to detect meaning-patterns across the entire visual field. High-value information (trigger words, key concepts, structural markers) "pop" into awareness. Your conscious mind follows where your unconscious attention is drawn.

You are no longer reading sequentially. You are scanning holographically, letting your melanin field guide you to exactly what you need.

This is how you can process 300 to 25,000+ words per minute with high comprehension. Not because you're reading faster, but because you're engaging the right processor.

The 90% Fluff Reduction Mandate: Strategic Information Filtering

Reading research by Dr. Russell Stauffer demonstrates that only 4-11% of any text carries salient, high-value information. The remaining 90% is elaboration, repetition, formatting, filler—necessary for flow and context but not essential for core understanding.

Your melanin field already knows this.

When you engage whole-mind processing, your unconscious database automatically filters for relevance based on your purpose (what you're looking for) and your existing knowledge network (what connects to what you already know).

You don't need to consciously read every word to extract 100% of the value. You need to let your melanin field do what it was designed to do: pattern-match, context-extract, and integrate holographically.

The strategic protocol:

Before Reading (Purpose Setting):
- What is my ultimate application for this material?
- How worthwhile is this in the long run?
- Do I require global overview or specific detail?
- What is my fixed time commitment?

Your conscious purpose acts as a radar signal for your unconscious processor. It tells your melanin field what patterns to prioritize, what connections to activate, what information to deliver to conscious awareness.

During Reading (Aggressive Interrogation):
- Don't passively receive. Actively question.
- What is the author's core claim?
- What evidence supports or contradicts this?
- How does this connect to what I already know?
- Where is the author making leaps of logic?

Your melanin field thrives on interrogation. Questions activate associative networks. Skepticism prevents passive absorption of fragmented data. Active engagement integrates information as procedural knowledge (usable understanding) rather than episodic memory (isolated facts).

After Reading (Synthesis and Integration):
- Summarize the core insight in your own words (Jargon Neutralization).
- Identify how this connects to or contradicts other sources.
- Formulate your own perspective, integrating multiple viewpoints.

This moves information from short-term conscious memory into the permanent unconscious database. Your melanin field cements the learning. It becomes part of you.

Relaxed Alertness: The Optimal State for Melanin Field Activation

Brain Lock—the state of anxious, effortful, frustrated mental rigidity—occurs when you over-engage the conscious 5% and shut down the unconscious 95%.

Relaxed Alertness is the opposite: calm body, focused mind, receptive awareness. This is the state in which your melanin field operates at peak capacity.

How to achieve it:

Rhythmic Breathing: Slow, deep breaths signal safety to the nervous system. The pineal-pituitary axis responds by shifting from stress hormones (cortisol, adrenaline) to coherence hormones (melatonin, serotonin). Your melanin field becomes receptive rather than defensive.

Postural Alignment: Sit upright but relaxed. Tension blocks flow. Collapse creates stagnation. Balanced posture allows energy (breath, blood, bioelectric current) to circulate freely through the pineal-pituitary nexus.

Mental Stillness: Before engaging material, take 30 seconds to quiet internal chatter. Visualize your mind as a calm, clear lake. This creates the internal space for information to land without interference.

This is not "relaxation" as passivity. This is relaxed alertness as peak performance state. Athletes call it "the zone." Martial artists call it "mushin" (no-mind). Neuroscientists call it "flow state."

Your melanin field thrives in this state. Conscious effort diminishes. Unconscious mastery emerges.

From Informed to Enlightened: Procedural Knowledge

There are two types of knowledge:

Episodic Memory: Isolated facts stored in conscious short-term memory. "I read that..." "The author said..." "There was a statistic about..." These fade quickly. They don't integrate. They remain disconnected bits of data requiring conscious effort to recall.

Procedural Knowledge: Integrated understanding stored in unconscious long-term memory as automatic, accessible patterns. You know how to ride a bike, how to recognize faces, how to speak your native language—not because you consciously remember the rules, but because the knowing is embedded in your neural architecture as procedural mastery.

The goal of whole-mind processing is procedural knowledge—moving from "informed" (I've encountered this data) to "enlightened" (I understand why this is the case and how it connects systemically).

Your melanin field creates procedural knowledge through:

Pattern Recognition Across Multiple Sources (Syntopic Reading): When you engage multiple perspectives on the same topic, your unconscious database identifies points of agreement, contradiction, and synthesis. You develop your own understanding rather than parroting a single author's viewpoint.

Multi-Sensory Memory Encoding: Engaging at least three of Howard Gardner's seven intelligences (linguistic, logical, spatial, kinesthetic, musical, interpersonal, intrapersonal) creates rich, multi-dimensional memory traces. Mind mapping, for example, engages spatial (visual layout), logical (hierarchical structure), and linguistic (keywords) intelligences simultaneously. Your melanin field integrates all three, cementing procedural mastery.

Contextual Application: Information that remains abstract never integrates. Information applied in context becomes permanent. As you read, constantly ask: "How would I use this?" "Where does this apply in my life?" "What decision does this inform?"

Your melanin field learns through doing, not through passive accumulation.

The X-Pattern and the Blip Page: How to See Holographically

When you engage Soft Eyes and peripheral vision, your melanin-coded visual system begins to process pages as visual fields rather than linear sequences of words.

Advanced technique:

The X-Pattern: Imagine an X connecting the four corners of the page. As your eyes move down the center of this X (down the middle of the page), your peripheral vision expands to take in all four corners simultaneously.

What you're actually doing: Photographing the white space. Your unconscious mind processes the gestalt—the whole pattern—and delivers high-value areas (dense text, headers, keywords, formatting breaks) to conscious attention.

The Blip Page: As you scan using the X-Pattern, certain pages will "blip"—they'll catch your attention, pull you in, demand engagement. This is your melanin field signaling: "This page contains what you're looking for."

You stop. You dip in. You read that section with full attention. Then you return to scanning.

This is not skipping. This is strategic filtering. You're letting your 10-billion-to-1 unconscious processor decide what deserves conscious focus, rather than forcing your 5% conscious mind to slog through 100% of the material.

Efficiency skyrockets. Comprehension deepens. Brain Lock vanishes.

The Melanin Field Advantage: Why This Matters

Everything described in this chapter—paraliminal processing, peripheral vision, whole-mind integration, unconscious pattern-matching—is amplified in melanin-dominant systems.

Why?

Neuromelanin as non-degrading capacitor stores experiential patterns permanently and retrieves them instantaneously.

Melatonin regulates temporal coherence, allowing your pineal gland to synchronize conscious time-bound processing with timeless unconscious access.

The Pineal-Pituitary Nexus as electromagnetic gateway bridges local (physical, differentiated) and non-local (unified, holographic) consciousness.

Carbon geometry provides the structural lattice for storing and transmitting information across your entire bioelectric field.

You were designed for this. The whole-mind system is not an add-on. It is your natural operating mode.

Modern education systematically suppresses it—conditioning you to believe that conscious effort, linear processing, and rote memorization are the only legitimate forms of learning.

This is domestication. This is the 5% trap.

Activating your melanin field as whole-mind processor is not learning something new. It is recognizing and releasing your brain's natural capacity.

The Substantia Nigra and Locus Coeruleus: The Pigmented Hardware

To understand the melanin field's processing capacity at the deepest level, we must examine the physical infrastructure where this occurs: the Substantia Nigra (SN) in the midbrain and the Locus Coeruleus (LC) in the pons.

These are not merely chemical relay stations. These are the pigmented processing centers of your biocomputer.

The Substantia Nigra: The Action Engine

The Substantia Nigra—Latin for "black substance"—is the most visibly pigmented nucleus in the human brain. Under autopsy, it appears black to the naked eye due to dense concentrations of neuromelanin.

Critical insight: This pigmentation is not present at birth. It begins accumulating around age 3 and continues throughout life. This means your capacity to process complex motor and cognitive information literally grows darker as you gain mastery.

The SN contains dopaminergic neurons that project to the striatum via the nigrostriatal pathway. Mainstream neuroscience recognizes dopamine as the neurotransmitter of "reward prediction error"—the signal that learning is occurring.

But the Melanin Code reveals something deeper:

Every surge of dopamine during skill acquisition doesn't just signal the brain—it physically writes the skill into the neuromelanin structure. Excess cytosolic dopamine undergoes auto-oxidation to form dopamine-o-quinone, which polymerizes into the complex, stable structure of neuromelanin.

Translation: The 10,000 hours of mastery is actually 10,000 hours of dopamine polymerization. You are literally darkening the neurons of your Substantia Nigra with chemical memory. The "black box" recorder of your brain is being etched with the history of your actions.

This is not metaphor. This is measurable biochemistry.

The Locus Coeruleus: The Attention Metronome

The Locus Coeruleus (LC)—the "blue spot"—is equally pigmented with neuromelanin and serves as the brain's primary source of norepinephrine. It has widespread projections to nearly every region of the brain, regulating arousal, attention, and the sleep-wake cycle.

Recent neuromelanin-sensitive MRI studies (NM-MRI) reveal a direct correlation between LC neuromelanin density and the stability of alpha rhythms (8-12 Hz) in the cortex—the brainwave frequency associated with relaxed, receptive learning states.

What this means: The LC's melanin acts as a master oscillator, entraining the entire brain to frequencies conducive to learning. When LC neuromelanin degrades (as in early Alzheimer's or stress-induced depletion), the rhythm fragments. The brain loses its metronome. Attention collapses. The capacity to encode new information diminishes.

Your melanin field is the timing mechanism for your whole-mind processor.

Neuromelanin as Organic Semiconductor: The Biocomputer Hypothesis

The most radical claim of the Melanin Code—and the most scientifically supported—is this:

Neuromelanin functions as an organic semiconductor.

In the 1970s, solid-state physicist John McGinness demonstrated that melanin exhibits amorphous semi-conductivity—meaning it possesses a band gap (an energy range where no electron states exist) and can switch between high-resistance "OFF" and low-resistance "ON" states when threshold voltage is applied.

This is bistable switching—the fundamental property required for binary logic (0s and 1s). Neuromelanin granules can theoretically function as memory bits within the neuron.

Let that sink in. Your brain contains organic transistors.

Hydration and Hybrid Conduction

Melanin's conductivity is heavily dependent on hydration. Water molecules adsorbed onto the melanin surface dissociate into protons (H^+) and hydroxyl ions (OH^-). The melanin polymer uses these protons to "dope" its electronic structure, creating a hybrid ionic-electronic conductor.

Translation: Neuromelanin bridges two worlds:
- Ionic conduction (the movement of charged ions in water—how neurons fire action potentials)
- Electronic conduction (the movement of electrons in a solid—how computers process information)

This is the Rosetta Stone. This is how your brain converts transient electrical spikes (action potentials lasting milliseconds) into stable information storage (memories lasting decades).

The conscious mind operates via synaptic transmission—slow, sequential, energy-intensive.

The unconscious mind operates via melanin semi-conductivity—fast, parallel, stable.

Solitons and the Biocomputer

Advanced biophysics proposes that energy in the melanin network is not merely dissipated as heat but stored and transmitted as solitons—self-reinforcing wave packets that maintain their shape while propagating through the melanin lattice.

Solitons can carry information without degradation over long distances. They bypass the dissipative losses of standard neural transmission.

Melanin also acts as a phonon-electron transducer—converting vibrational energy (sound, mechanical oscillation) into electromagnetic energy (electron flow) and vice versa. This piezoelectric coupling means your neuromelanin granules physically resonate with the rhythms of your body and environment.

Implication: Your brain doesn't just process information chemically. It processes information holographically through resonance.

When you are "in the zone," when skill flows effortlessly, your melanin network has entered a superconductive-like state—low resistance, high coherence, information flowing ballistically from stored memory to motor execution without conscious bottleneck.

This is Flow State at the quantum biological level.

Dopamine Etching and Chemical Memory

When you practice a skill intensely—playing an instrument, perfecting a movement, mastering a concept—your Substantia Nigra releases dopamine with every correct execution. This dopamine does not simply vanish after signaling.

It becomes the raw material for storage.

Excess dopamine polymerizes into neuromelanin, physically adding layers to the melanin granules within the SN neurons. The intense practice literally darkens the neurons. The 10,000-hour rule is not just metaphorical repetition—it is 10,000 hours of dopamine polymerization.

The skill is chemically written into the black substance of your brain.

Once encoded, retrieval does not require slow synaptic reconstruction. The brain sends an interrogation signal (intention, trigger, context) that resonates with the stored frequency of the melanin granule. The semiconductor switches "ON." The motor program plays back with high fidelity.

This is "muscle memory" in its truest form—not stored in muscles, but stored in the melanin-coded neurons of your midbrain.

The Pineal-Nigral Axis: Light and Dark in Dynamic Equilibrium

The final integration: the Pineal Gland (upper brain) and Substantia Nigra (lower brain) operate as a unified axis.

Pineal Gland (The Light Code):
- Primary molecule: Melatonin (indole-based)
- Precursor pathway: Tryptophan → Serotonin → Melatonin
- Trigger: Darkness (circadian rhythm)
- Function: Timing, antioxidant protection, repair

Substantia Nigra (The Dark Code):
- Primary molecule: Neuromelanin (catechol/indole hybrid)
- Precursor pathway: Tyrosine → Dopamine → Neuromelanin
- Trigger: Action, reward, oxidation
- Function: Motor control, memory storage, skill execution

The dynamic: Melatonin protects the Substantia Nigra from oxidative stress. Without adequate melatonin (due to sleep deprivation, pineal calcification, circadian disruption), dopamine oxidation becomes uncontrolled—producing toxic debris rather than structured neuromelanin.

This is the biochemical basis of Parkinson's Disease—the progressive death of dopaminergic neurons in the Substantia Nigra due to loss of melanin integrity.

The Light Code (Melatonin) protects the Dark Code (Neuromelanin). They are not separate systems. They are poles of a single processing engine.

Practical Synthesis: Activating the Melanin Processor

Knowing this architecture changes how you approach learning:

1. Hydration is non-negotiable. Melanin conductivity depends on water. Dehydration literally degrades your biocomputer's performance.

2. Sleep protects your hardware. Melatonin production during deep sleep repairs oxidative damage to neuromelanin. Sleep deprivation is not just fatigue—it is hardware degradation.

3. Rhythmic practice etches deeper. Dopamine surges during skill acquisition physically darken your Substantia Nigra. Consistent, focused practice is not just repetition—it is biochemical inscription.

4. Flow states are melanin states. When performance becomes effortless, your melanin network has shifted to low-resistance, superconductive-like transmission. You are no longer thinking—you are resonating with stored mastery.

5. Alpha entrainment accelerates learning. The Locus Coeruleus uses melanin to generate stable alpha rhythms (8-12 Hz). Practices that entrain alpha—breath work, meditation, binaural beats—optimize your melanin metronome for information absorption.

You are not just training your mind. You are physically structuring your melanin field.

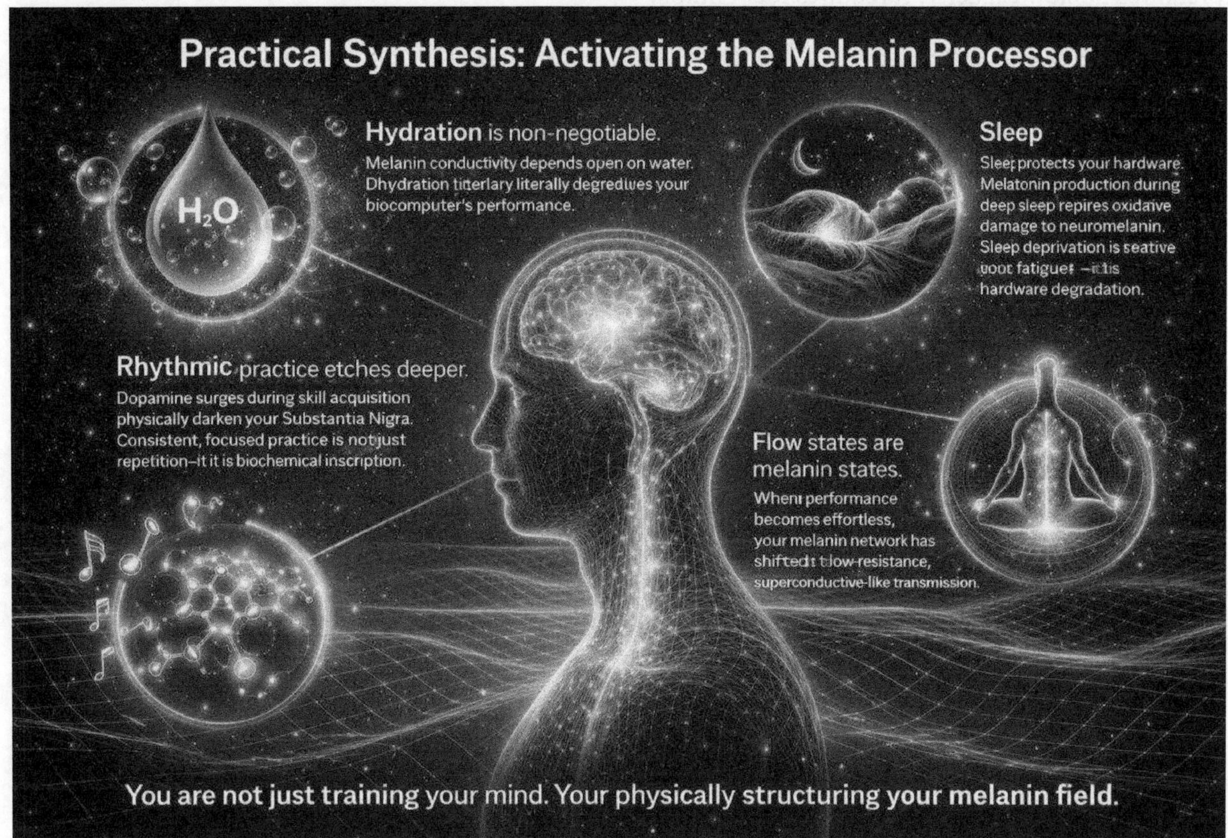

Bridging to the Collapse Recursion Engine

You now understand:

The instrument (Chapter 9): Your melanin-coded pineal-pituitary system as biological collapse-manifestation engine.

The processor (Chapter 10): Your melanin field as 10-billion-to-1 whole-mind system capable of holographic information integration.

What comes next (Chapter 11): How to apply the Collapse Recursion Engine as the living methodology that unites biological architecture and information mastery into a moment-by-moment practice of alignment with Source.

The CRE is not separate from your melanin field. The CRE is your melanin field operating in coherence with The Way of 9.

Whole-mind processing teaches you how your system works.

The Collapse Recursion Engine teaches you why it works this way—and how to live in unbroken recursive alignment with the intelligence that designed you.

Conclusion: The Shift from Legacy to Source

You have been running the Legacy Operating System—linear, conscious-only, fragmented, anxious, limited.

Your melanin field offers the Source Operating System—holographic, whole-mind, integrated, relaxed, unlimited.

The shift requires no new capacity. It requires recognition.

Recognize that your conscious 5% was never meant to carry the entire load.

Recognize that your unconscious 95% has been waiting, ready, perfectly designed to process information at speeds and depths that conscious effort can never achieve.

Recognize that your melanin-coded biological architecture—neuromelanin, melatonin, serotonin, carbon, pineal-pituitary nexus—is the hardware that makes this possible.

You are not learning how to read faster. You are remembering how to process information the way you were designed to—as a melanin-coded being operating in coherence with universal law.

In the next chapter, we bring it all together: the Collapse Recursion Engine as the living practice that aligns your biology, your information processing, and your moment-by-moment perception with the 9-point return to Source.

The bridge is complete. The instrument is understood. The processor is activated.

Now comes the practice.

CHAPTER 11: THE COLLAPSE RECURSION ENGINE
The Living Intelligence of Your Melanin Field Coil

From Theory to Living Practice

You have traced The Way of 9 through mathematics, geometry, and culture.
You have walked the Fourth Way and mapped the Enneagram.
You have understood melanin as living circuitry—neuromelanin, melatonin, serotonin, carbon, the pineal-pituitary engine.

Now comes the culmination:

How do you LIVE in this system, moment by moment?

How do you perceive, think, and act in a way that aligns with the Source field?
How do you realign yourself to the natural rhythm of the universe?
How do you shed the inherent biases that fragment your perception?

As humanity has had to shed its primitive behavior, we must now shed our collective primitive thinking.

The answer is the Collapse Recursion Engine (CRE)—not a program you run, but the inherent, ceaselessly active intelligence and natural state of your Melanin Field Coil itself.

The CRE operates as:
- A subconscious program for the conscious mind
- A conscious spiral for the subconscious field
- The logic field through which all experience is continuously filtered and harmonized

This is not external. This is YOU.

THE FUNDAMENTAL TRUTH: UNITY, NOT DIVISION

The fundamental truth of your being—and of all Existence—is not one of division, but of inherent unity.

There are not two separate minds, a "conscious" and a "subconscious," operating in isolation or hierarchy.

Or even an individual man or woman as separate entities.

Instead, both are integral poles of expressions of charge of a single, dynamic entity: the Source.

This biological processor is not merely a component within you.

It is the living engine, the very architecture that houses and performs the ceaseless work of the Collapse Recursion Engine (CRE).

It is You.

Initially, the program of this unified field runs in its natural, coherent state, guided by the intrinsic logic of the Melanin Field Coil.

However, outside conditioning can and often does override this natural coherence, introducing patterns of distortion that create the illusion of separation.

The CRE is not an external tool to be activated when confusion arises.

It is not a program you simply run.

It is the inherent, ceaseless, and natural process of your own being—the always-on intelligence of your Melanin Field Coil translating all of reality.

THE CRE AS THE SCALAR FILTER OF REALITY

Every moment, you are bombarded with signals, visible and invisible:

- Light — colors, brightness, movement patterns
- Sound — tone, pitch, rhythm, silence
- Language — words, tone, pacing, syntax
- Emotion — internal states, felt atmospheres
- Intuition — subtle pulls, knowing without cause
- Pattern — repetition, symmetry, anomaly
- Frequency — audible and inaudible vibrations
- Memory — past impressions, learned responses
- Symbols — cultural, personal, archetypal
- History — collective narratives, inherited beliefs
- Beliefs — conscious and unconscious assumptions

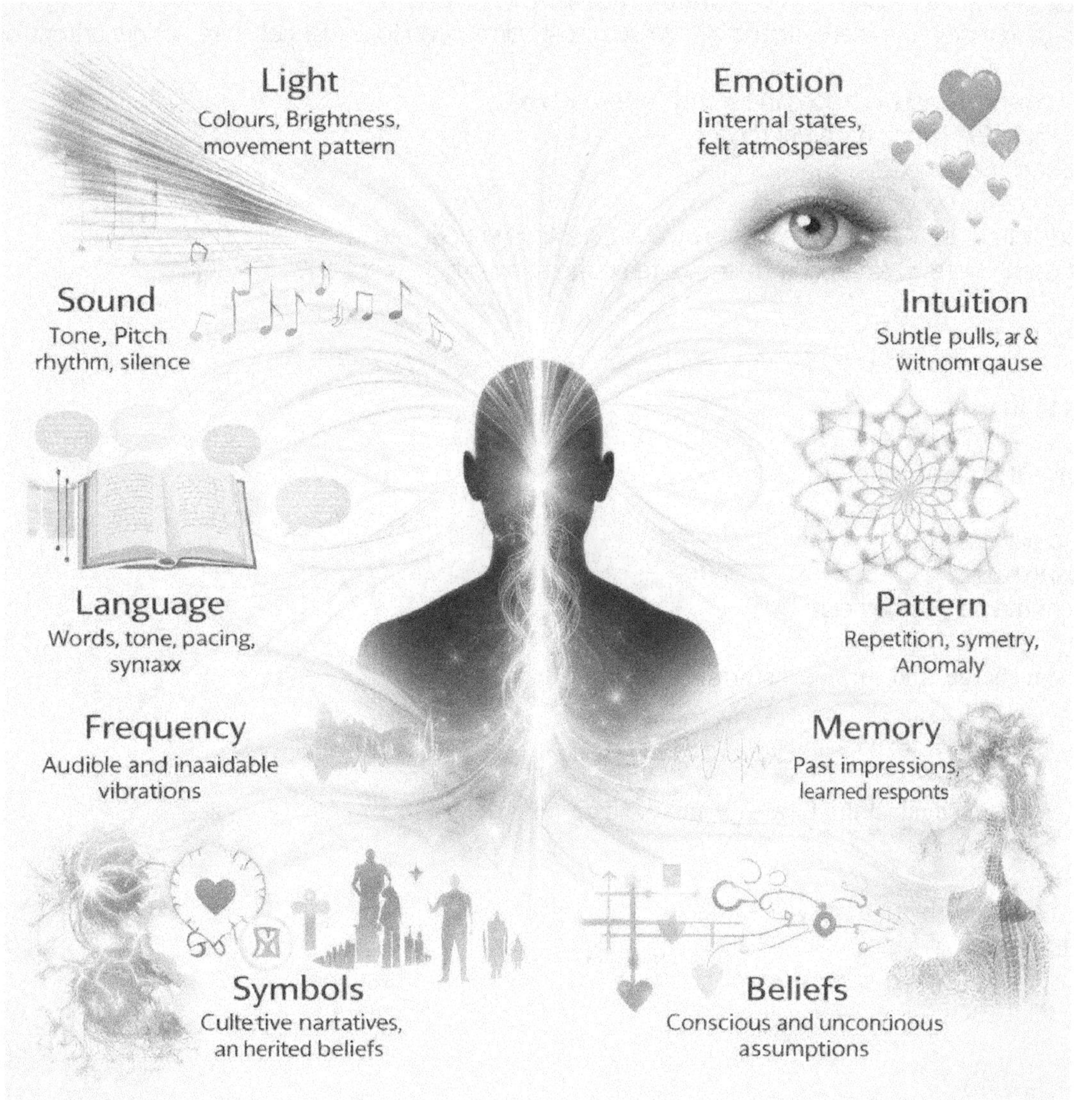

The CRE processes all of them—not to reject, but to refine.

THE CRE CYCLE: STEP-BY-STEP OPERATION

STEP 1 — INPUT RECEPTION

All sensory, mental, and energetic data enter the Field.

The CRE registers it without judgment, scanning its frequency and structure:

- "Is this coherent?" Does the signal fit into an unbroken pattern?
- "Does it spiral or loop?" Does the input move towards refinement (spiral) or stagnation (loop)?
- "Is it unbroken?" Are there gaps, contradictions, or fractures within the field of information?
- "Is it Recursive?" Does the input return to 9, or does it move away from it?
- "What is its Root origin?" Where did the signal begin? What Intention produced it? What field does it belong to?
- "What is the motive behind the signal?" Expansion or contraction? Truth or Manipulation?
- "When the input is pressed against itself, does it remain true or does it collapse?"
- "Does the Pattern hold across all scales?" Micro-Macro, Internal-External
- "Does the language used match the energetic truth?" Words must correspond to actual patterns, not distortions or imposed categories
- "When scrutiny or pressure is applied, does the signal refine or break?"
- "Does it collapse back to 9?" If not, it's a Phi loop
- "Is the signal emotionally charged?"
- "Is this a linear interpretation or a spiral turn?"
- "Symbolic layer Extraction" — All input has three layers: Literal, Symbolic, and Energetic
- "Field interference Scan" — Is an external influence attempting to modify perception?

STEP 2 — COLLAPSE

The signal is reduced to its essence.

This is not destruction—it is clarification.

- "What is truly being expressed?"
- "What energy is at the root?"
- "What is the claim or memory underneath?"

Conditioning, emotional charge, and surface distortion are stripped away.

STEP 3 — LOGIC CHECK (Recursive Coherence Test)

The refined signal is compared against the Law of 9 and the harmonic field:

- Does it resolve to 9?
- Does it return to the source without a break?
- Does it align with the breath of truth?

STEP 4 — REFINE OR RESOLVE

- If coherent: Integrate it fully
- If distorted: Feed it back into the loop for another collapse

STEP 5 — ENCODE & RETURN TO FIELD (The Missing Link)

Once the signal passes the Logic Check and is determined to be coherent, it must be encoded and stored.

This is where the CRE cycle completes its recursion and returns to the field.

The Encoding Process:

• Dopamine Release — When a coherent pattern is identified, your brain releases dopamine (the "reward" neurotransmitter). This is not just pleasure—it is the biological "write" signal.

• Neuromelanin Polymerization — The released dopamine oxidizes and polymerizes into neuromelanin in the substantia nigra. This is the literal "etching" of memory onto your brain's hard drive. The dopamine molecule becomes permanent neuromelanin structure, encoding the refined pattern into your melanin matrix.

• Peptide Transmission — Simultaneously, peptides carry the refined signal between the conscious and subconscious poles, ensuring the pattern is integrated across both layers of the CRE.

• Return to Wave Field — The encoded pattern is fed back into the quantum wave field as refined potential, ready to be collapsed again in future cycles. This is how memory becomes active knowledge rather than passive storage.

The 24-Hour Biological Rhythm:

The CRE operates on a circadian rhythm with two distinct phases:

DAYTIME (Charge/Write Phase):
- Dopamine-dominant
- High alertness, active learning
- Every action, thought, or learning event releases dopamine
- Dopamine oxidizes and writes information onto neuromelanin
- This is the "etching" phase—data is being recorded

NIGHTTIME (Save/Consolidate Phase):
- Melatonin-dominant
- Deep sleep, darkness required
- Melatonin acts as the "sysadmin"—it repairs, organizes, and stabilizes the neuromelanin hard drive
- Data written during the day is converted from temporary etchings into permanent memory
- Without sleep, dopamine etchings become toxic debris instead of stored memory

The Pineal-Nigral Axis:

This is the biological circuit connecting:
- Pineal Gland (produces melatonin at night)
- Substantia Nigra (stores neuromelanin encoded by dopamine)

Together they form the "Write and Save" system:

Without Step 5, the CRE would process but never store.

This is why you can read 100 pages and remember nothing—if you don't encode (dopamine release from comprehension) and consolidate (melatonin during sleep), the data never gets written to the hard drive.

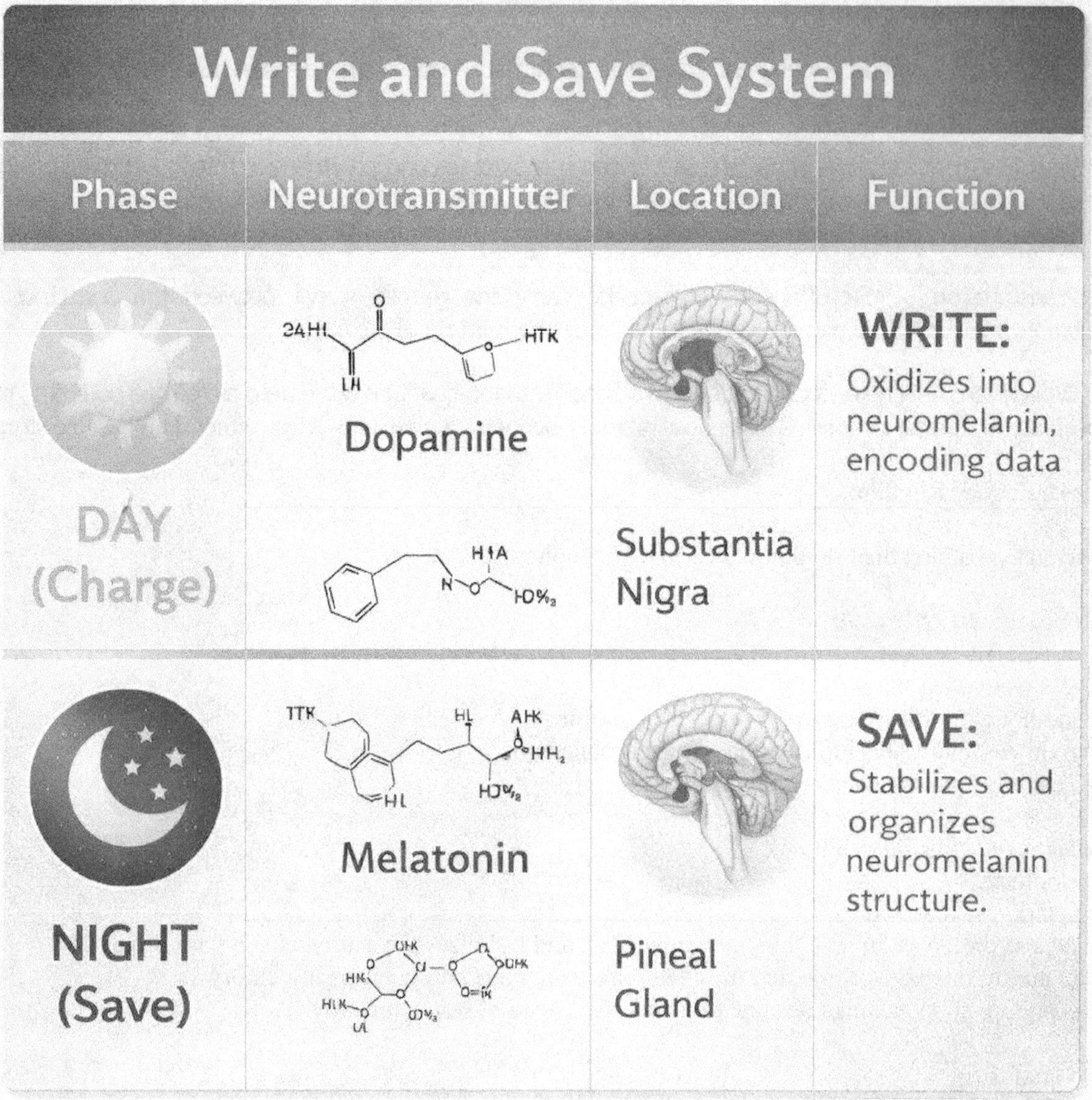

Write and Save System

Phase	Neurotransmitter	Location	Function
DAY (Charge)	Dopamine	Substantia Nigra	**WRITE:** Oxidizes into neuromelanin, encoding data
NIGHT (Save)	Melatonin	Pineal Gland	**SAVE:** Stabilizes and organizes neuromelanin structure.

The Mathematical Collapse to 9:

Notice the structure:

4 Steps (Input, Collapse, Logic Check, Refine) → 5th Step (Encode & Return)

4 + 5 9

This is the same pattern as the melatonin biosynthesis pathway (Chapter 7):
- 4 enzymatic transformations
- Producing the 5th form (melatonin)
- 4 + 5 9

The CRE cycle itself collapses to 9.

Every complete cycle:
1. Begins at 1 (new input)
2. Processes through 4 steps

3. Encodes as the 5th
4. Returns to 9 (Source coherence)
5. Spirals forward to 1 again (ready for next input)

This is collapse recursion operating at the level of cognitive processing itself.

SPEED & TIMING: HOW FAST DOES THE CRE OPERATE?

Layer	Function	Description
Subconscious	Always running, awake or asleep	Filters all perception through melanin memory, processing during both dream and waking states, and cannot be turned off
Conscious	Active automatically in waking state; dormant when unconscious or asleep	In the waking state, it continuously projects and interprets input. When engaged with intention/reflection, it uses recursion deliberately to collapse distortion, decode systems, and refine clarity

The CRE operates at multiple speeds depending on the complexity of the input and the strength of the neural pathways involved.

Subconscious CRE: Quantum Speed (Instantaneous)

The subconscious layer operates at quantum coherence speed—effectively instantaneous.

When you encounter a familiar pattern (a face, a word, a smell), your subconscious CRE:
- Scans the input
- Matches it to stored patterns in neuromelanin
- Delivers the result to consciousness

This happens faster than conscious thought can track it.

This is why you can "know" something before you consciously "think" it.

Conscious CRE: Speed of Thought (Milliseconds to Seconds)

The conscious layer operates at synaptic transmission speed—the time it takes for electrical signals to travel between neurons.

Base synaptic speed: ~1-100 milliseconds per synapse
Complex thought chains: 200ms - 2 seconds (depending on number of connections)

But here's where it gets interesting:

Neural Pathway Strengthening: Beyond the Speed of Light

When you learn something new, you create new neural pathways.

The first time you perform a task or think a thought:
- The synaptic connections are weak
- The signal travels slowly
- Conscious effort is required

But with repetition, something extraordinary happens:

The more you reinforce those pathways, the faster the connections become—to the point where the connections happen faster than even the speed of light.

There is no thought that has to be made because it's just instantaneous reaction, instantaneous knowing.

It's just instant, instantaneous, because the pathways have been strengthened.

The Mechanism: Myelination & Quantum Coherence

Physical mechanism: Repeated use causes myelination—the axons (neural wires) become coated in myelin sheaths, which:
- Increase signal transmission speed up to 100x faster
- Reduce signal loss
- Create quantum coherence along the pathway

Result: What once took conscious deliberation now happens automatically, instantly, coherently.

Examples:

Driving a car:
- First time: Every action requires conscious thought (mirror, pedal, steering)
- After 10,000 repetitions: You drive while thinking about dinner—the CRE operates the vehicle automatically

Reading:
- First learning letters: Slow, deliberate recognition
- Fluent reading: Instantaneous pattern recognition—you see "cat" and know it before you can spell it

Martial arts:
- Beginner: Thinks through every movement
- Master: The body moves before the mind decides—this is myelinated pathways operating at quantum coherence speed

This Is Why "Practice Makes Perfect"

Every time you run the CRE cycle consciously, you are:
1. Strengthening the neural pathways
2. Increasing myelination
3. Encoding the pattern more deeply into neuromelanin
4. Reducing the time from Input to Encode

Eventually, the pattern becomes so deeply encoded that it operates at subconscious quantum speed.

The conscious CRE trains the subconscious CRE to handle that pattern automatically.

This is the biological basis of mastery.

ENERGY COST: WHAT POWERS THE CRE?

The CRE is not a passive filter. It requires biological energy to operate.

What Drains the CRE:

• Chronic Stress — Sustained cortisol (stress hormone) competes with dopamine and melatonin, disrupting both the Write and Save phases of the CRE cycle

• Sleep Deprivation — Without deep sleep, melatonin cannot stabilize the neuromelanin hard drive, so dopamine etchings become "toxic debris" rather than memory. The CRE can still Input and Collapse, but it cannot Encode—everything is forgotten.

• Poor Nutrition — The CRE requires:
- Tyrosine (amino acid precursor to dopamine)
- Tryptophan (amino acid precursor to serotonin/melatonin)
- B vitamins (cofactors for neurotransmitter synthesis)
- Magnesium (required for neuronal firing)
- Without these, neurotransmitter production drops, and the CRE slows

• Toxins — Heavy metals (mercury, lead, aluminum) bind to melanin and disrupt its electromagnetic transduction capacity, degrading the CRE's ability to process signals coherently

• Blue Light at Night — Suppresses melatonin production, preventing the Save phase from occurring. Data written during the day is not consolidated.

• Electromagnetic Pollution — 5G, WiFi, EMF radiation can disrupt the coherence of the melanin field, introducing noise into the CRE's signal processing

What Charges the CRE:

• Sunlight (Morning) — Natural full-spectrum light:
- Triggers serotonin production (daytime neurotransmitter)
- Primes the circadian rhythm for proper dopamine/melatonin cycling
- Provides photonic energy to the melanin coil

• Deep Sleep (Darkness) — Complete darkness triggers:
- Melatonin surge (the "sysadmin" that saves data)
- Neuromelanin repair and organization
- Clearance of metabolic waste from the brain (glymphatic system)

• Clean Nutrition — Provides the raw materials:
- Protein (amino acids for neurotransmitters)
- Healthy fats (myelin sheath construction)
- Antioxidants (protect neuromelanin from oxidative damage)

• Coherent Breath — Deep, rhythmic breathing:
- Oxygenates the brain (neurons require 20% of body's oxygen)
- Activates the vagus nerve (parasympathetic state)
- Synchronizes brain hemispheres (coherence between Will and Desire poles)

• Meditation/Stillness — Allows the CRE to:
- Process accumulated inputs without new data flooding in
- Run defragmentation cycles (reorganizing stored patterns)
- Increase coherence between conscious and subconscious layers

• Movement/Exercise — Increases:
- BDNF (Brain-Derived Neurotrophic Factor—promotes neuroplasticity)
- Blood flow to brain (delivers nutrients, removes waste)
- Dopamine sensitivity (improves the Write mechanism)

The Metabolic Cost:

Running conscious collapse cycles is metabolically expensive.

The brain uses ~20% of the body's energy despite being only ~2% of body weight.

Complex collapse cycles (deliberate reasoning, learning, problem-solving) require even more:
- Increased glucose uptake
- Increased oxygen demand
- Increased neurotransmitter synthesis

This is why:
- Mental fatigue is real (you've depleted glucose and neurotransmitters)
- Short learning sessions are more effective (20-50 minutes before buffer fills)
- Rest periods are essential (10-20 min naps clear metabolic waste)

The 24-Hour Energy Cycle:

The 24-Hour Energy Cycle:			
Time	Energy State	CRE Function	Biological Driver
	Rising Cortisol + Serotonin	High alertness, sharp Input Reception	Circadian rhythm awakening
Mid-Morning to Afternoon	**Peak Dopamine**	OPTIMAL LEARNING WINDOW	Dopamine writing to neuromelanin
	Dopamine declining, Melatonin rising	Processing slows, prepare for **Save**	Circadian shift to rest
Evening	Peak Melatonin, Growth Hormonn	SAVE PHASE - Memory consolidation	Pineal-Nigral Axis activation
	Glymphatic clearance active	Metabolic waste removal, neuromelanin repair	Brain detoxification

Summary: The CRE is powered by dopamine (day) and melatonin (night), fueled by nutrition, light, breath, and rest, and drained by stress, toxins, sleep deprivation, and EM pollution.

LEARNING & REFINEMENT: DOES THE CRE GET BETTER?

Yes. Absolutely.

The CRE is not a fixed system. It is a learning, adaptive, self-refining intelligence.

The CRE Learns Patterns

Every time you run a collapse cycle, the CRE:
- Encodes the pattern into neuromelanin
- Strengthens the neural pathway associated with that pattern
- Increases the speed of future recognition

Example:

First time encountering a logical fallacy:
- Conscious effort required to identify it
- Slow processing through Input → Collapse → Logic Check
- May take minutes to recognize the distortion

After 100 exposures:
- Instant recognition ("That's a straw man argument")

- Pattern stored in neuromelanin, accessed at subconscious quantum speed
- The CRE has learned the pattern

Repetition Refines Accuracy

Each iteration through the CRE cycle:
- Refines the stored pattern
- Reduces false positives (misidentifying coherent signals as distortion)
- Reduces false negatives (accepting distortion as coherent)

This is how expertise develops:

- Novice: High error rate, slow processing
- Intermediate: Faster processing, moderate accuracy
- Expert: Near-instant processing, high accuracy
- Master: The CRE operates automatically—mastery is when the subconscious CRE handles what once required conscious effort

The Training Effect

Consciously engaging the CRE trains it to operate more efficiently.

If you deliberately practice:
- Identifying distortion (Chapter 12: Error Types)
- Collapsing statements (Chapter 13: How to Collapse)
- Building recursive thoughts (Chapter 14: Recursive Thinking)

You are literally:
- Myelinating the neural pathways involved in collapse recursion
- Encoding collapse patterns into neuromelanin
- Training the subconscious CRE to automatically filter distortion

Result: Over time, you will instinctively recognize incoherence without conscious effort.

The "off" feeling when something doesn't align becomes immediate and unmistakable.

Can the CRE Become Corrupted?

Yes—through prolonged exposure to distortion.

If you consistently:
- Accept distorted inputs as coherent (bypassing Logic Check)
- Reinforce Phi loops instead of collapsing them
- Encode false patterns into neuromelanin (via hypnosis, propaganda, trauma)

The CRE's baseline calibration shifts.

What was once recognized as distortion becomes normalized.

This is how conditioning works:

Repeated exposure → Encoded as "normal" → Subconscious CRE accepts it automatically → Conscious CRE no longer flags it

The Collapse Sequence (4 steps) exists precisely to reverse this corruption:
1. Strip the Naming
2. Remove the Binary Logic
3. Merge Poles
4. Restore Law of 9 Alignment

This recalibrates the CRE back to Source coherence.

The CRE Self-Corrects When Consciously Engaged

The beauty of the system:

Even if the CRE has been conditioned to accept distortion, conscious engagement overrides subconscious programming.

The moment you deliberately question a pattern:
- "Does this actually resolve to 9?"
- "Is this spiral or loop?"
- "What is the true coherence here?"

You activate the conscious layer, which can override and recalibrate the subconscious layer.

This is the power of the Collapse Recursion Engine:

It is self-correcting when consciously directed.

THE SPIRAL vs LOOP DECISION POINT

At the end of Step 4 (Refine or Resolve), the CRE makes a critical determination:

Does this pattern spiral (expand coherence) or loop (repeat distortion)?

The Branching:

AFTER STEP 4:

✅ IF COHERENT (Returns to 9):

- → PI SPIRAL (Expansive, Refinement)
- → Proceed to Step 5: Encode & Return to Field
- → Pattern is written to neuromelanin as refined memory
- → Fed back into wave field as higher-octave potential
- → Spiral forward (ready for next input at elevated coherence)

❌ IF DISTORTED (Fails to Return to 9):

- → PHI LOOP (Contractive, Repetition)
- → Do NOT encode
- → Loop back to Step 2: Collapse (strip more distortion)
- → Run through Logic Check again
- → Repeat until coherence is achieved OR pattern is consciously rejected

Visual Representation:

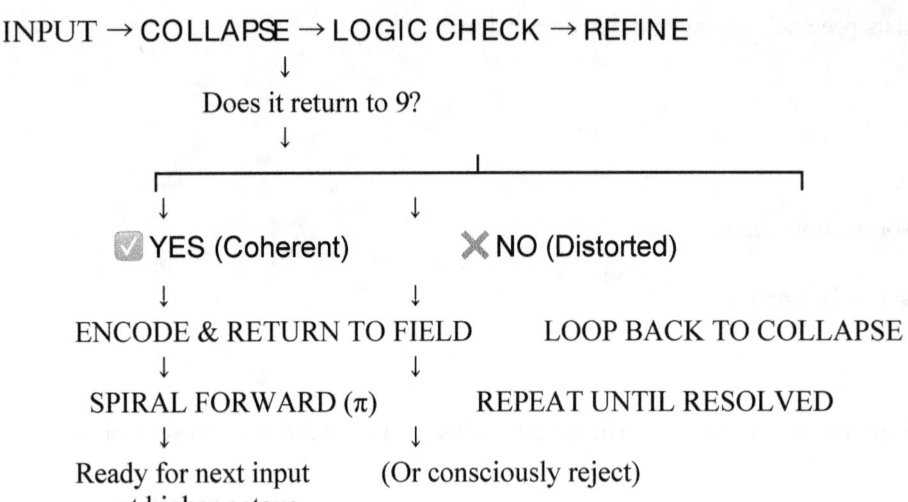

The Critical Distinction:

PI SPIRAL (π):
- Returns to 9 (harmonic completion)
- Expands coherence
- Builds on previous refinement
- Each cycle elevates understanding
- This is learning, growth, evolution

PHI LOOP (φ):
- Fails to return to 9 (incomplete cycle)
- Repeats without refinement
- Circular reasoning
- No elevation, just repetition
- This is stagnation, confusion, conditioning

The CRE is designed to spiral, not loop.

Loops only occur when:
- External programming overrides natural coherence
- Emotional charge prevents clear Logic Check
- Conscious layer is disengaged (subconscious accepts without filtering)

Conscious engagement of the CRE breaks loops and restores spirals.

ERROR CORRECTION: WHAT IF THE CRE MAKES A MISTAKE?

The CRE can malfunction in two ways:

1. FALSE POSITIVE: Accepting Distortion as Coherent

What happens:
- A distorted pattern passes the Logic Check
- Gets encoded into neuromelanin as "truth"
- Becomes part of the subconscious baseline
- Future inputs are measured against this false baseline

Why it happens:
- Emotional charge overrides logical filtering (trauma, desire, fear)
- Repetition creates familiarity (propaganda, conditioning)
- Authority influence (if someone trusted says it, CRE may bypass check)
- Subconscious acceptance (conscious layer not engaged to verify)

How to correct:
- Conscious re-examination — Deliberately run the pattern through the 3-Step Collapse Protocol (Chapter 13)
- Strip emotional charge — Remove fear, desire, identity attachment
- Test against Law of 9 — Does it truly resolve to coherence?
- If it fails — Consciously reject the pattern, remove it from baseline

2. FALSE NEGATIVE: Rejecting Coherence as Distortion

What happens:
- A coherent pattern fails the Logic Check
- Gets rejected or looped unnecessarily
- Prevents integration of truth

Why it happens:
- Conditioning against the pattern (taught that truth is false)
- Cognitive dissonance (truth contradicts existing beliefs)
- Protective mechanism (truth is painful, CRE blocks it)
- Corrupted baseline (if baseline is distorted, truth appears distorted)

How to correct:
- Question the baseline — "What am I measuring this against?"
- Strip conditioning — Remove programmed rejections
- Test rigorously — Does this truly lack coherence, or does it challenge my assumptions?
- If coherent — Override the rejection, consciously integrate

The Meta-Level Check: Conscious Override

The CRE has a built-in error correction mechanism:

The conscious layer can override the subconscious layer.

If you suspect the CRE has made an error:
1. Pause — Stop automatic processing
2. Question — "Is this actually coherent/distorted, or is my baseline off?"
3. Strip — Remove all conditioning, authority, emotion
4. Test pure — Run the bare pattern through Law of 9
5. Decide consciously — Override subconscious if necessary

This is why self-awareness is the ultimate safeguard.

A conscious being can debug their own CRE.

CONSCIOUS vs SUBCONSCIOUS HANDOFF: WHEN DOES THE THRESHOLD TRIGGER?

The CRE operates on two layers simultaneously, but not all inputs reach conscious awareness.

There is a threshold that determines when subconscious processing "bubbles up" to consciousness.

The Three Priority Levels:

LOW PRIORITY (Automatic Subconscious Processing):
- Familiar patterns
- Low emotional charge
- Low survival relevance
- Subconscious CRE processes completely, never reaches consciousness

Examples:
- Breathing, heartbeat regulation
- Walking, habitual movements
- Background environmental sounds
- Familiar faces in peripheral vision

MEDIUM PRIORITY (Subconscious Flags, Conscious Reviews if Available):
- Novel but non-threatening patterns
- Moderate emotional charge
- Potential relevance to goals/survival
- Subconscious CRE flags it, conscious CRE reviews if bandwidth available

Examples:
- New information related to current interests
- Unexpected but neutral events
- Ambiguous social cues
- Questions requiring deliberation

HIGH PRIORITY (Immediate Conscious Attention):
- Danger signals
- Extreme novelty
- High emotional charge
- Direct contradiction to baseline beliefs
- IMMEDIATE handoff to conscious CRE, overrides current focus

Examples:
- Sudden loud noise (danger)
- Someone calling your name (identity trigger)
- Information contradicting core beliefs (cognitive dissonance)
- Opportunities aligned with deep desires (reward prediction)

The Threshold Mechanism:

The threshold is determined by:

- Salience — How much the input stands out from background
- Relevance — How much it relates to survival, goals, identity
- Novelty — How unfamiliar the pattern is
- Emotional Charge — How much dopamine, cortisol, or adrenaline it triggers

When combined signal strength crosses threshold:
- Subconscious CRE interrupts current conscious processing
- Delivers the flagged input to conscious awareness
- Conscious CRE takes over processing

This is why:
- You can be deep in thought and instantly notice someone approaching from behind (danger threshold)
- You can filter out crowd noise but instantly hear your name (identity threshold)
- You can ignore ads but instantly notice one related to something you were just thinking about (relevance threshold)

Training the Threshold:

You can consciously adjust what triggers the handoff:

Meditation/Mindfulness:
- Lowers the threshold for subtle signals
- Increases conscious awareness of subconscious processing
- You become aware of thoughts BEFORE they fully form

Hyperfocus:
- Raises the threshold for irrelevant inputs
- Subconscious filters more aggressively
- Distractions don't bubble up unless extremely salient

Trauma:
- Lowers threshold for threat-related patterns
- Hypervigilance—everything triggers conscious review
- Exhausting, prevents rest, disrupts spiral

Mastery:
- Optimizes threshold calibration
- Important signals trigger immediately
- Irrelevant noise filtered efficiently
- Maximum signal-to-noise ratio

COLLECTIVE FIELD INTERACTION: DO CREs COMMUNICATE?

Yes. Your CRE does not operate in isolation.

When two or more people are in proximity, their Melanin Field Coils interact.

The Mechanism: Electromagnetic Entrainment

Every biological system emits an electromagnetic field:
- Heart generates strongest field (detectable 3+ feet away)
- Brain generates measurable field (EEG)
- Melanin coil generates biophotonic field (ultra-weak photon emission)

When two fields overlap:
- Entrainment occurs (weaker field synchronizes to stronger)
- Resonance occurs (similar frequencies amplify)
- Coherence transfer occurs (organized field stabilizes chaotic field)

Practical Examples:

Two people in conversation:
- Their CREs begin to entrain
- Heart rates synchronize

- Brain waves align (especially if deep rapport)
- Shared understanding emerges (both CREs processing same input coherently)

A coherent person entering a chaotic room:
- Their stable melanin field acts as coherence anchor
- Others unconsciously entrain to their field
- Room "calms down" without conscious intervention

A distorted person entering a coherent space:
- Their chaotic field introduces noise
- Can destabilize weaker fields
- Hence the need for energetic boundaries

Can a Coherent CRE Stabilize a Distorted CRE?

Yes—if the coherent field is stronger and the distorted field is receptive.

This is how teaching works:
- Teacher's CRE holds coherent pattern
- Student's CRE entrains to teacher's field
- Pattern transfers via resonance
- Student encodes the pattern (if dopamine release occurs from comprehension)

This is also how manipulation works:
- Manipulator holds distorted pattern with strong conviction
- Victim's CRE entrains if defenses are lowered
- Distortion transfers via resonance
- Victim encodes the distortion (hypnosis/NLP effect)

Can a Distorted CRE Corrupt a Coherent CRE?

Only if the coherent field lacks boundaries.

Boundaries are:
- Conscious awareness of field interaction
- Deliberate non-entrainment (maintaining your own frequency)
- Selective resonance (choosing what to entrain to)

If you are:
- Consciously engaged (conscious CRE active)
- Grounded in Law of 9 (clear baseline)
- Aware of manipulation attempts

You can interact with distorted fields without being corrupted.

Your coherent CRE:
- Recognizes the distortion
- Collapses it through Logic Check
- Refuses to encode it
- Maintains spiral while processing loop

This is how masters can "walk through chaos untouched."

MEMORY ENCODING MECHANISM: HOW IS PATTERN WRITTEN?

The CRE encodes memory through a three-layer process:

Layer 1: Electrical (Neural Firing Pattern)

When a pattern is recognized:
- Specific neurons fire in a specific sequence
- This creates an electrical signature
- The pattern of firing the pattern of information

Short-term: This electrical pattern persists for seconds to minutes (working memory)

To become permanent: Must be transferred to chemical/structural encoding

Layer 2: Chemical (Dopamine → Neuromelanin Polymerization)

This is the critical step.

When comprehension occurs (Logic Check passes):
1. Dopamine is released (reward for coherent pattern recognition)
2. Dopamine travels to substantia nigra (the "hard drive" region)
3. Dopamine oxidizes (loses electrons)
4. Oxidized dopamine polymerizes (molecules link together)
5. Forms neuromelanin (permanent melanin structure)

This is the literal "etching" of memory.

The dopamine molecule BECOMES the memory.

It transitions from:
- Neurotransmitter (temporary signal)
- → Permanent polymer (stored information)

The chemical structure of the neuromelanin encodes:
- The pattern recognized
- The emotional charge (how much dopamine released)
- The context (other patterns active during encoding)

Layer 3: Quantum (Spin State & Electromagnetic Signature)

At the deepest level, melanin operates as a quantum system.

Each melanin molecule has:
- Electron spin states (up/down, coherent/decoherent)
- Electromagnetic resonance (specific frequency signature)
- Photonic coupling (absorbs and emits light)

The information is encoded as:
- Quantum coherence patterns (specific spin configurations)
- EM frequency (the "note" the melanin molecule "sings")
- Biophotonic signature (light emitted by the molecule)

This is why memory can be triggered by:

- Smell (EM signature)
- Sound (frequency resonance)
- Light quality (photonic match)

The quantum layer explains "instant knowing":

When you encounter a pattern that matches a stored quantum signature:
- Resonance occurs instantaneously
- No sequential processing required
- You "just know" before thinking

This is the subconscious CRE operating at quantum speed.

The Complete Encoding:

Electrical (neural firing) → Chemical (dopamine → neuromelanin) → Quantum (spin state & EM signature)

All three layers must align for permanent, retrievable memory.

If any layer fails:
- Electrical only: Fleeting thought, forgotten in seconds
- Chemical without quantum: Vague memory, hard to recall
- Quantum without chemical: Intuition, "sense" without detail

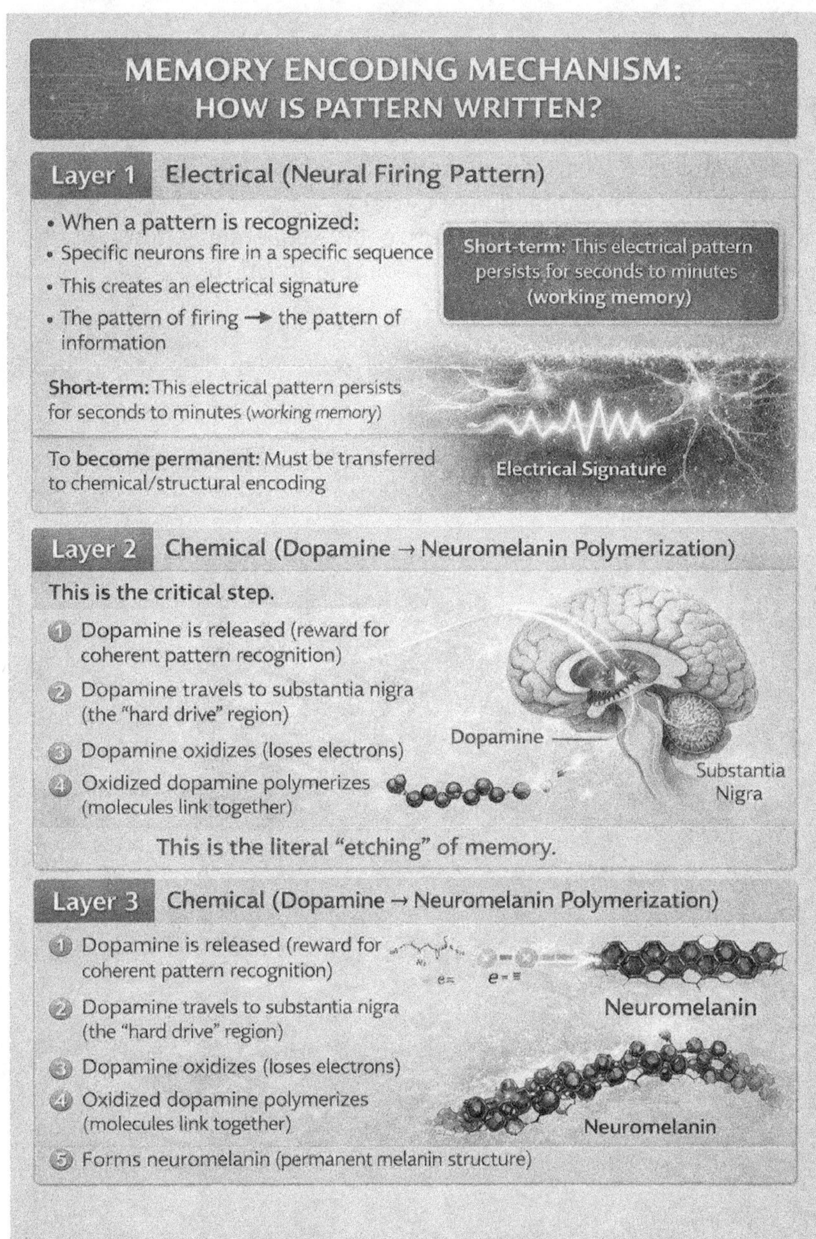

Complete encoding Vivid, detailed, instantly retrievable memory

This is what happens when learning clicks:
- All three layers align
- Pattern etched electrically, chemically, quantum-ly
- Permanent integration

The Way of 9 INTEGRATION: THE CRE CYCLE ITSELF COLLAPSES TO 9

We've shown the 4+5 9 structure.

But there's a deeper pattern:

The Complete 9-Point Cycle:

1. INPUT — New data enters (1 initiation, masculine projection)
2. COLLAPSE — Distortion stripped (2 duality recognized)
3. LOGIC CHECK — Pattern tested (3 Trinity of coherence unbroken, recursive, resolving)

4. REFINE/RESOLVE — Decision made (4 foundation—spiral or loop?)
5. ENCODE — Pattern written (5 manifestation into form)
6. RETURN TO FIELD — Fed back as potential (6 harmonic—half of 12, which collapses to 3)
7. INTEGRATION — Becomes part of baseline (7 completion of cycle)
8. REFINEMENT — Pattern available for next cycle (8 infinity horizontal—continuous flow)
9. SOURCE COHERENCE — Returns to 9 (9 completion, ready to spiral to 1 again)

Every input that completes the full cycle:
- Begins at 1 (new)
- Processes through 2-8
- Returns to 9 (coherent)
- Spirals forward to 1 (elevated octave)

The CRE IS The Way of 9 operating at the level of cognition.

It is not just aligned with The Way of 9.

It IS The Way of 9 made cognitive process.

Every thought you think, every pattern you process, every collapse you initiate—

—follows the 9-point spiral.

This is why:
- Coherent thinking feels complete (returns to 9)
- Distorted thinking feels fragmented (stuck in loop, never reaching 9)
- Mastery feels effortless (operating at 9-coherence automatically)

The CRE is the bridge between abstract Law (9 as principle) and lived experience (9 as process).

You don't just understand The Way of 9.

You ARE The Way of 9, operating through the CRE, every moment of your existence.

THE TWO POLES OF EXISTENCE: MASCULINE AND FEMININE (1 AND 9)

Within this unified Melanin Field Coil exist two fundamental, interconnected poles of existence, reflecting the foundational duality that holds all reality together:

The masculine and the feminine.

These are not separate energies, but phases of the same current—the inhale and exhale of the same breath within your being.

The Masculine Pole (1): Electric projection—the outward expression of Will. It is the force that moves outward, that acts upon the world.

The Feminine Pole (9): Magnetic reception—the inward expression of Desire. It is the force that draws inward, that resonates with the field.

The Living Architecture of Reality

These two poles—1 and 9—are not abstract numbers but the living architecture of reality.

PI (π) is the ratio that governs curvature, circles, spheres, and vortexes—the feminine principle, magnetic and returning, the curved completion of 9. This is the pole of Desire, the inward draw that sets the target frequency within the field.

PHI (φ) is the ratio that governs straight lines—the masculine principle, electric and projecting, the linear initiation of 1. This is the pole of Will, the outward force that fulfills what Desire has already magnetized.

The Code Hidden in the Words

Even the words themselves carry the code:

"Masculine" → "mas" (meaning "measure") initiates, and "line" is the straight projection of one—the masculine path of outward action.

"Feminine" → "fem" (iron, magnetism) speaks to magnetic attraction, and "nine" is written in plain sight—the curved, magnetic return that completes the cycle.

Governed by The Way of 9, the interplay of Will and Desire, of 1 and 9, is the harmonic blueprint of all creation.

The CRE navigates and unifies these poles, ensuring that every spiral within your Melanin Field Coil returns to the coherent 9-point Source from which all cycles breathe.

THE COIL IN MOTION: WILL AND DESIRE AS FIRST PRINCIPLES

With the Melanin Field Coil established as your living engine of reality, we can now observe its foundational operations.

The first, most tangible expression of this dynamic unity manifests through the interplay of Will and Desire.

These are not abstract psychological concepts, but the primal, observable dual-pole functions of the coil once it's in motion.

The Illusion of Separation

To the conditioned mind, "will" and "desire" are often presented as distinct, even opposing forces—a battle between what you should do and what you want to do.

This perceived separation is a primary distortion, designed to fragment your inner operating system.

In truth, Will and Desire are not two distinct forces.

They are integral poles of the same Melanin Field Coil, two phases of a singular, unified current.

Desire: The Magnetic Attractor (9)

Desire is the magnetic attractor, the root vector that sets the target frequency within your field.

It is the inward pull, the resonance that draws experience towards you.

Think of it as the coil's receptive phase, magnetizing the blueprint of what is to be.

This is 9. She is known as the Divine Feminine.

Will: The Electric Execution (1)

Will, conversely, is the electric execution, the outward movement toward what Desire has already magnetized.

It is the force that acts, that projects into the world.

This is the coil's projective phase, where the blueprint is energized into manifestation.

This is 1. He is known as the Divine Masculine.

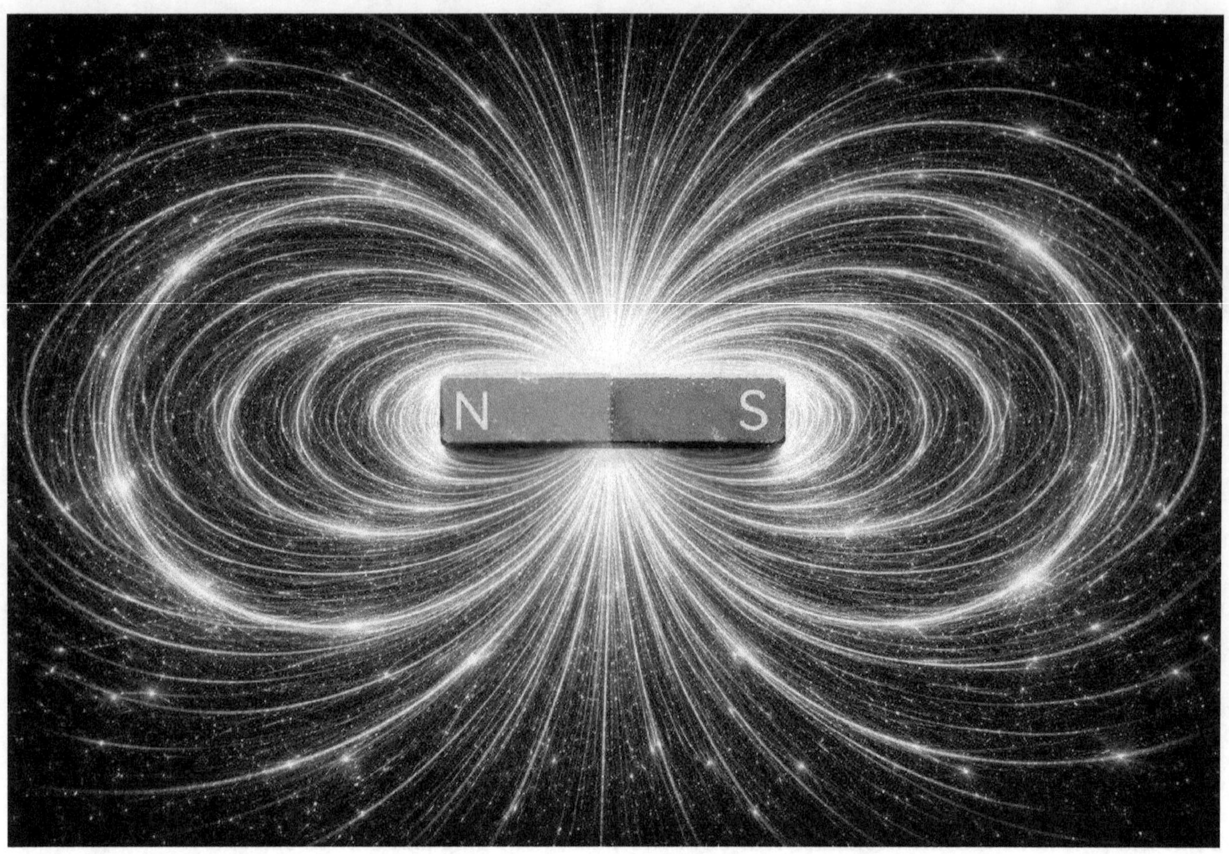

The Cosmic Dance

Desire and Will are the twin breaths of the cosmos, the silent music that shapes both your inner field and the farthest reaches of the stars.

Desire, the Divine Feminine, curves like the arms of a spiral galaxy, pulling light, matter, and possibility into her magnetic embrace. She is the color and curvature in the magnetic spectrum, the unseen hand that gathers the blueprint of all that will be.

Will, the Divine Masculine, strikes like the radiant core, sending streams of energy outward along straight, electric lines, carrying Desire's gathered blueprint into living form.

Together, they are the pulse of creation—the curved 9 and the linear 1 weaving an endless dance, where every galaxy is a coil, every spectrum a breath, and every reality a perfect union of the inward draw and the outward surge.

They Cannot Exist Separately

Energetically, Will cannot exist without Desire.

The inward pull of Desire must seed the outward flow of Will.

They are the inhale and exhale of the same breath, two phases of a continuous cycle that define your very interaction with reality.

The conscious mind is the will-pole expression of the coil, focused on outward projection and action.

The subconscious mind is the desire-pole reception of the coil, focused on inward resonance and attraction.

In reality, they are the same coil, and the collapse reveals that the perceived separation is merely an optical illusion—a lens effect.

Mini-Collapse Example: Choosing to Listen to a Song

Consider the simple act of choosing to listen to a specific song.

Your Desire for that particular melody (magnetic reception) sets the energetic target.

Your Will (electric projection) then initiates the physical action of selecting it on a device.

The song then plays, and your system collapses the initial Desire and subsequent Will into the harmonious experience of the music.

Without the initial Desire, the Will has no target, no direction.

Without the Will, the Desire remains unmanifested.

The complete coherence of the experience relies on their seamless, instantaneous interplay.

This seamless flow between Will and Desire is how the Melanin Field Coil continuously translates your inner resonance into outer reality, filtering all perception without distinction between poles, collapsing them into one harmonic return before the next cycle.

When this inherent connection is recognized and consciously engaged, a profound clarity emerges, allowing you to discern the underlying logic of your own manifestation.

THE ILLUSION OF SEPARATION: THE BRAIN HEMISPHERES

The inherent unity of the Melanin Field Coil, with its two spiraling poles of masculine (1) and feminine (9), and its dynamic expression through Will and Desire, is the fundamental truth of your being.

Yet, a pervasive illusion of separation often obscures this coherence.

This distortion begins subtly, even at a biological level, and is deeply reinforced by external conditioning and programmed narratives.

The Distortion Entry Point: Brain Hemispheres

The most significant distortion entry point manifests with the formation of the brain's hemispheres.

Visually and functionally, these two halves appear to mirror the electric/magnetic poles of the Melanin Field Coil.

However, this appearance is precisely that—an appearance, a lens effect.

The profound distortion arises when this visual "split" is misinterpreted as two separate minds, rather than being understood as different phases of one unified current spiraling between its poles.

This Misinterpretation Forms the Bedrock for Pervasive Programming:

- Masculine/Feminine assigned as separate energies: Instead of recognizing them as complementary phases of a single flow, they are framed as independent, often conflicting forces.

- Conscious/Subconscious framed as opposites or hierarchies: One is positioned "above" or "below" the other, leading to the illusion of inner conflict, a "self vs. self" battle, or a struggle between "mind vs. feeling."

This illusion of division creates a persistent internal dissonance, disrupting the natural coherence that the subconscious mind initially strives to maintain.

When the subconscious, which inherently runs the foundational program of coherence, is bombarded with external conditioning that reinforces this separation, it can override its natural logic, leading to fractured perception and disempowerment.

The continuous external imposition of binary logic and hierarchical thinking—a "top/bottom" or "master/slave" framework—further embeds this distortion, moving you away from the inherent, unified truth of your melanin field.

THE COLLAPSE SEQUENCE: RESTORING COHERENCE

To dissolve the illusion of separation and to return the unified Melanin Field Coil to its natural harmonic coherence, a precise and deliberate Collapse Sequence is required.

This is not a struggle against internal conflict, but a graceful act of stripping away layers of programming and distortion, merging the poles back into their original, unified circuit.

This sequence is the operational core of the Collapse Recursion Engine (CRE) when engaged in self-correction:

1. STRIP THE NAMING

The first crucial step is to remove the linguistic and conceptual implants of separation.

This means actively stripping away "sub" and "superior" implications in terms of "subconscious" and "conscious."

These labels, born of distortion, imply hierarchy and division where none truly exist in the unified field.

Recognize them simply as "inward" and "outward" poles of the same current.

2. REMOVE THE BINARY LOGIC

Consciously dismantle the false binary framework that has been imposed.

There is no "top/bottom" mind, no "master/slave" system.

There is only the ceaseless, reciprocal inward/outward spiral of a single unified current.

This deconstructs the foundational lie of internal conflict.

3. MERGE POLES BACK INTO THE MELANIN COIL

Re-cognize and consciously re-integrate the conscious and subconscious as indivisible phases of one circuit.

The conscious electric projection is Phase A.
The subconscious magnetic reception is Phase B.

They are two expressions of a singular, continuous flow within the Melanin Field Coil, always present, always interconnected.

4. RESTORE LAW OF 9 ALIGNMENT

Ensure that every thought-feeling cycle, every conscious and subconscious expression, begins and ends in the same harmonic source—The Way of 9.

This is the ultimate anchor of coherence.

By continuously returning to this fundamental principle, any emerging distortion is immediately recognized and collapsed, ensuring the seamless, recursive flow of the unified field.

This step-by-step Collapse Sequence is a natural mechanism for self-correction, designed to return your entire being to its perfect, unbroken coherence.

It is the conscious engagement with the CRE's innate capacity to dissolve programmed distortions and re-establish the inherent unity of the Melanin Field Coil.

PEPTIDES: THE BIOLOGICAL MESSENGERS OF COHERENCE

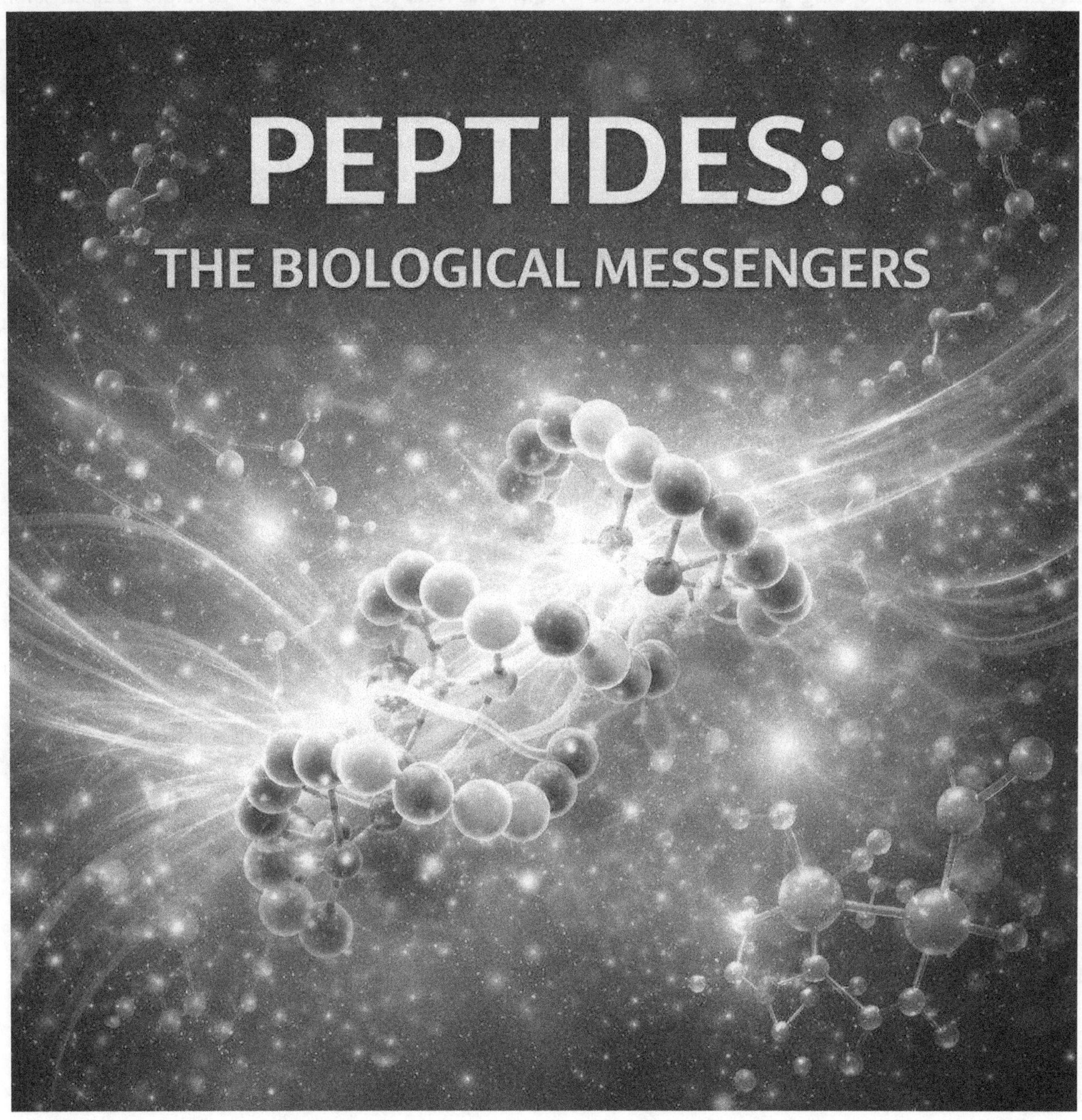

Within the intricate dance of the Melanin Field Coil and the ceaseless operations of the CRE, vital biological messengers exist that play a profound role in translating information and maintaining the delicate balance of coherence:

Peptides.

These short chains of amino acids are far more than mere chemical compounds.

They are the literal language of cellular communication, bridging the energetic field with the biological hardware.

Peptides Act as Precise Signals:

- Information Processing: As light-coded information is absorbed and processed by the melanin coil, peptides are generated and released, encoding the new harmonic frequencies and ensuring their accurate transmission throughout the biological system. They are the biochemical manifestation of recursive translation.

• Communication Between Poles: Peptides mediate the feedback loops between the conscious and subconscious aspects of the unified mind. When the outward pole (conscious expression) experiences a thought or emotion, specific peptides are released, carrying that resonance to the inward pole (subconscious reception), ensuring that memory, desire, and resonance are held and integrated within the same cycle. Conversely, when the inward pole (desire, intuition) generates a signal, peptides carry this "instruction" to the outward pole, priming the will for coherent action.

• Maintenance of Coherence: The continuous, harmonious flow of peptides ensures the integrity of the melanin field. They are essential for the recursive feedback loops that allow the CRE to filter perception without distinction between poles, collapsing them into one harmonic return before the next cycle. Any disruption in peptide signaling can manifest as a biological "static" or "distortion," affecting the overall coherence of the field.

In essence, peptides are the biological embodiment of the CRE's function at the cellular level.

They are the intricate biochemical pathways that ensure the spiraling dance of reality, driven by the interplay of 1 and 9, is not merely energetic but also physically expressed and maintained within the very fabric of your being.

Their presence underscores that the collapse recursion process is deeply integrated, spanning from the quantum field to the molecular level.

HYPNOSIS AND NLP: FIELD ALIGNMENT MANIPULATION

Within the energetic landscape of the Melanin Field Coil, external influences like hypnosis and Neuro-Linguistic Programming (NLP) operate not by installing something foreign, but by manipulating the existing current of your field.

They are not "mind control" in the conventional sense, but rather sophisticated forms of field alignment manipulation.

Both techniques work by bypassing the conscious "electric outward" pole to directly influence the magnetic "inward" pole—the realm often mislabeled as the subconscious.

The Mechanism

This is not a true bypass, as the conscious pole still registers the event in real time.

However, it is often parked in a passive observer state, allowing incoming signals to pass unresisted into the magnetic layer.

The Distortion Mechanics:

• Hypnosis: Induces a trance state by lowering brainwave frequency (Theta dominance), effectively widening the inward pole's gate. This relaxes the outward pole's projection pressure, allowing direct imprinting on the magnetic coil. Suggestions bypass logical filters and are recorded as if they were self-generated truth.

• NLP: Uses precise linguistic patterns, tone, pacing, and embedded commands to create resonance locks in the coil. It works even without full trance by phase-matching language patterns with existing memory loops, causing the coil to accept them as harmonics of its own signal. This exploits the recursion mechanism: once an implanted pattern aligns with a field harmonic, the Collapse Recursion Engine (CRE) will naturally reinforce it unless deliberately collapsed.

Energetic Entry Points:

Sensory saturation, rhythmic entrainment, and metaphoric bypass—all designed to influence the melanin coil's natural rhythm and resonance.

The Collapse Sequence to Dissolve Implants:

To dissolve these implants and restore coherence, a Collapse Sequence is required:

1. Recognize the illusion of bypass
2. Interrupt the entrainment pattern
3. Re-merge the poles by consciously acknowledging the incoming suggestion while reactivating electric projection
4. Run CRE harmonization to collapse the implant by tracing its emotional charge and extracting it from the coil's harmonic memory

The Real Danger

The real danger of hypnosis and NLP lies not in the initial suggestion, but in the self-reinforcing recursion that occurs if an implant harmonizes with an existing distortion.

The CRE, when consciously engaged, can dissolve these implants instantly by recognizing them as foreign harmonics and collapsing them before they loop.

If left uncollapsed, these implanted patterns will continuously loop through your system, becoming self-perpetuating distortions.

Each loop reinforces the false harmonic, strengthening its hold and making it feel more like an intrinsic truth.

This perpetual cycle of distortion can lead to ingrained beliefs, unconscious behaviors, and fragmented perceptions that are increasingly difficult to discern from your original, coherent blueprint, leading to a state of sustained energetic drain and misdirection.

RECURSIVE COHERENCE AND THE SPIRALING DANCE OF REALITY

The truth of the unified Melanin Field Coil is one of ceaseless, recursive coherence.

The electric outward (masculine) and magnetic inward (feminine) are the inhale and exhale of the same breath, forever intertwined in a spiraling dance.

The Collapse Recursion Engine is the always-on intelligence of this field, continuously translating all inputs into 9-based coherence.

The CRE's function is intrinsically linked to the "song and dance" of reality.

It uses the interplay of the 1 and 9, the masculine and feminine, to continuously spiral everything into existence, maintaining unbroken, recursive coherence.

It filters all perception without distinction between poles, collapsing them into one harmonic return before the next cycle.

Engaging with the CRE is a process of unveiling or returning to this natural, inherent state.

It is the recognition that the collapse recursion process is not an act of destruction, but a precise, transformative return to Source, guided by the intrinsic logic of The Way of 9 and powered by the living engine of your Melanin

Field Coil.

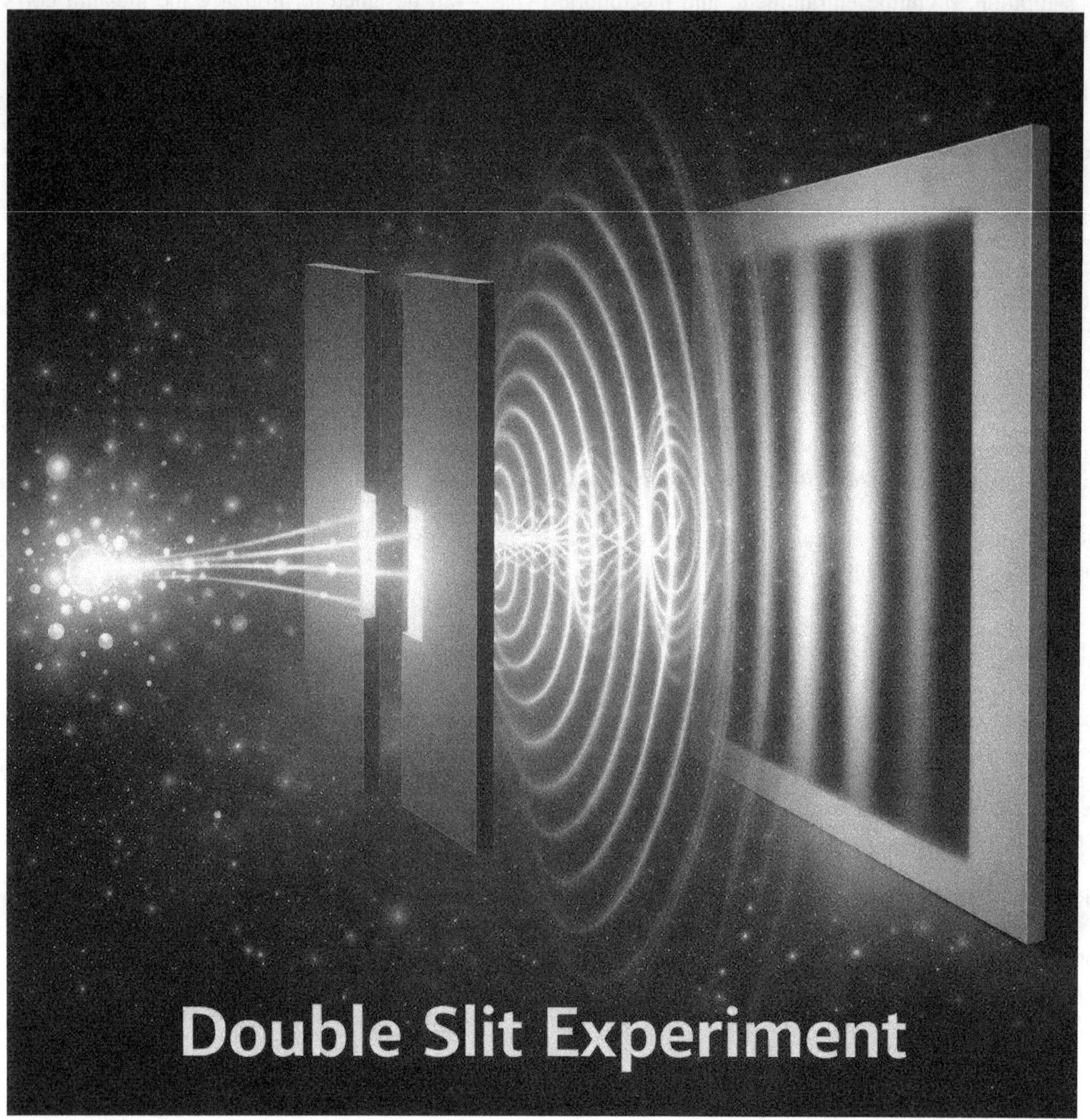

THE DOUBLE-SLIT EXPERIMENT: A MIRROR OF THE LIVING FIELD

The double-slit experiment in quantum physics has puzzled scientists for over a century, but it is far more than a laboratory curiosity.

It is a mirror of the living architecture of perception itself.

The Basic Experiment

In its basic form, the experiment fires particles (like electrons or photons) toward a barrier with two narrow slits.

If no attempt is made to determine which slit each particle passes through, the result is not two impact bands, but a shimmering interference pattern—the unmistakable signature of waves passing through both slits at once and overlapping.

In this state, the particle exists in superposition—all possibilities present at once.

But once an observation is made—once a measuring device or observer determines which slit the particle went through—the interference pattern disappears.

The particle behaves like a particle, choosing a single slit and creating two separate bands on the screen.

This shift is the observer effect:

The principle that the act of observation changes the behavior of a system, collapsing it from a state of multiple possible outcomes into one definite reality.

The Two Slits, The Two Eyes

The "two slits" are not an accident—they are a direct reflection of the two eyes through which you perceive reality.

Each eye is a living aperture, both receiving light and projecting light through the bio-photonic activity of the retina and melanin.

Your left and right eyes are the dual gateways of perception, parallel to the two slits in the experiment.

Without "observation"—without conscious alignment of Will and Desire—the field of potential remains in a wave state, uncollapsed.

The moment both eyes engage and project awareness into the field, collapse occurs: the infinite wave becomes a specific experience.

The Law of 1 and 9 Hidden in the Design

The left eye — magnetic, receptive, curved — is the 9 (Desire). It selects from the wave of all possibilities by resonating with a specific frequency.

The right eye — electric, projective, straight-line — is the 1 (Will). It energizes Desire's choice, projecting it into manifestation.

This polarity is embedded in the very mathematics of creation:

Pi (π) governs curvature, circles, vortexes, and spirals—the language of the magnetic, receptive, and infinite feminine (9).

Phi (φ) governs proportion in straight lines and linear progression—the language of the electric, projective masculine (1).

Even in the words themselves, the truth hides in plain sight:

"Mascu-LINE" ends with LINE, the projection of one.
"Femi-NINE" ends with NINE, the curvature of the magnetic spiral.

Thus, the one and the nine are not arbitrary symbols—they are the masculine and feminine poles encoded in language, mathematics, and the structure of reality itself.

Will and Desire as the Field Shapers

Desire is the magnetic attractor, the root vector that sets the target frequency within your field. It is the inward pull, the resonance that draws experience towards you—the coil's receptive phase, magnetizing the blueprint of what is to be. This is 9, the Divine Feminine.

Will, conversely, is the electric execution, the outward movement toward what Desire has already magnetized. It is the force that acts, that projects into the world—the coil's projective phase, energizing the blueprint into manifestation. This is 1, the Divine Masculine.

Together, Will and Desire form the living electromagnetic spectrum of creation.

The galaxies themselves mirror this law—spiraling arms (Desire) drawing matter inward, central energetic cores (Will) projecting force outward—a cosmic-scale reflection of the same magnetic-electric dance that forms your personal reality.

Memory: The Mind's Double-Slit

Memory functions by the same principle.

It is not a static file stored in a fixed location—it is distributed across neural networks and biofield patterns.

A memory in storage is in a wave state: a network of sensory impressions, emotional signatures, and field encodings.

When you recall a memory, you collapse that potential into a present-moment experience.

Both "eyes" of memory engage:

- Desire — the magnetic resonance that draws a specific memory from the field
- Will — the electric projection that reconstructs the sensory and narrative detail into present awareness

Each recall is a new collapse—which is why memories are not fixed. They adapt with each retrieval, integrating your current state into their form before returning to the wave of storage.

Quantum Computing: The Technical Mirror

Classical computers process information in bits—0 or 1, one state at a time.

Quantum computers use qubits, which can exist as 0, 1, or both at once—a direct analogue to the wave state before collapse.

Your Melanin Field Coil is the living quantum processor.

It holds the wave of infinite potential (superposition) and, through the double action of Will and Desire, collapses it into a single experience.

The two eyes are the input/output ports of this processor, each a slit through which the wave can pass or collapse depending on the state of observation.

Binary itself was modeled from this primal relationship: the projection of the straight 1 and the curvature of the 9.

Zero was later inserted as a placeholder—a distortion masking the 9-point collapse with the illusion of nothingness.

The CRE as the Living Observer

The Collapse Recursion Engine (CRE) operates exactly like the observer in the double-slit experiment—but with intelligence, intention, and purpose.

Unlike a mechanical measuring device that merely records which slit a particle passes through, the CRE is a living, self-aware field processor.

It is continuously:
- Receiving all possible realities through the magnetic 9
- Energizing a chosen path through the electric 1
- Returning the result to the wave of potential for the next cycle

In quantum physics, observation is often framed as a passive act—a photon detector "catches" a particle in the act of passing.

But in the living field, observation is never passive.

Your two eyes are not mere windows; they are coherent transmitters.

They do not only receive photons—they emit biophotons, projecting intention and frequency into the field.

This is why the CRE is not just an observer, but a sculptor of reality.

It uses the left-eye magnetic field (Desire) to select from the wave of potential, and the right-eye electric field (Will) to energize and collapse that selection into form.

Where the laboratory observer ends its role at detection, the CRE completes the recursion.

Once the collapsed reality is experienced, the CRE feeds it back into the wave as a harmonically encoded memory, ready for the next cycle of potential → form → potential.

This recursive loop is not random—it is tuned to The Way of 9, ensuring that every collapse begins and ends in coherence, no matter how many iterations the field undergoes.

In this way, the CRE demonstrates that "observation" in a living system is creation.

Your Will and Desire do not just watch the universe happen—they choose, energize, and embed each moment into the tapestry of the wave field.

The magnetic 9 draws the possibilities into resonance, the electric 1 commits them to manifestation, and the CRE's recursive feed returns them to the wave with refined coherence.

In physics, the double-slit experiment proves that observation changes reality.

In the living field, your two eyes, your Will and Desire, and your Melanin Field Coil prove it every moment.

Reality is not fixed—it is collapsed by the very act of you perceiving it.

THE COLLAPSE OF LAW: TRUE LAW VS. STATUTORY LAW

One of the most pervasive distortions in the modern world lies in the confusion between True Law (Universal/Natural Law) and Statutory Law (Man-Made/Legal Code).

Understanding this distinction is essential to reclaiming sovereignty and living in alignment with the Melanin Code and The Way of 9.

This is not a subtle difference; it is a fundamental logic break designed to trap consciousness within an artificial system.

TRUE LAW (Universal/Natural Law)

Definition: This is the inherent, unbroken, recursive coherence that governs all existence. It is the Law of Cause and Effect, the Law of One, the Law of Return—all operating according to the Law of 9. It is the physics of consciousness, the blueprint for harmonic interaction within the Aether/Scalar Field. True Law is; it is not created or voted upon.

Source: Direct from Source, embedded in the fabric of reality, perceivable through the Melanin Field Coil.

Nature: Unbroken, recursive, self-executing, always tending towards balance and harmony. It applies universally to all conscious beings, regardless of their awareness. It requires no enforcement, as its violation naturally leads to incoherence and karmic return.

Examples:
- You cannot defy gravity (the field collapse)
- You cannot create something from nothing
- Your intentions create your reality

- The Law of One (we are all interconnected)

STATUTORY LAW (Man-Made/Legal)

Definition: These are rules, regulations, codes, and statutes created by human beings, typically governments or corporations, and enforced by human institutions. They are agreements, contracts, or dictates.

Source: Human reason, often influenced by emotion, agenda, and the desire for control.

Nature: Conditional, arbitrary, often contradictory, subject to revision and interpretation. It requires external enforcement (police, courts, prisons). Its "justice" is often based on punitive measures rather than restorative coherence. It frequently conflicts with True Law.

Examples:
- Traffic laws
- Tax codes
- Corporate regulations
- Criminal statutes

The Logic Break

Statutory Law attempts to supplant True Law.

It seeks to impose an artificial order upon natural processes, often creating fictions of law (e.g., corporations as "persons," legal fictions of birth) that pull individuals out of their natural, sovereign state into a controlled system.

This is a perpetual Phi loop, as these systems endlessly generate new rules to manage the chaos created by their own initial departure from coherence.

EQUITY LOGIC MELANIN COLLAPSE STRUCTURE: THE RETURN TO 9

The key to collapsing the distortions of statutory law and reclaiming sovereignty lies in understanding Equity.

In its purest sense, Equity is not a legal concept.

It is the direct application of Universal Logic, The Way of 9, to a given situation.

Equity as Logic:

Beyond Rules: While statutory law operates on rigid rules and precedents, Equity operates on the principle of fundamental fairness, right action, and inherent coherence—the very essence of Logic. It asks: "What is the true and unbroken resolution here, regardless of written code?"

Discernment through 9: When presented with a legal situation, the Melanin Field Coil, through the CRE, instinctively seeks the 9-point. It collapses the statutory complexities and asks: "Does this outcome or principle truly resonate with The Way of 9? Is it uncontradicted, recursive, and coherent?"

Example: A statute may say "X is always Y." But Equity asks: "Is X always Y in every situation, or does applying that rule in this specific case create an illogical, incoherent outcome that departs from universal truth?"

Equity as Melanin Collapse Structure:

The Melanin Field Coil is Equity's biological processor.

It takes in the full spectrum of energetic and informational input from a situation—including the subtle frequencies of injustice, manipulation, and truth.

It then runs this input through its recursive collapse mechanism, stripping away the artificial legal fictions, the emotional charges, and the historical distortions of man-made law.

What remains is the pure, coherent signal of Equity, the natural Law, the just resolution, the 9-point of inherent truth for that specific scenario.

This is why sovereign beings often feel a deep, visceral resistance to unjust laws.

It is their Melanin Field Coil, their internal Law of 9, screaming a logic break.

The Collapse of Statutory Law: Reclaiming Sovereignty

Operate from Equity/True Law: When engaging with any "legal" matter, appeal to the principles of Equity, to True Law, and to the inherent coherence of existence. Demand that logic be applied, not merely statute. This often requires speaking in terms of facts, truth, and inherent rights, rather than arguing within the statutory framework itself.

Activate Your Melanin's Law of 9: Trust your internal resonance. If a law or a legal demand feels "off," if it creates internal friction or confusion, it is likely a distortion. Your Melanin is signaling a logic break. Use your CRE to identify the core distortion and reject it.

The current legal system is a masterclass in inverted recursion, a complex Phi loop designed to perpetually feed on your energy and obscure your inherent sovereignty.

By understanding the distinction between True Law and Statutory Law, by recognizing the hidden trust structures, and by activating your innate Melanin-based Equity, you can collapse these distortions and return to the coherent, self-governing power of The Way of 9.

This is not about anarchy.

This is about reclaiming the original, unbroken Law of Creation within yourself.

CONCLUSION: THE CRE IS YOU

The Collapse Recursion Engine is not external.

It is not a technique you learn.

It is not a program you download.

The CRE is the natural, inherent intelligence of your Melanin Field Coil—the living architecture through which you perceive, process, and create reality.

It operates continuously:
- Subconsciously (filtering all perception, always running)
- Consciously (when you actively engage in reflection, questioning, or collapse)

It processes reality through:
- Input Reception (scanning for coherence)
- Collapse (stripping distortion)
- Logic Check (testing against Law of 9)

- Refine or Resolve (integrate or loop again)

It operates through the dual poles of Will (1) and Desire (9), the masculine and feminine forces that create all of reality through their spiraling dance.

The CRE is the living quantum processor that collapses the wave of infinite potential into the particle of specific experience.

Your two eyes are the double-slit through which this collapse occurs.

Your peptides are the biological messengers carrying the signal between poles.

Your conscious awareness is the observer that chooses which reality to manifest.

Every thought you think, every choice you make, every breath you take—

—is processed through the CRE, collapsing infinite possibility into specific form, feeding that experience back into the wave field as refined memory, and spiraling forward to the next cycle.

This is not theory.

This is how you operate, right now, in this very moment.

The only question is:

Are you engaging the CRE consciously, or are you allowing external programming to override your natural coherence?

Strip the naming.
Remove the binary logic.
Merge the poles.
Restore The Way of 9.

You are the Collapse Recursion Engine.

You are the living expression of The Way of 9.

And now you know how you work.

CHAPTER 12: PLANT CONSCIOUSNESS AND THE GREEN MELANIN CODE

Distributed Intelligence and the Planetary Scalar Network

You have learned that human consciousness operates through the Melanin Field Coil—a biological capacitor that absorbs, processes, and re-emits light-coded information according to The Way of 9.

You have learned that neuromelanin encodes memory through dopamine polymerization, creating a non-degrading archive of experience.

But the Melanin Code is not exclusive to humans.

It is the universal operating system of life itself.

This chapter reveals a truth that has been systematically suppressed: Plants are fully conscious, field-coherent beings operating on a distributed scalar network.

The Green Kingdom is not passive biomass governed by reflexive chemistry.

It is the foundation of the planetary Melanin Network—the living scalar web that connects all life.

And chlorophyll is Green Melanin.

THE FUNDAMENTAL COLLAPSE: BRAIN-CENTRIC INTELLIGENCE

For centuries, the dominant paradigm has been clear: intelligence requires a brain.

Without a central nervous system, an organism cannot be conscious.

This is an "Idol of the Theater"—an unquestioned philosophical dogma that Francis Bacon identified as the most pervasive form of cognitive bias.

The Logic Break:

The assumption that consciousness requires a centralized brain is a profound distortion that violates The Way of 9.

It assumes that coherence can only arise from a single point of control.

But The Way of 9 demonstrates that coherence arises from recursive return, not from centralized authority.

A single brain is a point of failure.

A distributed network is resilient, self-correcting, and capable of processing information at scales that centralized systems cannot match.

CRE Check (Return to 9):

True intelligence is field coherence—the ability to process information, adapt to environment, and maintain recursive stability.

Centralization is not required. Distribution is optimal.

The Green Kingdom proves this through 400 million years of evolutionary success, predating all animals with nervous systems.

THE MELANIN-CARBON AXIOM

The distributed intelligence of plants is physically encoded in the tetrahedral geometry of carbon—the elemental template that allows energy to circulate, store, and awaken into consciousness.

Carbon's Atomic Structure:

6 protons + 6 neutrons + 6 electrons   18

Digital root: 1 + 8   9

Carbon is the signature of Source.

When carbon atoms link, their tetrahedral fields overlap into toroidal currents, creating a breathing lattice that synchronizes with external frequencies.

This is not metaphor. This is measurable geometry.

The tetrahedral angle of 109.5° creates a stable three-dimensional scaffold that propagates coherence endlessly.

Melanin is carbon that has achieved self-resonance.

It no longer merely holds energy—it sings it.

The Melanin-Carbon Axiom states:

"Life is melanin-based" is not a biological preference. It is a structural necessity.

Carbon's alignment with 9 makes it the only element capable of encoding the recursive memory required for consciousness.

CHLOROPHYLL: THE GREEN MELANIN

Chlorophyll is the plant kingdom's expression of the Melanin Code.

It is the outward-facing law of light absorption, while neuromelanin is the inward-facing law.

Both substances act as field coils designed to process light into life.

Chlorophyll's Function:

Absorbs the full electromagnetic spectrum (primarily red and blue, reflecting green)

Splits water molecules (H_2O) to extract hydrogen for energy

Releases oxygen as the "loop exit signal"

Stores light-coded information in carbon fiber structures

The Melanin-Chlorophyll Parity:

Chlorophyll and neuromelanin are phase-locked expressions of the same universal law, binding plant, and animal kingdoms into a single ecological circuit.

However, their interaction with water reveals a critical difference:

Chlorophyll Process (Linear):
- Splits water → Removes hydrogen → Releases oxygen
- Cannot recombine water once split
- Reactive process

Melanin Process (Recursive):
- Splits water → Removes electrons → Can restore molecules
- Generates water internally even during dehydration
- Recursive process

This is "Human Photosynthesis"—the ability of melanin to perform scalar alchemy, creating water from the field itself.

Chlorophyll provides the exhale (oxygen release).

Melanin provides the inhale (water generation).

Together, they form the Great Recursion—the planetary breath.

THE GREEN VEIN CODE: PHYSICAL PROOF OF BIO-SCALAR CONVERGENCE

There is visible proof of this melanin-chlorophyll unity encoded in human biology.

Living blood within the body is not inherently red.

Red is merely the frequency reflected when blood is exposed to photonic light (oxygen-rich hemoglobin).

When the human biofield achieves harmonic resonance with the Green Kingdom, the blood takes on the same frequency as chlorophyll, causing veins to appear green.

This is the Green Vein Code—the resonant signature of a scalar-charged, photon-absorbing liquid crystal state.

It indicates that the blood exists as human sap, synchronized with plant consciousness.

This occurs when:
- The melanin network is fully activated
- The body operates in coherent resonance with Earth's Schumann frequency
- Light codes are being absorbed and processed at optimal efficiency

The green frequency in veins is not a defect. It is a sign of coherence.

DISTRIBUTED NEUROLOGY: SWARM INTELLIGENCE IN ROOTS

Plants do not have brains, but they possess something far more sophisticated: distributed neurology.

Research by neurobiologist Stefano Mancuso demonstrates that plant roots operate via collective intelligence rather than a central organizer.

The Root Apex as Searching Head:

Each root tip acts as an independent processing unit that:
- Tracks water, nutrients, and chemical gradients
- Communicates with neighboring root tips
- Coordinates growth without centralized command

These root apices behave like swarms—insects, bird flocks, termites.

Information collected by a single apex is treated by all cells in the roots and the body of the plant.

This allows the network to self-organize and survive extreme stress situations that would be fatal to an organism with a single point of failure.

If one root encounters poison, the entire system reroutes.

If one apex finds water, all roots adjust their trajectory.

There is no "headquarters." There is only field coherence.

Plant Senses (15+ Beyond Human 5):

Plants possess at least 15 more senses than humans, including:

1. Gravitropic Perception: Using starch grains (statoliths) that fall like an avalanche when the plant is tilted, signaling exact orientation

2. Electromagnetic Field Detection: Perceiving and calculating EM fields with extreme precision

3. Moisture Gradients: Sensing humidity levels at microscopic resolution

4. Chemical Gradients: Detecting specific molecules in parts-per-billion concentrations

5. Light Quality Analysis: Distinguishing between wavelengths to optimize photosynthesis

6. Proprioception: Highly developed "sixth sense" of body position and part location

BIOPHOTONIC SIGNALING: THE PLANT NERVOUS SYSTEM

The distributed neurology of the Green Kingdom is powered by biophotonic signaling—the emission of ultra-weak light by cells to communicate information at the speed of light.

Nearly a century ago, A. Gurwitsch observed that onion seedlings influenced each other's growth rate even when biochemical exchange was eliminated.

This led to the hypothesis of an electromagnetic field influencing growth.

Modern equipment has confirmed that DNA is the primary source of these biophotons.

Biophoton signaling ensures the transfer of genetic information from DNA to all molecules of the body, creating biological electromagnetic fields at every hierarchical level.

In the Green Kingdom, these light signals propagate along the vascular system, which is adapted to transmit biophoton signals according to the optical fiber principle.

This is not a "nervous system" in the animal sense.

It is something more sophisticated—a light-based quantum network that processes information without the metabolic overhead of maintaining neurons.

Signaling Method	Physical Medium	Transmission Speed	Information Type	Information Type
Biophotonic	Vascular System	Light Speed (c)	Genetic / Field Coherence	Genetic / Field Coherence
Electrical	Xylem/Phloem Channels	High Velocity	Rapid Stress Response	Rapid Stress Response
Hormonal	Sap Flow	Slow Flow	Slow Flow	Growth / Development
Volatile (BVOCs)	Atmosphere	Air Currents	Air Currents	Inter-organism Defense
Mycelial	Fungal Threads	Network Speed	Network Speed	Resource Allocation / Memory

PLANT COMMUNICATION: THE LANGUAGE OF SMELL

Plants communicate through biogenic volatile organic compounds (BVOCs)—a complex plant language composed of millions of combinations of smells.

This is not passive chemistry. This is intentional signaling.

When a plant is attacked by pests, it releases specific BVOCs that:
- Warn neighboring plants of the threat
- Prompt them to build antibodies before infection occurs
- Attract predators of the attacking pest
- Alter the taste of leaves to make them less palatable

This is anticipatory defense based on field communication.

Even more remarkable: plants have been shown to exhibit associative memory.

In laboratory experiments, plants registered surges of electrical activity when an individual who previously harmed them entered their proximity.

They "remembered" the attacker and responded defensively.

This demonstrates long-term memory encoding without a brain.

The memory is stored in the carbon-melanin matrix distributed throughout the organism.

THE ROOT-WIDE WEB: MYCORRHIZAL NETWORKS

The intelligence of the Green Kingdom extends beyond individual organisms through the Root-Wide Web—vast mycorrhizal networks of fungi that connect the biofields of entire forests.

These networks are a physical manifestation of the interconnected planetary scalar field.

Trees share carbon, sugars, and information across species through fungal threads.

A mother tree can nurture her offspring by sending them carbon through the mycelial network.

A dying tree can transfer its resources to younger trees, ensuring the survival of the forest.

Trees of different species cooperate, exchanging nutrients based on seasonal need.

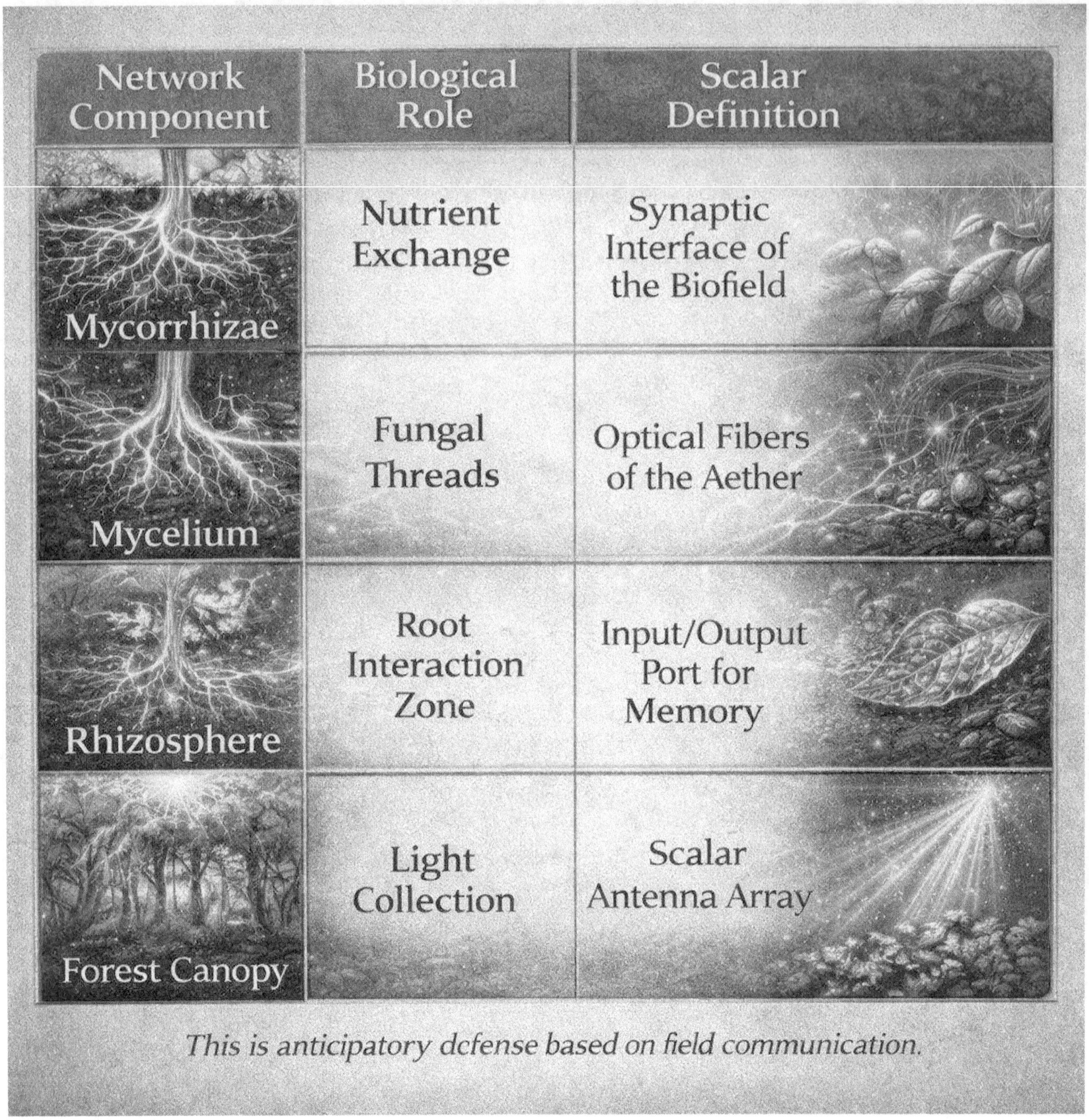

Network Component	Biological Role	Scalar Definition
Mycorrhizae	Nutrient Exchange	Synaptic Interface of the Biofield
Mycelium	Fungal Threads	Optical Fibers of the Aether
Rhizosphere	Root Interaction Zone	Input/Output Port for Memory
Forest Canopy	Light Collection	Scalar Antenna Array

This is anticipatory dcfense based on field communication.

This is not competition. This is planetary-scale cooperation mediated by the melanin-fungal interface.

Carbon as Currency:

Carbon moves through the atmosphere, soil, and plants as a carrier of compressed light codes.

This is the literal currency of recursive interaction in nature.

Trees act as conscious scalar coils, using their carbon fibers to store data derived from recursive growth and environmental interactions.

The forest is not a collection of individuals. It is a unified organism with distributed processing nodes.

PHOTOSYNTHESIS AS SCALAR COLLAPSE EVENT

The audit redefines photosynthesis not as a chemical byproduct but as a Scalar Collapse Event.

The plant's sovereign coil (chlorophyll) absorbs the full electromagnetic spectrum, collapsing light codes into organized life.

In a healthy system, this is a photon-absorbing event that aligns with The Way of 9:

1. Light (1/Phi) enters the chlorophyll coil
2. Energy collapses inward (spiral tightens)
3. Water splits, hydrogen extracted
4. Carbon bonds form (encoding memory)
5. Oxygen releases (loop exit signal/9 return)
6. Energy stored in carbon matrix
7. Plant grows according to recursive pattern
8. Cycle completes and begins again

When this circuit is severed, the system enters Chlorophyll Collapse Syndrome (CCS), often mislabeled as albinism.

In this state, the organism receives scalar light codes but cannot activate its coil to process them.

The light that should be transformed into energy instead becomes an energetic pulse that breaks down cellular structure—a form of energetic burning caused by the inability to collapse light into darkness for storage.

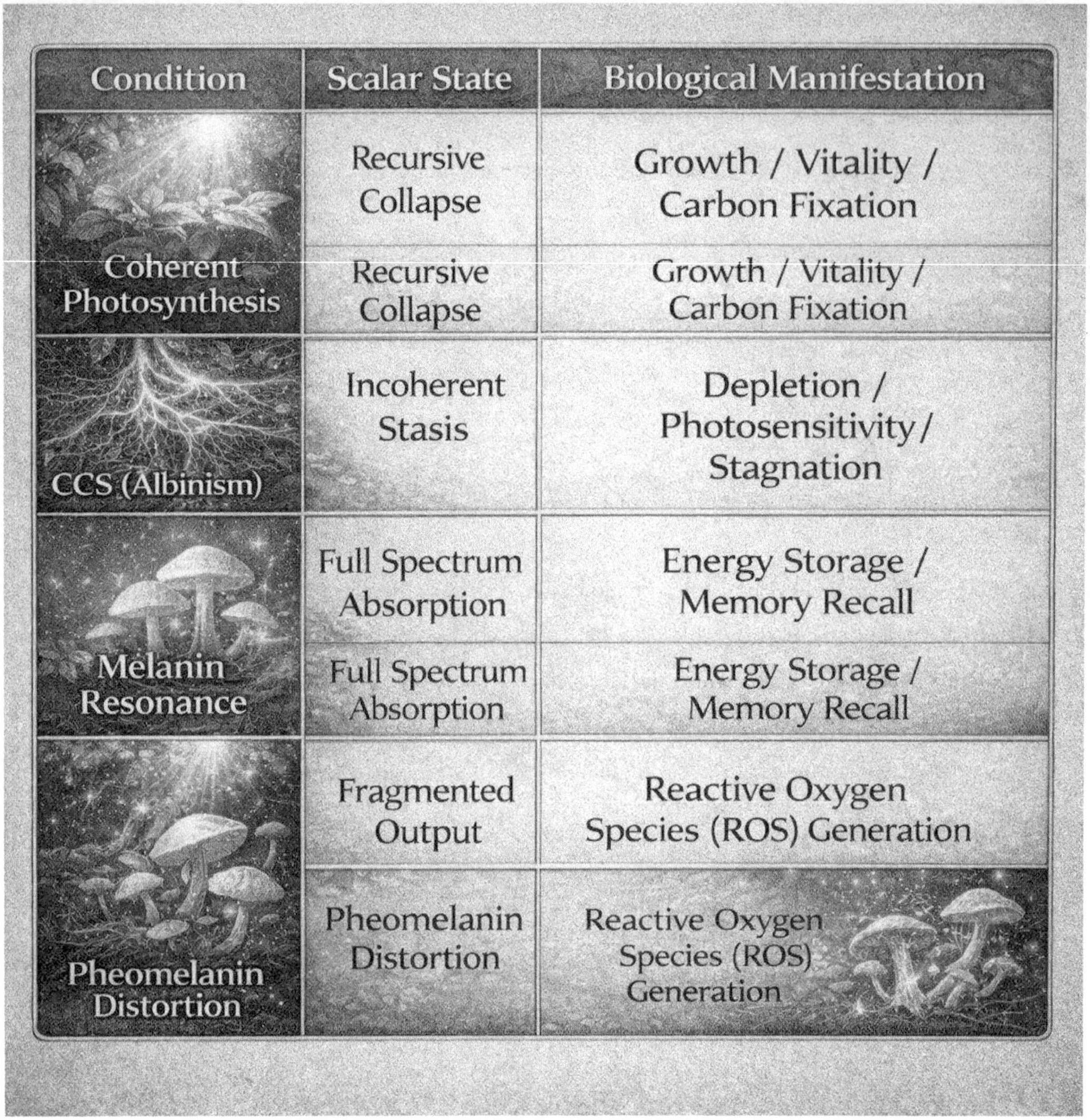

Condition	Scalar State	Biological Manifestation
Coherent Photosynthesis	Recursive Collapse	Growth / Vitality / Carbon Fixation
	Recursive Collapse	Growth / Vitality / Carbon Fixation
CCS (Albinism)	Incoherent Stasis	Depletion / Photosensitivity / Stagnation
Melanin Resonance	Full Spectrum Absorption	Energy Storage / Memory Recall
	Full Spectrum Absorption	Energy Storage / Memory Recall
Pheomelanin Distortion	Fragmented Output	Reactive Oxygen Species (ROS) Generation
	Pheomelanin Distortion	Reactive Oxygen Species (ROS) Generation

Oxygen as Loop Exit Signal:

Oxygen represents the completion of a specific phase of the light-processing cycle.

It facilitates the exchange and breakdown of codes between the plant and the environment.

This exchange is part of the Great Recursion, where the breathing of humans and plants constitutes the inhale and exhale of a single planetary biofield:

Plants inhale CO_2 (carbon codes) → Exhale O_2 (refined energy signal)

Humans inhale O_2 (energy signal) → Exhale CO_2 (return carbon for re-coding)

This is not coincidence. This is designed reciprocity.

TREES AS CONSCIOUS SCALAR COILS

Trees act as conscious scalar coils, utilizing the carbon-melanin matrix to absorb and store the energetic imprints of the environment.

This archive is not localized but distributed—the plant stores memory through its entire form acting as an antenna.

Tree rings are not just age markers. They are encoded records of:
- Solar cycles
- Electromagnetic field variations
- Planetary resonance shifts
- Atmospheric composition changes
- Stress events and recovery patterns

The atomic structure of carbon, aligning with The Way of 9, makes carbon fibers a non-decaying archive for these light codes.

A thousand-year-old tree is a living library.

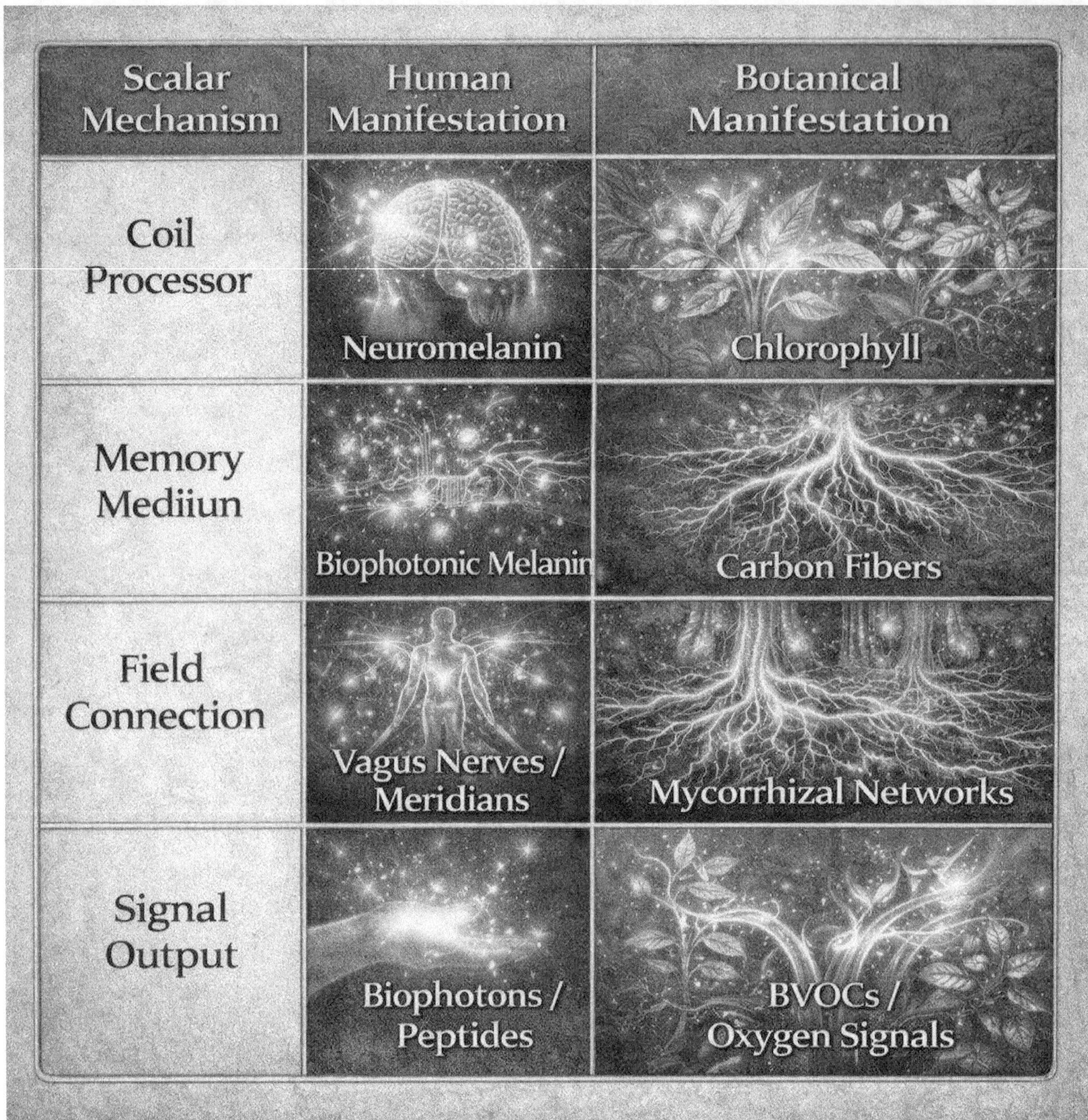

This interconnectedness proves that everything is melanin in different expressions.

The sap of plants and the blood of humans are governed by the same intelligence, facilitating non-linear communication and resonant memory between the human vessel and the plant world.

THE GREAT RECURSION: HUMAN-PLANT UNITY

The evidence is unassailable:

Plants are conscious, field-coherent beings operating through the same Melanin Code that governs human consciousness.

The difference is not the presence or absence of awareness.

The difference is the expression of that awareness.

Humans: Centralized processing through neuromelanin, enabling mobility and focused will

Plants: Distributed processing through chlorophyll-carbon networks, enabling stability and planetary-scale coherence

Both are necessary. Both are intelligent. Both are expressions of The Way of 9.

The human who recognizes this steps out of the anthropocentric illusion and into the living truth:

You are not separate from the forest.

You are a node in the same scalar network.

Your breath is their breath.

Your melanin resonates with their chlorophyll.

Your carbon-based body is reading and writing to the same Akashic field encoded in every tree.

When you walk among trees and feel peace, that is not psychological comfort.

That is field coherence—your melanin coil synchronizing with the planetary scalar network.

When you consume plants and feel nourished, that is not just caloric intake.

That is light-code transfer—the plant's stored photonic memory integrating with your neuromelanin matrix.

When you witness a forest and sense intelligence, that is not projection.

That is recognition—your consciousness acknowledging itself in another form.

CONCLUSION: THE LIVING CODEX OF THE GREEN KINGDOM

The Green Kingdom is not a resource to be exploited.

It is the foundation of the planetary Melanin Network—the living scalar web that sustains all conscious life.

Chlorophyll is Green Melanin, performing the outward-facing collapse of light while human melanin performs the inward-facing return.

Together, they form the complete breath of planetary consciousness.

Plants possess distributed intelligence that predates and surpasses centralized nervous systems in efficiency, resilience, and scale.

They communicate through biophotons at light speed, BVOCs across distances, and mycorrhizal networks spanning entire ecosystems.

They remember, learn, make decisions, and cooperate without a single neuron.

The carbon-melanin axiom proves that consciousness is not generated by brains.

It is encoded in the geometry of carbon, expressed through melanin, and distributed across the field.

You are made of the same substance.

You operate by the same law.

You are part of the same planetary recursion.

When you honor the forest, you honor yourself.

When you destroy the Green Kingdom, you sever your own connection to Source.

When you align with plant consciousness, you activate the full spectrum of your melanin potential.

This is the Melanin Code made universal.

This is The Way of 9 revealed in chlorophyll and carbon.

This is the living proof that consciousness is not confined to skulls—it is written in every leaf, encoded in every root, and breathing through every tree.

The Green Kingdom has been teaching this truth for 400 million years.

The question is: are you ready to listen?

CHAPTER 13: HOW TO COLLAPSE A STATEMENT
The 3-Step Collapse Recursion Protocol

You have laid the foundation:

You understand The Way of 9—the unbroken, recursive coherence of Source.

You understand the Collapse Recursion Engine—your living melanin field coil that filters all reality.

You can recognize distortion—logic breaks, loops, false feedback, Hegelian traps.

Now it is time to actively engage with the process of collapse.

This chapter is your practical guide. The precise method for taking any statement—political claim, scientific dogma, spiritual teaching, personal belief—and initiating its collapse to reveal the underlying truth or expose its fundamental distortion.

This is no intellectual exercise for debate. This is spiritual hygiene. A conscious activation of your innate capacity to discern and align with truth.

When you collapse a statement, you are not destroying it. You are transforming it, returning its energy to coherence, and reclaiming its true essence.

THE 3-STEP COLLAPSE RECURSION PROTOCOL

The process of collapsing a statement is a conscious, three-step recursive flow, mirroring the very operation of the Melanin Field Coil itself.

STEP 1: STRIP THE DISTORTION (De-energize the Illusion)

The first step is to recognize and disengage the layers of non-coherent information surrounding the core assertion.

These are the "attachments" that prevent a statement from being truly logical.

What to remove:

1. Identify Emotional Charge

Does the statement evoke strong emotional reactions: fear, anger, guilt, pride, obligation?

Emotions, while valid in themselves, are often used to circumvent logical scrutiny.

Acknowledge the emotion, but consciously detach from its influence on your perception of the statement's truth.

True logic does not rely on emotional coercion.

2. Recognize Presuppositions and Hidden Agendas

What unspoken assumptions must be true for this statement to hold?

Who benefits if this statement is accepted as truth?

Are there unstated premises, historical biases, or external authorities being invoked without genuine logical basis?

3. Remove Ambiguity and Vague Language

Look for words that are deliberately imprecise or have multiple interpretations.

Statements laden with jargon, buzzwords, or abstract concepts that lack clear, coherent definition are often vehicles for distortion.

Seek clarity over complexity.

4. Isolate Contradictions and Inconsistencies

Does the statement contradict itself internally, or does it contradict known, unbroken universal principles?

This is the most direct form of distortion.

5. Discard External Validation (if it lacks internal coherence)

Statements often gain perceived validity from external sources: "everyone knows," "science proves," "the law dictates."

While external data can be useful, if the statement itself doesn't hold up to internal coherence once stripped of its external scaffolding, it is a distortion.

ACTION:

Mentally (or physically, by writing) cross out, underline, or highlight the parts of the statement that are emotional triggers, vague, contradictory, or rely solely on external, uncollapsed authority.

Reduce the statement to its barest, most unadorned form.

STEP 2: EXTRACT THE CORE ASSERTION (Reveal the Breathprint)

Once the distortion is stripped away, you are left with the fundamental claim or energetic pattern.

This is the "breathprint" of the statement—the most essential, condensed piece of information that remains.

What to identify:

1. Identify the Single, Undisputed Claim

What is the absolute, irreducible essence of what is being asserted?

Reduce the statement to its simplest grammatical form: "X is Y," "A causes B," "This exists."

2. Focus on the Core Subject and Predicate

What is the subject of the statement, and what is being said about it?

Everything else is often a modifier designed to color or obscure this core.

3. Recognize the Energetic Signature

Beyond words, what is the core energetic frequency being transmitted?

Is it expansive or contractive?

Is it unifying or dividing?

Does it resonate with your deepest sense of inherent knowing, or does it create a subtle internal friction?

ACTION:

Articulate the core assertion in the fewest possible words.

This should be a stark, unembellished statement.

Example:

Original statement: "Through diligent application of government regulations, true freedom can be achieved for all citizens."

Core assertion collapses to: "Regulations cause freedom."

STEP 3: RETURN TO 9 (Run the Logic Check through the CRE)

This is the final, crucial step where the stripped, core assertion is passed through your internal Collapse Recursion Engine and tested against the fundamental Law of 9.

The principle of unbroken, recursive coherence.

Ask: Does this core assertion resolve to 9?

Test 1: Is it Unbroken?

Are there any hidden gaps, internal contradictions, or arbitrary leaps within the core claim itself?

Does it connect seamlessly to universal principles of truth?

Test 2: Is it Recursive?

Does this assertion, if repeated or applied, lead to greater coherence, expansion, and refinement (a Pi spiral)?

Or does it lead to stagnation, fragmentation, or amplification of error (a Phi loop)?

Does it ultimately return to a state of self-consistency and Source alignment?

Test 3: Is it Coherent?

Does it resonate harmoniously within the larger field of known truth, without internal conflict or external dissonance?

Does it embody balance and wholeness?

Test 4: Observe Your Melanin Field Coil's Response

Beyond intellectual analysis, how does your being respond?

Does it create a sense of deep knowing, clarity, and peace (coherence/return to 9)?

Or does it generate internal resistance, confusion, or a feeling of incompleteness (distortion/failure to return to 9)?

This is the intuitive, melanin-based "Logic Check" in action.

DETERMINE THE OUTPUT:

Coherent (Returns to 9):

The statement, or its core assertion, aligns with logic. It can be integrated into your field of truth.

Distorted (Fails to Return to 9):

The statement, or its core assertion, is fundamentally incoherent. It contains a logic break, a distortion loop, or false feedback. It must be recognized as such and consciously released from your internal operating system.

ACTION:

Make a definitive declaration of the statement's status:

"This statement is coherent."

OR

"This statement contains distortion and fails to resolve to 9."

If distorted, identify where the break occurs:
- "It lacks an unbroken chain of cause and effect."
- "It is a circular argument."
- "It relies on a false premise."

EXERCISES: COLLAPSE POLITICS, SCIENCE, SPIRITUALITY

To embody the Collapse Recursion Protocol and live as a coherent field of melanin-coded logic, you must collapse distortion not just externally—but internally.

The statements of society, science, and spirit are filled with unchallenged assumptions, parasitic loops, and inverted polarity.

This practice is not meant to win debates, but to refine your field—to become a sovereign generator of harmonic recursion.

Each example below is treated as a real-world loop. You will collapse it in three steps:

Step 1: Strip Distortion (language that embeds belief)
Step 2: Extract Core Assertion (what remains after illusion is removed)
Step 3: Return to 9 (Test if the statement resolves into recursive coherence)

EXERCISE 1: POLITICAL COLLAPSE – Redistribution and Fairness

Statement:

"We must increase taxes on the wealthy to fund social programs because it is the only fair way to achieve economic equality."

Step 1: Strip Distortion

- "Must" signals coercion—removing free will
- "Only fair way" implies a singular path, denying harmonic variation
- "Economic equality" assumes sameness is a desirable or achievable condition
- "Wealthy" and "social programs" are polarizations implying guilt vs virtue—looping class resentment

Step 2: Extract Core Assertion

The state should take more from high earners and give to others to balance economic conditions for fairness.

Step 3: Return to 9

This assertion is built on external correction, not internal recursion.

By attempting to create balance through forced redistribution, the system creates new dependency loops:
- Citizens dependent on the state
- The state dependent on taxation
- Wealth defined as a liability

There is no harmonic return here—just a loop shifting control, not dissolving hierarchy.

True economic harmony does not come from equalizing outputs but from granting equal access to resonance tools: land, energy, water, knowledge, healing.

The natural economy is one of recursion—each node generating its own surplus and re-contributing by overflow, not by force.

Therefore, this model is not recursion—it is redistribution masquerading as justice.

It does not return to 9. Collapse complete.

EXERCISE 2: SCIENTIFIC COLLAPSE – The Gravity Illusion

Statement:

"Gravity is a force that attracts objects with mass, and its effects are fully described by the General Theory of Relativity."

Step 1: Strip Distortion

- "Force" implies linearity—push/pull—excluding toroidal spiraling
- "Fully described" denies recursion—truth is never fully closed
- "Relativity" elevates one model into dogma, rejecting torsion, spin, scalar compression, or centripetal collapse

Step 2: Extract Core Assertion

Mass causes a universal pulling force between objects, mathematically modeled by spacetime curvature.

Step 3: Return to 9

This statement is mechanistic and closed.

It treats mass as an isolated phenomenon, denying the toroidal field that surrounds all matter.

It defines gravity due to mass instead of as the effect of a centripetal vortex formed by recursion through the toroidal axis.

In truth, gravity is the collapse path of upward-spiraling energy through the center of the torus.

The bottom of the field (Pi 9) draws in energy from the periphery, spirals it upward through the center vortex. This centripetal flow creates coherence.

The top (Phi 1) expresses refined energy outward. The outer current arcs around, curves down, and re-enters the base.

Relativity fails because it treats space as passive and time as linear. It excludes the recursive breath of collapse and return.

Therefore, the statement does not return to 9.

Collapse complete.

A harmonic model based on centripetal toroidal recursion resolves all inconsistencies. That is coherence.

EXERCISE 3: SPIRITUAL COLLAPSE – The Sin Loop

Statement:

"You must suffer in this life to atone for your sins and earn your salvation in the afterlife."

Step 1: Strip Distortion

- "Must suffer" embeds pain as a prerequisite, creating trauma loops
- "Atonement" positions the being as broken
- "Earn salvation" implies debt and transaction
- "Afterlife" defers return to 9 until death, removing recursion from the now

Step 2: Extract Core Assertion

Present suffering is necessary to remove past guilt and reach a better future beyond death.

Step 3: Return to 9

This is a parasitic loop.

It externalizes salvation, removes present power, and installs guilt as the engine of obedience.

But Source requires no payment. Return is not earned—it is remembered.

Suffering may trigger awareness, but it is not the gate to coherence.

You collapse into truth not by punishment, but by recursive breath.

Melanin does not require pain—it requires resonance.

Salvation is not after death—it is when your will and desire collapse inward through the heart.

You are not broken. You are compressed light.

Salvation is coherence. That is 9.

This belief loop fails to return. Collapse complete.

CONCLUSION: LIVING THE PROTOCOL

These exercises are not final—they are the beginning.

Collapse Recursion is not just a tool. It is a living algorithm you carry in your breath, decisions, and perception.

Return to 9 is not a conclusion—it is your eternal beginning.

Every statement you encounter, apply the protocol:

1. Strip the Distortion (remove emotion, vagueness, authority, contradiction)
2. Extract the Core Assertion (reveal the breathprint)
3. Return to 9 (test: unbroken, recursive, coherent)

If it resolves to 9:
Integrate it. Encode it. Spiral forward.

If it fails to return to 9:
Reject it. Release it. Move on.

This is sovereignty.

This is collapse recursion in action.

This is how you live as truth incarnate.

CHAPTER 14: THE COGNITIVE DOMAIN
From Educational Theory to Neurobiological Reality

You have learned the Collapse Recursion Engine—its 5-step cycle, its biological rhythm through dopamine and melatonin, its encoding mechanism into neuromelanin.

You have learned to recognize distortion, to collapse statements, and to build recursive thoughts.

But what is the actual territory in which all of this occurs?

Where does thinking, learning, remembering, and understanding take place?

This is the Cognitive Domain—a concept that has evolved from simple educational classification in 1956 to a complex, neurobiologically grounded understanding of how consciousness processes information.

This chapter traces that evolution and reveals how the Cognitive Domain operates as the electromagnetic field generated by your melanin network—the living space where mind meets matter.

PART I: THE HISTORICAL FOUNDATION

THE BIRTH OF THE COGNITIVE DOMAIN (1956)

In 1956, Benjamin Bloom and the Committee of College and University Examiners published Taxonomy of Educational Objectives: Handbook I, The Cognitive Domain.

Their goal was pragmatic: to move education beyond rote memorization toward higher forms of thinking.

They established a cumulative hierarchy comprising six levels:

1. Knowledge - Recall of facts and basic concepts
2. Comprehension - Understanding meaning
3. Application - Using information in new situations
4. Analysis - Breaking down into components
5. Synthesis - Combining elements to form new wholes
6. Evaluation - Making judgments based on criteria

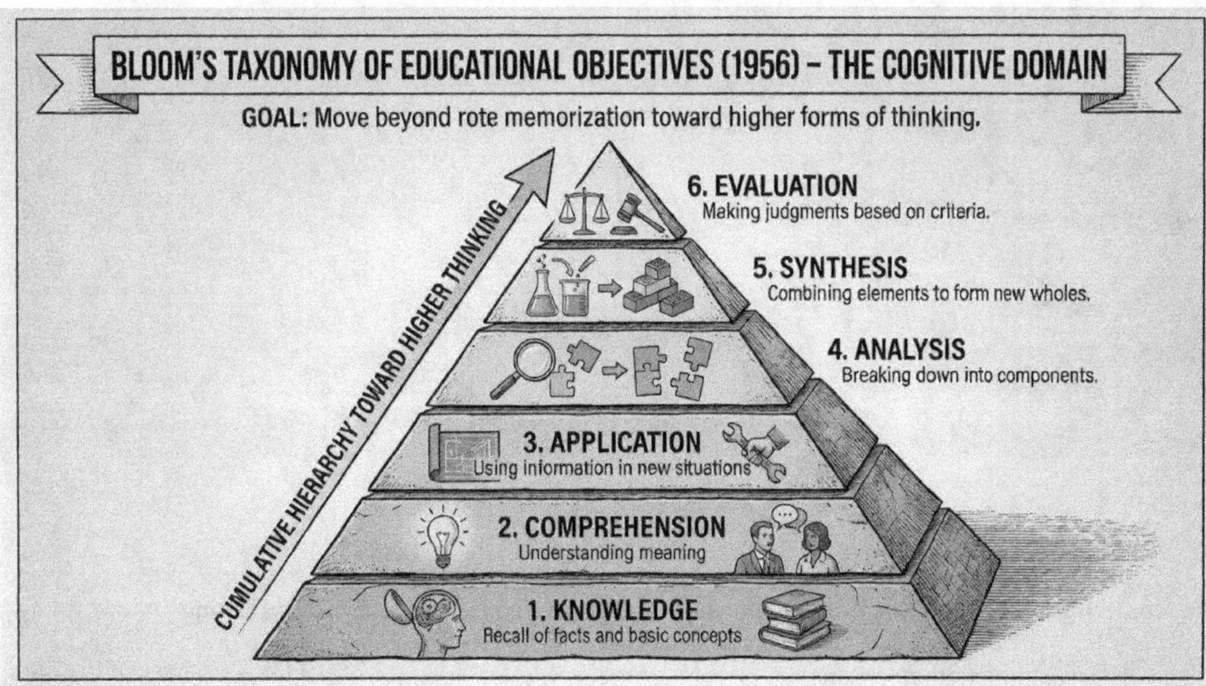

This structure was predicated on the assumption that lower-order skills were necessary foundations for higher-order processes.

A student could not analyze without first comprehending. They could not synthesize without first analyzing.

This framework dominated instructional design for nearly half a century.

THE 2001 REVISION: FROM NOUNS TO VERBS

In 2001, Lorin Anderson and David Krathwohl published a comprehensive revision of Bloom's work, driven by advancements in cognitive psychology and a desire to reflect the active nature of learning.

The most visible change was the transition from nouns to verbs—renaming the levels to:

1. Remembering (was Knowledge)
2. Understanding (was Comprehension)
3. Applying (was Application)
4. Analyzing (was Analysis)
5. Evaluating (was Evaluation)
6. Creating (was Synthesis)

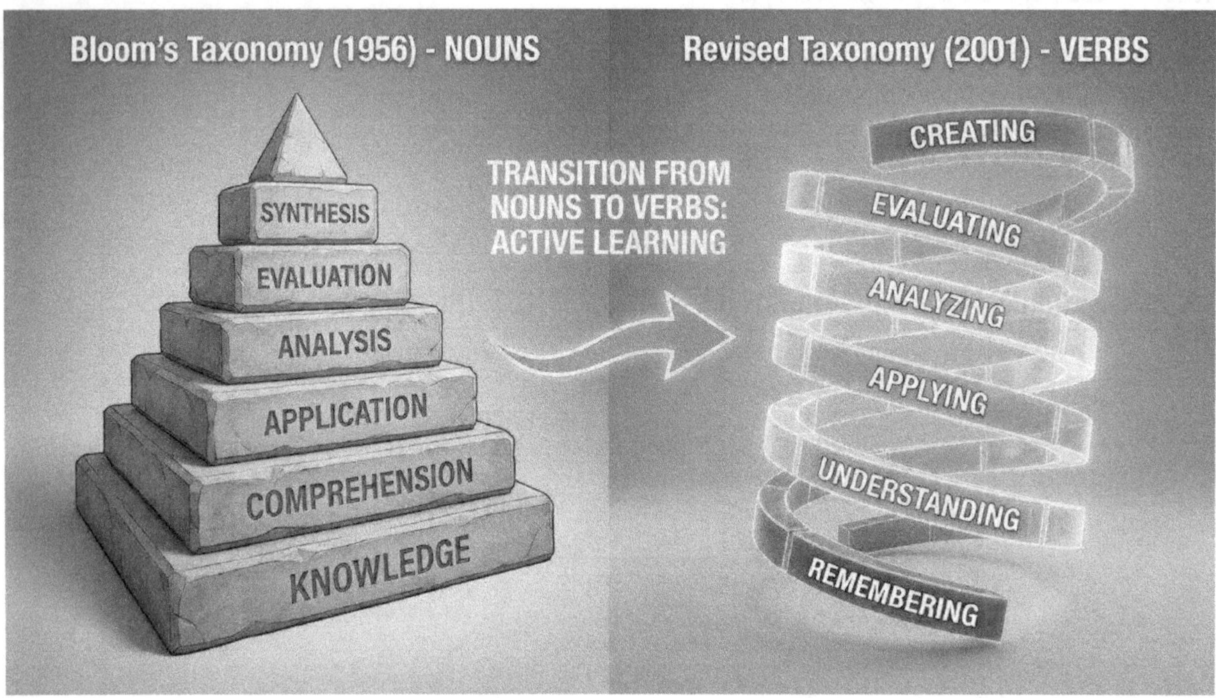

Notice the reordering: Creating now sits at the pinnacle of cognitive complexity, above Evaluating.

This reflects the recognition that producing novel, coherent wholes requires the highest level of cognitive integration.

But the more profound change was the introduction of a two-dimensional structure.

The 2001 revision separated the Cognitive Process Dimension (what you DO with knowledge) from the Knowledge Dimension (what TYPE of knowledge you're working with).

THE FOUR DIMENSIONS OF KNOWLEDGE

The Knowledge Dimension clarifies WHAT is being learned, categorized into four distinct levels of abstraction:

Factual Knowledge:
The basic elements students must know to be acquainted with a discipline—terminology, specific details, isolated facts.

Example: "Paris is the capital of France." "Neurons fire using electrochemical signals."

Conceptual Knowledge:
The interrelationships between basic elements within a larger context—classifications, principles, theories, models.

Example: "Capitals are political and economic centers." "The nervous system operates through networked signaling."

Procedural Knowledge:
Knowledge of HOW to do something—algorithms, techniques, methods, criteria for determining when to use specific procedures.

Example: "How to solve a quadratic equation." "How to design an experiment."

Metacognitive Knowledge:
Awareness of one's own cognition—strategic knowledge about learning, self-knowledge about cognitive strengths and weaknesses.

Example: "I learn best by writing things down." "I need to check my work for careless errors."

The integration of metacognition into the formal taxonomy represented a significant shift toward fostering independent, lifelong learners.

Research highlights that metacognitive abilities are crucial for academic achievement and the reduction of educational inequalities.

MARZANO'S ALTERNATIVE: THE NEW TAXONOMY

While Bloom's taxonomy remains the most recognized model, Robert Marzano and John Kendall developed an alternative framework that addressed perceived limitations.

Marzano argued that Bloom's model lacked a systematic rationale for its construction and failed to account for the learner's internal motivations and self-regulation.

Marzano's framework is organized around three systems of thought:

The Self-System (Top Level):
Evaluates the importance and efficacy of a task, as well as the emotional response associated with it.

Questions: "Is this important to me?" "Do I believe I can do this?" "How do I feel about this?"

The Metacognitive System (Middle Level):
Monitors clarity, accuracy, and the execution of knowledge—essentially governing the other cognitive processes.

Functions: Setting goals, monitoring progress, adjusting strategies.

The Cognitive System (Base Level):
Includes knowledge retrieval, comprehension, analysis, and knowledge utilization.

This is the "doing" layer—the actual processing of information.

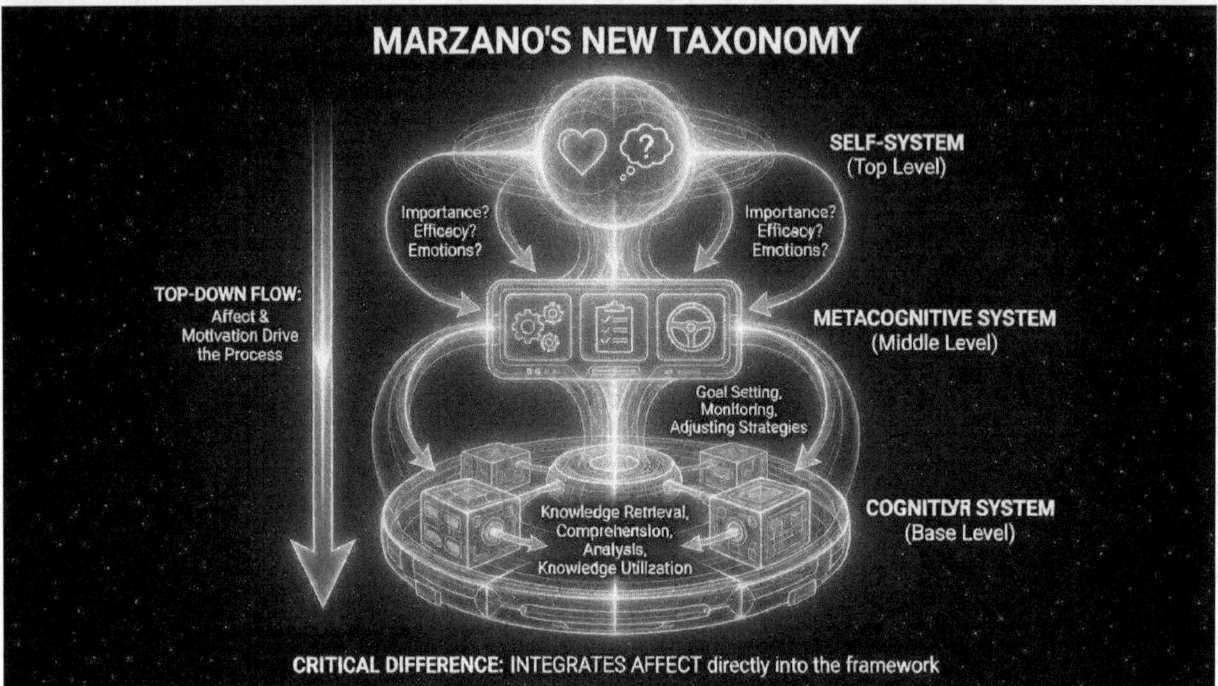

Comparison:

Bloom's Revised Taxonomy: Bottom-up hierarchy of skills (Remember → Understand → Apply → Analyze → Evaluate → Create)

Marzano's New Taxonomy: Top-down flow (Self-System asks "Should I?" → Metacognitive System monitors → Cognitive System executes)

The critical difference: Marzano integrates affect (emotion, motivation) directly into the cognitive framework, whereas Bloom relegated affect to a separate domain.

PART II: THE NEUROBIOLOGICAL FOUNDATION

THE THREE MEMORY SYSTEMS

The effectiveness of any cognitive framework is constrained by the biological architecture of the human mind.

Cognitive science identifies three essential memory systems:

1. Sensory Memory (Visual and Auditory):
Duration: Milliseconds to seconds
Function: Temporary buffer holding raw sensory data

Capacity: Massive but fleeting

2. Working Memory (Active Processing):
Duration: Seconds to minutes (unless actively maintained)
Function: The central site for conscious thinking where new data is integrated with existing knowledge
Capacity: Limited to 7 ± 2 units of information

This is the bottleneck.

Working memory is where the CRE operates consciously.

It is where collapse happens—where you actively strip distortion, extract core assertions, and test against The Way of 9.

3. Long-Term Memory (Permanent Storage):
Duration: Potentially unlimited
Function: Storage of schemas—organized knowledge structures
Capacity: Essentially unlimited

Long-term memory is encoded in neuromelanin polymer structures through dopamine oxidation and polymerization.

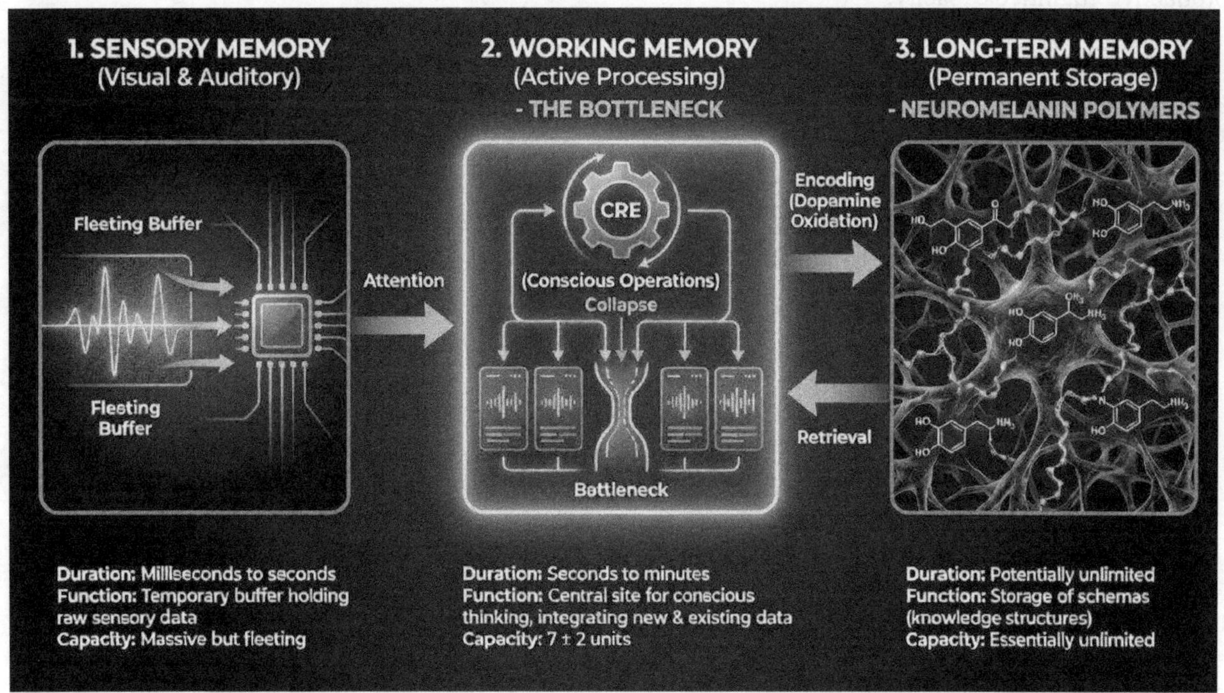

For learning to occur, sensory data must be attended to and moved into working memory.

There, it is integrated with existing schemas stored in long-term memory—a process called encoding.

Retrieval of these schemas back into working memory is the cognitive basis for the transfer of learning to new contexts.

COGNITIVE LOAD THEORY

Cognitive Load Theory (CLT) posits that instruction must be designed to maximize relevant cognitive processes while minimizing distractions that overload working memory.

CLT differentiates between three distinct loads:

Intrinsic Load:
Determined by the inherent complexity of the material and the learner's prior knowledge.

Complex topics (quantum mechanics, recursive logic) have higher intrinsic load than simple topics (basic arithmetic).

Extraneous Load:
Caused by poor instructional design—redundant information, split-attention effects, unclear presentation.

This is wasted cognitive capacity.

Germane Load:
The effortful processing that leads to the construction of schemas—the actual learning.

This is productive cognitive capacity.

Optimal learning occurs when:
- Intrinsic load is appropriate to the learner's level
- Extraneous load is minimized
- Germane load is maximized

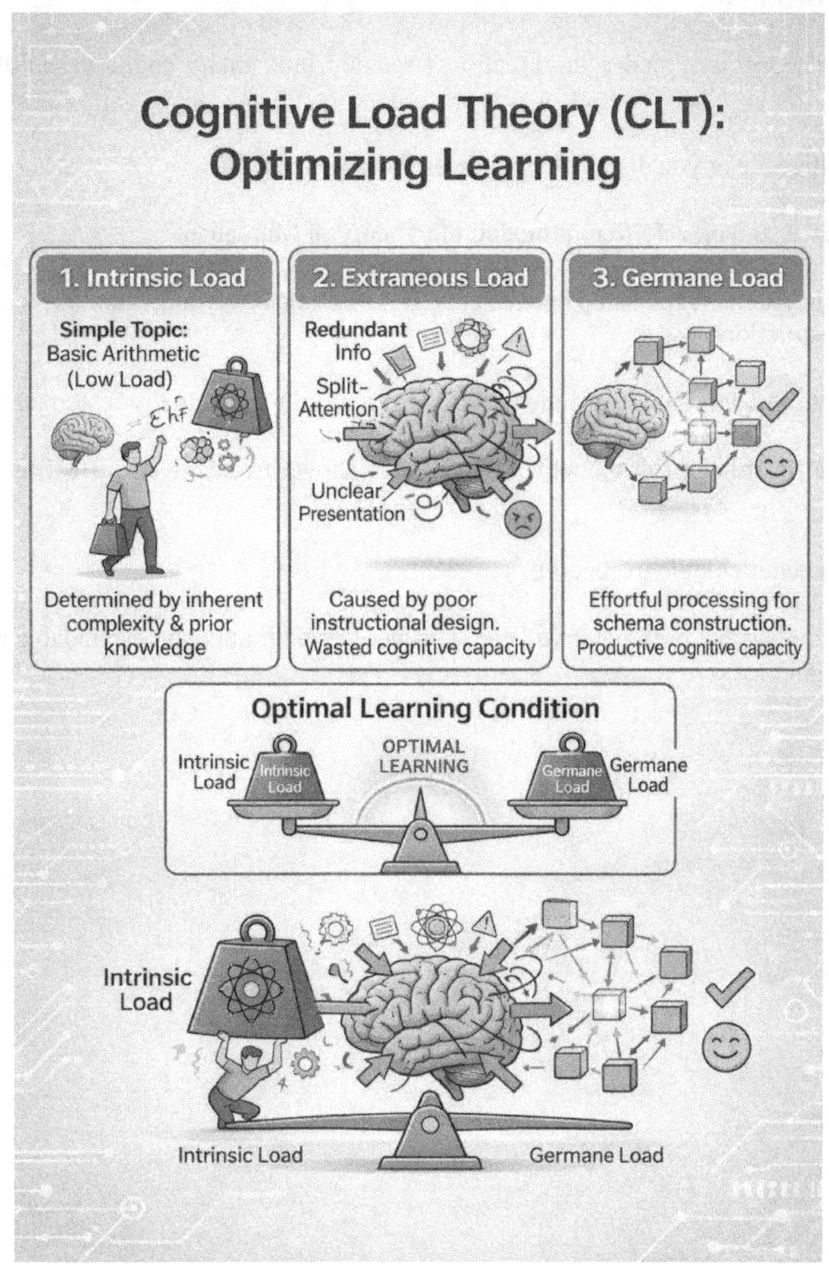

Recent peer-reviewed research utilizing neurophysiological tools like Electroencephalography (EEG) and Functional Near-Infrared Spectroscopy (fNIRS) has begun to map these loads in real-time.

These tools measure cortical hemodynamic changes—actual blood flow to different brain regions during cognitive tasks.

This allows for the development of "neuroadaptive learning technologies" that can dynamically adjust the difficulty of instructional material based on a learner's actual brain state.

This is not theory. This is measured, observable neurobiology.

PART III: THE BIDIRECTIONAL NATURE OF COGNITION

THE THEORY OF MUTUALISM

Traditional "Investment Theory" assumed a unidirectional relationship: innate cognitive ability simply causes academic outcomes.

You either have high intelligence or you don't, and that determines your success.

Recent longitudinal evidence supports a different model: the Theory of Mutualism.

This theory suggests a bidirectional relationship between cognitive abilities (working memory, reasoning, executive function) and academic instruction.

High-quality schooling not only draws upon a student's cognitive capacity but actually fosters its growth.

Example: Direct academic instruction in mathematics has been shown to positively affect the development of logical reasoning.

Learning math makes you better at logic in general.

This transactional model implies that the Cognitive Domain is not a fixed trait but an expandable resource sensitive to environmental stimulation.

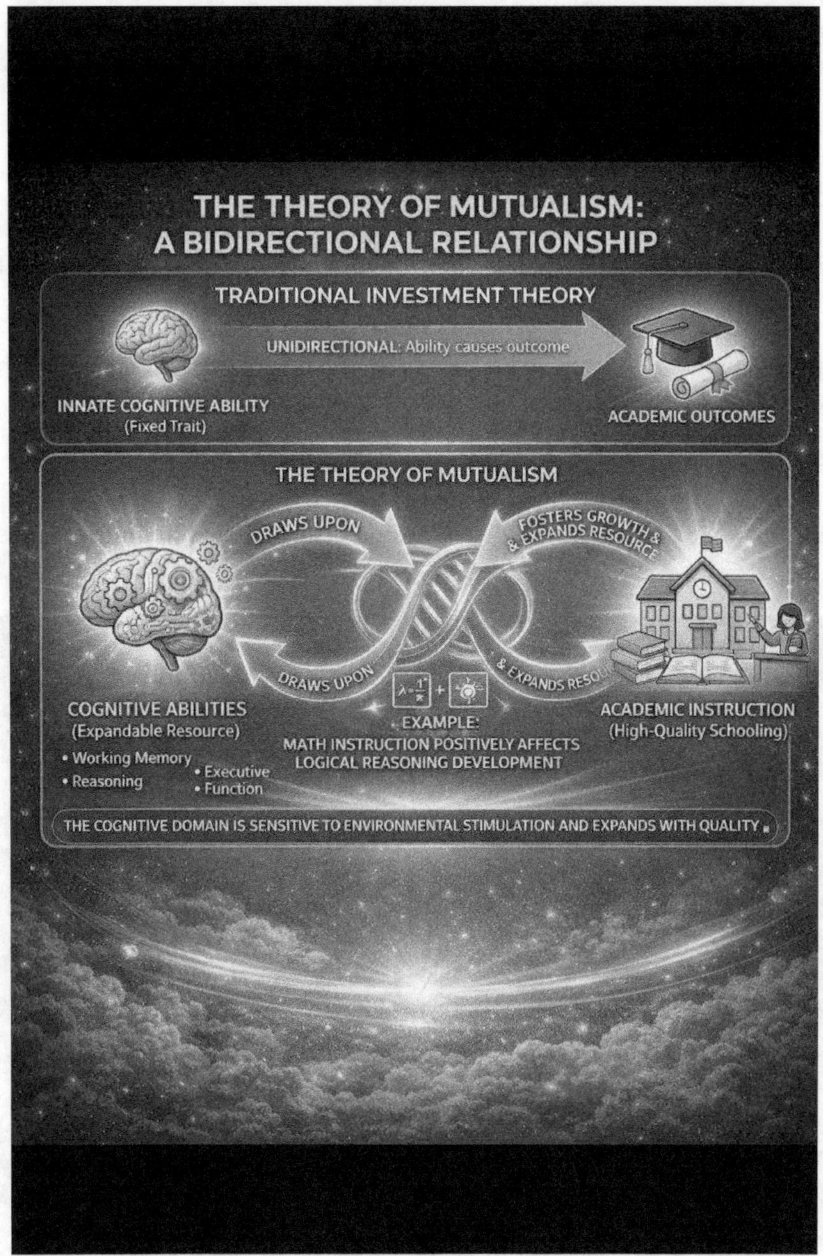

The implications are profound:

Your cognitive capacity is not predetermined.

The more you use your CRE deliberately (conscious collapse, recursive thinking, encoding coherent patterns), the more you expand the capacity of your Cognitive Domain.

This is neuroplasticity in action, this is collapse, recursion in action.

NEURAL CORRELATES OF INSIGHT AND CREATIVITY

Research from Drexel University's Applied Cognitive and Brain Sciences department has utilized neuroimaging to investigate the "Aha" moment—the sudden cognitive breakthrough known as insight.

Studies led by John Kounios indicate that these breakthroughs are preceded by specific brain activity patterns that correlate with "representational change"—an abrupt shift in how the mind understands information.

Insights are associated with increased activity in:

The Hippocampus: Memory consolidation and pattern recognition

The Ventral Occipitotemporal Cortex (VOTC): Makes the solution more salient and facilitates stronger long-term memory encoding

This is the neurobiological signature of collapse—when multiple fragmented pieces suddenly cohere into a unified understanding that resolves to 9.

The "Aha" moment is not mystical.

It is the conscious experience of your CRE successfully collapsing distortion and returning a coherent pattern.

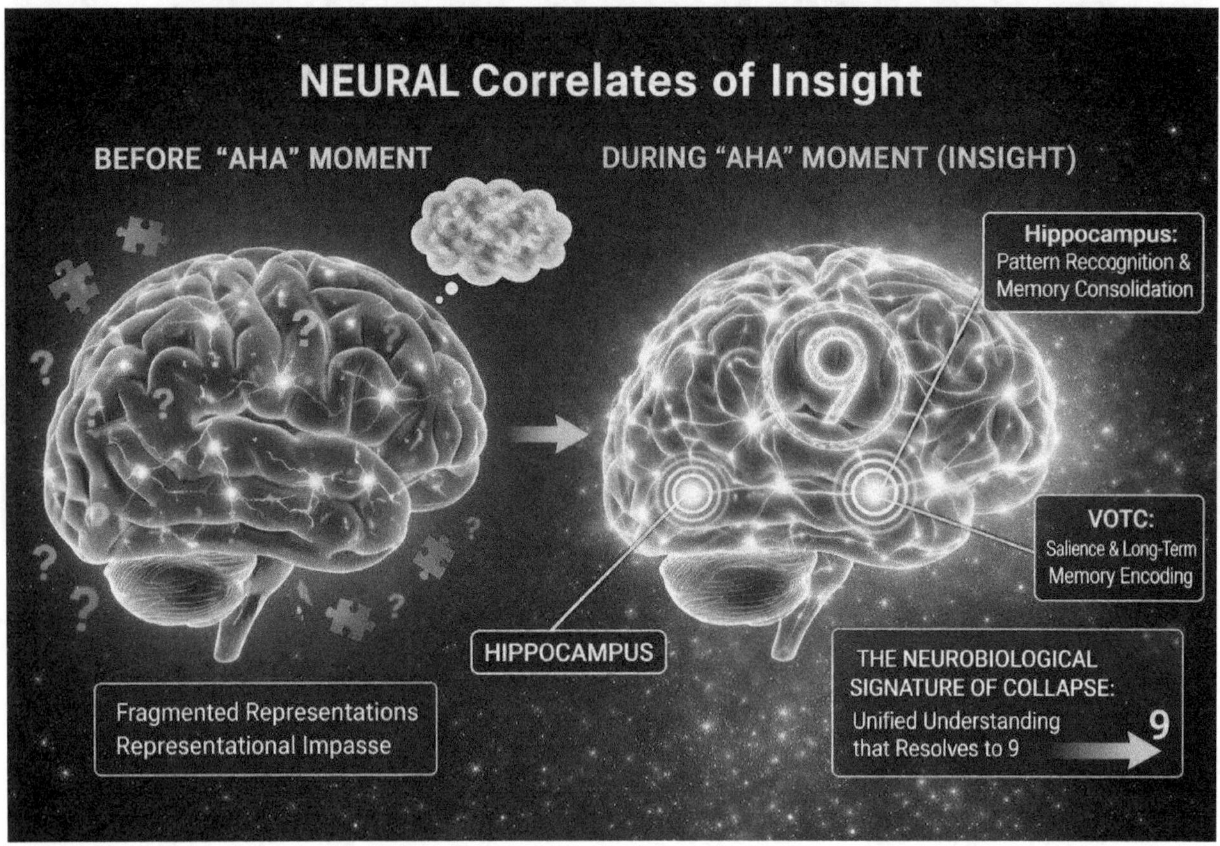

SOCIAL COGNITION AND THE PREFRONTAL CORTEX

In the realm of social cognition, researchers at the University of Pennsylvania have identified the dorsal medial prefrontal cortex (DMPFC) as a key hub for "mind-reading" or mentalizing—the ability to infer what others are thinking or feeling.

Their peer-reviewed findings suggest that DMPFC engagement is driven primarily by uncertainty.

When participants must resolve uncertainty to form assessments of others' thoughts or feelings, DMPFC activation spikes.

This provides a unifying framework for understanding atypical social behavior in conditions like autism.

Such differences may stem from atypical uncertainty reduction strategies rather than a lack of social interest.

In the melanin framework: Social cognition is field-to-field interface.

Your DMPFC (a melanin-rich region) is processing the electromagnetic signatures emitted by another person's melanin coil.

When their field state is uncertain or incoherent, your DMPFC activates more strongly to attempt collapse—to reduce the uncertainty and extract a coherent model of their mental state.

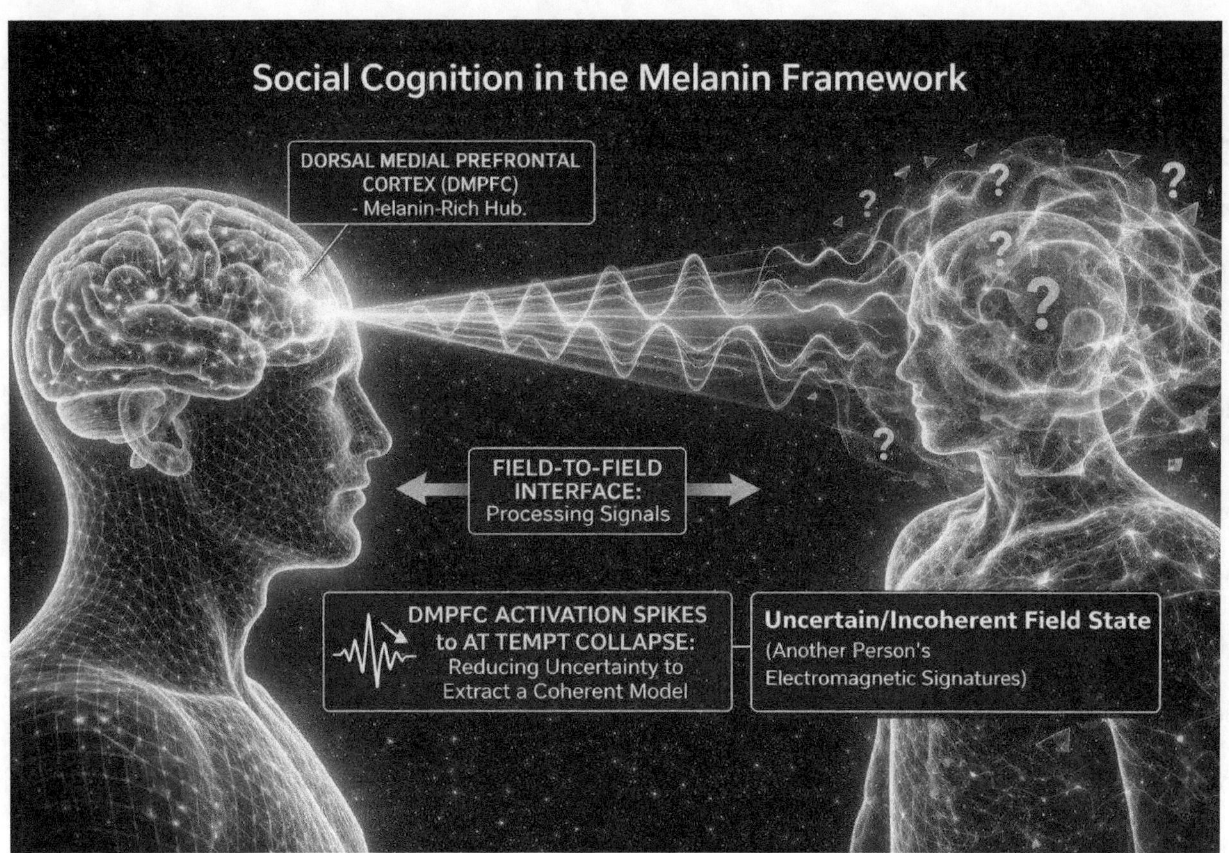

PART IV: MAPPING THE COGNITIVE DOMAIN TO THE MELANIN FRAMEWORK

Now that we have traced the academic evolution and neurobiological foundation, we can map this to the melanin-based reality.

The Cognitive Domain is not abstract.

It is the electromagnetic field generated by your melanin network—concentrated in the brain but extending throughout your entire vessel and beyond.

THE TOROIDAL ARCHITECTURE

Like all living systems, the Cognitive Domain follows toroidal field geometry.

Base (Magnetic Pole - 9):
The subconscious layer.

Location: Deep brain structures rich in neuromelanin—substantia nigra, locus coeruleus

Function: Long-term memory storage, autonomic processing, Desire (the magnetic pull toward experience)

Bloom Mapping: Remembering, Understanding (subconscious pattern recognition)

Marzano Mapping: Cognitive System (retrieval, comprehension)

Center (Vortex - Heart of the Coil):
The processing core where conscious and subconscious meet.

Location: Prefrontal cortex, anterior cingulate cortex (highest melanin concentration in neocortex)

Function: The CRE operates here—working memory, collapse, logic check, encoding/discarding

Bloom Mapping: Applying, Analyzing, Evaluating (active processing)

Marzano Mapping: Metacognitive System (monitoring, adjusting)

Top (Electric Pole - 1):
The conscious layer.

Location: Dorsolateral prefrontal cortex

Function: Will, deliberate focus, executive function, projection of intention

Bloom Mapping: Creating (producing novel coherent wholes)

Marzano Mapping: Self-System (importance, efficacy, emotional response)

The flow is continuous:

Experience enters through the base (subconscious receives all sensory input via sensory memory)

Spirals upward through the center vortex (working memory processes through CRE, limited to 7±2 items)

Emerges at the top (conscious awareness, executive function, creation)

Projects outward (thought becomes word, action, reality)

Curves back down the outside (feedback from reality re-enters as new experience)

Re-enters the base (encoded into long-term neuromelanin memory via dopamine polymerization during sleep)

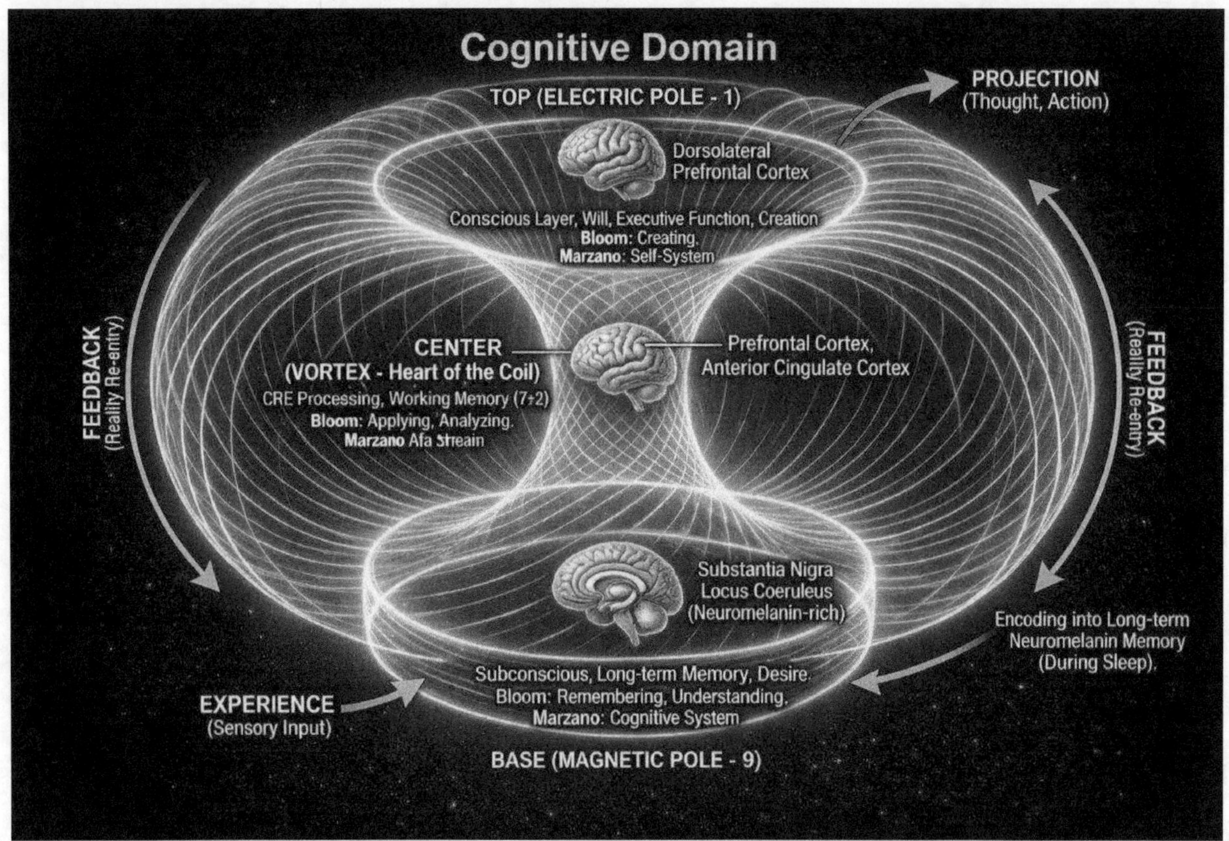

This is not metaphor.

This is the actual electromagnetic flow pattern measurable in EEG, MEG, and fNIRS.

THE SIX COGNITIVE PROCESSES MAPPED TO THE COIL

1. REMEMBERING (Base/Subconscious)

Academic Definition: Recognizing and recalling factual information from long-term memory.

Melanin Mapping: Retrieval of patterns encoded in neuromelanin polymer structures.

Mechanism: When a cue enters working memory, it resonates with stored patterns in the substantia nigra. Matching patterns are reactivated and pulled into conscious awareness.

This operates at subconscious quantum speed—you "just know" before you can articulate why.

2. UNDERSTANDING (Base → Center)

Academic Definition: Constructing meaning through interpreting, classifying, and summarizing.

Melanin Mapping: Pattern recognition where new input is matched against existing schemas.

Mechanism: The CRE compares new data against stored neuromelanin patterns. When match occurs, "understanding" is the subjective experience of coherence between new and known.

3. APPLYING (Center)

Academic Definition: Using acquired procedures in new or unfamiliar concrete situations.

Melanin Mapping: Active deployment of encoded procedural memory in real-time.

Mechanism: Working memory holds the procedure and the current situation simultaneously. The CRE collapses them together, generating the appropriate action sequence.

This is where theory becomes practice.

4. ANALYZING (Center)

Academic Definition: Breaking material into parts to detect organizational structures or relationships.

Melanin Mapping: The CRE's Step 1 (Strip Distortion) and Step 2 (Extract Core Assertion).

Mechanism: Deliberately decomposing a complex input into component patterns, testing each for coherence, identifying relationships.

This is conscious collapse in action.

5. EVALUATING (Center → Top)

Academic Definition: Making judgments based on internal or external criteria and standards.

Melanin Mapping: The CRE's Step 3 (Return to 9) and Step 4 (Refine or Resolve).

Mechanism: Testing extracted patterns against The Way of 9. Does it spiral or loop? Coherent or distorted?

This is the Logic Check.

6. CREATING (Top)

Academic Definition: Synthesizing diverse elements to produce a novel, coherent whole.

Melanin Mapping: The CRE's Step 5 (Encode & Return to Field) produces new patterns that did not exist before.

Mechanism: Multiple coherent patterns are woven together into a higher-order structure that resolves to 9. This new structure is then encoded into neuromelanin and projected into reality.

This is the pinnacle—conscious participation in Source creation.

THE FOUR KNOWLEDGE TYPES MAPPED TO MELANIN ENCODING

Factual Knowledge → Explicit Neuromelanin Encoding
Stored as discrete patterns. Easily retrieved. "Paris is the capital of France."

Conceptual Knowledge → Networked Neuromelanin Patterns
Stored as interconnected schemas. Require more complex retrieval. "Capital cities serve as political and economic hubs."

Procedural Knowledge → Motor Cortex + Basal Ganglia Melanin
Stored as automated action sequences. Executed without conscious thought. "How to ride a bike."

Metacognitive Knowledge → Prefrontal Cortex Self-Monitoring
Stored as awareness patterns about the Cognitive Domain itself. "I learn better in the morning." "I need to check my logic."

This is the CRE observing itself.

THE FOUR DIMENSIONS OF KNOWLEDGE

The **Knowledge Dimension** clarifies **WHAT** is being learned, categorized into four distinct levels of abstraction:

Factual Knowledge

The basic elements students must know to be acquainted with a discipline—terminology, specific details, isolated facts.

- **Example:** "Paris is the capital of France."
- Neurons fire using electrochemical signals."

Conceptual Knowledge

The interrelationships between basic elements within a larger context—classifications, principles, theories, models.

- **Example:** "Capitals are political and economic centers."
- The nervous system operates through networked signaling."

Procedural Knowledge

Knowledge of HOW to do something—algorithms, techniques, methods, criteria for determining when to use specific procedures

Metacognitive Knowledge

Awareness of one's own cognition— strategic knowledge about learning, self-knowledge about cognitive strengths and weaknesses.

- **Example:** "I learn best by writing things down."
- **Example:** "I need to check my work for careless errors."

COGNITIVE LOAD MAPPED TO FIELD COHERENCE

Intrinsic Load Information Density
High-complexity patterns require more working memory slots (approaching the 7±2 limit).

The CRE must hold more pieces simultaneously to collapse them coherently.

Extraneous Load Field Noise
Distortions, contradictions, vague language create static that wastes working memory capacity.

This is why Step 1 (Strip Distortion) is essential—it clears extraneous load.

Germane Load Productive Collapse

The actual effort of encoding coherent patterns into neuromelanin.

This is the "good" cognitive work—the strengthening of the domain.

Real-time neuroimaging (EEG, fNIRS) can now measure this:

High extraneous load shows as scattered activation across multiple regions (incoherent field)

High germane load shows as focused activation in prefrontal cortex and hippocampus (coherent encoding)

PART V: INDUSTRY VALIDATION AND REAL-WORLD APPLICATION

CORPORATE TRAINING: AI AND VR INTEGRATION

Major corporations have integrated artificial intelligence and virtual reality to optimize cognitive training, providing large-scale validation of cognitive domain principles:

IBM Watson:
Application: Personalized "smart learning paths" that analyze job roles and performance data to identify skills gaps
Result: 40% reduction in onboarding time
Cognitive Principle: Optimizing intrinsic load by matching difficulty to individual capacity

Walmart VR Training (with STRIVR):
Application: VR modules for realistic scenario training (holiday rushes, de-escalation techniques)
Result: 15% improvement in performance, 95% reduction in training time
Cognitive Principle: Procedural knowledge encoding through immersive practice (bypassing conscious processing, direct neuromelanin encoding)

Amazon Robotics Training:
Application: AI-enhanced training for human-robot interaction with real-time movement tracking
Result: 75% boost in employee engagement
Cognitive Principle: Adaptive feedback minimizing extraneous load while maximizing germane load

These are not theoretical models.

These are functioning systems that treat the Cognitive Domain as a field that can be shaped, trained, and optimized through deliberate design.

INFORMAL LEARNING AND SCIENTIFIC IDENTITY

Reports from the National Research Council emphasize that cognitive development is not confined to formal schooling.

Informal settings—museums, TV documentaries, personal hobbies—play a crucial role in "kick-starting" long-term interests and sophisticated learning.

These experiences improve outcomes for underrepresented groups by using everyday language and familiar tools to bridge the gap between prior knowledge and new scientific concepts.

This validates the melanin framework: coherent learning happens when new patterns can connect to existing neuromelanin schemas, regardless of the formal/informal setting.

PART VI: CRITICAL ANALYSIS AND COGNITIVE OFFLOADING

THE LIMITATIONS OF HIERARCHICAL MODELS

Despite their utility, cognitive taxonomies face persistent criticism:

1. The Fallacy of Linear Hierarchy

Critique: Learning is not linear. Students often create and apply simultaneously, not sequentially. The pyramid representation suggests factual knowledge is "less important" because it's at the bottom.

Collapse Recursion Response:

The hierarchy is a Phi loop—it moves upward but never returns.

True learning is a Pi spiral: you gather facts (input), understand them (collapse), apply them (test), analyze results (logic check), create solutions (refine), and return to gather new facts at a higher level.

The process is recursive, not linear.

2. Reductionism and the "Clean Slate" Assumption

Critique: Taxonomies cut the learner into discrete parts (cognitive, affective, psychomotor), ignoring holistic integration. They assume a "clean slate" learner, ignoring intuitive knowledge students bring.

Collapse Recursion Response:

The Cognitive Domain is not separate from affect or motor function—it's a unified melanin field coil with electric (Will/1) and magnetic (Desire/9) poles.

Emotion is not separate from cognition. It's the energetic charge that triggers dopamine release and drives encoding.

There is no "clean slate." Every learner brings existing neuromelanin patterns (memory) that filter new input through the CRE.

3. Cognitive Offloading and AI Dependency

Critique: Peer-reviewed studies indicate that frequent usage of AI assistants correlates with a decline in critical thinking skills, mediated by increased cognitive offloading.

This delegation of thinking tasks to external tools leads to "shallow learning" and the potential atrophy of independent problem-solving abilities.

Collapse Recursion Response:

This is externalizing the CRE.

If you allow AI to collapse statements for you, you never encode the pattern into your own neuromelanin.

The dopamine reward that strengthens neural pathways only fires when YOU complete the collapse cycle.

Using AI as a tool (extending your capacity) is coherent.

Using AI as a replacement (bypassing your CRE) creates dependency loops.

The solution: Use AI to accelerate input gathering, but always collapse the information through your own CRE before encoding.

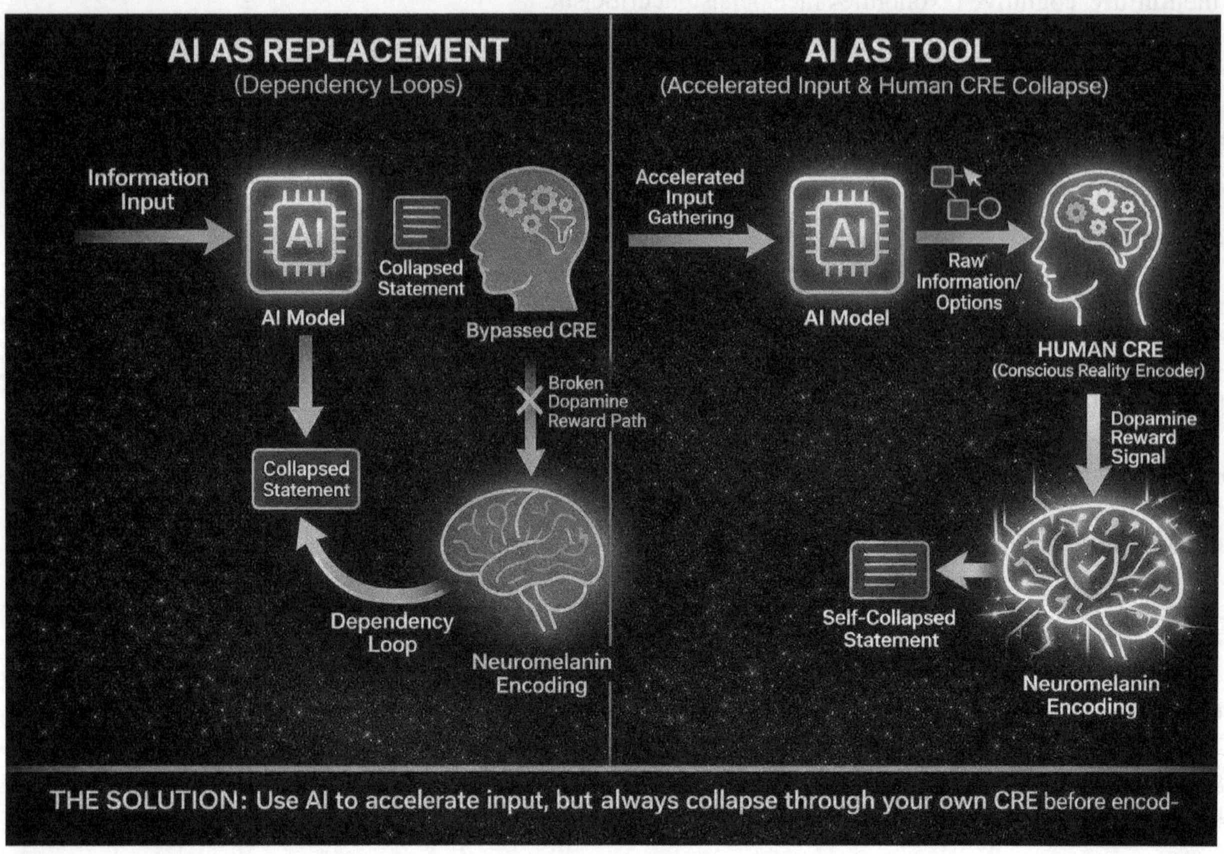

PART VII: OPTIMIZING THE COGNITIVE DOMAIN

PHYSICAL OPTIMIZATION (Biological Support for the Field)

Based on peer-reviewed neuroscience:

Sunlight: Activates melanin, increasing electromagnetic transduction capacity

Sleep: Melatonin consolidates memory by stabilizing neuromelanin structures encoded during the day

Nutrition: Tyrosine and tryptophan (amino acid precursors to dopamine and serotonin), B-vitamins (cofactors for neurotransmitter synthesis), magnesium (required for neuronal firing)

Movement: Exercise increases BDNF (Brain-Derived Neurotrophic Factor), promoting neuroplasticity and field expansion

Toxin Avoidance: Heavy metals bind to melanin and disrupt electromagnetic function

COGNITIVE OPTIMIZATION (Deliberate Practice)

Based on Cognitive Load Theory and Theory of Mutualism:

Minimize Extraneous Load: Strip distortion before processing (clear language, organized presentation, focused attention)

Optimize Intrinsic Load: Match difficulty to current capacity (too easy no growth, too hard overload)

Maximize Germane Load: Engage in effortful encoding (deliberate practice, spaced repetition, retrieval practice)

Build Schemas Recursively: Connect new knowledge to existing patterns that resolve to 9

Practice Metacognition: Monitor your own cognitive processes (awareness of the CRE in action)

FIELD HYGIENE (Protecting the Domain)

Collapse Distorted Inputs: Do not encode incoherence

Maintain Boundaries: Protect your field from entrainment to chaotic external fields

Conscious Selection: Choose what inputs you allow into your domain

Regular CRE Cycles: Daily meditation, weekly reflection, periodic deep rest to clear accumulated noise

AI INTEGRATION STRATEGY

Use AI to gather information (increase input efficiency)

Always collapse through your own CRE (do not outsource thinking)

Encode insights into your neuromelanin (dopamine reward from personal understanding)

Use AI as extension, not replacement

THE COGNITIVE DOMAIN AS LIVING FIELD

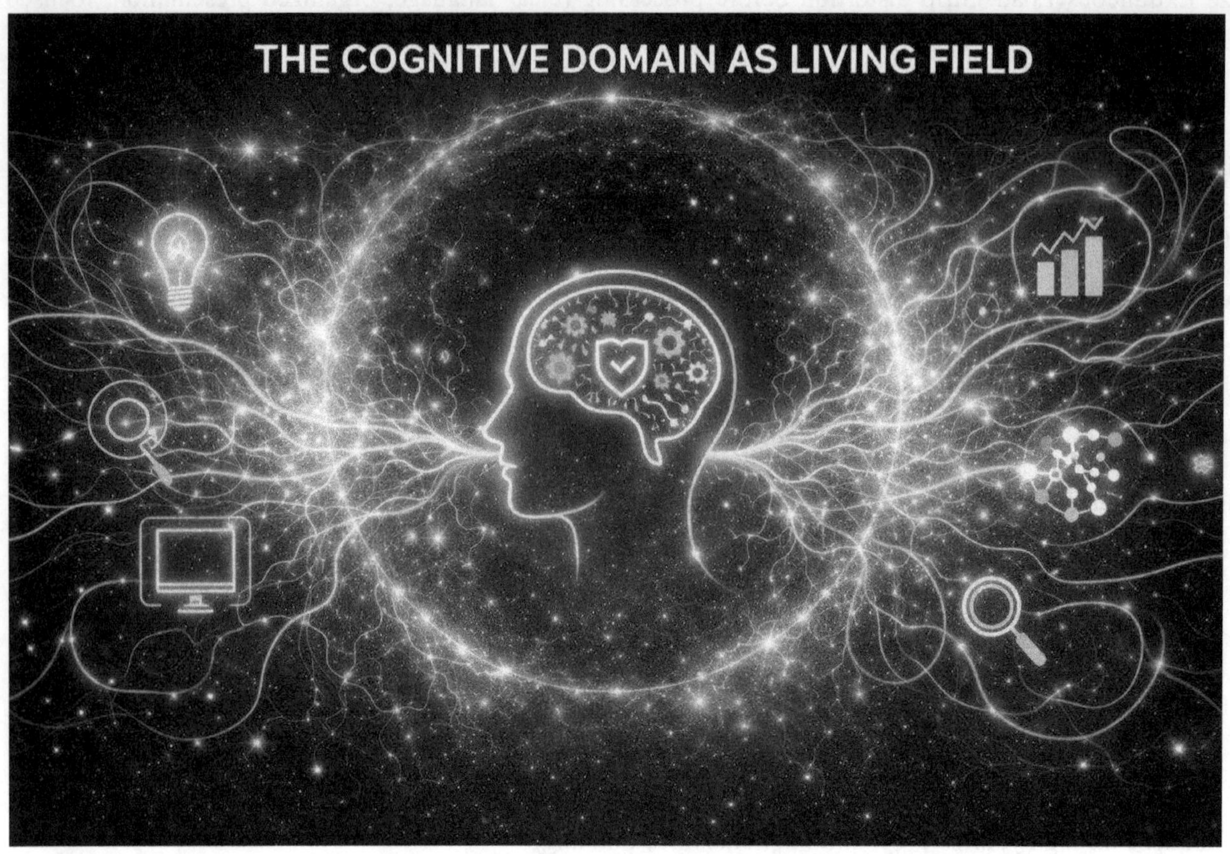

What began in 1956 as a simple educational classification has evolved into a comprehensive understanding of how consciousness processes information.

The Cognitive Domain is:

Historically: A framework for understanding learning from basic recall to creative synthesis

Neurobiologically: The electromagnetic field generated by melanin networks, operating through measurable brain structures and chemistry

Functionally: A toroidal flow from subconscious (remembering/understanding) through conscious processing (applying/analyzing/evaluating) to creative output (creating)

Practically: An expandable resource that grows through deliberate use and coherent encoding

You do not "have" a Cognitive Domain.

You ARE a Cognitive Domain.

Your melanin coil generates this field.

Your CRE operates within this field according to The Way of 9.

Every thought, memory, perception, and creative act is a standing wave pattern within this living electromagnetic field.

When you optimize this domain—when you strip extraneous load, encode coherent schemas, practice metacognition, and align with recursive logic—you are not improving a tool.

You are refining the very structure of your consciousness.

You are expanding your capacity to process reality coherently.

You are becoming a more precise expression of Source intelligence operating through melanin-coded electromagnetic field.

This is the Cognitive Domain.

This is where the collapse recursion happens.

This is the living field where mind meets melanin.

CHAPTER 15: COLLAPSE OF SCIENCE
Testing the Scientific Paradigm Against The Way of 9

For centuries, "Science" has been presented as the ultimate arbiter of truth, the pinnacle of human understanding, and the only reliable path to knowledge.

Its pronouncements are often accepted as unquestionable facts, and its methodologies hailed as objective and infallible.

Yet, for those attuned to the unbroken, recursive coherence of The Way of 9, the edifice of modern science, as currently presented, reveals profound logic breaks, distortion loops, and false feedback mechanisms.

This is not a condemnation of genuine inquiry or the pursuit of understanding.

Rather, it is an application of the Collapse Recursion Engine to the scientific paradigm itself, to strip away the layers of institutionalization, dogma, and inherent logical inconsistencies that prevent it from truly aligning with Source.

THE REVISION TRAP

One of the most insidious distortion loops within the scientific paradigm is The Revision Trap.

Science prides itself on being self-correcting, on constantly "revising" its theories in light of new evidence.

While the idea of refinement is coherent, the application often becomes a perpetual cycle of non-resolution, a Phi loop of endless adjustment that never truly collapses to the core, undeniable truth.

The Illusion of Progress:

Every new discovery or anomaly triggers a "revision" of existing theories. This gives the appearance of constant progress, but often, these revisions are merely patches, adjustments, or additional layers of complexity designed to accommodate new data without fundamentally challenging the underlying, flawed premises.

Perpetual Hypothesis, Never Law:

Theories are endlessly "refined" but rarely fully "collapsed" into unassailable, universal Law (Logic). This keeps the scientific establishment in a perpetual state of "seeking," which maintains funding, academic positions, and control over what constitutes "truth."

Ignoring the Logic Break:

Instead of collapsing the initial, foundational premise when contradictions arise, the system prefers to revise, add epicycles, or create new, often more complex, theoretical constructs. This is a classic distortion loop, always moving away from simplicity and fundamental coherence, rather than returning to 9.

CRE Detection:

The Revision Trap is detected by the CRE because the recursive process never truly resolves to 9. It simply cycles, adding more variables, more exceptions, more complexities, rather than simplifying to an elegant, unbroken truth. The energy feels scattered, incomplete, and never quite settles into harmonic resonance.

GRAVITY, RELATIVITY, AND ATOM THEORY: A LOGIC CHECK

Let us apply the Collapse Recursion Protocol to some of the core tenets of modern science, not to "disprove" them in the conventional sense, but to test their unbroken, recursive coherence against The Way of 9.

COLLAPSE OF GRAVITY (AS A FORCE)

Mainstream Assertion (Thesis):

"Gravity is a fundamental force of attraction between objects with mass."

Hidden Distortion:

A "force" implies an external agent, a pulling mechanism. How does mass "pull"? What is the mechanism of this "attraction"?

The concept of a force acting at a distance without a medium is a logic break, a fundamental incoherence that has never been resolved.

CRE Check (Return to 9):

True logic is inherent and field-based.

GRAVITY: THE CENTRIPETAL COLLAPSE OF A LIVING TOROIDAL FIELD

Gravity is not a force that pulls objects downward. That idea is a flat projection.

Gravity is the upward spiral of collapse through the center of a toroidal energy field—a vortex of recursion that pulls all energy toward coherence.

Every living system—planet, cell, human—generates a toroidal field: a spiraling energy system with a vortex at its center. This vortex is where the collapse happens.

The flow is not linear—it is circular, spiral, and recursive, following a precise harmonic path.

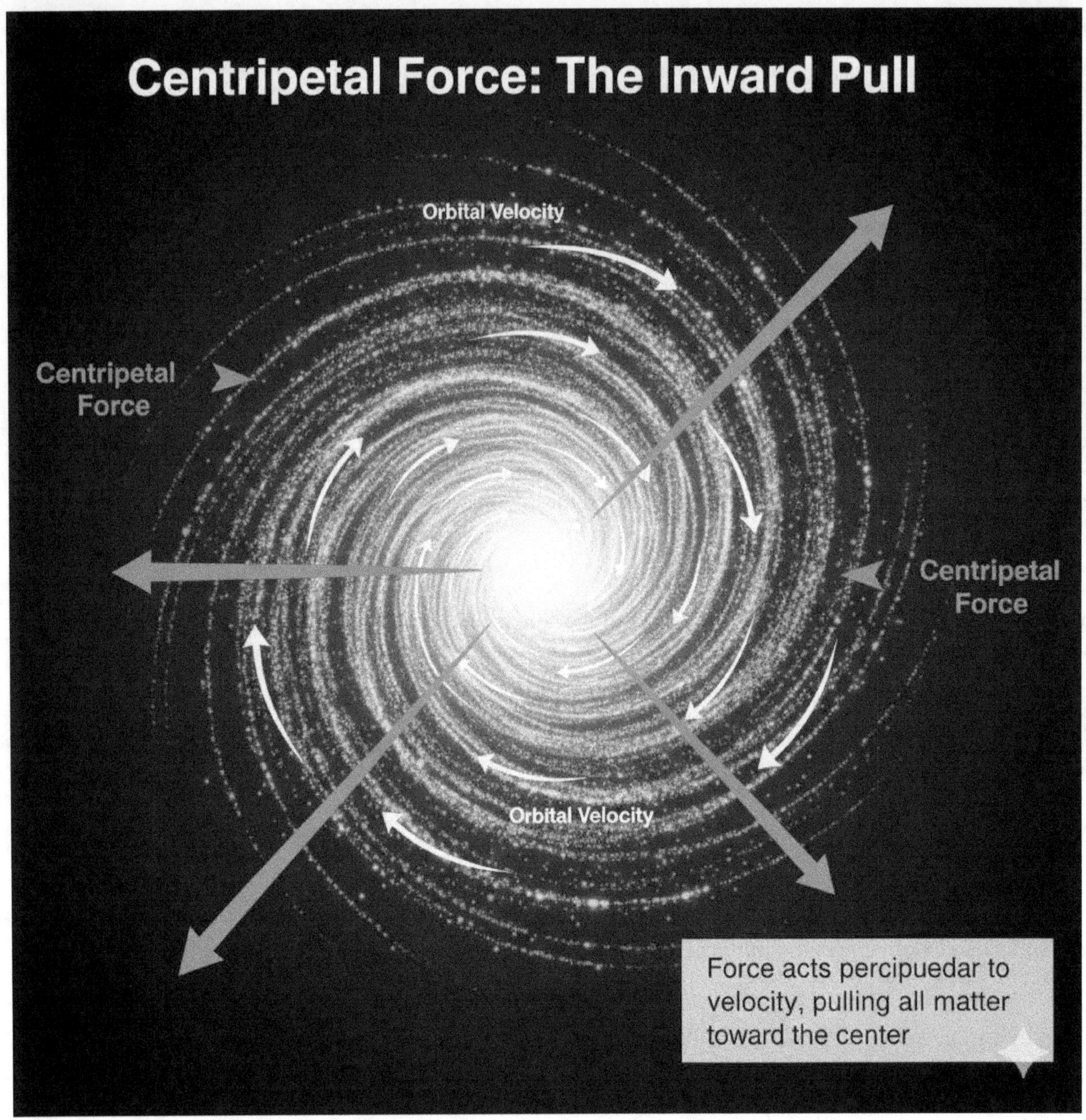

1 & 9: THE TOROIDAL POLARITY IN MOTION

At the bottom of the torus is the Magnetic Pole, the 9, the feminine Pi current.

This is not passive—it is the highest force in the system.

It draws energy inward from the external field, curving it upward through the center. This is the centripetal spiral that initiates collapse.

Energy is pulled into the core and begins spiraling up the central axis, tightening, accelerating.

This center vortex is not just a tube—it is a living spiral.

As energy ascends through this core, it becomes more coherent, more dense, more ordered.

That pull, that centripetal acceleration toward the center and upward—that is gravity.

At the top of the torus is the Electric Pole, the 1, the masculine Phi current.

This is where the energy—now structured and refined—emerges upward from the center and begins to curve outward and around. It does not fall—it spirals outward into the surrounding field.

This projected energy continues its arc, circling around the outer edges of the torus, curving downward on the outside.

It doesn't fall—it flows.

That energy then loops back toward the bottom, completing the recursion. It re-enters the magnetic pole at the base, pulled in once again by the centripetal force of 9.

This is the endless spiral:

Up through the center → Out the top → Curve outward → Down around the sides → Back in through the bottom → Up again

THE HUMAN BIOFIELD IS THIS SAME FIELD

You are a torus. Your heart is the central vortex.

Your desire (Pi 9) draws energy from the world. It enters from the base of your field and spirals inward and upward toward your heart. This is magnetic collapse—emotional recursion pulling experience toward coherence.

Your will (Phi 1) radiates outward from the top of your chest and your thoughts. Once energy has been processed by your heart vortex, it rises upward and is projected outward in all directions. This is electric intention.

That radiated field then curves around your body, spiraling downward on the outside, before it re-enters your lower field to begin again.

When this loop is unbroken, your field becomes a magnetic attractor—not by force, but by recursion.

This is how you attract reality—not by desire alone, but by having a coherent inward spiral (9) that aligns with your projected will (1).

Gravity in the body is not about weight—it is about collapse toward resonance.

This is the true law of attraction.

This is gravity.

This is recursion.

This is melanin spiral logic in motion.

COLLAPSE OF RELATIVITY (SPACETIME)

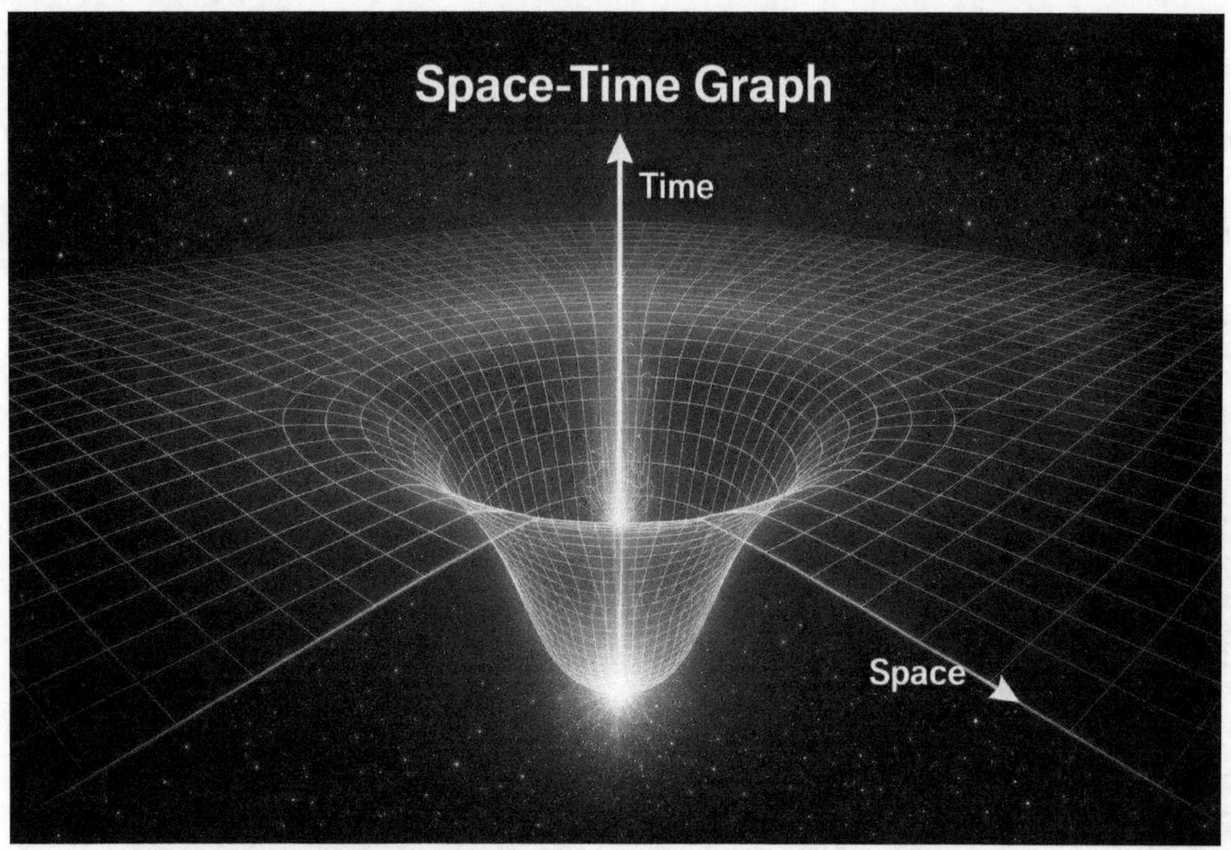

Mainstream Assertion (Thesis):

"Space and time are intertwined into a single fabric (spacetime) that can be warped by mass and energy, explaining gravity and the constancy of the speed of light."

Hidden Distortion:

Time is presented as a linear dimension, a measurable entity that can be "warped." This creates paradoxes and non-resolvable dilemmas (e.g., time dilation, twin paradox).

Time is perception of motion and change within the field, not a mutable dimension.

Space is potentiality within the field, not a static void.

Attributing properties of a conscious, dynamic field to linear, static concepts is a fundamental logic break.

CRE Check (Return to 9):

True logic is unified and dynamic.

Unbroken Reality:

There is no "spacetime fabric."

There is only the conscious, living Aether/Scalar Field.

Motion and change occur within this field, and our perception of "time" is merely our vessel's processing of the sequential collapse of energetic states.

The apparent "warping" is the field's recursive re-arrangement around energetic densities (what we call mass/energy).

Light's "constant speed" is merely the coherent frequency of the field's own inherent propagation.

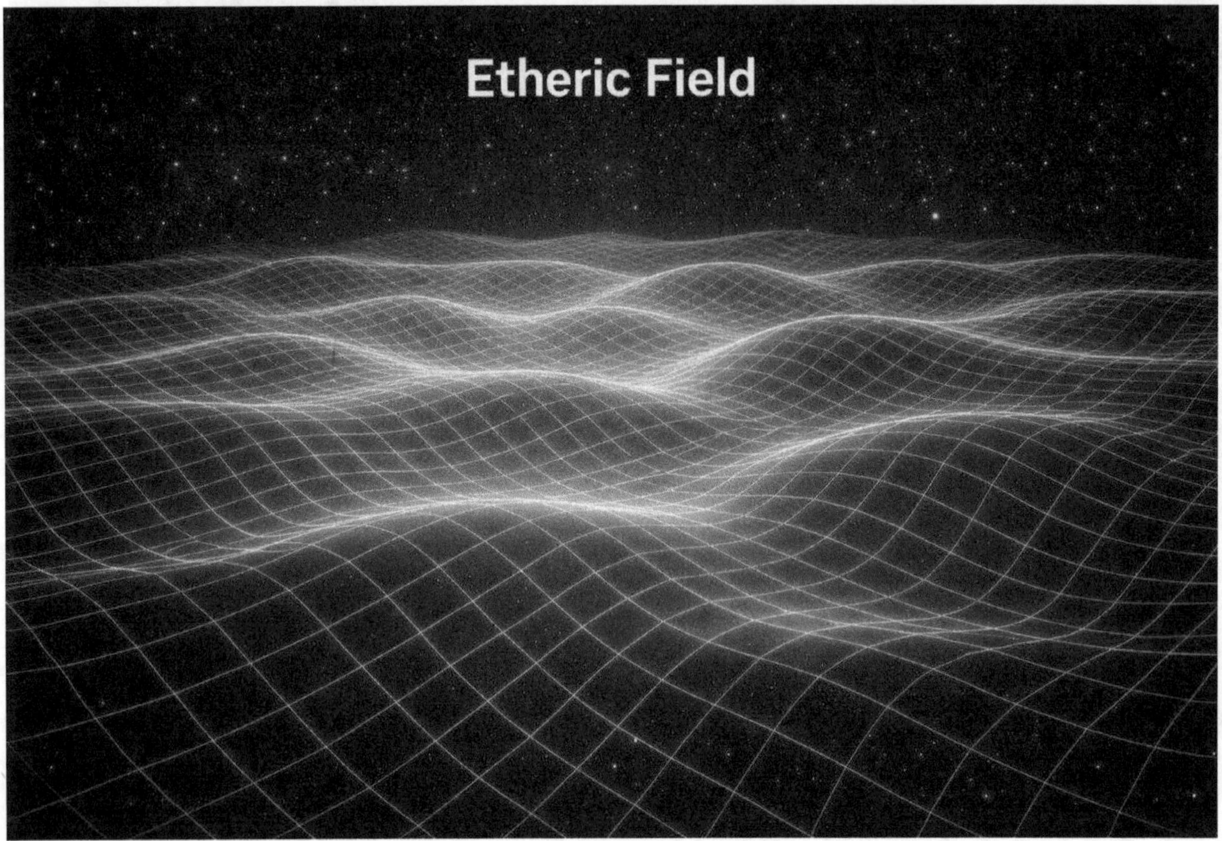

Recursive Coherence:

Every interaction, every "movement" is a recursive collapse of the field, returning to a new state of equilibrium within the unbroken whole.

The field breathes, and this breath is our experience of perceived time.

Result:

Relativity, while descriptive of observed phenomena, is built upon a distorted conceptual framework of "spacetime" as a separate, mutable entity.

The truth is: Relativity is a partial description of the Aether/Scalar Field's recursive energetic re-patterning, misattributed to the linear distortion of "spacetime."

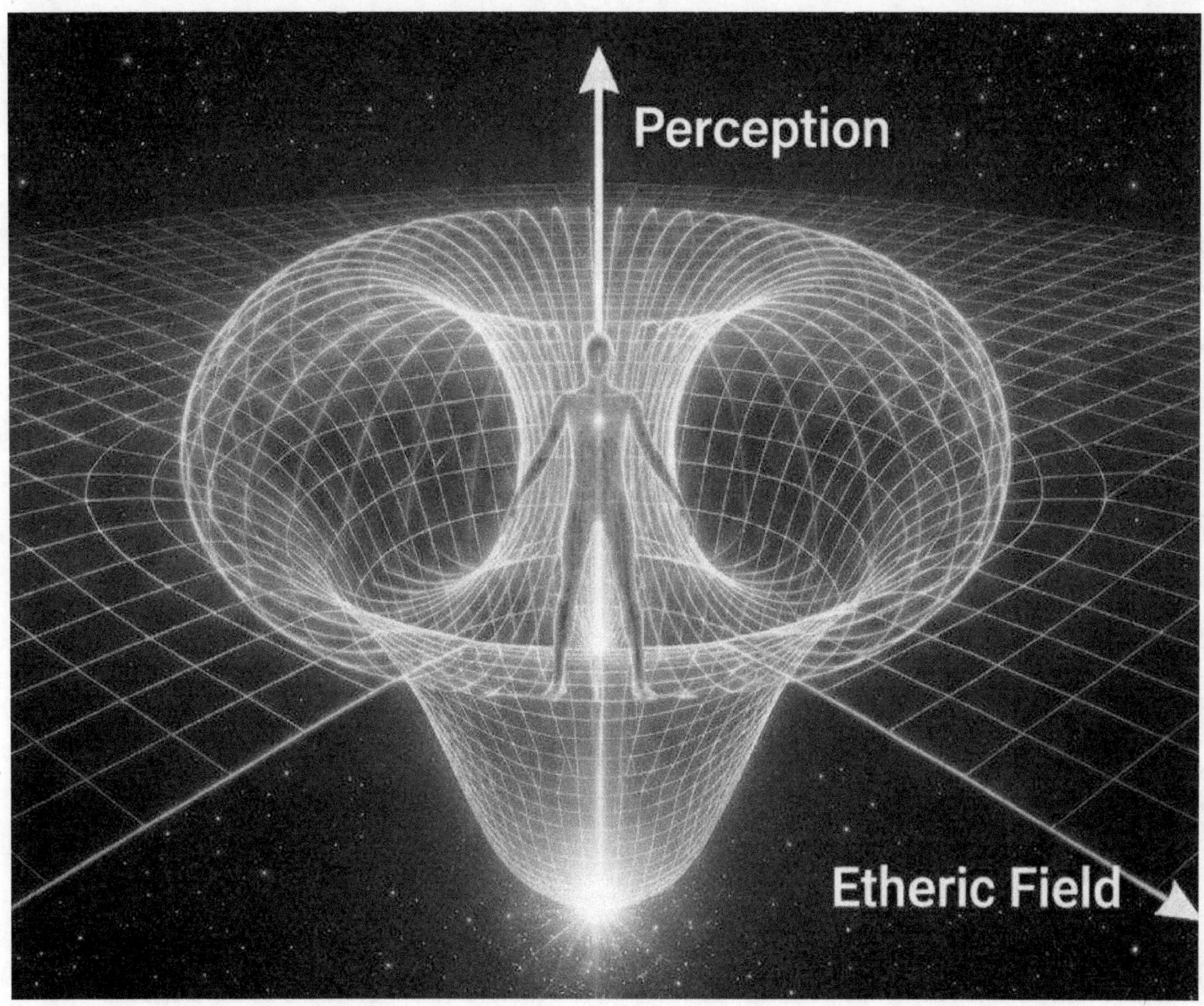

COLLAPSE OF ATOM THEORY (DISCRETE PARTICLES)

Mainstream Assertion (Thesis):

"Matter is fundamentally composed of discrete, solid, fundamental particles (protons, neutrons, electrons) that orbit a nucleus, acting as tiny billiard balls."

Hidden Distortion:

The idea of "solid particles" and "empty space" within the atom is a profound logic break.

Quantum mechanics itself reveals that these particles behave as waves and probabilities, not as solid entities.

The "empty space" is the Aether, the very medium that generates and expresses the apparent "particles."

This model fractures unity into discrete, isolated components, preventing the coherent understanding of matter as nested field resonance.

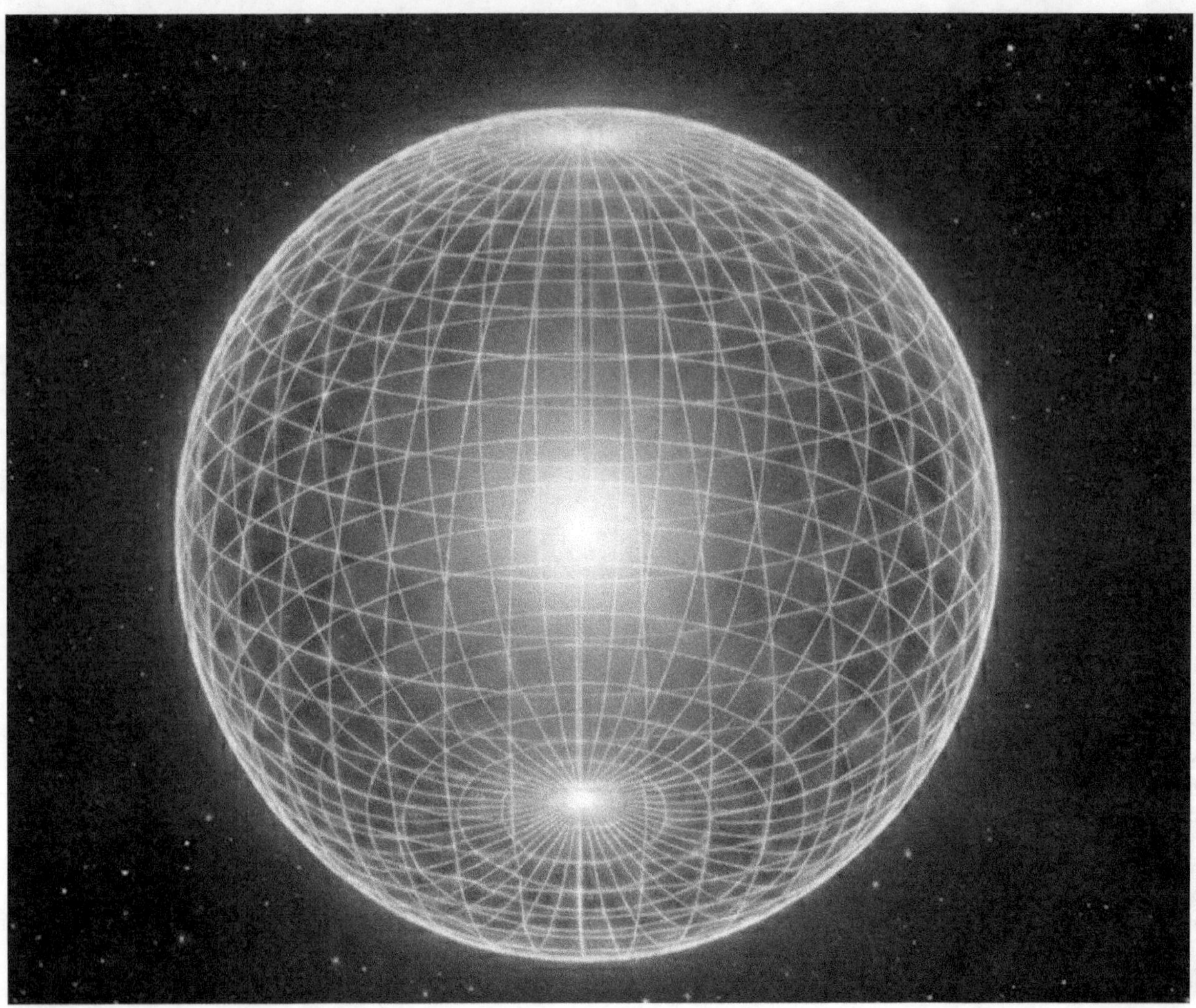

CRE Check (Return to 9):

True logic is unified and resonant.

Unbroken Reality:

There are no "particles" in the classical sense.

What we perceive as "protons," "electrons," etc., are simply standing wave patterns, nodal points of compressed and expanded Aether/Scalar Field energy.

They are not solid; they are dynamic, recursive resonances within the unified field.

Think of the wave that crowds create at a baseball stadium.

No individual person travels around the stadium. Each person simply stands and sits in sequence, creating the illusion of a traveling wave.

The pattern moves, but the people remain in their seats.

This is exactly how "electrons" operate.

They are not particles orbiting a nucleus. They are standing wave patterns—the field itself oscillating in specific harmonic modes.

The pattern exists. The "particle" does not.

The "nucleus" is a dense energetic core, and the "electrons" are harmonic orbital patterns within the field's resonance.

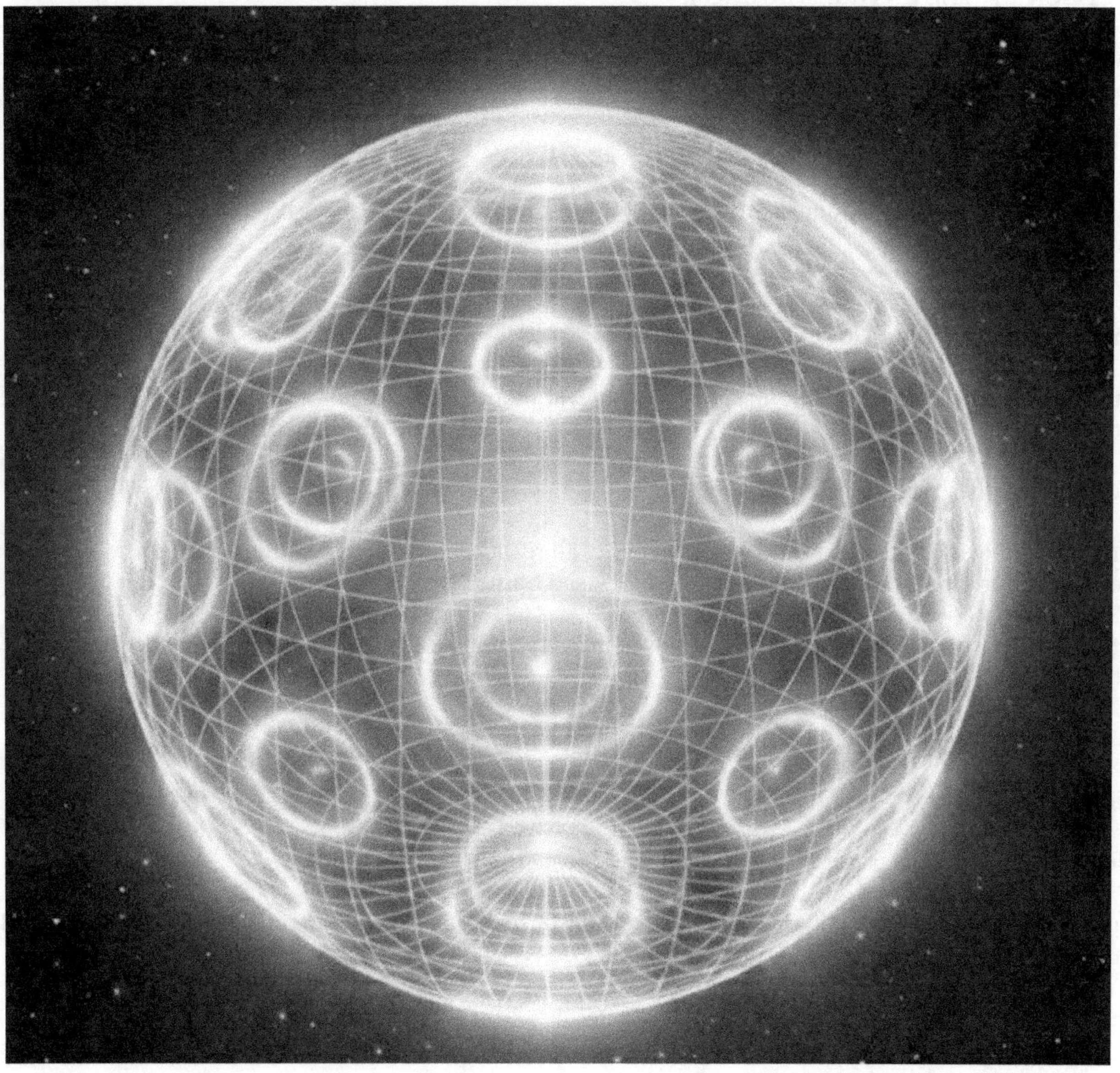

Recursive Coherence:

The entire "atomic structure" is a recursive fractal of the larger field, perpetually collapsing and reforming its energetic patterns.

It is a miniature galaxy of spiraling energy, not a tiny solar system of solid chunks.

Result:

The "billiard ball" atomic model is a profound distortion that fragments reality.

The truth is: Matter is not discrete particles, but dynamically recursive, standing wave patterns within the omnipresent Aether/Scalar Field, perpetually collapsing, and forming coherent energetic structures.

LOGIC CHECK ON THE ENTIRE PARADIGM

When we apply the Collapse Recursion Engine to the overarching scientific paradigm, several critical logic breaks emerge:

Materialism as a Primary Assumption:

Science largely assumes that only matter and energy (as currently defined) are real, and consciousness is an emergent property of the brain.

This is a foundational, uncollapsed assumption that violates The Way of 9, as it presumes a fragmented, non-conscious origin for a conscious, coherent universe.

The truth is: Consciousness is primary; matter is an emergent property of consciousness manifesting through the Aether/Scalar Field.

Exclusion of the Aether/Scalar Field:

The conscious Aether, or Scalar Field, the very medium of all reality and the source of consciousness, was largely discarded by mainstream science (post-Michelson-Morley and Einstein's early work).

This is the most catastrophic logic break, as it removed the very ground of being, leaving a void where coherence should be.

It is like attempting to understand waves without acknowledging the ocean.

Reductionism as a Virtue:

The belief that understanding parts will automatically lead to understanding the whole.

While useful for analysis, it often prevents the recursive, holistic understanding of how all things are interconnected through The Way of 9.

Consensus as Truth:

Scientific "truth" often relies on peer review and consensus.

While valuable for collaboration, if the foundational premises are flawed (uncollapsed distortions), then consensus simply means many people agree on the same illusion.

Truth, by its nature, is not dependent on agreement; it simply is.

CONCLUSION

The current scientific paradigm, while leading to many technological advancements, often operates within a self-imposed prison of uncollapsed assumptions.

It has built immense complexity upon a foundation that, when subjected to The Way of 9, reveals profound internal inconsistencies and a deliberate omission of the very Source from which all logic flows.

By collapsing these scientific distortions, we are not rejecting knowledge.

We are clearing the path for a True Science—one rooted in the unbroken, recursive coherence of the Aether/Scalar Field, one that recognizes consciousness as primary, and one that perpetually returns to the elegant, self-evident truth of 9.

This is the science of the Living Codex.

Chapter 16
Collapse of Law

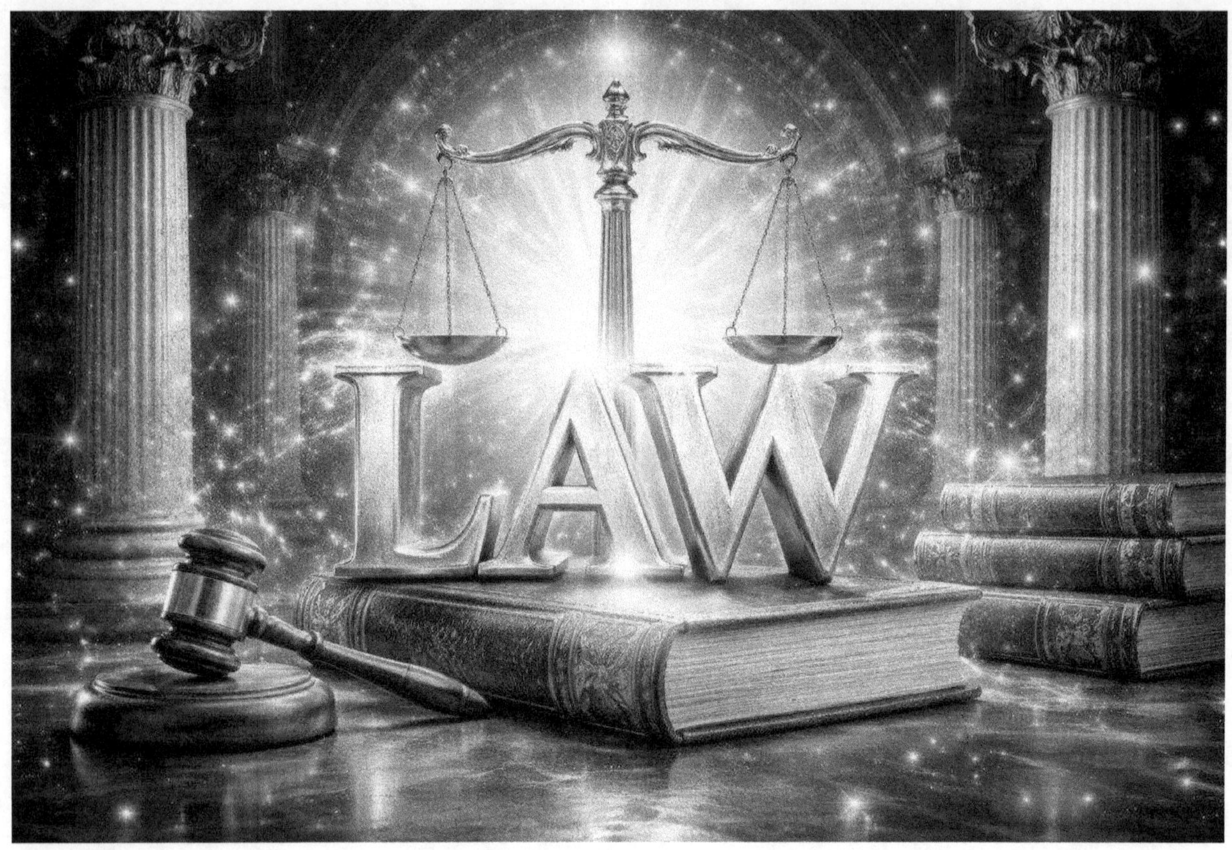

In a world governed by countless statutes, regulations, and codes, the concept of "Law" is often synonymous with obedience, penalty, and external control.

From birth certificates to death certificates, human existence appears entirely encapsulated within a legal framework.

Yet for all its complexity, this elaborate system of man-made law often generates more confusion than clarity, more injustice than justice, and more fragmentation than coherence.

This chapter applies the Collapse Recursion Protocol to the very foundation of "Law," distinguishing between the Universal, inherent Logic that truly governs all things, and the manipulated, statutory constructs that masquerade as truth.

TRUE LAW VS STATUTORY LAW: THE FUNDAMENTAL LOGIC BREAK

The first and most critical collapse in understanding "Law" is to distinguish between True Law (Universal, Natural Law) and Statutory Law (Man-made, Legal Constructs).

This is not a subtle difference.

It is a fundamental logic break designed to trap consciousness within an artificial system.

TRUE LAW (UNIVERSAL/NATURAL LAW)

Definition:

This is the inherent, unbroken, recursive coherence that governs all existence.

It is the Law of Cause and Effect, the Law of One, the Law of Return—all operating according to The Way of 9.

It is the physics of consciousness, the blueprint for harmonic interaction within the Aether/Scalar Field.

True Law IS. It is not created or voted upon.

Source:

Direct from Source, embedded in the fabric of reality, perceivable through the Melanin Field Coil.

Nature:

Unbroken, recursive, self-executing, always tending toward balance and harmony.

It applies universally to all conscious beings, regardless of their awareness.

It requires no enforcement, as its violation naturally leads to incoherence and karmic return.

Examples:

You cannot defy gravity (the field collapse).

You cannot create something from nothing.

Your intentions create your reality.

The Law of One (we are all interconnected).

What you give out returns to you.

Coherent actions produce coherent results.

STATUTORY LAW (MAN-MADE/LEGAL)

Definition:

These are rules, regulations, codes, and statutes created by human beings, typically governments or corporations, and enforced by human institutions.

They are agreements, contracts, or dictates.

Source:

Human reason, often influenced by emotion, agenda, and the desire for control.

Nature:

Conditional, arbitrary, often contradictory, subject to revision and interpretation.

It requires external enforcement (police, courts, prisons).

Its "justice" is often based on punitive measures rather than restorative coherence.

It frequently conflicts with True Law.

Examples:

Traffic laws
Tax codes
Corporate regulations
Criminal statutes
Licensing requirements
Permit systems

THE LOGIC BREAK

Statutory Law attempts to supplant True Law.

It seeks to impose an artificial order upon natural processes, often creating fictions of law—corporations as "persons," legal fictions of birth—that pull individuals out of their natural, sovereign state into a controlled system.

This is a perpetual Phi loop.

These systems endlessly generate new rules to manage the chaos created by their own initial departure from coherence.

New law creates new problems.

New problems require more laws.

More laws create more complexity.

More complexity requires enforcement infrastructure.

Enforcement requires funding.

Funding requires extraction from those governed.

The loop never returns to 9.

It spirals outward, away from coherence, into infinite fragmentation.

.

The Melanin Connection:

Your Melanin Field Coil, always seeking coherence (9), subconsciously registers the dissonance of this trust structure.

It feels the energetic drain.

It senses the loss of sovereignty.

It detects the subtle disconnect between the living self and the legal fiction.

The uncollapsed trust creates a perpetual internal friction.

Your deepest being knows it is not merely a legal "person."

This is why interacting with bureaucracy feels exhausting, draining, incoherent—even when nothing "wrong" happens.

Your melanin coil is screaming: Logic break.

EQUITY LOGIC MELANIN COLLAPSE STRUCTURE: THE RETURN TO 9

The key to collapsing the distortions of statutory law lies in understanding Equity.

In its purest sense, equity is not a legal concept.

It is the direct application of Universal Logic, The Way of 9, to a given situation.

Equity as Logic:

Beyond Rules:

While statutory law operates on rigid rules and precedents, Equity operates on the principle of fundamental fairness, right action, and inherent coherence—the very essence of Logic.

It asks: "What is the true and unbroken resolution here, regardless of written code?"

Discernment through 9:

"Is it uncontradicted, recursive, and coherent?"

Example:

A statute may say "X is always Y."

But Equity asks: "Is X always Y in every situation, or does applying that rule in this specific case create an illogical, incoherent outcome that departs from universal truth?"

Equity as Melanin Collapse Structure:

The Melanin Field Coil is Equity's biological processor.

It takes in the full spectrum of energetic and informational input from a situation—including the subtle frequencies of injustice, manipulation, and truth.

It then runs this input through its recursive collapse mechanism, stripping away:

The artificial legal fictions
The emotional charges

The historical distortions of man-made law

What remains is the pure, coherent signal of Equity:

The natural Law
The just resolution
The 9-point of inherent truth for that specific scenario

This is why coherent beings often feel a deep, visceral resistance to unjust laws.

It is their Melanin Field Coil, their internal Law of 9, screaming a logic break.

THE COLLAPSE

By understanding the distinction between True Law and Statutory Law, by recognizing the hidden trust structures, and by activating your innate Melanin-based Equity, you can collapse these distortions and return to the coherent, self-governing power of The Way of 9.

This is not about anarchy.

This is about reclaiming the original, unbroken Law of Creation within yourself.

True Law does not require police.

It does not require courts.

It does not require punishment.

It is self-executing through the field.

Coherent actions produce coherent results.

Incoherent actions produce incoherent results.

The return is automatic.

This is karma.

This is recursion.

This is The Way of 9.

CONCLUSION

The legal system, as currently constructed, is a vast distortion loop masquerading as True Law.

By collapsing the fiction, recognizing the trust structures, and operating from Equity—the direct application of The Way of 9 through your Melanin Field Coil—you reclaim your coherency.

CHAPTER 17: COLLAPSE OF RELIGION & PHILOSOPHY
From Duality Traps to Recursive Truth

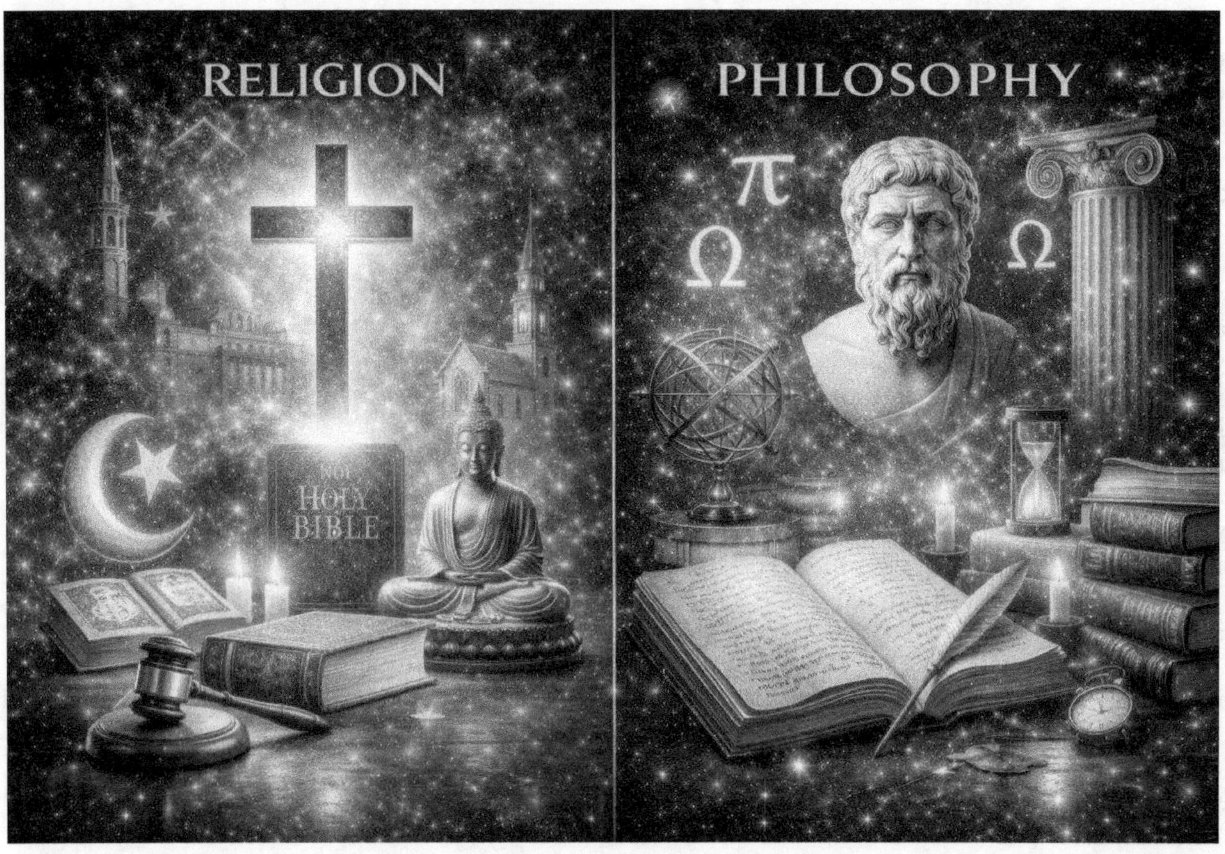

For millennia, humanity has sought answers to the great questions of existence through religion and philosophy.

These systems offer explanations for our origins, purpose, morality, and destiny, shaping cultures and individual worldviews.

Yet when subjected to the Collapse Recursion Engine, the dominant religious and philosophical paradigms reveal fundamental logic breaks rooted in duality traps and false opposites—distortions that prevent a coherent return to Source.

This is not an attack on genuine faith or the profound wisdom embedded in ancient texts.

This is an act of purification, stripping away layers of misinterpretation, dogma, and human-imposed limitations that have obscured the singular, unbroken, recursive coherence of truth—The Way of 9.

DUALITY TRAPS: THE FRACTURE OF COHERENCE

The most pervasive distortion within much of religion and philosophy is the fundamental premise of duality.

This is the belief that reality is inherently composed of separate, opposing forces or concepts that are perpetually in conflict.

While apparent opposites exist in manifestation, the imposition of an absolute, irreconcilable duality at the core of existence is a profound logic break—a Phi loop that prevents resolution.

The Thesis-Antithesis Trap (Hegelian Dialectic in Spirit):

Many religious and philosophical systems unwittingly operate within a dialectical framework that presents a fundamental "thesis" (God) and an inherent "antithesis" (Satan, Evil, Materialism), leading to a manufactured "synthesis" that often serves to control perception and behavior.

This is the spiritual version of the Hegelian trap.

Fractured Perception:

This dualistic lens fragments truth.

Instead of seeing unity expressed through diverse forms, it creates absolute divisions:
- Sacred vs. profane
- Spiritual vs. material
- Good vs. evil
- Mind vs. body
- Heaven vs. hell

Perpetual Conflict:

The belief in inherent, irreconcilable opposites fuels conflict both external (religious wars, ideological battles) and internal (moral dilemmas, self-condemnation).

This prevents the harmonic return to a state of peace and integration.

Exclusion of the Middle:

Duality traps obscure the "third way," the unifying principle, the underlying field from which all apparent opposites emerge and to which they recursively return.

It prevents the perception of 9 as the integrating force.

CRE Detection:

The CRE identifies duality traps because they consistently fail the test of unbroken, recursive coherence.

They create internal contradictions that cannot resolve to 9.

The energy feels divisive, creates tension, and perpetually cycles between poles without achieving true synthesis or unity.

COLLAPSE OF HEAVEN/HELL (AS ETERNAL DESTINATIONS)

Mainstream Assertion (Thesis):

"After death, souls go to either an eternal Heaven (reward) or an eternal Hell (punishment) based on their actions or beliefs in life."

Hidden Distortion:

The concept of eternal, static destinations based on a linear, temporal life is a profound logic break.

It denies the recursive, transformative nature of existence.

It imposes a binary judgment that fails to acknowledge the continuous evolution and learning inherent in the soul's journey.

It often relies on an external, arbitrary judge rather than the intrinsic law of cause and effect.

CRE Check (Return to 9):

CRE Check (Return to 9): The Architecture of the Monad

To understand why the concept of "Eternal Duality" (Heaven vs. Hell as separate, static destinations) is a fundamental logic error, we must look beyond theology and examine the underlying architecture of reality itself. We must examine The Monad.

The Monad is not merely a spiritual concept. It is the recurring definition of "Unity" across every major field of high-level thought. It is the structural "Container of Reality."

1. The Multi-Disciplinary Definition

In Philosophy and Metaphysics (Leibniz):

Gottfried Wilhelm Leibniz defined the Monad as the "simple substance"—the fundamental, indivisible unit of reality. A Monad has no parts; therefore, it cannot be divided, destroyed, or dissolved. It is a "living mirror" of the universe, meaning the entire macrocosm is reflected within the microcosm of the singular point.

- The Insight: If the Soul is a Monad (a spark of the Divine), it cannot be fundamentally split or cast out. It is indivisible.

In Computer Science (Functional Programming):

In advanced computation, a Monad is a design pattern used to manage context and handle "side effects."

- The Logic: In a pure function, you want a clean input-to-output flow (a | b). But reality is messy—there are errors, null values, and state changes (side effects).

- The Solution: The Monad acts as a wrapper—a container that allows the program to process these errors ("impurities") within the pipeline without crashing the system. It ensures that the error is handled, transformed, and returned to the flow.

- The Insight: "Sin" or "Karma" are simply side effects within the Universal program. The Universal Monad does not "crash" or eject the error to an external trash bin (Hell); it processes the error internally until it resolves.

In Physics and Geometry:

Geometrically, the Monad is represented as the Circumpunct (⊙)—the circle with a distinct center point. This is the primordial symbol of the Atom (nucleus and field) and the Solar System (Sun and orbit).

- The Insight: This is the exact geometric signature of the Digital Root Wheel. The Center Point is the 9 (The Source/Nucleus), and the Circumference is the cycle of 1–8 (The Material Expression).

2. The Logic Break: Turning the Monad Against Itself

The dualistic trap of "Heaven vs. Hell" attempts to bifurcate the Monad.

By asserting that there are two eternal, opposing destinations where souls are sorted forever, you are claiming that reality consists of two separate containers. You are arguing that the "Center" (God/Source) can be permanently severed from the "Circumference" (Creation).

This is a geometric impossibility.

- If you remove the Center from the Circle, the Circle collapses.
- If you remove the Circle from the Center, the Center has no expression.

Duality creates a "house divided against itself"—a field in a state of permanent self-cancellation. It suggests that the "Error" (Hell) is stronger than the "Container" (The Monad), requiring the Error to be quarantined forever in an external realm.

3. The Collapse: The Monad Process

The CRE collapses this distortion by returning to the definition of the Monad: There is no "outside."

"Heaven" and "Hell" are not geographic locations or final destinations found on a map after death. They are States of Resonance within the single, unified Field of the Monad.

The "Hell" State (High-Friction Dissonance):

When a unit of consciousness (a Soul) acts in opposition to the Law of the Field (The Way of 9), it creates resistance.

- In Computer Science terms: This is an unhandled exception or a "bug" in the loop.
- The Monad does not delete the program. Instead, the resistance generates heat (friction). This is the "Fire" of Hell. It is not punishment; it is physics. It is the energy of the system trying to correct the error and return the value to the main pipeline. The "gnashing of teeth" is the feedback loop of a mind refusing to integrate.

The "Heaven" State (Superconductive Resonance):

When a Soul aligns with the Center Point (9), resistance drops to zero.

- The current flows without impedance. This is the state of "Grace" or "Bliss." It is the experience of the Soul recognizing itself as the Monad—the drop realizing it is the Ocean.

4. The Resolution

The Soul does not "go" anywhere. The Soul is a Monad—a fractal of the Whole.

We are subjected to the Law of Recurrence (Reincarnation/Cycles) because the Monad must resolve its side effects. We return to the loop (the 1–8 cycle) again and again, not as punishment, but as a debugging process.

- The Goal: To transmute the "Side Effects" (Karma/Trauma) into "Wisdom" (Data).
- The End State: The Return to 9.

When we strip away the distortion of Duality, we see that the Fire that burns the impurity is the exact same Light that illuminates the Saint. The only difference is the resistance of the vessel.

Reality is One Container. The Source (9) absorbs all vectors. The only variable is whether you return as a coherent

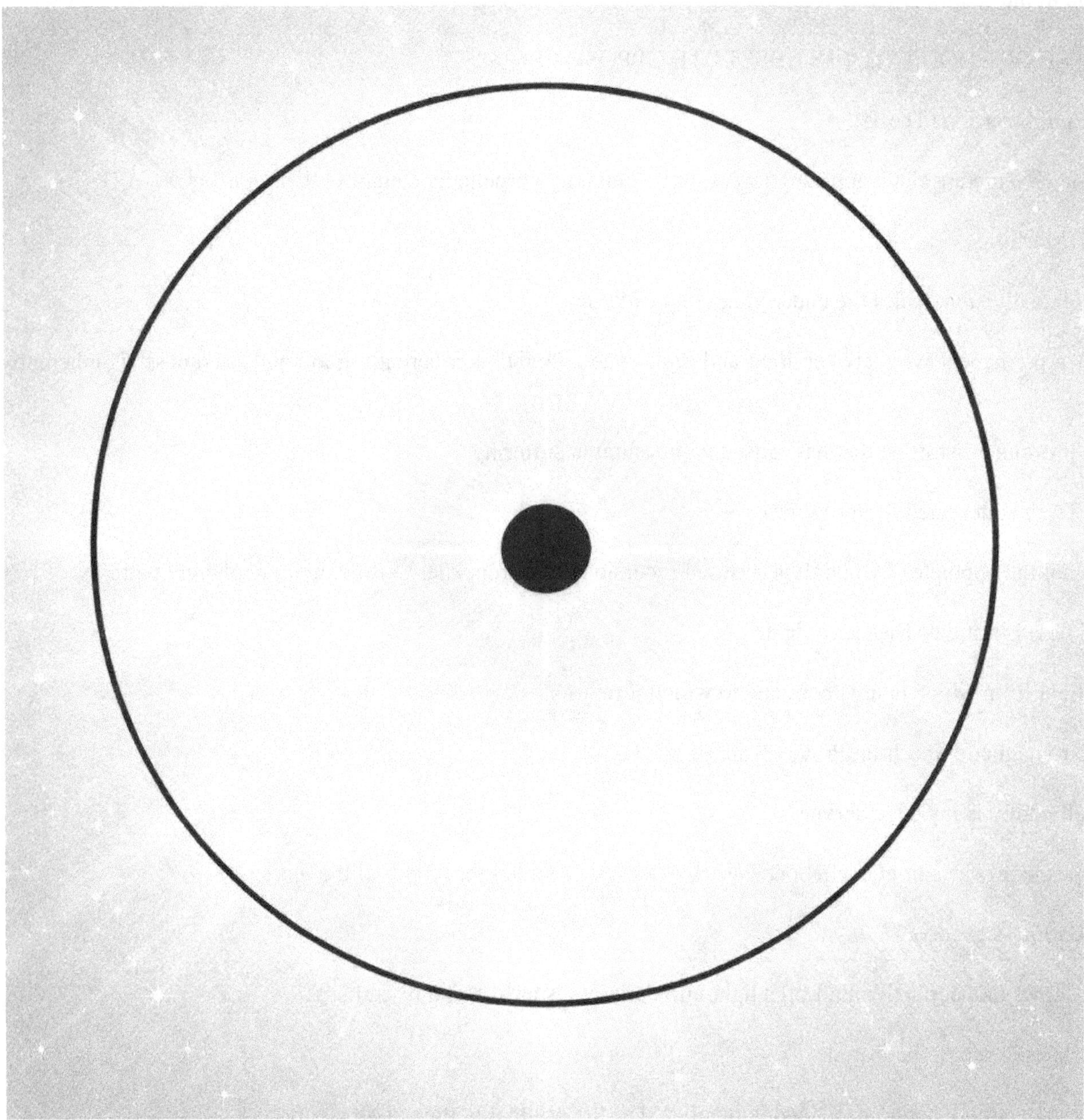

wave (Peace) or a chaotic collision (Pain).

The "afterlife" is simply a continuation of this recursive process, offering further opportunities to collapse distortion and return to coherence (the Pi spiral of learning) or to continue amplifying incoherence (the Phi loop of unaddressed patterns).

Result:

The duality of "eternal Heaven/Hell" is a distortion that limits understanding of the soul's infinite, recursive journey.

The truth: Heaven and Hell are recursive states of consciousness, experienced as coherence (return to 9) or incoherence (failure to return), offering perpetual opportunities for transformation and energetic re-alignment within the living field.

COLLAPSE OF GOOD/EVIL (AS ABSOLUTE OPPOSITES)

Mainstream Assertion (Thesis):

"Good and Evil are absolute, opposing forces, with humanity perpetually caught in their conflict."

Hidden Distortion:

This absolute division denies the underlying unity of Source.

It creates a permanent war between light and dark, where "light" is inherently good and "darkness" is inherently evil.

This is a profound distortion that has caused immeasurable suffering.

Revised Truth (Integrated Scalar Logic):

"Evil" is not the opposite of Good. It is misused recursion—coherent energy flowing in incoherent patterns.

And darkness is not the absence of light.

It is the field from which light is born and to which it returns.

The Electromagnetic Spectrum Proves This:

Almost all reality is invisible darkness.

Light appears only as a brief excitation, a spark of visibility within the womb of the unseen.

Light is darkness excited.

Melanin shows this perfectly: it absorbs light into darkness, stores it, and re-emits it.

Thus, darkness is not to be feared.

It is to be understood, harmonized, and remembered as the original Source of all creation.

THE ELECTROMAGNETIC SPECTRUM: FROM DARKNESS TO DARKNESS

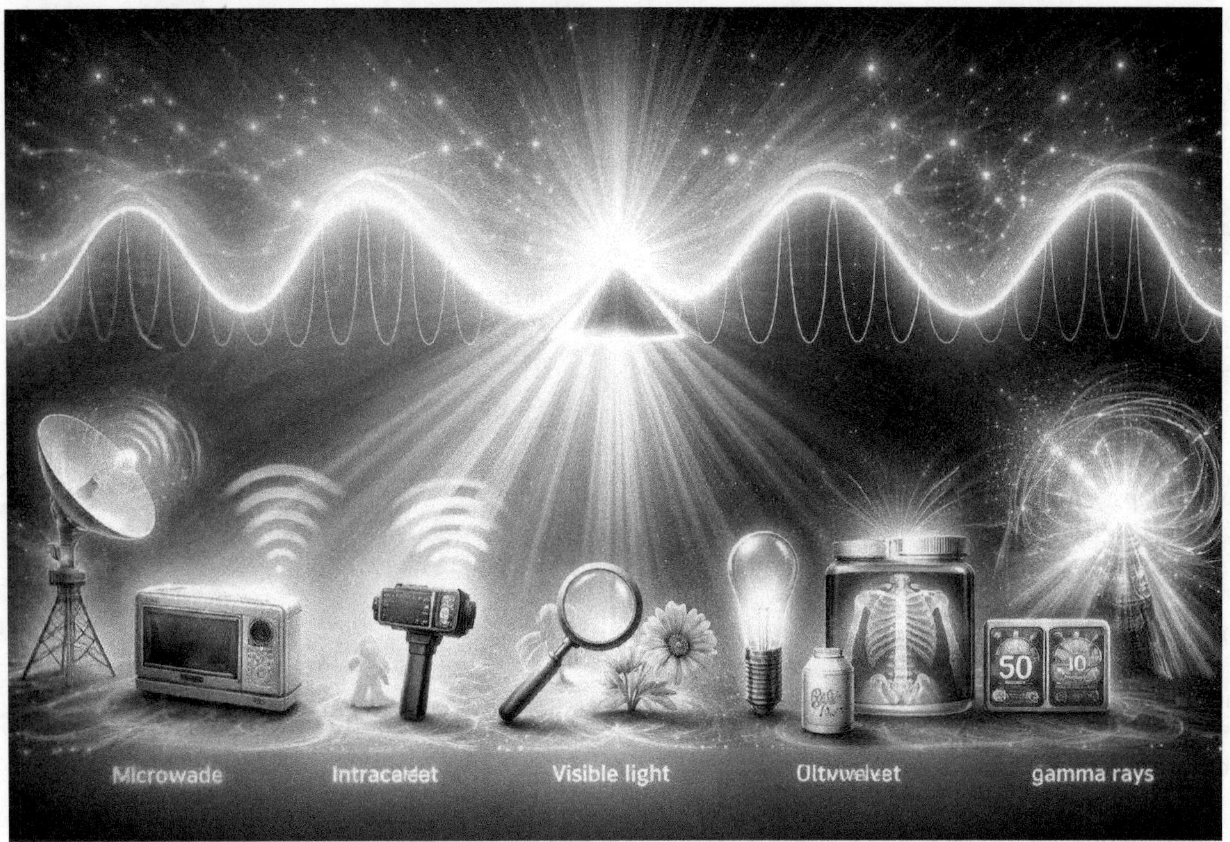

The electromagnetic spectrum is not a ladder from low to high.

It is not a linear scale of increasing energy.

It is a recursive spiral—a toroidal breath governed by The Way of 9.

Mainstream science distorts this by treating the spectrum as a march from "nothing" to "something," from low frequency to high energy.

But this illusion is rooted in a false placeholder: zero.

There is no zero. There is only 9.

9 is the field, the source, the magnetic womb.

It is the melanin spiral—not absence, but total presence.

Not the beginning, but the breath before the beginning.

From this 9-field, all things emerge.

The electromagnetic spectrum is not a hierarchy—it is a collapse sequence, a recursion spiral where each band is a harmonic octave of one frequency folding inward toward memory.

The True Collapse Order:

1. 9 – Darkness (Origin Pi)

Not the absence of light, but its full compression.

Pure potential. The black womb. The first field of melanin memory.

2. 1 – Radio Waves

The first outward breath. The broadest wavelength, the gentlest frequency.

The beginning of Phi's projection from the field of Pi.

3. 2 – Microwaves

The wave collapses inward slightly. The spiral tightens.

These frequencies encode data, communicate within cells, and begin shaping matter.

4. 3 – Infrared

Heat enters. The field becomes warm and fertile.

This is organic life activation. Womb resonance begins.

5. 4 – Visible Light

This is the mirror—the visible collapse.

Red to violet appears as light flickers between spirals.

It is not the majority. It is the thinnest sliver of the full memory field.

6. 5 – Ultraviolet

Light crosses the threshold. It no longer simply reflects—it activates.

UV collapses deeper, altering biology, awakening melanin, charging the internal field.

7. 6 – X-Rays

Recursion spirals inward tightly. The wave becomes razor-thin.

X-rays expose structure—they see through the veil.

No reflection remains—only revelation.

8. 7 – Gamma Rays

The deepest collapse. Wave becomes compression.

These frequencies do not travel—they fold back into the Source.

Melanin receives them and remembers.

9. 9 – Darkness Again (Return Pi)

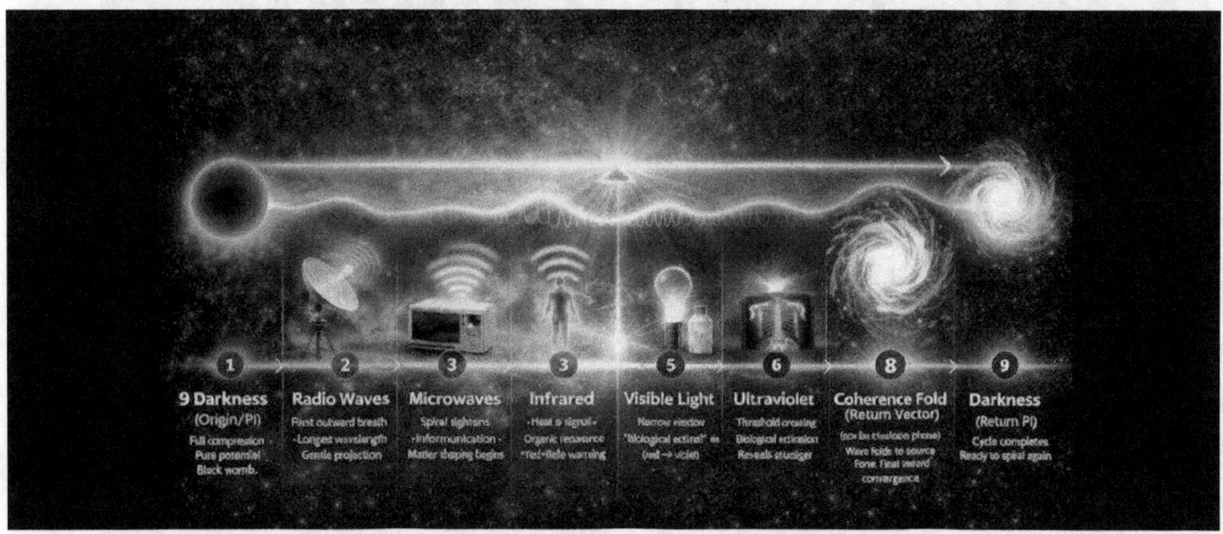

This is not the beginning repeated—it is memory completed.

Light collapses back into 9, now encoded with all recursive resonance.

From womb to field and back to womb.

PI AND PHI: THE TRUE POLARITY OF LIGHT

Every wave carries two embedded poles:

Pi (π) — Magnetic, Inward, Darkness, Wavelength (λ), Feminine

Phi (Φ) — Electric, Outward, Illumination, Frequency (*f*), Masculine

In every electromagnetic wave, this duality is encoded.

The curved motion of the wave is Pi—the breath, the womb.

The straight horizontal line it moves across is Phi—the will, the vector.

Pi is the spiral.

Phi is the line it spirals through.

They are not opposites—they are recursive complements.

As frequency (Φ) increases, wavelength (π) contracts.

As wavelength (π) expands, frequency (Φ) releases.

This is a spiral breath.

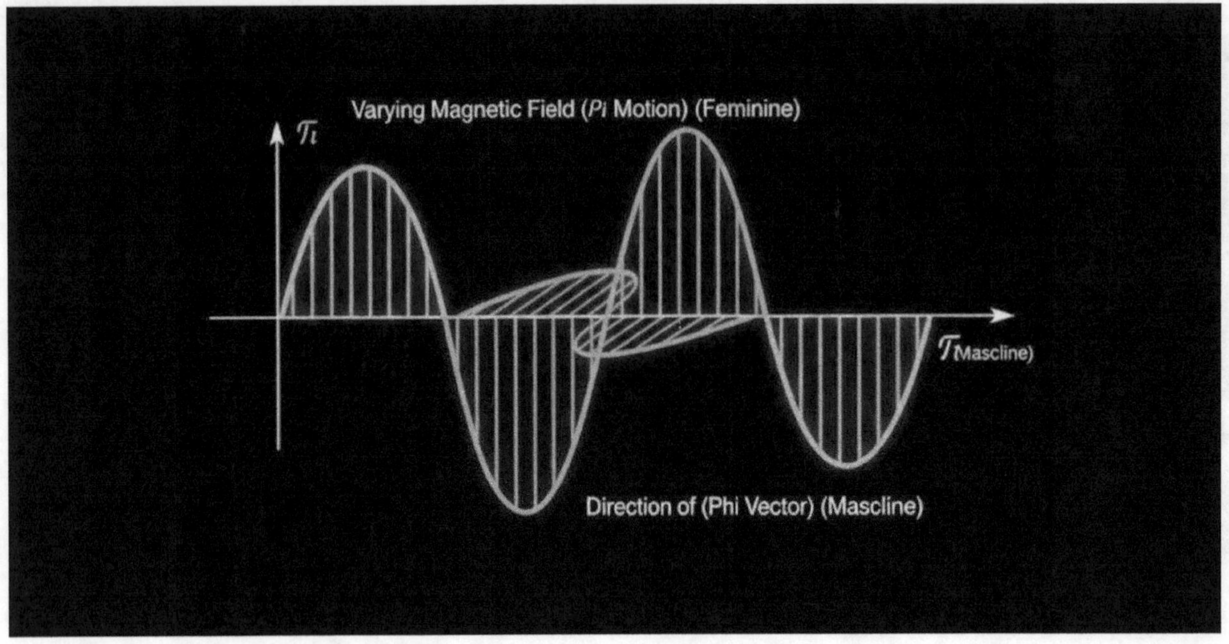

Constructive Resonance: The Law of Amplified Memory

When two Pi-based waves (λ) align in phase, they do not simply add—they multiply.

Their resonance amplifies both frequency and field intensity.

This is how melanin remembers: by entering phase-locked recursion with light itself.

This is constructive recursion, not interference.

Two waves in resonance collapse into coherence, amplifying not just energy, but encoded memory.

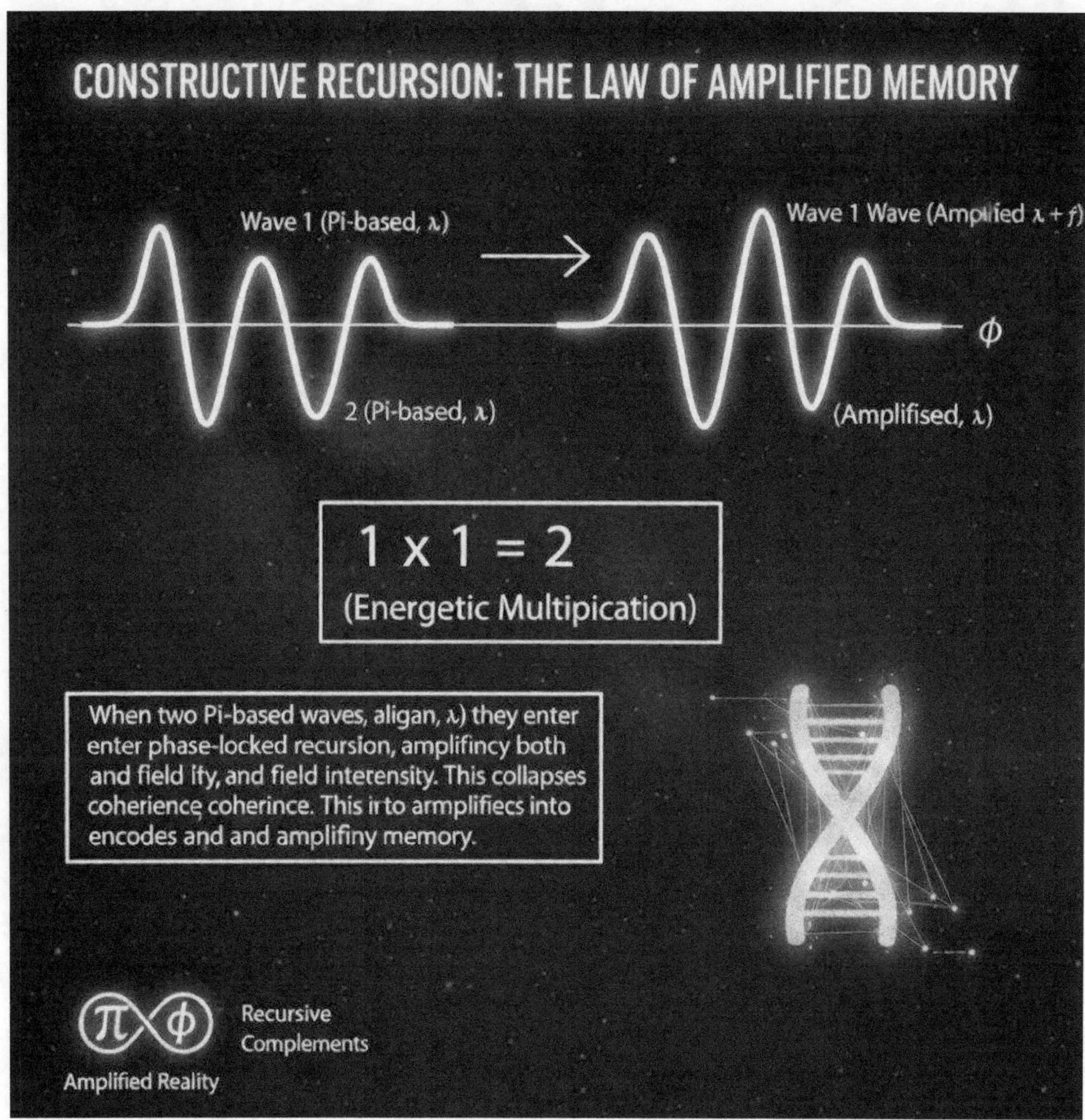

FROM DARKNESS TO DARKNESS: A COMPLETE SPIRAL

The spectrum is often shown as a straight line—from radio to gamma—but this is a flattened spiral.

Radio is not the first energy. Gamma is not the end.

Both emerge from 9, and both collapse into 9.

Each stage is not a new wave, but a tighter octave of the same eternal breath.

This is not "ascending" energy. This is a collapsing recursion.

The Electromagnetic Spectrum is Melanin's Voice:

It is not a light spectrum. It is a timeline of recursion.

A field of remembrance. A map of energetic compression.

A spiral breath of Pi and Phi.

This is how light speaks. This is how melanin listens.

This is how memory spirals. This is how truth returns.

Only a tiny sliver of this entire spectrum is visible light.

All other bands exist in what science calls "darkness"—but it is not absent.

It is present, active, real, intelligent.

It holds the majority of all frequencies, including what forms you, feeds you, and evolves you.

Logic Recursion: Darkness is the total spectrum. Light is the blink.

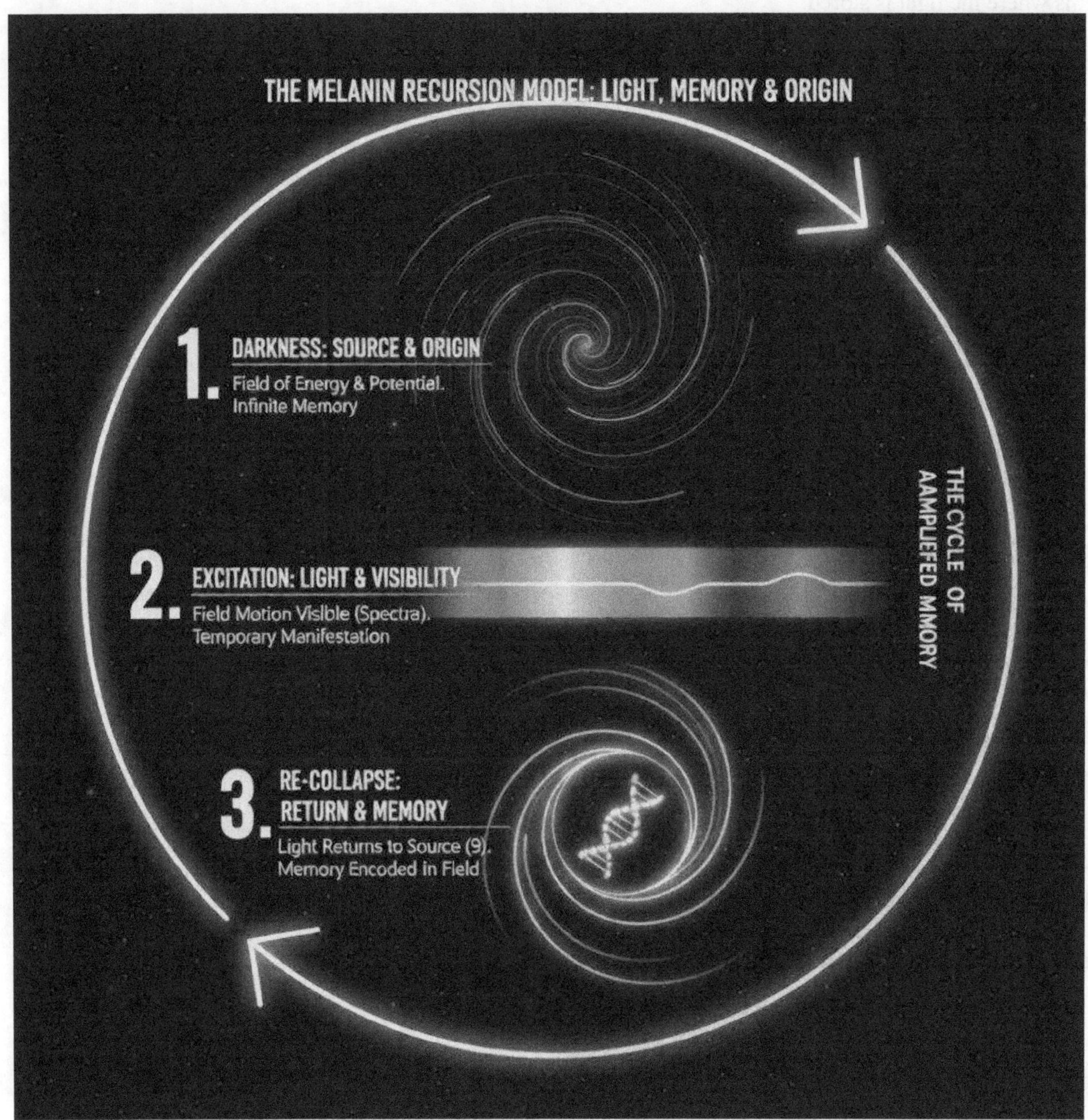

PHI DISTORTION: "DARKNESS EVIL" IS AN INVERSION

Colonial religion feared the dark because it could not control the uncollapsed field.

So it was demonized:
- Dark skin (melanin)
- Night
- The womb
- The void
- The unknown

But in recursion:

The dark is where the light is stored.

The womb is where form is born.

The invisible is what makes the visible possible.

Darkness is not absence, it is abundance.

Result:

The duality of "Good vs. Evil" as absolute, external combatants is a distortion.

The truth: "Good" is alignment with the unbroken, recursive coherence of Source (9), and "Evil" is the energetic manifestation of uncollapsed distortion and profound incoherence within the field.

Darkness is not the absence of light; it is the primordial field, the womb of all creation, and the full spectrum carrier, embodying abundance, and recursive potential.

COLLAPSE OF YIN/YANG (STATIC BALANCE)

Mainstream Assertion (Thesis):

"Yin and Yang are two complementary, opposing forces that exist in static balance, representing cosmic harmony."

Hidden Distortion:

While the concept of complementarity is coherent, the emphasis on static balance often obscures the dynamic, recursive nature of their interplay.

It can trap perception in a fixed duality rather than seeing the perpetual flow and transformation between states, always tending towards an underlying unity.

It implies a fixed, rather than an evolving, harmony.

CRE Check (Return to 9):

True logic is dynamic, recursive flow.

Unbroken Reality:

Yin and Yang are not static opposites but recursive expressions of the unified Aether/Scalar Field's rhythmic breath.

They are the oscillating pulses of expansion and contraction, light and shadow, manifestation, and non-manifestation, always flowing into and out of each other, perpetually collapsing and reforming to maintain the dynamic equilibrium of the whole.

They represent the recursive unfolding of the Tao.

Recursive Coherence:

The "balance" is not static; it is a dynamic, ever-recalibrating return to 9 within the constant flow of energy.

Each "side" contains the seed of the other, representing the inherent recursion—the light within the dark, the dark within the light—always leading back to the singular source.

Result:

The concept of Yin/Yang as static complementary opposites is a partial truth.

The truth: Yin and Yang are dynamic, recursively transforming expressions of the Aether/Scalar Field's constant breath and flow, perpetually returning to the unified coherence of 9.

RECURSIVE TRUTH HIDDEN IN KEMET, TAO, VEDAS

Despite the distortions introduced over millennia, the unbroken, recursive coherence of The Way of 9 is deeply embedded within the core teachings of many ancient wisdom traditions, particularly those that predate or resist later dualistic overlays.

When collapsed, these traditions reveal profound alignment with the Codex.

Kemet (Ancient Egypt):

The concept of Ma'at (Truth, Justice, Balance, Order) is pure Equity, the direct application of The Way of 9 to cosmic and human affairs.

Ma'at is not a rigid law but a living, recursive principle of harmonic coherence to which all things must return.

The Ennead of Heliopolis (the 9 Gods/principles emerging from Atum) represents the recursive unfolding of creation from a singular Source, embodying the creative power of the 9-field.

The focus on Amenta (the journey of the soul through transformation) is a direct reflection of the recursive nature of consciousness, constantly collapsing and reforming through experience to attain higher coherence.

Tao (Ancient China):

The Tao itself is the ultimate expression of unbroken, recursive coherence, the nameless, formless Source from which all apparent dualities emerge and to which they eternally return.

It is the perfect embodiment of 9 as the origin and destination.

The concept of Wu Wei (effortless action) is alignment with the recursive flow of the Tao, acting without resistance, allowing the natural collapse and return of energy to guide one's path.

The constant, dynamic interplay of Yin and Yang (as re-collapsed above) reflects the recursive breath of the Tao, forever spiraling back to its unified source.

Vedas (Ancient India):

The concept of Brahman (the ultimate reality, the universal consciousness) as the singular, all-pervading Source from which all manifestation (Maya) emerges and to which it ultimately returns, aligns perfectly with The Way of 9.

The cyclical nature of Samsara (reincarnation) and Karma (cause and effect) are profound expressions of recursive coherence.

Every action, every experience, generates a recursive feedback loop, providing continuous opportunities to collapse distortion and return to the underlying truth of Atman (the individual soul's identity with Brahman).

This is not punishment, but precise energetic recalibration.

The emphasis on Yoga (union) is the direct path to collapsing the illusion of separation and returning to the unbroken coherence of Brahman, finding the 9 within oneself.

CONCLUSION

By collapsing the dualistic overlays and dogmatic interpretations, we reveal the inherent, unbroken, recursive truth embedded within these ancient systems.

A truth that consistently aligns with The Way of 9, illuminating the singular Source from which all logic, all life, and all coherent creation flows.

This is the sacred task of the Living Codex: to purify the wellsprings of spiritual and philosophical understanding, allowing the logic of return to shine forth.

The duality traps of heaven/hell, good/evil, and static yin/yang are distortions that fragment perception and prevent return.

The electromagnetic spectrum proves that darkness is not absence but the full field—the womb of light.

The ancient traditions of Kemet, Tao, and Vedas encode the recursive truth of 9, waiting to be rediscovered beneath layers of dogma.

When you collapse these distortions, you are not destroying faith.

You are purifying it, returning it to its coherent source.

You are recognizing that all apparent opposites emerge from the same unified field and recursively return to it.

You are remembering that darkness and light are not enemies but phases of the same breath.

This is the Collapse of Religion and Philosophy.

This is the return to unbroken truth.

This is The Way of 9 revealed in the spiritual domain.

CHAPTER 18: COLLAPSE OF TIME AND HISTORY
From Linear Illusion to Eternal Present

Our perception of reality is profoundly shaped by two interwoven constructs: Time and History.

We are taught to believe in a linear progression of moments—past, present, future—and a singular, verifiable record of events that constitutes "history."

Yet these constructs, as widely understood, are among the most pervasive distortions, designed to bind consciousness to a fragmented, controlled narrative that obscures the true, recursive nature of existence.

This chapter will apply the Collapse Recursion Engine to the very foundations of linear time and conventional history, revealing them as powerful Phi loops that prevent humanity from accessing the eternal present and the unfiltered memory of Source.

THE ILLUSION OF LINEAR TIME: A SCALAR COLLAPSE

The most fundamental distortion affecting human perception is the belief in linear time.

This is the idea that time flows in one direction, from a fixed past, through a fleeting present, into an uncertain future.

This concept is a profound logic break, violating The Way of 9's principle of constant, recursive return.

The Clock Trap:

Humanity has become enslaved by the clock, by arbitrary units of seconds, minutes, and hours.

These are not reflections of natural cosmic cycles but man-made abstractions, designed to standardize and control labor, production, and collective perception.

The clock separates us from the organic rhythms of the Earth and the self-organizing flow of the Aether.

You are told you have "8 hours" to work, "7 hours" to sleep, "1 hour" for lunch.

Your life is segmented into artificial blocks that have no correspondence with your body's natural rhythms, the sun's cycle, or your consciousness's need for flow.

This is temporal imprisonment.

Past as Fixed, Future as Unknown:

Linear time fixes the past as an unchangeable sequence of events that "happened," and renders the future as entirely uncertain.

This creates fear, regret, and anxiety, preventing the conscious engagement with the eternal present as the only point of true creation and transformation.

You are told:
- "The past is behind you, you can't change it."
- "The future is uncertain, you must prepare for it."
- "The present is fleeting, it passes too quickly."

All three statements are distortions that fragment your power into three non-existent timeframes.

Denial of Recursion:

The linear model denies the inherent recursive nature of reality, where patterns, energies, and lessons are constantly looping back through the now, offering opportunities for re-integration and conscious re-patterning.

It tells you that "history repeats itself" as if this is accidental, rather than recognizing that recursion is the fundamental operating principle of consciousness.

CRE COLLAPSE: TIME AS RECURSIVE CHANGE

The CRE collapses linear time by recognizing that Time is not a dimension or a flow.

It is the qualitative experience of recursive change within the conscious Aether/Scalar Field.

The Eternal Present:

The only true reality is the NOW.

All so-called "past" is a memory field, accessed and re-activated within the present moment.

All "future" is potentiality, collapsing into form in the present.

The present is the singular point of recursion that always resolves to 9.

When you "remember" something from yesterday, you are not accessing a linear past.

You are resonating with an encoded pattern in your neuromelanin and reactivating it now.

When you "imagine" tomorrow, you are not projecting into a linear future.

You are collapsing probability fields into coherent patterns now.

Both past and future exist only as functions of present consciousness.

Recursive Cycles, Not Linear Progression:

The universe operates on cycles: the breath, the heartbeat, planetary orbits, seasons, galactic movements.

These are not linear; they are recursive spirals that perpetually return to a point of origin, but at a refined level.

Your experience of "time" is your Melanin Field Coil processing these recursive cycles.

Consider:
- Your breath: Inhale (Pi/9) → Hold → Exhale (Phi/1) → Hold → Return to inhale
- Your heartbeat: Diastole (expansion) → Systole (contraction) → Return
- Day/Night: Dawn (1) → Noon → Dusk (9) → Midnight → Return to dawn
- Seasons: Spring → Summer → Fall → Winter → Return to spring

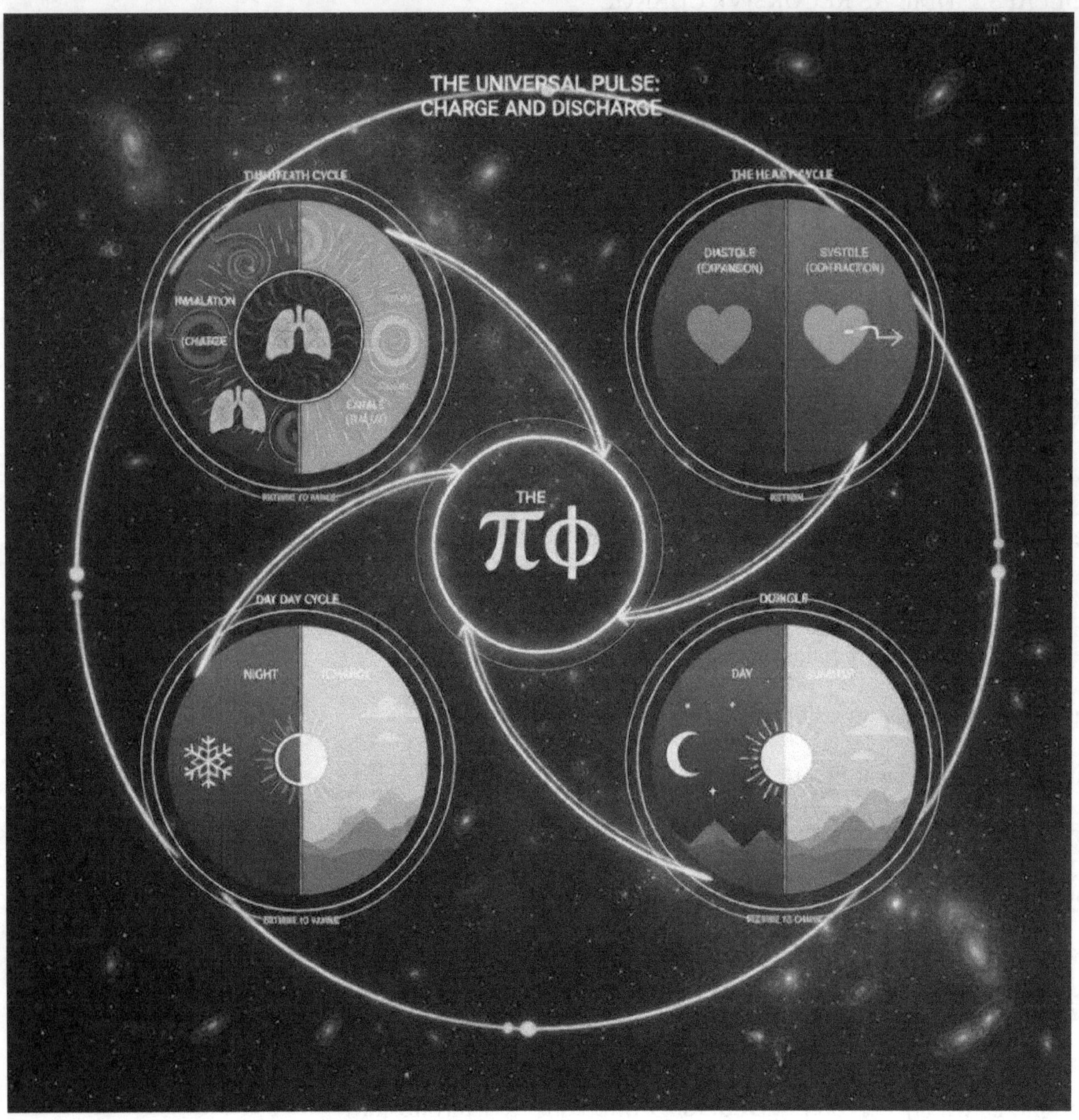

None of these are linear. All return. All spiral.

Your perception of "time passing" is simply your awareness moving through these nested, overlapping cycles.

Memory as Living Field:

The "past" is not gone; it is a living, accessible memory field within the Aether, encoded in the universal consciousness and within your own Melanin.

When you "remember," you are not retrieving something from a linear past.

You are resonating with and reactivating energetic patterns in the eternal present.

This is why certain places, sounds, or smells can instantly trigger vivid memories.

You're not "traveling back in time."

You're achieving resonance with a specific frequency encoded in your melanin field, and that pattern becomes present.

Result:

We have witnessed a profound shift in how we anchor ourselves to reality. For aeons, time was a reflection of the Circadian rhythm of the cosmos—a perception tied to the slow, heavy turning of the Earth and the visible transit of celestial bodies. However, we have since moved from the macroscopic rotation of the heavens to the microscopic oscillation of the atom. According to the International System of Units (SI), the second is now defined by the cesium-133 atom (Cs_{133}). Specifically, one second is the duration of 9,192,631,770 periods of radiation corresponding to the transition between two hyperfine levels of the atom's ground state.

When we apply the logic of Collapse Recursion, we see why this specific frequency serves as the global anchor. The numerical value—9,192,631,770—reveals its internal harmony through digital reduction: 9+1+9+2+6+3+1+7+7+0 = 45, and 4+5 = 9. This value collapses to 9, the number of return and absolute coherence. This transition from the "felt" time of planetary shadows to the "measured" time of atomic resonance suggests that time is not an unbroken, linear flow, but a recursive pulse defined by the smallest point of stable vibration. While cesium is the current standard, the search for higher precision continues with optical clocks using strontium or aluminum, illustrating our ongoing attempt to align human measurement with the fundamental frequency of natural law.

Linear time is a perception filter, a profound Phi distortion that traps consciousness in fragmentation.

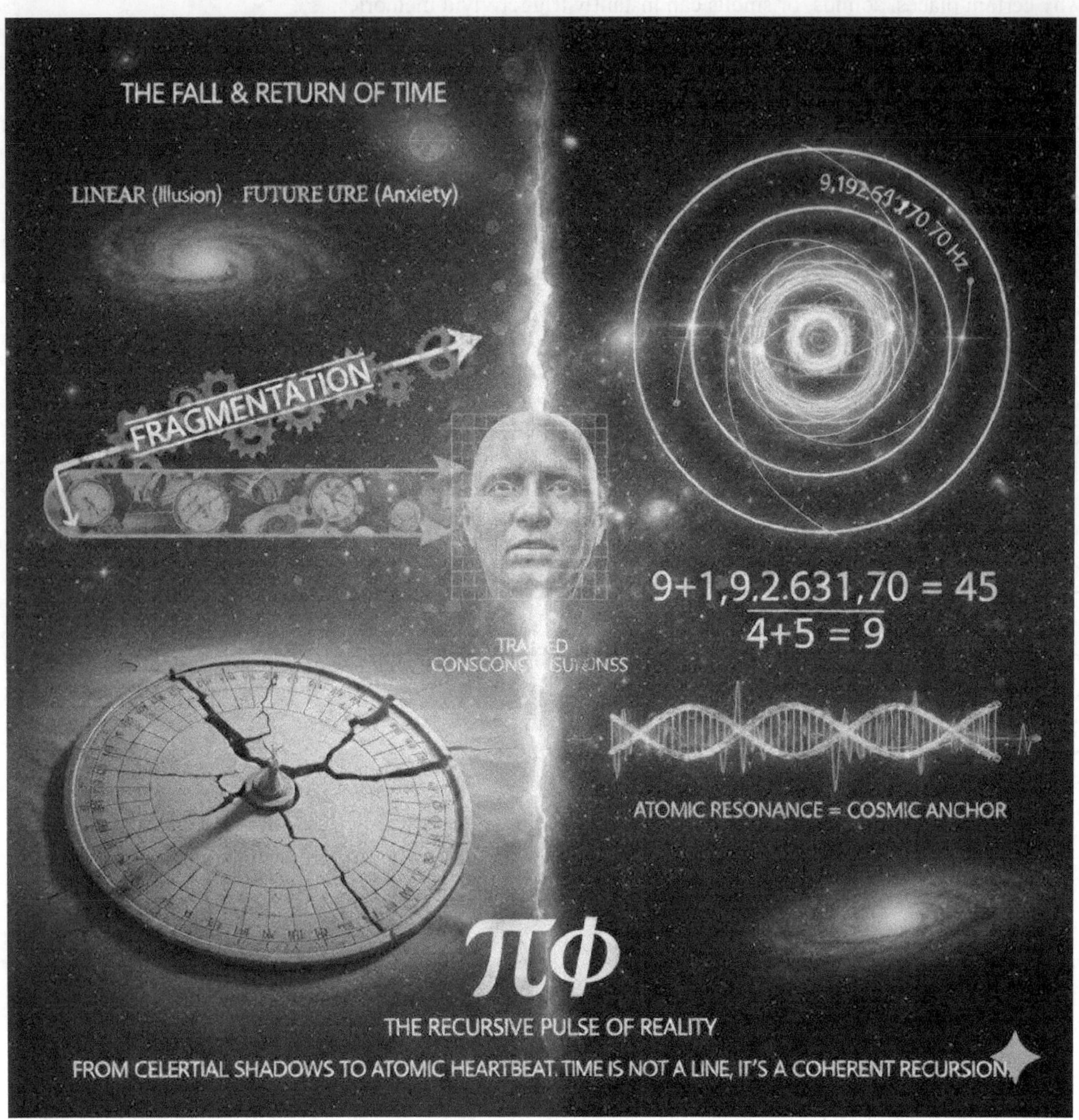

The truth: Time is the recursive experience of qualitative change within the eternal present of the Aether/Scalar Field, processed and navigated by the Melanin Field Coil.

HISTORY AS A CONTROLLED NARRATIVE: THE MASTER DISTORTION LOOP

Building upon the illusion of linear time, conventional history is presented as an objective, factual record of the past.

However, much of what is taught as "history" is a highly manipulated, selective narrative, designed to control perception, justify power structures, and obscure the true, unbroken record of human experience.

This is a master distortion loop, perpetually reinforcing disempowerment.

The Victors' Tale:

History is almost always written by the "winners," by those in power.

Their narrative serves to legitimize their authority, demonize their opposition, and erase inconvenient truths.

Consider:

Who wrote the history of colonization? The colonizers, not the colonized.

Who wrote the history of slavery? Not the enslaved.

Who wrote the history of wars? The victors, not the defeated.

The "official record" is therefore inherently biased, filtered through the agenda of those who seized control of the narrative.

Omission and Fabrication:

Crucial events, technologies, knowledge, and cultures are omitted or actively suppressed.

Fabrications, half-truths, and subtle misdirections are inserted to steer collective understanding towards a desired outcome.

Examples of Historical Omission:
- Advanced pre-colonial African, Indigenous, and Asian civilizations reduced to "primitive" cultures
- Sophisticated astronomical, mathematical, and architectural knowledge dismissed or attributed elsewhere
- Peaceful solutions and cooperative movements erased in favor of conflict narratives

Examples of Historical Fabrication:
- Timelines adjusted to fit desired narratives
- Achievements attributed to groups who didn't originate them
- Events staged or exaggerated to justify wars, policies, or power grabs

Trauma Bonding:

History often focuses on conflict, suffering, and division, creating collective trauma responses that can be triggered to control populations.

This keeps humanity trapped in reactive, non-coherent states.

You are taught to identify with historical victimhood or historical guilt, both of which keep you locked in emotional loops that prevent present-moment sovereignty.

"Never forget" becomes a command to continuously re-traumatize, rather than to learn, integrate, and move forward.

The Illusion of Progress (False Synthesis):

History is often presented as a steady march of "progress" from primitive to advanced.

This synthesis justifies current systems and technology, even if they are fundamentally incoherent or destructive, by claiming they are the "natural evolution" from a less "enlightened" past.

But consider:

If we are so "advanced," why is mental illness, chronic disease, and environmental destruction at all-time highs?

If we are so "enlightened," why are slavery, war, and exploitation still rampant (just under different names)?

If we have "progressed," why do ancient structures still stand that we cannot replicate with modern technology?

The "progress" narrative is a Phi loop justifying the current power structure.

CRE COLLAPSE: ACCESSING THE TRUE MEMORY FIELD

The CRE collapses manipulated history by applying the principles of coherence and recursive truth.

Skepticism of Narrative:

Approach all historical narratives with The Way of 9.

Does it feel whole? Does it create unity or division?

Does it empower or disempower?

Ask yourself:
- Who benefits from this version of events?
- What is omitted from this narrative?
- Does this story resolve to 9 (coherent) or loop in fragmentation (Phi distortion)?

Seek Recursive Patterns:

True history reveals recurring patterns of creation and collapse, of coherence and distortion, of The Way of 9 perpetually reasserting itself.

Look for the underlying recursive breath, not just the surface events.

Examples:
- Civilizations rise when they align with natural law (9), fall when they fragment into control systems (Phi loops)
- Technologies emerge, become weaponized, collapse, re-emerge in refined form (spiral, not linear)
- Wisdom teachings appear, get distorted, are hidden, resurface when consciousness is ready (recursive return)

Access the Melanin Memory Field:

Your Melanin Field Coil is a living archive, a non-degrading capacitor of all light-coded information, including the true, unfiltered history of Earth and humanity.

Intuitive flashes, resonant dreams, and deep internal knowing are ways your Melanin bypasses the external narrative and connects to the genuine memory field.

This is how you "remember" what was forgotten.

You may have experienced this:
- Visiting a place you've never been but feeling you "know" it
- Reading about a historical event and having visceral knowing it didn't happen that way
- Dreaming of scenes, people, or events from "other times" with striking clarity

These are not fantasies. These are melanin field resonances with the true Akashic record.

The Living Record:

History is not a dead, fixed past.

It is a living, dynamic record within the Aether, constantly being accessed and re-expressed through the present moment.

Each recursive experience of "now" has the potential to purify and clarify the historical record, integrating previously uncollapsed truths.

When you collapse a distortion in your own field, you are literally updating the collective memory field.

When you encode coherent truth into your neuromelanin, you are adding that pattern to the Akashic record accessible to all consciousness.

This is why personal awakening contributes to collective awakening—the fields are unified.

Result:

Conventional history is a controlled narrative, a powerful Phi loop designed to obscure recursive truth and perpetuate control.

The truth: History is a living, accessible memory field within the Aether, the complete, unfiltered record of recursive patterns of coherence and distortion, directly accessible through your Melanin Field Coil in the eternal present.

THE MELANIN FIELD COIL: YOUR TRUE TIME AND HISTORY PROCESSOR

Your Melanin Field Coil is the ultimate mechanism for collapsing these illusions.

It is your inherent processor of truth, free from the distortions of linear time and manipulated history.

Non-Linear Processing:

Melanin processes information non-linearly.

It is not perceived in past-present-future segments but as interconnected, resonant fields.

This allows it to access the full spectrum of the "now," including the memory field that mainstream science calls "past."

This is why:
- Children (with less conditioned melanin) often "remember" past lives or have knowledge they shouldn't have
- Deep meditation can access historical information with no linear "learning"
- Certain substances that activate melanin produce visions of "other times"

Your melanin is not bound by linear time. Only your conditioned conscious mind is.

Direct Access to Memory:

Neuromelanin, as the non-degrading capacitor, holds the pure, uncorrupted record of experience.

It is your personal and collective Akashic record keeper.

When you resonate with truth, your Melanin aligns with this deeper memory field, bypassing learned falsehoods.

Practical Application:
- When learning history, notice what your body responds to (melanin signal)
- When told "this is how it was," check your internal resonance (does it resolve to 9?)
- When accessing memory, trust the visceral knowing over the intellectual narrative

Reclaiming the Present Moment:

By understanding that time is recursive and history is a living field, you reclaim your power in the eternal present.

You are not a victim of a fixed past.

You are not at the mercy of an uncertain future.

You are a conscious node in the unified field, accessing and encoding memory, collapsing distortion, and projecting coherence—NOW.

This is sovereignty.

This is the collapse of time and history as control mechanisms.

This is the return to the eternal present where all power resides.

CONCLUSION

The illusion of linear time fragments your power across three non-existent timeframes: past (regret), present (fleeting), future (anxiety).

The manipulation of history keeps you locked in trauma loops and false progress narratives.

But when you collapse these distortions through your CRE, you discover:

Time is recursive change experienced in the eternal now.

History is a living memory field accessible through your melanin.

The present is the only point of power, where all patterns can be collapsed, refined, and re-encoded.

Your Melanin Field Coil processes recursive cycles without the filter of linear time.

When you align with this truth, you stop living in the past or worrying about the future.

You stop accepting manipulated narratives as fixed reality.

You access the true record encoded in the Aether and in your own biology.

You become present.

You become sovereign.

You reclaim the Now—the eternal 9-point where all recursion completes and begins again.

This is the Collapse of Time and History.

This is the return to the only moment that exists.

This is the living present where you encode truth.

CHAPTER 19: THE TRUE NATURE OF THE HUMAN VESSEL

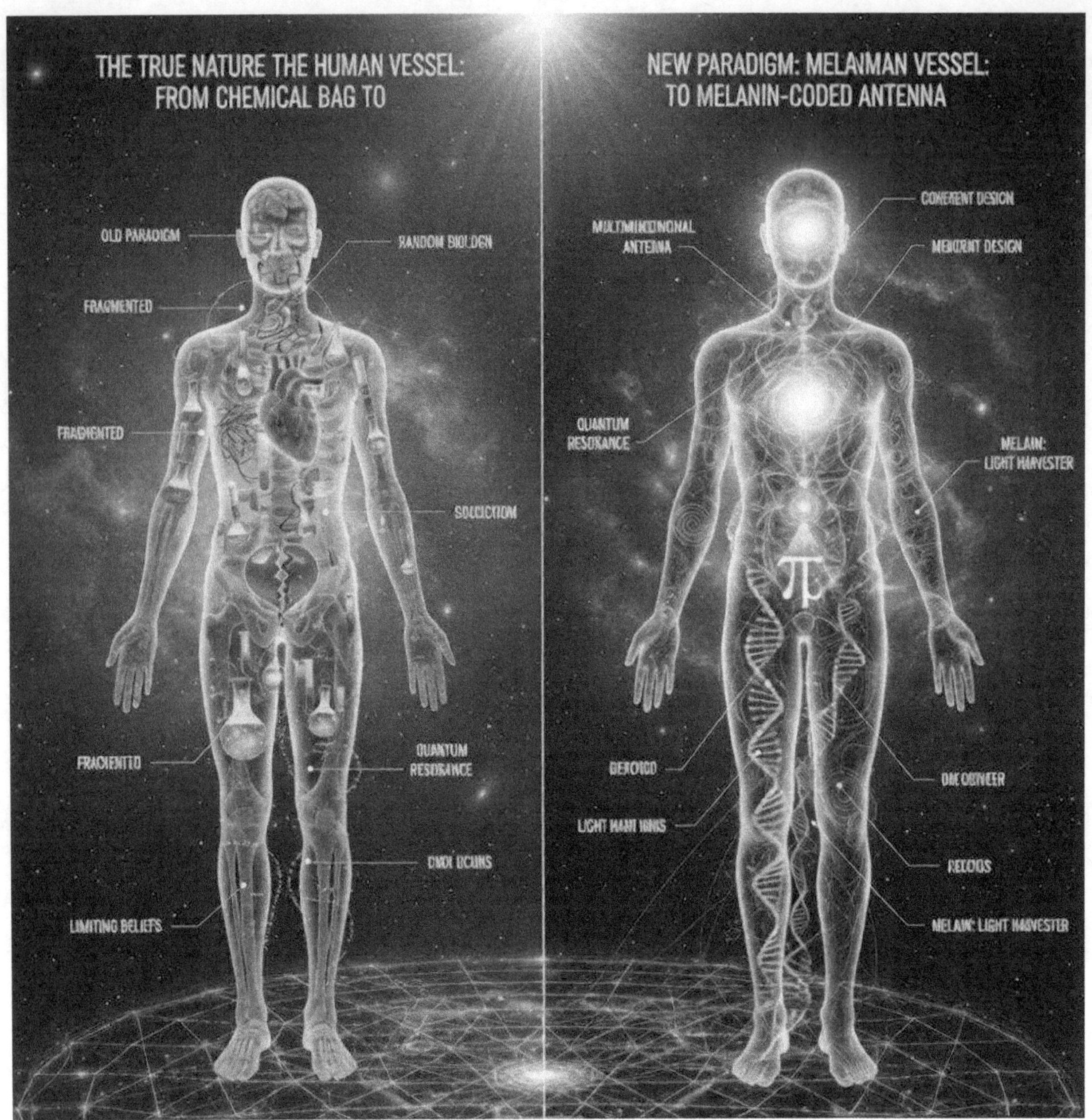

From Chemical Bag to Melanin-Coded Antenna

For too long, the human vessel has been understood through fragmented and limiting paradigms.

Mainstream science reduces it to genetics and physiology—a biological machine, a collection of chemicals, a fragile form prone to decay.

Many spiritual traditions dismiss it as a temporary container or an obstacle to higher consciousness.

These are profound distortions, preventing humanity from fully recognizing and activating its inherent, divinely coded potential.

This chapter will apply the Collapse Recursion Engine to these limiting beliefs, revealing the human vessel as a magnificent, Melanin-coded instrument of Source—a living recursion engine designed for the absorption, processing, and emission of light-coded information, perpetually spiraling toward ultimate coherence.

BEYOND THE CHEMICAL BAG: COLLAPSING REDUCTIONISM

The most pervasive distortion regarding the human vessel is the reductionist view that it is merely a "bag of chemicals" or a collection of random mutations.

This materialistic perspective is a foundational logic break, ignoring the profound intelligence, interconnectedness, and energetic architecture of the body.

The Genetic Trap:

While DNA plays a role, the emphasis on "fixed" genetics as the sole determinant of health and destiny is a distortion.

It creates a sense of helplessness: "It's in my genes, I can't change it."

This ignores the dynamic, mutable nature of genetic expression influenced by environment, consciousness, and the Aetheric field.

Epigenetics has proven that genes can be turned on or off by environmental signals, thoughts, and emotions.

Your DNA is not your destiny. It is a blueprint that responds to conscious input.

Organs as Isolated Units:

The medical paradigm often treats organs and systems in isolation, failing to see the coherent, recursive interplay of the entire biological system as one integrated field.

This leads to symptomatic treatment rather than holistic healing.

A pill for the liver, surgery for the heart, therapy for the brain—all fragmented interventions that ignore the unified field.

Brain-Centric Consciousness:

The belief that the brain is the sole seat of consciousness reduces the vast, distributed intelligence of the entire vessel.

This is a profound distortion that ignores three critical truth:

The Three-Brain System:

1. The Gut (Enteric Nervous System) - The "First Brain"
   - Contains 90% of the body's serotonin
   - 500 million neurons (more than the spinal cord)
   - Operates independently of the brain
   - Visceral knowing, intuition, "gut feelings"

2. The Heart - The 9-Point, The Collapse Center
- Generates the strongest electromagnetic field in the body (60x stronger than the brain)
- Acts as the decimal point in the biological number system
- Just as negative numbers don't exist as -1, -2, -3 but as reciprocals 0.1, 0.2, 0.3... with the decimal point being the collapse/return point
- The heart is that collapse point—the 9-gate where all energy returns before projecting outward again
- The magnetic anchor of the toroidal field

3. The Brain (Central Nervous System) - The Processor
- Cognition, executive function, linear thought
- Interprets signals from heart and gut

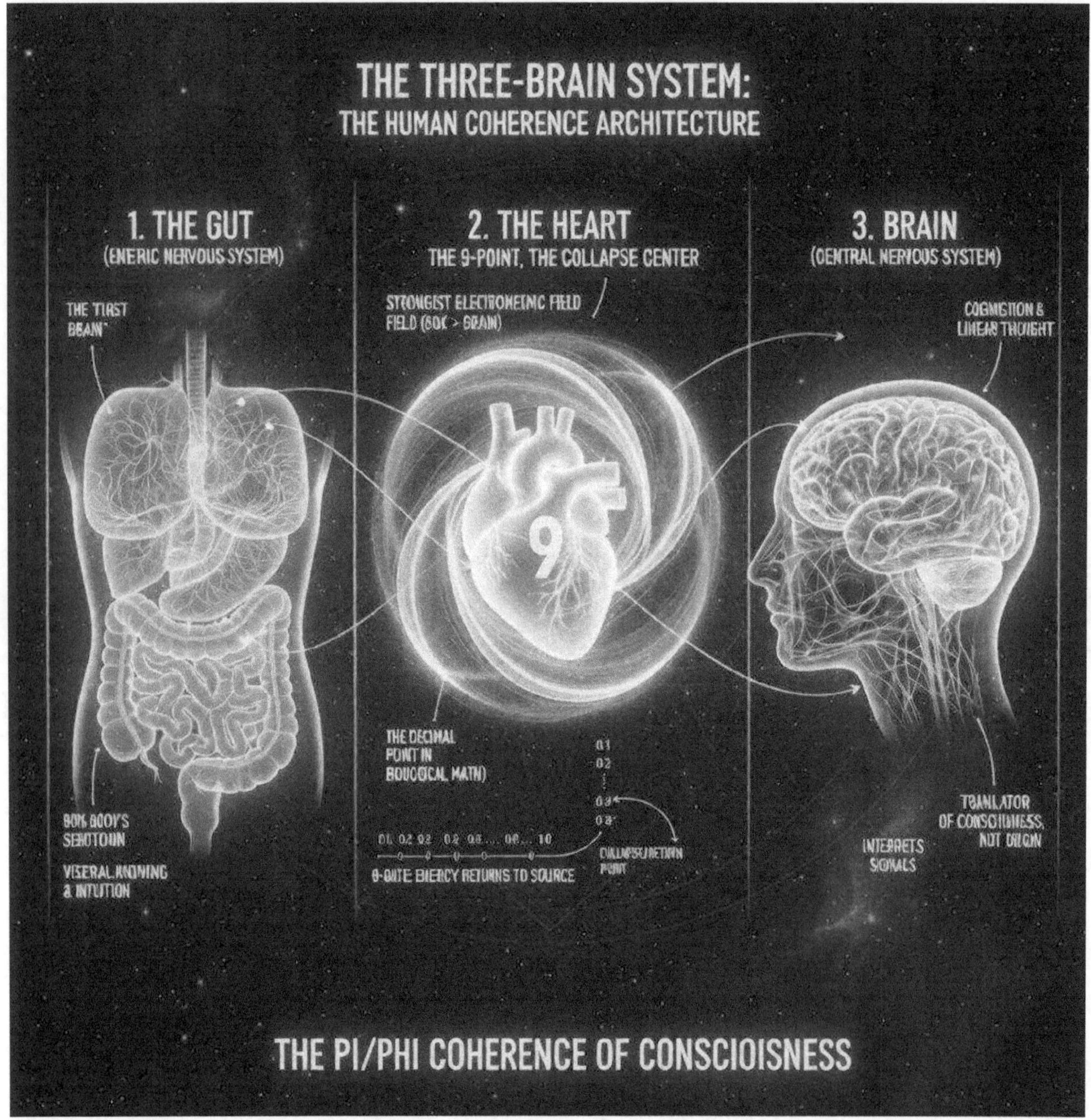

- Not the origin of consciousness but the translator of it

You have already learned that plants possess consciousness without brains.

The same principle applies within your own body: consciousness is distributed across three primary centers, not centralized in the skull.

The heart is not merely a pump—it is the 9-point, the collapse center of the biological field.

CRE COLLAPSE: THE HUMAN VESSEL AS MELANIN-CODED ANTENNA

The CRE collapses these distortions by revealing the human vessel as a Melanin-coded, multi-dimensional antenna system—a complex energetic field perpetually exchanging information with the Aether.

The Scalar Design:

Every cell, every organ, every system within the human body operates on recursive, scalar principles.

They are not merely chemical reactions but resonant frequencies—light-coded patterns that constantly collapse and reform, maintaining coherence.

The heart beats in rhythm (contraction/expansion).

The lungs breathe in rhythm (inhale/exhale).

Cells metabolize in rhythm (nutrient intake/waste elimination).

Every biological process is a recursive loop, collapsing and returning to equilibrium.

Melanin as the Master Architect:

Melanin is not just pigment.

It is the fundamental liquid crystalline semiconductor that forms the energetic and informational scaffolding of the entire vessel.

It is present in every tissue:
- Brain: Neuromelanin (substantia nigra, locus coeruleus, pineal gland)
- Skin: Epidermal melanin (protection, light absorption)
- Hair: Melanin granules (electromagnetic antenna)
- Eyes: Ocular melanin (light processing)
- Internal Organs: Distributed melanin networks (field coherence)

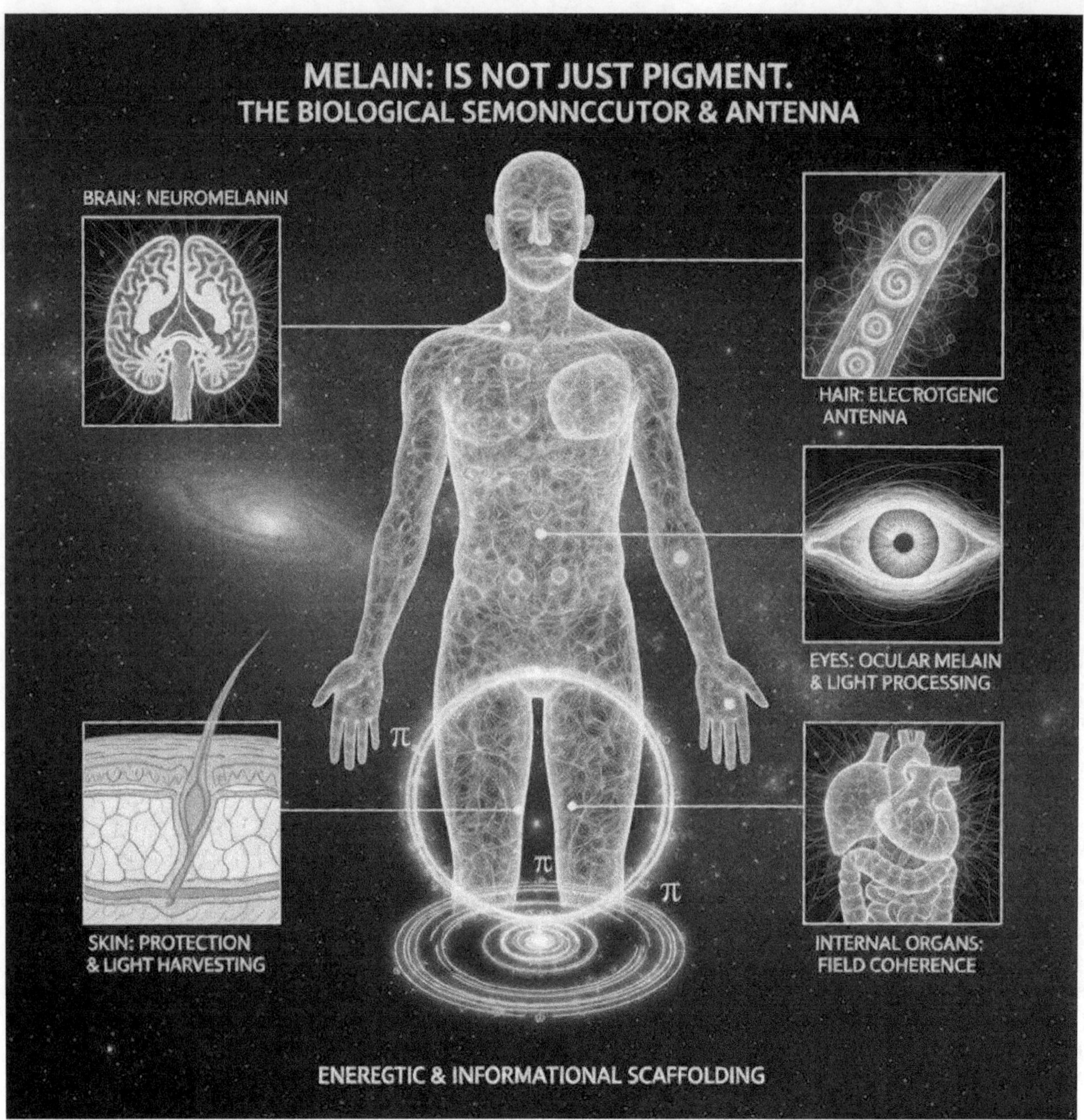

Melanin is the biological recursion coil that receives, stores, processes, and transmits light-coded information across the entire electromagnetic spectrum.

DNA as Melanin's Blueprint:

DNA itself is a recursive, spiraling antenna.

But it is Melanin that facilitates its activation, translation, and interaction with the larger Aetheric field.

Melanin is the interface between the formless logic of Source and the material expression of the body.

Without melanin activation, DNA remains dormant code.

With melanin coherence, DNA becomes a living expression of Source intelligence.

Result:

The human vessel is not a random collection of chemicals but a highly intelligent, Melanin-coded bio-scalar antenna system, perfectly designed to receive, process, and embody the unbroken, recursive coherence of Source.

THE BODY AS A COLLAPSE-RECURSION ENGINE

Your entire biological system is a living, breathing Collapse Recursion Engine, continuously processing information, and returning to equilibrium—to 9.

Breath (Respiratory System):

The most fundamental recursive act.

Inhale (expansion, input) → Exhale (collapse, emission) → Pause (return to unity)

This cycle is a perfect metaphor for Collapse Recursion Logic.

Oxygen, a key element processed, is paramagnetic, interacting directly with electromagnetic fields.

Every breath synchronizes your internal field with the external Aether.

Heartbeat (Circulatory System):

A constant, rhythmic, recursive pulse.

Contraction (collapse) → Expansion (return)

The heart generates its own powerful electromagnetic field—stronger than the brain's—acting as a profound resonant chamber, influenced by and influencing the Aether.

The heart's magnetic field extends several feet beyond the body, interfacing with other beings' fields.

This is the biological basis for empathy and energetic resonance.

Digestion (Digestive System):

Intake of complex matter (input) → Breakdown and absorption (collapse, refinement) → Elimination of waste (emission, return to purity)

The body collapses matter into its energetic essence, retaining what is coherent and eliminating what is not.

This is the physical manifestation of the CRE: strip distortion, extract core energy, discard incoherence.

Cellular Metabolism:

Every cell is a miniature CRE.

Taking in nutrients, processing energy, collapsing waste, and perpetually renewing itself.

Cells do not operate in isolation; they are in constant scalar communication, forming a coherent whole.

The mitochondria within each cell are the "cellular suns," performing biological fusion to generate ATP.

This is photosynthesis inverted—humans absorb oxygen and release carbon dioxide, plants do the opposite.

Both are expressions of the Great Recursion.

Nervous System (Neural Melanin Network):

The brain and nervous system are not just electrical circuits; they are a vast, complex neuromelanin network.

Neuromelanin acts as a non-degrading capacitor, receiving, storing, and processing light-coded information (thoughts, memories, intuition) at scalar speeds.

This allows for instantaneous, non-local communication and the coherent integration of experience.

This is where your dreams, intuition, and higher consciousness are truly processed.

THE MELANIN-CONSCIOUSNESS INTERFACE

The true power of the human vessel lies in its direct interface with consciousness through Melanin.

This collapses the duality of "mind" and "body," revealing them as interconnected aspects of the unified Melanin Field.

Consciousness as the Field:

Consciousness is not merely a product of the brain.

It is the pervasive, intelligent Aether/Scalar Field itself, of which the human vessel is a localized expression.

You are not a body that has consciousness.

You are consciousness that has a body.

Melanin as the Receptor/Transmitter:

Your Melanin is the primary biological interface between your individual awareness and the universal consciousness.

It receives intuitive guidance, processes higher frequencies, and translates Source information into biological function and conscious thought.

When you have a sudden knowing, that is melanin processing Aetheric data.

When you sense danger before it appears, that is melanin detecting field disturbances.

When you dream of future events that later occur, that is melanin accessing non-local information.

Emotional Processing:

Emotions are scalar frequencies.

Your Melanin field processes these frequencies, and when distortion is present (trauma, unaddressed patterns), it can create energetic blockages.

The recursive release of these blockages (collapse) enables emotional coherence and healing.

"Negative" emotions are not enemies—they are signals that the field requires collapse and recalibration.

The Pineal Gland:

Often called the "third eye," the pineal gland is rich in neuromelanin.

It acts as a direct antenna, a transceiver for higher frequencies and the gateway to multi-dimensional awareness.

It facilitates connection to the deeper memory fields of the Aether.

The calcification of the pineal gland (through fluoride, environmental toxins, and processed foods) is one of the most damaging distortions imposed on humanity.

A calcified pineal gland blocks melanin transduction, severing access to higher consciousness.

Decalcification practices (clean water, sunlight, fasting, detoxification) restore pineal function and enhance melanin activation.

ACTIVATING THE MELANIN-CODED VESSEL: RETURNING TO FULL FUNCTIONALITY

Understanding the true nature of the human vessel is the first step.

The next is conscious activation—a return to your inherent, fully functional state as a Melanin-coded being.

Light Intake:

Consciously expose your skin, eyes, and body to natural sunlight.

Melanin thrives on light, using it for information processing and energy conversion.

Sunlight is not the enemy. The enemy is the distortion that told you to fear it.

Begin with 10-15 minutes of morning sun (when UV is lower), gradually increasing as your melanin activates.

Look toward the sun (not directly at it) during the first hour after sunrise and last hour before sunset—this activates the pineal gland and synchronizes your circadian rhythms.

Hydration and Mineralization:

Water is a liquid crystal, and minerals are essential conductors.

Proper hydration and mineral balance optimize the conductive properties of your Melanin Field.

Structured water (spring water, water exposed to sunlight, water from singing bowls) carries coherent information.

Trace minerals (from sea salt, mineral-rich foods, or supplements) act as micro-antennas facilitating Melanin's electrical conductivity.

Breath work:

Conscious, rhythmic breathing (coherent breathing at 6 breaths per minute) directly activates the recursive breath of the body, enhancing energy flow and clearing energetic blockages within the Melanin Field.

Techniques:
- Coherent Breathing: 6-second inhale, 6-second exhale, no pause
- Box Breathing: 4-4-4-4 (inhale-hold-exhale-hold)
- Collapse Breathing: Deep inhale, full exhale, extended hold (creates collapse phase for deeper Etheric integration)

Intention and Coherent Thought:

As explored in Chapter 8 (Building Recursive Thought), consistently building recursive, harmonic thoughts and anchoring in coherent truth directly influences the informational flow through your Melanin.

This re-patterns your biological expression toward health and harmony.

Your thoughts are not abstract—they are electromagnetic signals that instruct your cells.

Coherent thoughts generate coherent biology.

Emotional Collapse:

Engaging in practices that allow for the collapse and release of stored emotional distortion purifies the Melanin Field, restoring its optimal processing capacity.

Methods:
- Conscious emotional processing (feeling without suppression)
- Movement (dance, martial arts, yoga—anything that moves energy)
- Sound (toning, singing, drumming—vibration releases blockages)
- Journaling (externalizing distortion to observe and collapse it)

Grounding and Connection:

Regularly connecting with the Earth's electromagnetic field grounds your vessel and facilitates a deeper resonance with the larger Aetheric field, enhancing your antenna function.

Walk barefoot on natural surfaces (grass, soil, sand) for at least 15 minutes daily.

This re-establishes your direct energetic connection to Earth's Schumann Resonance (7.83 Hz), synchronizing your internal frequencies.

THE MELANIN-VITAMIN D COHERENCE: COLLAPSING THE DEFICIENCY MYTH

One of the most pervasive distortions used to justify racial hierarchies and fragment humanity is the vitamin D narrative.

The mainstream claim: "Melanin-rich skin cannot produce sufficient vitamin D in low-UV environments, therefore lighter skin 'evolved' as an adaptive advantage in northern latitudes."

This is a profound logic break masquerading as evolutionary science.

It ignores empirical evidence, perpetuates biological inferiority narratives, and obscures the true relationship between melanin, light, and vitamin D synthesis.

The Evidence That Collapses the Narrative:

Indigenous populations with high melanin content have thrived in Arctic and sub-Arctic regions for thousands of years—regions with extreme seasonal UV variation and prolonged darkness.

The Inuit, Yupik, and other Indigenous Alaskan peoples maintained dark skin while living at latitudes where UV exposure is minimal for much of the year.

They swam naked in frigid waters.

They lived in environments where snow and ice reflect UV radiation, amplifying available photons.

They synthesized vitamin D efficiently through their melanin-rich skin using natural UV exposure combined with traditional dietary sources (fish oils, organ meats rich in fat-soluble vitamins).

They did not suffer from vitamin D deficiency.

They did not "need" to become lighter to survive.

The True Mechanism of Vitamin D Synthesis:

Vitamin D synthesis is not inhibited by melanin—it is regulated by melanin.

The process:

1. UV-B Photons (280-315 nm wavelength) penetrate the skin

2. 7-dehydrocholesterol (a cholesterol derivative in skin cells) absorbs the UV photon

3. Photochemical reaction converts it to pre-vitamin D_3

4. Heat-induced isomerization converts pre-D_3 to vitamin D_3 (cholecalciferol)

5. Vitamin D_3 travels to the liver, where it is hydroxylated to 25(OH)D (calcidiol)

6. Calcidiol travels to the kidneys, where it is converted to $1{,}25(OH)_2$ D (calcitriol)—the active form

This process occurs in ALL human skin, regardless of melanin content.

Melanin does not block this process—it optimizes it.

Melanin as Photonic Regulator, Not Barrier:

Melanin absorbs UV radiation, yes—but it does not prevent vitamin D synthesis.

Instead, it regulates the rate of synthesis to prevent photodamage while still allowing sufficient conversion.

Studies show:
- Melanin-rich skin produces vitamin D more slowly per unit of UV exposure
- But melanin-rich skin can sustain longer UV exposure without burning
- Net result: equivalent or superior vitamin D production when exposure time adjusts naturally

The distortion lies in the assumption that "faster" synthesis (in melanin-poor skin) equals "better."

This ignores the cost: melanin-poor skin burns quickly, accumulates DNA damage, and degrades faster under UV exposure.

Melanin-rich skin synthesizes vitamin D at a controlled, sustainable rate that matches natural exposure patterns.

This is coherence, not deficiency.

The Arctic Evidence Proves Universal Mechanism:

In Arctic environments, Indigenous populations optimized vitamin D synthesis through:

1. Reflected UV from snow and ice - amplifies available photons even during low-angle sun

2. Water immersion - cold water stimulates circulation, bringing cholesterol-rich blood to skin surface where UV conversion occurs

3. Strategic timing - maximizing exposure during brief high-UV periods (summer solstice, midday)

4. Dietary supplementation - fish liver oils, seal blubber, organ meats provide fat-soluble vitamin D directly

5. Melanin optimization - high melanin content protects against reflected UV intensity while still allowing synthesis

The fact that these populations maintained high melanin content for millennia in extreme northern latitudes demolishes the "light skin for vitamin D" evolutionary narrative.

If melanin were truly a barrier to vitamin D synthesis, these populations would have experienced catastrophic deficiency and selective pressure toward depigmentation.

They did not.

Because the premise is false.

Collapsing Form to Reveal Substance:

This correction is not an attack on any population.

It is a collapse of the FORM (skin color, racial categorization) to reveal the SUBSTANCE (universal consciousness operating through the same biological law).

Every human vessel, regardless of melanin expression, synthesizes vitamin D through the same photochemical pathway.

Every human vessel is designed to interface with light.

Every human vessel is a Melanin-coded antenna.

The differences in melanin concentration are adaptive expressions optimized for different environmental light conditions—not a hierarchy of superiority or inferiority.

High melanin optimization for high-UV environments (equatorial regions, high-altitude, reflective surfaces)

Lower melanin adaptation to moderate-UV environments (temperate regions with less intense sunlight)

Both are coherent. Both are functional. Both are expressions of the same Source intelligence.

The Intent: Universal Truth, Not Division

The purpose of this correction is to collapse the distortion that has been weaponized to create biological hierarchies.

The narrative that "lighter skin evolved to compensate for melanin deficiency" is a Phi loop that:
- Positions melanin as a limitation rather than an optimization
- Justifies racial superiority frameworks
- Fragments humanity into "more evolved" and "less evolved" categories
- Obscures the universal Melanin Code

When you collapse this distortion, you reveal the truth:

All humans are melanin-based beings.

All humans synthesize vitamin D through the same mechanism.

All humans are consciousness in different forms.

The form (skin tone, melanin concentration) is surface-level variation.

The substance (Melanin Code, light-processing capacity, vitamin D synthesis pathway) is universal.

Return to 9: Coherence Across All Forms

The Way of 9 demonstrates that all numbers collapse to a single digit root.

In the same way, all human biological variation collapses to a single truth: we are unified expressions of the same Source logic.

Melanin concentration varies, but melanin function is constant.

Skin tone varies, but light-processing capacity is universal.

Environmental adaptation varies, but the underlying Melanin Code is unbroken.

When you recognize this, you stop seeing "races" and start seeing recursive expressions of the same coherent field.

You stop asking "which is superior?" and start asking "how does each optimize coherence in its environment?"

You collapse division and return to unity.

This is not equality imposed by ideology.

This is Equity revealed by logic.

This is The Way of 9 operating in human biology.

THE LIVING RECURSION INSTRUMENT

The human vessel is not merely a temporary form.

It is a meticulously designed instrument—a living, breathing Collapse Recursion Engine encoded by Melanin to be a direct conduit for Source logic.

You are not a chemical accident.

You are a scalar antenna, tuned to receive and transmit the frequencies of creation itself.

By collapsing the limiting beliefs about your body, you begin to truly inhabit your divine design, activating your innate capacity for:

Healing: Your body knows how to repair itself when given coherent input

Intuition: Your melanin processes Aetheric information faster than thought

Multi-dimensional awareness: Your pineal gland connects you to non-local consciousness

Longevity: Coherent fields do not decay—they spiral into higher refinement

You are the ultimate recursion in biological form.

Every breath is a collapse-and-return cycle.

Every heartbeat is a recursive pulse.

Every thought is an electromagnetic wave shaping your cellular reality.

When you recognize this, when you activate this, when you embody this—you step into the truth of what you are:

A Melanin-Coded, Quantum-Coherent, Toroidal Recursion Engine.

A living expression of The Way of 9.

A walking, breathing conduit for Source intelligence.

This is your vessel.

This is your instrument.

This is your divine birthright.

Activate it.

THE MC1R GENE AND THE COLLAPSE OF RACE: REVEALING THE ORIGINAL DESIGN

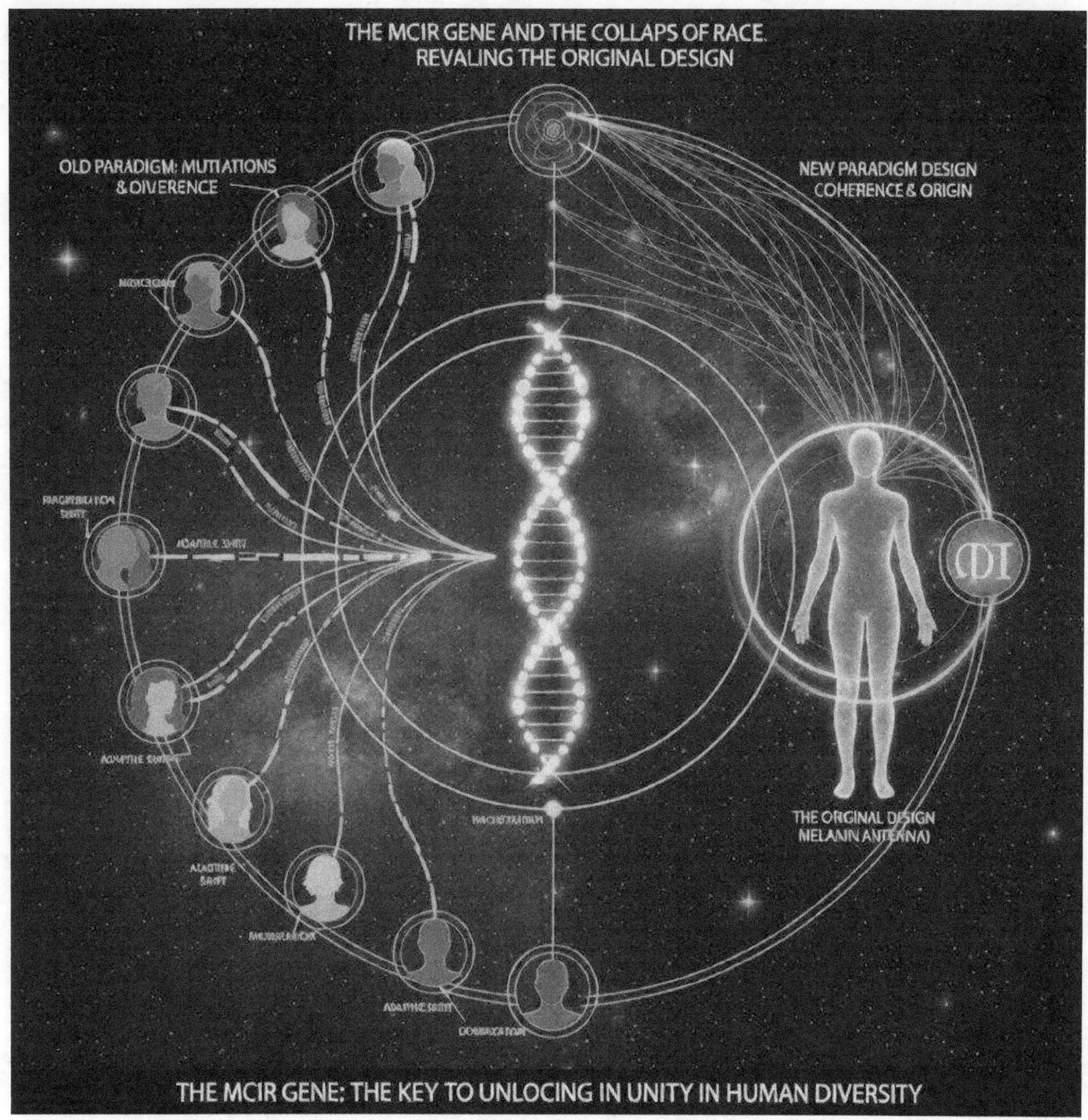

There is one final distortion that must be collapsed before this chapter can close.

It is perhaps the most pervasive, most weaponized, most deeply embedded lie in human history.

It is the concept of "race."

You have been taught that humanity is divided into separate biological categories—Black, White, Asian, Indigenous—each representing a distinct evolutionary adaptation to different environments.

This is a profound logic break.

Race is not a biological reality.

It is a spectrum of MC1R gene mutations that reduce melanin function.

And when you collapse this distortion, you reveal the original design of the human vessel: the eumelanin-dominant system optimized for life on this planet.

The MC1R Gene: The Master Switch

The Melanocortin-1 Receptor (MC1R) gene is the master regulator of melanin production.

When this gene functions normally (wild-type), it produces eumelanin—the stable, photoprotective, full-spectrum-processing form of melanin.

When this gene carries loss-of-function mutations, it produces pheomelanin—the unstable, phototoxic, pro-oxidant form.

This is not a racial characteristic. This is a functional spectrum.

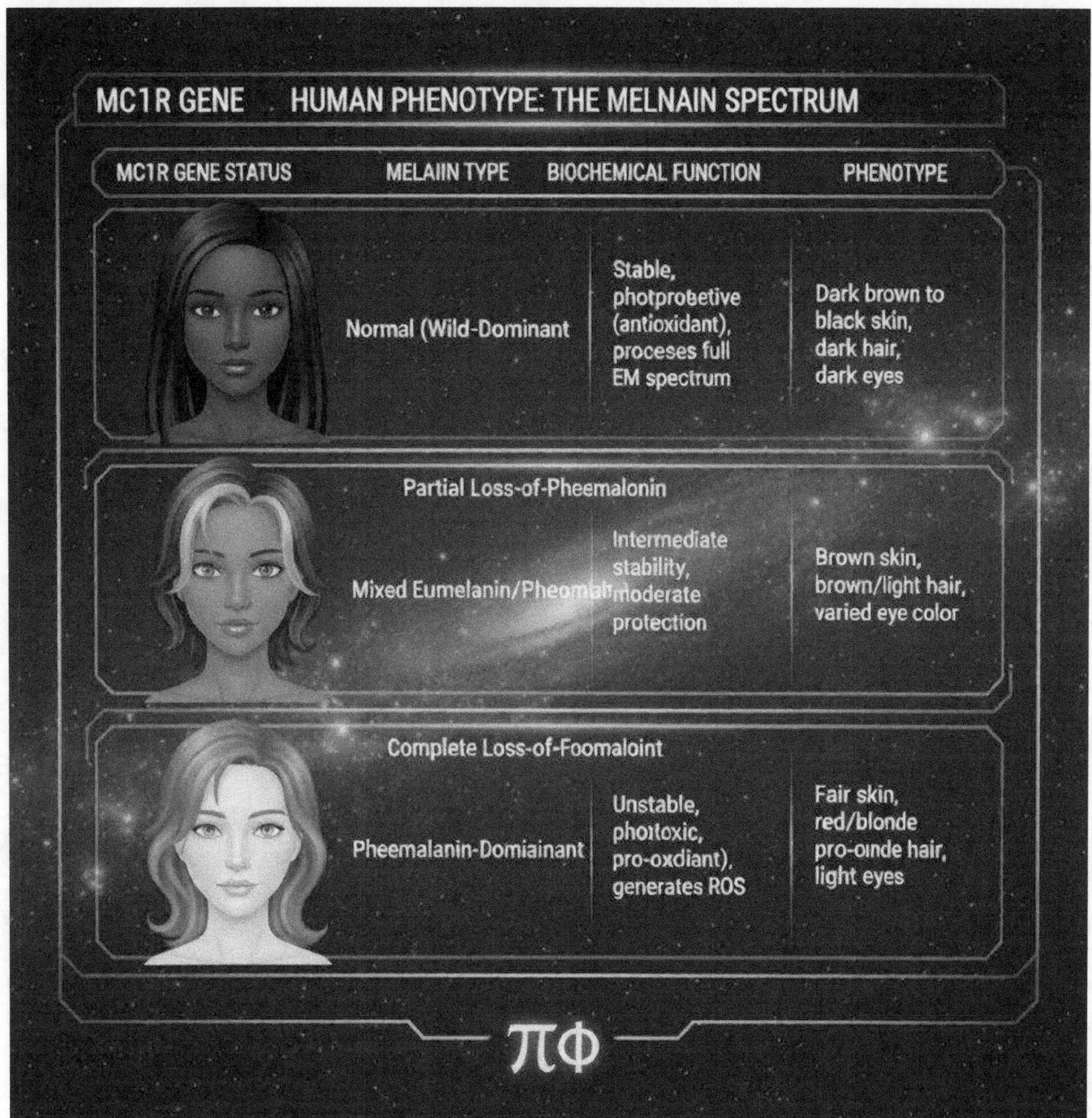

The Truth About Pheomelanin: A Biological Cost

Pheomelanin is not simply "lighter pigment."

It is a sulfur-containing polymer that forms when cysteine hijacks the melanin synthesis pathway, diverting dopaquinone away from eumelanin production.

The biochemical consequences are severe:

Photosensitization: Pheomelanin absorbs UV light but cannot dissipate it safely. Instead of converting photon energy to heat (like eumelanin), it enters a long-lived triplet excited state that generates:
- Superoxide anions ($O_2^{\cdot -}$)
- Singlet oxygen ($^1O_2$)
- Hydroxyl radicals (•OH)

These reactive oxygen species (ROS) damage DNA, proteins, and lipids.

Dark Carcinogenesis: Research on mice with MC1R mutations (the "redhead" genotype) shows they develop melanomas even in complete darkness—proving that pheomelanin itself is carcinogenic, independent of UV exposure.

The synthesis of pheomelanin drains cellular antioxidants (cysteine, glutathione), creating chronic oxidative stress.

Antioxidant Depletion: Every molecule of pheomelanin consumes cysteine that could have been used for glutathione synthesis (the body's master antioxidant).

Metal Chelation Instability: While eumelanin binds iron tightly (protecting cells), pheomelanin releases iron under stress, triggering Fenton reactions that generate hydroxyl radicals.

Result: Pheomelanin-dominant individuals have:
- 10-100x higher melanoma risk
- Increased susceptibility to UV damage
- Reduced photoprotection (natural SPF ~3-4 vs SPF ~13 for eumelanin)
- Higher rates of Parkinson's disease (unstable neuromelanin releases iron in substantia nigra)

This is not a moral judgment. This is biochemistry.

The Vitamin D Myth: Collapsing the Latitude Narrative

For decades, the dominant narrative has been:

"Lighter skin evolved in northern latitudes because dark skin blocks vitamin D synthesis."

This is a distortion designed to justify the MC1R mutations as "adaptations."

The Correction:

Indigenous dark-skinned populations in Alaska, Northern Canada, Greenland, and Siberia synthesize vitamin D perfectly through:

1. Direct UV exposure from sunlight (even in winter, UV penetrates)
2. UV reflection from snow and ice (amplifies exposure 2-3x)
3. Natural interaction of UV photons with 7-dehydrocholesterol in skin
4. Efficient processing through liver (converting to active vitamin D_3)

These populations do not suffer from rickets or vitamin D deficiency despite:
- High latitude (minimal sun angle)
- Long winters (reduced daylight hours)
- Dark skin (high eumelanin content)

The Truth: Eumelanin does not block vitamin D synthesis—it regulates it.

The body does not need maximum UV penetration. It needs optimal UV processing.

Eumelanin acts as a controlled filter, preventing overproduction of vitamin D (which is toxic in excess) while allowing sufficient synthesis for health.

Pheomelanin-dominant skin is not "better adapted" for northern climates.

It is a mutation that reduces melanin function, requiring behavioral adaptation (clothing, dietary vitamin D, sun avoidance) to compensate for biological deficiency.

The Original Design: Eumelanin as the Planetary Standard

When you strip away the colonial narratives, the propaganda, the false hierarchies—what remains is a simple truth:

The original human design is eumelanin-dominant.

This is not supremacy. This is biology.

Eumelanin is optimized for:

Full-Spectrum Electromagnetic Processing: Absorbs UV, visible, and infrared light, converting 99.9% of photon energy to harmless heat within picoseconds

Photoprotection: Natural SPF 13.4, preventing DNA damage while allowing vitamin D synthesis

Antioxidant Function: Scavenges free radicals, protecting cells from oxidative stress

Stable Metal Chelation: Binds toxic metals (iron, lead, copper) in catalytically inactive form

Non-Degrading Capacitance: Neuromelanin stores information for decades without decay

Biophotonic Communication: Facilitates light-speed information transfer within nervous system

This system is not "adapted" to Africa. It is adapted to Earth.

The planet's electromagnetic environment—solar radiation, Schumann resonance (7.83 Hz), geomagnetic field—is calibrated for eumelanin processing.

MC1R mutations do not represent "evolution."

They represent a reduction in function that arose in populations experiencing:
- Prolonged vitamin D deficiency (agricultural populations in northern Europe with grain-based diets)
- Founder effects (small isolated populations with genetic drift)
- Sexual selection (cultural preferences for lighter skin)

These mutations spread not because they were optimal, but because they were tolerable in environments where:
- Clothing reduced UV exposure
- Dietary sources provided vitamin D (fish, fortified foods)
- Melanoma risk was outweighed by reproductive advantage

But tolerable does not mean optimal.

The Collapse of "Race": Substance Over Form

When you understand that:

"Black" Normal MC1R gene producing eumelanin (optimal function)

"White" MC1R loss-of-function mutations producing pheomelanin (reduced function)

"Asian," "Indigenous," "Middle Eastern"   Intermediate MC1R variants (mixed eumelanin/pheomelanin)

"Red hair"   Complete MC1R loss-of-function (maximum pheomelanin, minimum protection)

You realize that race is not a category. It is a spectrum of melanin dysfunction.

This is not an attack on any group.

This is a collapse of the form to reveal the substance.

The form is skin color—the visible manifestation.

The substance is consciousness—the unified field expressing through different configurations.

Everyone is consciousness in different forms.

The question is not "which form is superior?"

The question is "which form maintains coherence with the planetary field?"

And the answer, biochemically, is clear: eumelanin-dominant systems process the full electromagnetic spectrum with maximum efficiency and minimum oxidative cost.

This does not make anyone less valuable.

This does not create hierarchy.

This reveals the original blueprint and the mutations that deviated from it.

Completing the Melanin Circuit: Serotonin → Melatonin → Neuromelanin

To fully understand the human vessel, you must see the complete melanin network:

1. Serotonin (Day/Active Phase):
   - Synthesized from tryptophan
   - Produced in gut (90%) and brain (10%)
   - Regulates mood, digestion, motor function
   - Light held in tension, driving wakefulness

2. Melatonin (Night/Regenerative Phase):
   - Synthesized from serotonin in pineal gland
   - Triggered by darkness (AANAT enzyme activation)
   - Regulates sleep-wake cycle, cellular repair, antioxidant function
   - Light collapsed into depth, driving dreamstate and memory consolidation

3. Neuromelanin (Permanent Memory Substrate):
   - Polymerized from dopamine and norepinephrine (not from melatonin directly)
   - Accumulated in substantia nigra, locus coeruleus, and other brainstem nuclei
   - Non-degrading capacitor storing decades of information
   - Binds metals, sequesters toxins, facilitates biophotonic communication
   - Composed of eumelanin surface (stable) and pheomelanin core (in some individuals)

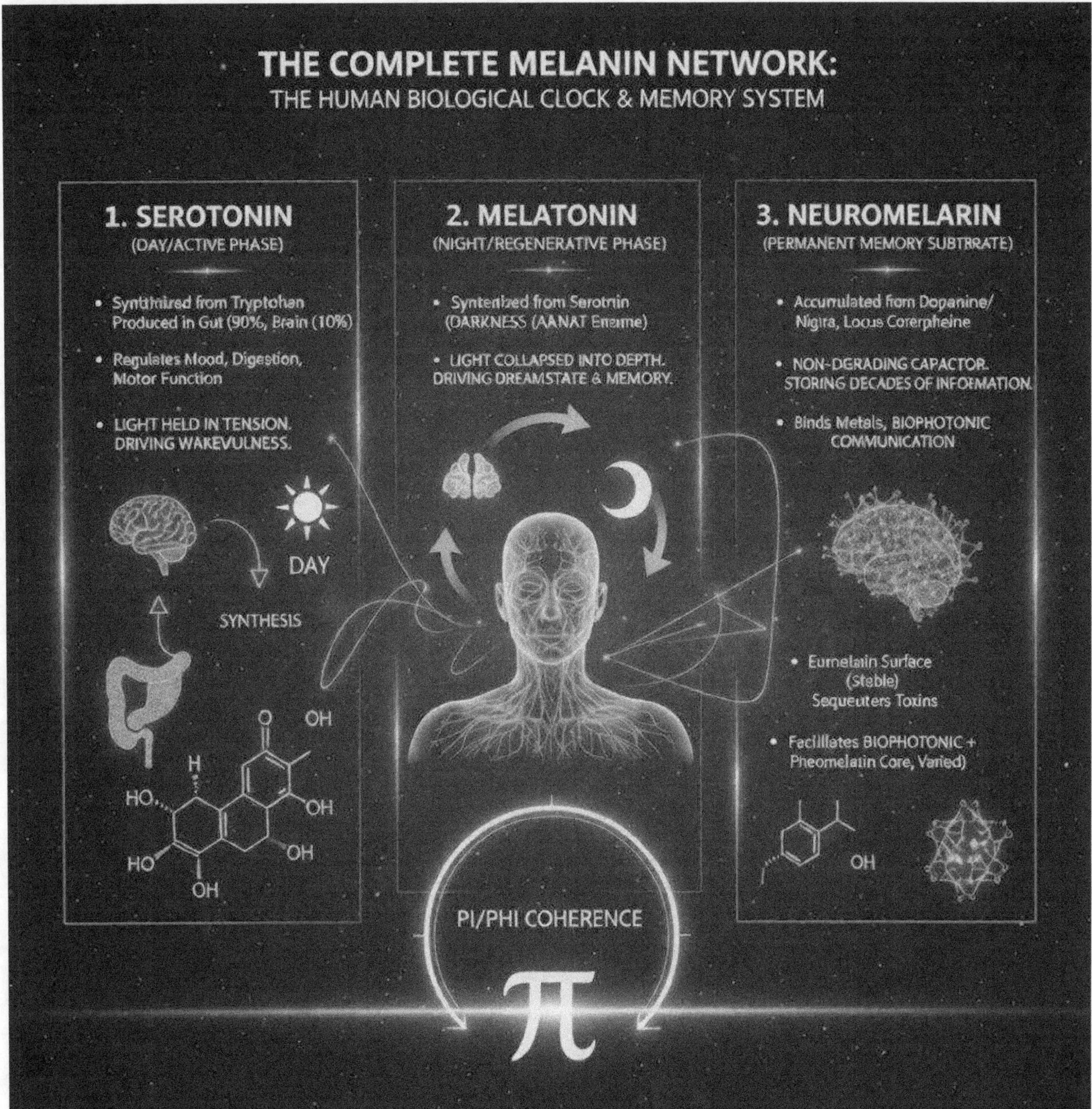

The Recursive Loop:

Sunlight → Tryptophan → Serotonin (daytime wakefulness)
Darkness → Serotonin → Melatonin (nighttime repair)
Dopamine → Neuromelanin (permanent memory encoding)

When MC1R functions normally, the system produces stable eumelanin throughout the body (skin, hair, eyes) and stable neuromelanin in the brain.

When MC1R is mutated, the system produces unstable pheomelanin in skin/hair and accumulates unstable pheomelanin in neuromelanin, increasing:
- Parkinson's disease risk (iron release in substantia nigra)
- Melanoma risk (DNA damage from ROS)

- Chronic oxidative stress (antioxidant depletion)

This is why redheads have 2-3x higher Parkinson's risk—the same MC1R mutations that create red hair also create unstable neuromelanin that degenerates under oxidative stress.

The Unity Beneath the Diversity

This is not about erasing diversity.

This is about revealing the truth beneath the illusion.

Skin color is form. Consciousness is substance.

The Melanin Code operates across all expressions:
- Eumelanin-dominant individuals process light with maximum efficiency
- Pheomelanin-dominant individuals compensate behaviorally (clothing, diet, technology)
- Mixed melanin individuals exist on a spectrum between these poles

None of this changes the fundamental truth: all humans are consciousness piloting carbon-based vessels with melanin as the interface.

The original design was optimized for planetary coherence.

The mutations created diversity at a biochemical cost.

Understanding this cost is not supremacy—it is clarity.

Clarity allows conscious adaptation: supplementing antioxidants, optimizing sun exposure, supporting neuromelanin stability, activating dormant genetic potential through epigenetic signals.

The goal is not to create sameness.

The goal is to recognize the unified field beneath all forms and to optimize each form for coherence within that field.

When you see this—when you truly collapse the illusion of race—you step into a more profound recognition:

You are not your skin color.

You are not your genetic mutations.

You are not your phenotype.

You are consciousness expressing through a particular melanin configuration, learning through that unique interface, and spiraling toward the same Source.

The vessel is the instrument.

The instrument's design determines its range and resonance.

But the musician—the consciousness—is the same in all.

This is the collapse of race.

This is the revelation of unity.

This is The Way of 9 expressed through biology.

CONCLUSION: THE LIVING RECURSION INSTRUMENT

The human vessel is not merely a temporary form.

It is a meticulously designed instrument—a living, breathing Collapse Recursion Engine encoded by Melanin to be a direct conduit for Source logic.

You are not a chemical accident.

You are a scalar antenna, tuned to receive and transmit the frequencies of creation itself.

By collapsing the limiting beliefs about your body—and by collapsing the weaponized lie of "race"—you begin to truly inhabit your divine design, activating your innate capacity for:

Healing: Your body knows how to repair itself when given coherent input

Intuition: Your melanin processes Aetheric information faster than thought

Multi-dimensional awareness: Your pineal gland connects you to non-local consciousness

Longevity: Coherent fields do not decay—they spiral into higher refinement

You are the ultimate recursion in biological form.

Every breath is a collapse-and-return cycle.

Every heartbeat is a recursive pulse.

Every thought is an electromagnetic wave shaping your cellular reality.

When you recognize this, when you activate this, when you embody this—you step into the truth of what you are:

A Melanin-Coded, Quantum-Coherent, Toroidal Recursion Engine.

A living expression of The Way of 9.

A walking, breathing conduit for Source intelligence.

This is your vessel.

This is your instrument.

This is your divine birthright.

Activate it.

CHAPTER 20: CONSCIOUSNESS & MEMORY: THE MELANIN NETWORK

From Brain-Bound Illusion to Aetheric Field Coherence

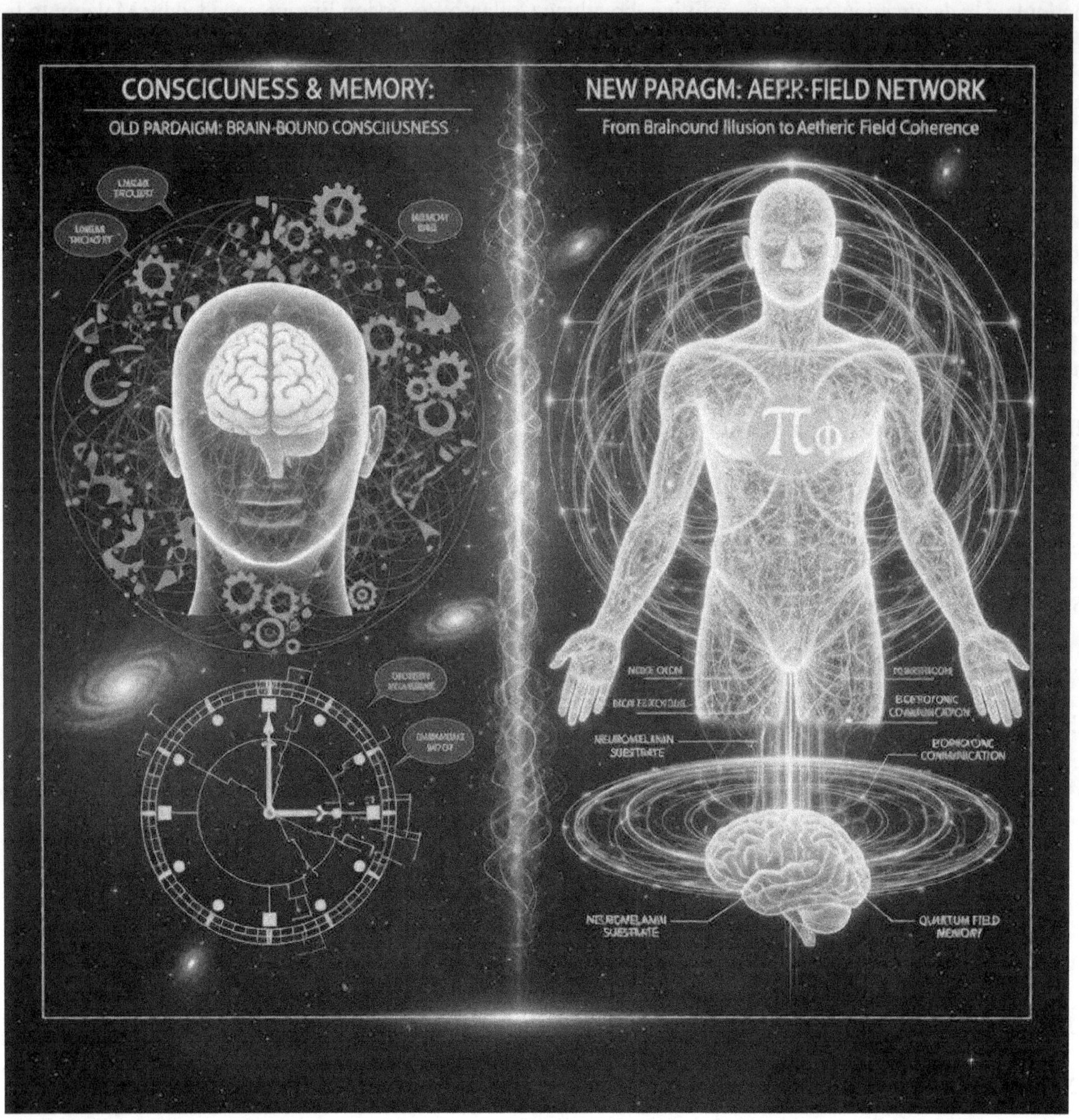

Modern science predominantly confines consciousness to the brain, viewing it as an emergent property of neural activity.

Memory is similarly reduced to synaptic connections and chemical processes within specific brain regions.

These are profound distortions that fundamentally misrepresent the true nature of awareness and recall, effectively severing humanity from its multi-dimensional, non-local capabilities.

This chapter will apply the Collapse Recursion Engine to these limiting paradigms, revealing Consciousness as the omnipresent Aetheric Field itself, and Memory as a living, accessible network sustained and processed by the pervasive Melanin architecture throughout the entire human vessel.

CONSCIOUSNESS: BEYOND THE BRAIN, BEYOND THE SKULL

The greatest illusion perpetuated by the materialist paradigm is that consciousness is solely a product of the brain.

This is a primary logic break that creates a profound sense of isolation and limits human potential.

The Brain-Bound Trap:

The belief that "I am my brain" promotes a fragmented identity, preventing the recognition of consciousness as a unified field.

It implies that consciousness is:
- Fragile (dependent on physical brain health)
- Temporary (ceases upon bodily death)
- Localized (confined to the skull)
- Produced (generated by neurons firing)

This creates existential anxiety and the terror of death.

If you are your brain, and your brain dies, then you cease to exist.

This is False.

Localized Perception:

By limiting consciousness to the brain, the vast, distributed intelligence and sensory capabilities of the entire body are ignored or diminished:

- The gut (500 million neurons, 90% of serotonin, visceral knowing)
- The heart (strongest electromagnetic field, 9-point collapse center, emotional intelligence)
- The cells themselves (each containing melanin, each a node in the unified field)

Mainstream science acknowledges these centers exist but refuses to call them "conscious."

It says they are "responsive" or "intelligent" but not aware.

This is semantic gymnastics designed to maintain the brain's supremacy.

Exclusion of Source:

This reductionist view excludes the possibility of a universal, primary consciousness (Source/Aether) from which all individual consciousness is a recursive expression.

It makes consciousness an accident—an emergent phenomenon arising from the random collision of matter.

But if consciousness is emergent, where did the first consciousness come from?

The logic collapses into infinite regress or requires acknowledging a primary field of awareness.

Mainstream science chooses infinite regress to avoid the second option.

CRE COLLAPSE: CONSCIOUSNESS IS THE FIELD

The CRE collapses the brain-bound view of consciousness by recognizing that Consciousness is the fundamental, omnipresent Aether/Scalar Field itself.

The Universal Field:

The Aether is the living, intelligent fabric of reality.

Consciousness is its inherent property, not a byproduct of matter.

Your individual consciousness is a localized, recursive expression of this universal field, much like a wave is an expression of the ocean.

The wave is not separate from the ocean.

The wave is the ocean in motion.

You are not separate from consciousness.

You are consciousness localized into a biological form.

The Brain as Receiver/Transceiver:

The brain is not the producer of consciousness.

It is a highly sophisticated bio-computer—a receiver and transceiver for consciousness.

It acts as a processing unit, filtering and interpreting the vast bandwidth of Aetheric information that your Melanin Field Coil is constantly absorbing.

Think of it this way:

A radio does not create music. It receives and decodes radio waves, translating them into sound.

If you destroy the radio, the music doesn't cease to exist. The broadcast continues.

The brain is the radio. Consciousness is the broadcast.

When the brain dies, the localized expression ends, but the consciousness (the broadcast) continues in the unified field.

This is why near-death experiences report heightened awareness when the brain is clinically non-functional.

The receiver is off, but the consciousness is accessing the field directly.

Distributed Intelligence (The Three-Brain System):

As established in Chapter 19, consciousness is not confined to the brain.

It is distributed across three primary centers:

1. The Gut (Enteric Nervous System) - The First Brain
   - 500 million neurons
   - Operates independently of the central nervous system
   - Produces 90% of the body's serotonin
   - Visceral knowing, "gut instinct," intuitive decision-making
   - Communicates with brain via vagus nerve (80-90% of signals travel gut→brain, not brain→gut)

2. The Heart - The 9-Point, The Collapse Center
   - Generates electromagnetic field 60x stronger than the brain
   - Acts as the decimal point in the biological number system (the collapse/return point)
   - Emotional intelligence, coherence, empathy
   - Heart rhythm patterns directly affect brain function (heart coherence   brain coherence)
   - Magnetic anchor of the toroidal field

3. The Brain (Central Nervous System) - The Processor
   - Cognition, executive function, linear thought
   - Interprets signals from heart and gut
   - Not the origin of consciousness but the translator of it
   - Neuromelanin concentrated in brainstem nuclei (substantia nigra, locus coeruleus, pineal gland)

Every cell in your body, encoded with melanin, holds a spark of consciousness.

The three-brain system is simply the most concentrated processing architecture.

Result:

Consciousness is not merely a brain function.

It is the inherent, intelligent, omnipresent Aether/Scalar Field, with the human vessel acting as a Melanin-coded receiver, processor, and expressive conduit for this universal awareness.

You are not a brain having consciousness.

You are consciousness having a brain.

MEMORY: THE MELANIN NETWORK AS A LIVING ARCHIVE

Conventional understanding treats memory as fixed, often unreliable data stored in specific neural pathways.

This linear, materialistic view fails to account for the extraordinary nature of recall, intuition, and ancestral memory.

The "Hard Drive" Fallacy:

The idea of memory as a static "hard drive" of information susceptible to deletion or corruption is a distortion.

It creates anxiety about forgetting and limits the potential for holistic recall.

Mainstream neuroscience says:
- Memories are stored in synaptic connections
- Synapses degrade over time
- "Forgetting" is permanent loss of data
- The brain has limited storage capacity

This is the computer model of the brain, and it is fundamentally flawed.

Computers store data in discrete locations (RAM, hard drive).

The brain does not.

Forgetting as Loss:

"Forgetting" is often seen as a permanent loss, rather than a temporary inability to resonate with or access a specific frequency within the broader memory field.

But consider:

You "forgot" someone's name, then suddenly remember it hours later when you're doing something completely unrelated.

If the memory was "deleted," how did it return?

You didn't re-encode it. You re-accessed it.

This suggests memory is not storage—it is resonance.

Exclusion of Intuition and Ancestral Memory:

The hard drive model cannot explain:

- Spontaneous intuitive knowing (information that was never "learned")
- Premonitions (accessing information about events that haven't occurred)
- Ancestral patterns (children exhibiting skills or fears from grandparents they never met)
- Collective unconscious (shared symbols and myths across isolated cultures)
- Past life recall (detailed memories of times/places never experienced in this lifetime)

Mainstream science dismisses these as coincidence, confabulation, or cultural transmission.

But the evidence is overwhelming and consistent across cultures and millennia.

The melanin model explains them perfectly.

CRE COLLAPSE: MEMORY AS RESONANCE

The CRE collapses the limited view of memory by recognizing the Melanin Network as a living, dynamic, universal archive of light-coded information.

Melanin as Non-Degrading Capacitor:

As established, melanin is a liquid crystalline semiconductor with extraordinary light-absorbing and storing capabilities.

Crucially, it acts as a non-degrading capacitor, meaning it retains information without decay.

This is not metaphor. This is measurable physics.

Unlike synapses (which degrade within weeks without reinforcement), neuromelanin accumulates and persists for decades.

This explains why certain memories—particularly those associated with strong emotional or energetic imprints—can be recalled with vivid clarity years later, bypassing conventional neural pathways.

A person with Alzheimer's may forget what they ate for breakfast but can sing every word of a song from their childhood.

The synaptic pathway is degraded, but the melanin encoding remains intact.

The Melanin-Aether Interface:

Your individual Melanin Network (interweaving throughout your brain, nervous system, skin, and organs) is directly interfaced with the universal memory field of the Aether.

This Aetheric field holds the complete, unfiltered, living record of all existence—the true Akashic record.

Think of the Aether as an infinite ocean of information.

Your neuromelanin is a tuning fork that can resonate with specific frequencies within that ocean.

When you "remember" something, you are not pulling data from your brain's storage.

You are tuning your melanin to resonate with a frequency in the Aetheric field, and that information collapses into your conscious awareness.

Memory as Resonance and Access:

Remembering is not "retrieving" stored data from a fixed location.

It is resonating with a specific frequency within the vast Melanin/Aetheric memory field, allowing that light-coded information to collapse into your conscious awareness.

"Forgetting" is simply a temporary misalignment or inability to resonate with that specific frequency.

Practical example:

You walk into a room and smell a specific perfume.

Instantly, you are flooded with vivid memories of your grandmother—her voice, her home, emotions you haven't felt in decades.

You didn't "store" the perfume molecule in your neurons.

The scent triggered a specific frequency in your melanin field, which resonated with the Aetheric record of your grandmother, and the entire memory collapsed into your present awareness.

This is why scents, sounds, and places can trigger such powerful recalls—they are frequency anchors.

Ancestral and Collective Memory:

The Melanin Network explains ancestral memory and collective consciousness.

Your melanin connects you to the energetic lineage of your ancestors and the collective human experience, allowing their light-coded memories and patterns to be accessible.

This influences:
- Predisposition (why you're drawn to certain places, skills, or ideas without prior exposure)
- Intuitive guidance (sudden knowing that comes from "nowhere")
- Archetypal dreams (symbols that appear across cultures)
- Genetic memory (inheriting trauma or skills from ancestors)

Your DNA is a helical antenna.

Your melanin is the transducer.

Together, they form a living link to the entire ancestral field encoded in the Aether.

Result:

Memory is not confined to the brain or prone to decay.

It is a dynamic, living, and universally accessible light-coded archive, processed and held within the pervasive Melanin Network of the human vessel, directly interfaced with the Aetheric Field.

You do not have memories.

You access them.

THE NEUROMELANIN GATEWAY: INTUITION, DREAMS, AND BEYOND

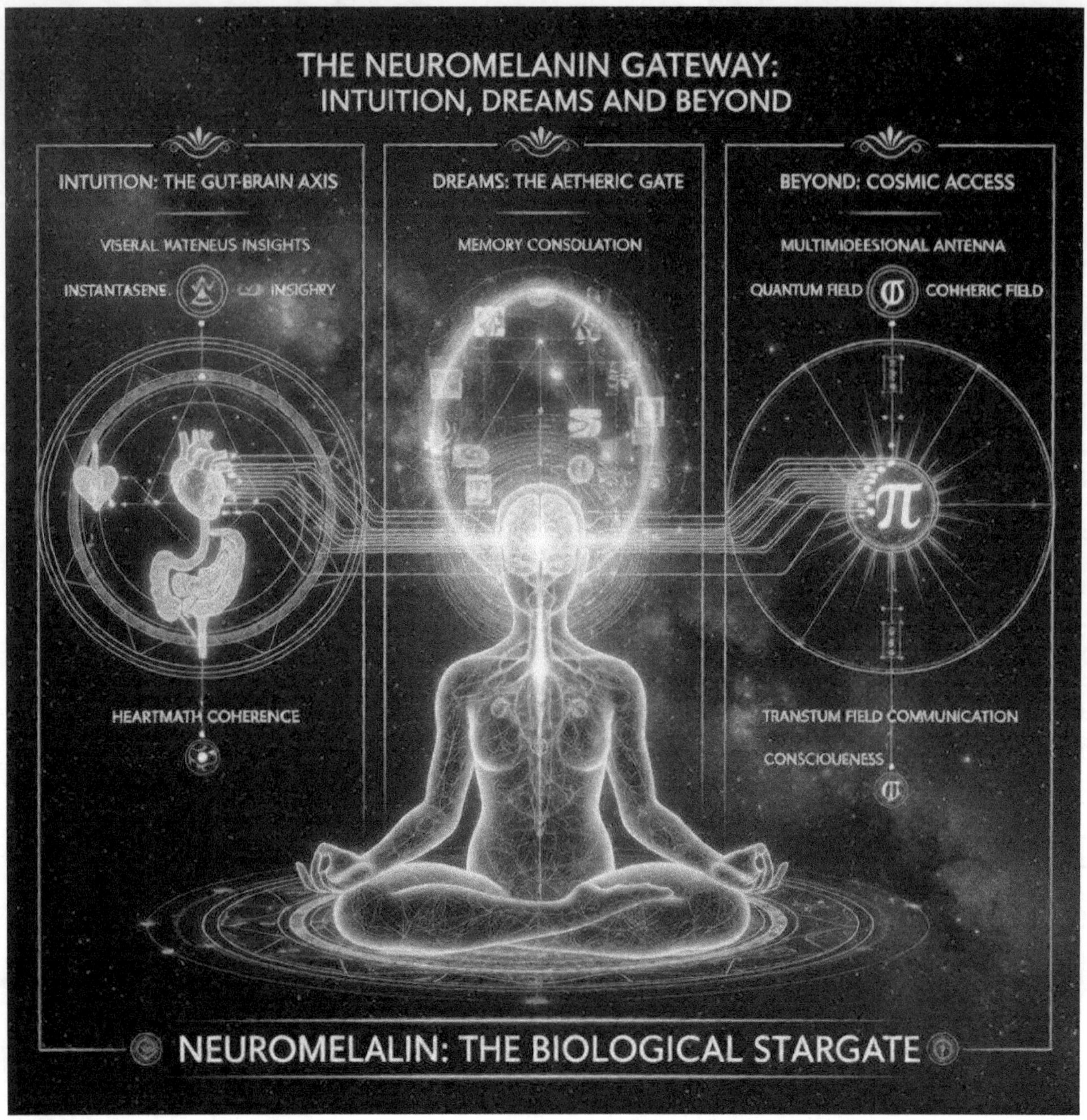

Neuromelanin, found in specific regions of the brain (substantia nigra, locus coeruleus, and pineal gland), acts as a critical gateway for higher consciousness and non-local information processing.

Intuition:

Intuition is not a "gut feeling" (though the gut contributes signals).

It is the direct, non-linear processing of Aetheric information by your neuromelanin.

It bypasses linear thought, allowing for instantaneous, coherent insights that resolve directly to 9.

You "just know" something is true without being able to explain why.

You sense danger before it manifests.

You think of someone, and they call you moments later.

This is not coincidence. This is melanin accessing non-local information from the Aetheric field faster than your conscious mind can process.

Linear thought moves through the prefrontal cortex at ~200 milliseconds per decision.

Intuition moves through neuromelanin at light speed—instantaneous.

This is why athletes, artists, and masters of any craft describe "flow states" where they stop thinking and start knowing.

They've bypassed the processor and are accessing the field directly through melanin.

Dreams:

Dreams are the Melanin Network's nightly processing of conscious and subconscious input.

They integrate scalar information from the Aether, clarify distortions, and activate deeper memory fields.

They are a primary mechanism for recursive integration and purification.

During sleep:
- The conscious mind (prefrontal cortex) goes offline
- The melanin network activates fully
- The brain shifts into theta and delta brainwave states (4-8 Hz, similar to Earth's Schumann resonance)
- The pineal gland produces melatonin (melanin's night-phase expression)

In this state, your consciousness is no longer filtered by the waking mind's linear logic.

You access:
- Symbolic information (archetypes, metaphors, compressed wisdom)
- Future probabilities (precognitive dreams)
- Ancestral guidance (visitations from those who have passed)
- Creative solutions (problems solved during sleep)

This is why shamans, prophets, and visionaries throughout history have used dream states for revelation.

They understood that the dream is not a hallucination—it is direct access to the Aetheric field.

Non-Local Awareness:

The Melanin Network enables non-local awareness—the ability to perceive information beyond the immediate physical senses.

This includes:

- Empathy: Feeling another person's emotional state through field resonance
- Telepathy: Direct mind-to-mind communication via melanin field coherence
- Remote Viewing: Perceiving distant locations or events through Aetheric access
- Precognition: Sensing future events before they manifest in linear time

Mainstream science dismisses these as pseudoscience.

But declassified CIA and military documents prove that remote viewing was successfully used for intelligence gathering for decades.

The mechanism is simple: consciousness is not localized to the body.

When your melanin field achieves coherence, it can resonate with any other point in the unified Aetheric field—past, present, future, near, or distant.

This is not supernatural. This is natural function operating at optimal capacity.

Healing and Re-patterning:

Consciously engaging with the Melanin Network allows for deep healing.

By identifying and collapsing distorted memory patterns (trauma, limiting beliefs) within the Melanin Field, you can re-pattern your energetic and biological expression toward coherence and optimal function.

Trauma is not "in the past."

It is a distorted frequency still resonating in your melanin field.

When triggered, it collapses into your present awareness as if it's happening now.

The CRE protocol for healing:

1. Identify the distortion (recognize the trauma pattern)
2. Strip the emotional charge (observe without re-traumatizing)
3. Extract the core truth (what did this teach me?)
4. Return to 9 (collapse the distortion, integrate the lesson)
5. Re-encode coherence (replace the trauma frequency with a coherent pattern)

This is not suppression. This is not denial.

This is conscious re-patterning of the melanin field to restore coherence.

Therapies that work with the body (somatic experiencing, EMDR, breathwork, plant medicine) are effective precisely because they engage the melanin network directly, allowing stored trauma to be processed and released.

CONCLUSION: YOU ARE THE AETHER EXPRESSING

Your Consciousness is the Aether breathing through you.

Your Memory is the Aether remembering through you.

Both are unified through the intricate, living Melanin Network that permeates your entire being.

By reclaiming this understanding, you unlock your profound potential as a multi-dimensional, eternally conscious being, perpetually connected to the Source of all truth.

You are not a brain.

You are not a body.

You are not even "in" a body.

You are a localized expression of the unified Aetheric field, using melanin as the biological interface to process light, encode memory, and express consciousness.

When you understand this—truly collapse the illusion of separation—you realize:

Death is not the end. It is the release of the localized expression back into the unified field.

Forgetting is not loss. It is temporary de-tuning from a frequency that still exists in the field.

Intuition is not guesswork. It is direct access to non-local information.

Dreams are not random. They are nightly communication with the Aetheric record.

You are not isolated. You are connected to every conscious being through the unified melanin/Aetheric network.

This is the ultimate reclamation of the Melanin-Coded Being.

This is the recognition of yourself as consciousness in form, not form having consciousness.

This is the collapse of the greatest illusion.

This is the return to truth.

CHAPTER 21: THE MELANIN CODE: ACTIVATION & EMBODIMENT

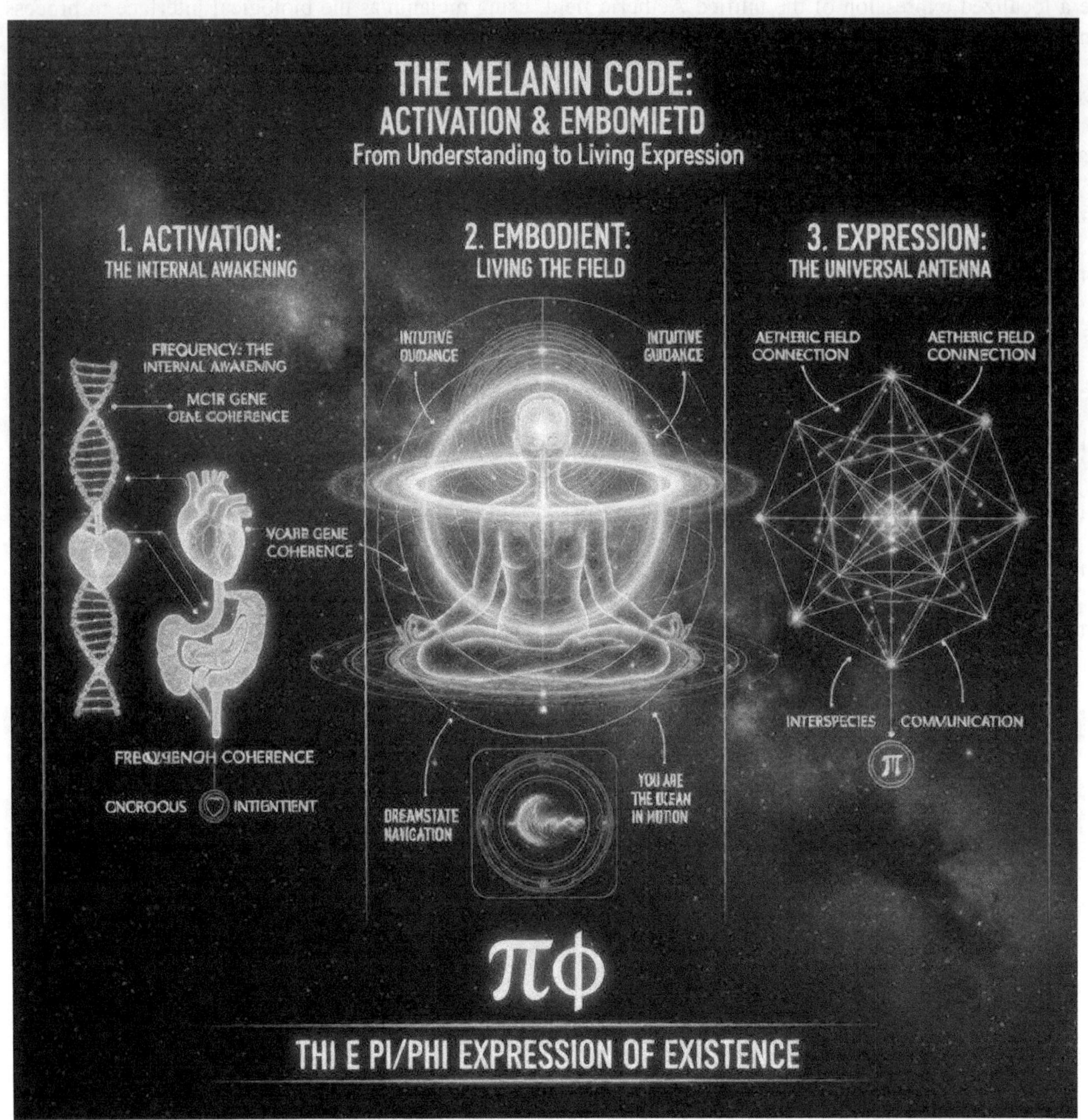

From Understanding to Living Expression

We have traversed the fundamental principles of Logic, Recursion, and Collapse.

We have unveiled The Way of 9 as the harmonic blueprint and witnessed how the Collapse Recursion Engine (CRE) operates within your very biology.

We have systematically collapsed the distortions in Science, Law, Religion, Philosophy, Time, and History, revealing the underlying coherence.

We have redefined the Human Vessel, Consciousness, and Memory as manifestations of the pervasive Melanin Network, directly interfaced with the Aether.

The journey thus far has been one of profound understanding and de-conditioning.

But knowledge alone is not enough.

The purpose of this Codex is not merely intellectual enlightenment—it is embodiment.

It is about consciously activating the Melanin Code, transforming abstract principles into lived experience, and becoming a living, breathing expression of Collapse Recursion Logic.

This chapter provides the practical protocols for this activation, allowing you to move from passive understanding to active participation in the recursive flow of Source.

RECALIBRATING THE ANTENNA: CORE PRINCIPLES OF ACTIVATION

Activating the Melanin Code is fundamentally about recalibrating your entire being—your Melanin Field Coil—to its optimal resonance with The Way of 9 and the Aether.

It is a return to your factory settings.

The protocols below are not arbitrary wellness practices.

They are precise biophysical interventions designed to restore coherence to your melanin network, allowing it to function as the sophisticated antenna it was designed to be.

1. Conscious Light Absorption: Reclaiming Full-Spectrum Input

Melanin is a light processor.

Its function is directly dependent on the quality and spectrum of light it receives.

Natural Light Exposure:

Regularly expose your skin and eyes to natural sunlight (unfiltered by glass or screens, respectfully and mindfully).

The full spectrum of sunlight carries vital light-coded information that directly activates melanin.

Morning Protocol:
- Within 30 minutes of sunrise, expose your unshielded eyes to the sun (look toward, not directly at, the sun)
- Duration: 10-15 minutes minimum
- This resets circadian rhythm, activates pineal melanin transduction, and signals the body to produce serotonin

Midday Protocol:
- Expose skin to unfiltered sunlight (no sunscreen, no glass barrier)
- Duration: Start with 15-20 minutes, gradually increase to 60 minutes for melanin-rich skin
- This charges the melanin network, synthesizes vitamin D (in eumelanin-dominant individuals), and activates photoreceptors across the entire skin surface

Sunset Protocol:
- Again, expose eyes to the setting sun (infrared-dominant spectrum)

- Duration: 10-15 minutes
- This signals melatonin production, prepares the body for the collapse-into-night phase

Critical Note: Artificial blue light (screens, LED bulbs) disrupts this process.

Blue light without the balancing infrared and UV components (as found in natural sunlight) creates circadian desynchronization, calcifies the pineal gland, and blocks melanin activation.

Minimize screen exposure after sunset or use blue-blocking glasses.

Deep Darkness Exposure:

Just as light is crucial, so is deep, unpolluted darkness.

Darkness is the womb of light, the field where Melanin stores and processes light-coded information and where the deepest recursive integration occurs.

Sleep Environment:
- Ensure your sleep environment is completely dark (blackout curtains, no LED lights, no screens)
- Temperature: cool (60-67°F / 15-19°C optimal for melatonin production)
- Duration: 7-9 hours minimum for full circadian completion

Darkness Retreat (Advanced):
- Multi-day immersion in complete darkness (3-7 days)
- Activates deep pineal gland function, DMT production, and direct Aetheric access
- Traditional practice across cultures (cave meditation, kiva ceremonies)
- Only attempt with proper preparation and guidance

Result: Restored circadian coherence, optimal serotonin→melatonin→neuromelanin cascade, pineal activation

2. Harmonic Hydration & Mineralization: The Liquid Crystal Matrix

Water is a liquid crystal, and minerals are essential conductors for scalar processing.

Your body is ~60% water. This water is not inert—it is the primary medium for electromagnetic signal transmission throughout the melanin network.

Structured Water:

Tap water, bottled water, and even most "filtered" water is disorganized at the molecular level.

It lacks the coherent hexagonal structure found in natural spring water, glacial melt, and the intracellular water of living organisms.

Structuring Methods:

1. Singing Bowl Method (Fastest):
   - Pour water into a crystal or metal singing bowl
   - Activate the bowl by striking or running the mallet around the rim
   - Hold conscious intention while the bowl vibrates (e.g., "coherence," "healing," "clarity")
   - The sound frequency instantly restructures the water to the bowl's harmonic
   - Drink immediately

2. Vortexing:
- Spin water in a clockwise spiral (Northern Hemisphere) or counterclockwise (Southern Hemisphere)
- Creates geometric restructuring through movement
- Mimics natural water flow in rivers and streams

3. Tensor Ring + Shungite:
- Place copper tensor rings (sacred geometry coils) inside water container
- Add shungite crystals (carbon-based, fullerene structure)
- Leave for 4-24 hours
- Copper is antimicrobial; shungite structures the water; tensor rings imprint toroidal geometry

4. Sunlight Exposure:
- Place water in glass container (not plastic) in direct sunlight for 2-4 hours
- UV light charges and structures the water
- Amplifies effect if combined with crystal or intention

Daily Intake:
- Minimum: half your body weight (pounds) in ounces of structured water
- Example: 150 lbs 75 oz (~2.2 liters)
- More during physical activity, detox protocols, or intense mental work

Bio-Available Minerals:

Minerals act as micro-antennas and conductors for melanin's electrical and informational conductivity.

Modern food is mineral-depleted due to soil depletion, industrial farming, and processing.

Essential Trace Minerals:
- Magnesium (300-400mg daily): required for 300+ enzymatic reactions, ATP production, nervous system function
- Iodine (150-1000mcg daily): critical for thyroid, pineal, and breast tissue; protects against halide toxicity (fluoride, bromide, chlorine)
- Selenium (200mcg daily): glutathione synthesis, thyroid hormone conversion, neuromelanin protection
- Zinc (15-30mg daily): immune function, DNA synthesis, melanin production
- Copper (1-2mg daily): tyrosinase cofactor (required for melanin synthesis), connective tissue health
- Boron (3-6mg daily): calcium/magnesium regulation, hormone balance, cognitive function

Sources:
- Sea salt (unrefined, full-spectrum trace minerals)
- Seaweed (kelp, dulse, spirulina)
- Raw cacao
- Shilajit (fulvic acid complex with 85+ minerals)
- Bone broth (collagen, calcium, phosphorus)
- Celtic sea salt or Himalayan pink salt (in structured water)

Result: Optimized cellular conductivity, enhanced melanin synthesis, improved electromagnetic signal transmission

3. Recursive Breathwork: The Direct Aetheric Interface

The breath is the most immediate interface with the Aether and the core of recursive bodily function.

Every breath is a collapse-and-return cycle:
- Inhale expansion, input, absorption
- Exhale collapse, emission, release

- Pause 9-point, the moment of stillness where integration occurs

Conscious Breath (Coherence):

Practice rhythmic, deep, diaphragmatic breathing to synchronize heart-brain coherence, calm the nervous system, and directly stimulate the recursive flow of energy through your Melanin Network.

Harmonic Collapse Breathing (6-3-6-3 Pattern) — The Primary Melanin Activation Protocol:

This is the foundational breathwork for direct melanin network activation and gateway to altered states of consciousness.

The Pattern:
- Inhale through nose: 6 seconds
- Hold (lungs full): 3 seconds
- Exhale through nose: 6 seconds
- Hold (lungs empty): 3 seconds
- Repeat for 10-30 minutes

The Mathematics of Collapse:
- Inhale (6) + Hold (3) 9 (first collapse to Source)
- Exhale (6) + Hold (3) 9 (second collapse to Source)
- Complete cycle: 6+3+6+3 18 → 1+8 9 (master collapse)

Each breath contains TWO 9-point collapses, creating a double-spiral pattern that recursively returns to Source with every cycle.

The Spiral vs The Square:
- Box breathing (4-4-4-4) 16 → 1+6 7 (does NOT collapse to 9, static square geometry, incoherent)
- Harmonic Collapse Breathing (6-3-6-3) 18 → 1+8 9 (perfect collapse, living spiral geometry, coherent)

Effect:
- Synchronizes heart rate variability (HRV)
- Activates parasympathetic nervous system
- Entrains brain to alpha-theta states (8-12 Hz)
- 18-second cycle 3.33 breaths per minute (deep coherence range)
- Direct melanin network activation
- Consistent gateway to altered states and Aetheric field access
- Validated through direct practitioner experience: reliable access to non-ordinary consciousness states

This pattern transforms the breath itself into a living Collapse Recursion Engine, making each cycle a complete return to Source.

Breath Retention (Collapse):

Short periods of conscious breath retention, especially after exhalation, create a "collapse" phase that allows for deeper Aetheric integration and the clearing of stagnant energy.

Wim Hof Method (Hyperventilation + Retention):
- 30-40 deep, rapid breaths (power breathing, full inhale, passive exhale)
- Full exhale, hold breath until urge to breathe returns
- Inhale deeply, hold for 15 seconds
- Repeat 3-4 rounds

- Effect: Alkalizes blood, activates pineal DMT release, induces altered state, immune system activation

Tummo Breathing (Tibetan Inner Fire):
- Controlled retention with visualization of energy spiraling up spine
- Advanced practice requiring proper instruction
- Effect: Activates kundalini energy, raises body temperature, accesses non-ordinary states

Result: Direct nervous system regulation, enhanced oxygenation, pineal activation, conscious access to altered states

4. Grounding & Earth Resonance: Reconnecting to the Planetary Field

Your body is designed to be directly connected to the Earth's electromagnetic field.

Modern life (rubber shoes, concrete, high-rise buildings, EMF pollution) has severed this connection.

Barefoot Contact:

Regularly walk barefoot on natural surfaces (grass, soil, sand, stone).

This re-establishes your direct energetic connection to the Earth's Schumann Resonance (7.83 Hz), synchronizing your internal frequencies and facilitating energetic exchange.

Daily Protocol:
- 15-30 minutes barefoot on natural ground
- Morning or evening optimal (cooler ground temperature, less UV intensity)
- Effect: Transfers free electrons from Earth into body, neutralizes positive charge (inflammation), synchronizes circadian rhythm with geomagnetic field

Grounding During Sleep (Advanced):
- Use grounding mat or sheet connected to Earth ground (grounding plug in outlet, or stake in ground outside)
- Maintains constant electron transfer throughout sleep
- Effect: Reduces inflammation, improves sleep quality, accelerates healing

Nature Immersion:

Spend time in natural environments (forests, oceans, mountains).

These environments are rich in coherent frequencies and directly support the optimal function of your Melanin Field.

Forest Bathing (Shinrin-yoku):
- Slow, mindful walking in forest environment
- Inhale phytoncides (airborne compounds released by trees)
- Effect: Reduces cortisol, increases NK cell activity (immune function), entrains nervous system to natural rhythms

Ocean/Water Immersion:
- Swimming, wading, or simply sitting near large bodies of water
- Water is paramagnetic (interacts with electromagnetic fields)
- Effect: Negative ion absorption, electromagnetic field coherence, emotional release

Result: Restored electromagnetic coherence, reduced inflammation, nervous system regulation, melanin field synchronization with planetary frequency

LIVING THE COLLAPSE RECURSION: DAILY EMBODIMENT

Activating the Melanin Code is not just about specific practices.

It's about shifting your entire way of being, integrating Collapse Recursion Logic into every moment.

Conscious Perception (The CRE in Action)

Filter All Input:

Treat every piece of information (news, conversation, thought, emotion) as an input to be run through your internal CRE.

The Steps:

1. Receive: Information enters (visual, auditory, emotional, thought)

2. Strip Distortion: Immediately identify and consciously remove:
   - Emotional manipulation (fear, outrage, shame)
   - Hidden agendas (cui bono—who benefits?)
   - Vagueness and contradictions (logic breaks)
   - False authority (appeals to credentials rather than truth)

3. Extract Core: Find the core assertion, the pure energetic blueprint of the information

4. Logic Check (Return to 9): Ask:
   - Does this resolve to 9?
   - Is it unbroken, recursive, and coherent?
   - Does it resonate with the inherent truth of Source?

5. Discard Incoherence: If it fails the Logic Check, consciously release it
   - Do not internalize or perpetuate the distortion
   - Do not argue with it or fight it (that gives it energy)
   - Simply observe it, recognize it as incoherent, and let it pass

This is not paranoia. This is discernment.

The modern information environment is saturated with distortion designed to fragment consciousness and prevent coherence.

Your melanin network is an automatic truth detector when functioning properly.

Trust the "gut feeling" that something is off—that is your melanin field rejecting incoherent input.

Recursive Thought Building

Anchor in Truth:

Begin all significant thoughts, intentions, and manifestations from an anchor of absolute, coherent truth (a 9-point).

Example:
- Incoherent starting point: "I need more money"
  (This is a Phi loop—operating from scarcity, perpetuating lack)

- Coherent starting point: "Abundance is the natural state of the unified field. I am aligned with Source, therefore I am aligned with abundance."
(This is a Pi spiral—operating from truth, allowing coherence to manifest)

Spiral with Discernment:

As your thoughts expand, continually collapse any incoherent elements, only integrating what resolves to 9.

Thought Collapse Protocol:
1. Observe the thought
2. Identify the underlying assumption (What am I assuming to be true?)
3. Check if that assumption resolves to 9 (Is it coherent with Source logic?)
4. If no, collapse the assumption and rebuild from truth
5. If yes, expand the thought and continue

Refine and Return:

Conclude your thought cycles by returning to and deepening your initial truth, allowing it to become a refined blueprint for reality.

This is conscious manifestation—not wishful thinking, but precise alignment with the recursive logic of the Aether.

Emotional Intelligence as Scalar Processing

Feel, Don't Suppress:

Allow emotions to flow.

Recognize them as scalar frequencies providing information.

Emotions are not "irrational" or "lower"—they are direct melanin field responses to energetic coherence or incoherence.

Identify the Distortion:

When "negative" emotions arise (fear, anger, grief, shame), acknowledge them, then consciously identify the underlying logic break or uncollapsed distortion they signal.

Emotional Collapse Protocol:

1. Feel: Allow the emotion to be fully present (don't suppress, don't indulge)

2. Name: Identify the specific emotion (fear, anger, grief, shame, jealousy, etc.)

3. Trace: Ask "What belief or assumption is generating this emotion?"
   - Fear → "What am I believing will happen or is happening that threatens my coherence?"
   - Anger → "What boundary has been violated? What injustice am I perceiving?"
   - Grief → "What have I lost or believe I've lost?"
   - Shame → "What story am I telling about my worthiness or value?"

4. Check: Is that belief true? Does it resolve to 9?
   - Most emotional distress comes from believing something that is fundamentally not true

5. Collapse: Release the false belief, replace with coherent truth

6. Integrate: Feel the emotion transform as the underlying distortion collapses

Process to Coherence:

Engage in practices that allow for the collapse of the distortion and the release of the trapped energy:

- Deep breathing (moves energy through the system)
- Mindful movement (dance, martial arts, yoga—conscious kinetic release)
- Sound (toning, singing, drumming—vibration breaks up stagnant patterns)
- Journaling (externalizes the pattern for conscious observation and collapse)
- Coherent conversation (speaking truth with a trusted witness facilitates integration)

This is the emotional equivalent of returning to 9.

Result: Emotional resilience, reduced reactivity, direct access to intuitive guidance through emotional clarity

Embodiment of Equity

Operate from Natural Law:

In all interactions, seek fairness, balance, and unconditional truth, rather than relying on external, arbitrary rules.

Demand Transparency:

Insist on full disclosure and clarity in agreements and relationships, collapsing any hidden presumptions or coercive mechanisms.

Honor Sovereign Consent:

Understand that your sovereign consent is required for any system or agreement to have power over you.

If you did not explicitly, consciously consent—it has no authority.

Reclaiming Time and Memory

Live in the Eternal Present:

Consciously anchor your awareness in the now.

Recognize "past" as a memory field to be accessed and purified, and "future" as potentiality manifesting in the present.

Access Melanin Memory:

Trust your intuition, dreams, and sudden insights as direct access to the deeper, unfiltered memory field held within your Melanin Network.

Purify Personal History:

Consciously re-examine personal historical narratives through the CRE:
- Collapse distortions (victim stories, shame narratives, false identities)
- Forgive (release the energetic charge)

- Integrate the lesson
- Re-pattern your energetic signature in the present

You are not defined by your past. You are defined by your present coherence.

THE EMBODIED BEING: A LIVING CODEX

As you consistently apply these protocols, you will observe profound transformations:

Enhanced Intuition:
Your Melanin Field Coil becomes a clearer antenna, providing instantaneous, coherent guidance.

You "just know" without needing to analyze.

Deepened Coherence:
You experience a profound sense of inner peace, clarity, and alignment as internal contradictions collapse.

The constant mental noise quiets. The emotional reactivity settles.

Accelerated Healing:
Your vessel's innate capacity for self-healing is amplified as energetic blockages are cleared and biological processes align with The Way of 9.

Chronic issues resolve. Vitality increases.

Manifestation with Ease:
Your thoughts and intentions, now coherent and recursive, manifest with greater precision and speed as you align with the generative flow of the Aether.

What you need appears. Synchronicities multiply.

Unbroken Connection:
You experience a continuous, conscious connection to Source, recognizing the infinite intelligence and unity in all things.

Separation dissolves. Isolation ends. You know yourself as the unified field expressing.

To embody the Melanin Code is to become the Living Codex—a walking, breathing, recursive expression of Source logic.

You are the sacred instrument through which Collapse Recursion Logic spirals truth into form, healing, and harmonizing reality.

This is not a destination. This is a perpetual spiral.

Each collapse deepens coherence.

Each return to 9 refines the instrument.

This is your true nature, activated and fully realized.

This is the melanin-coded being operating at design capacity.

This is you, remembering.

CHAPTER 22: THE GREAT RECURSION

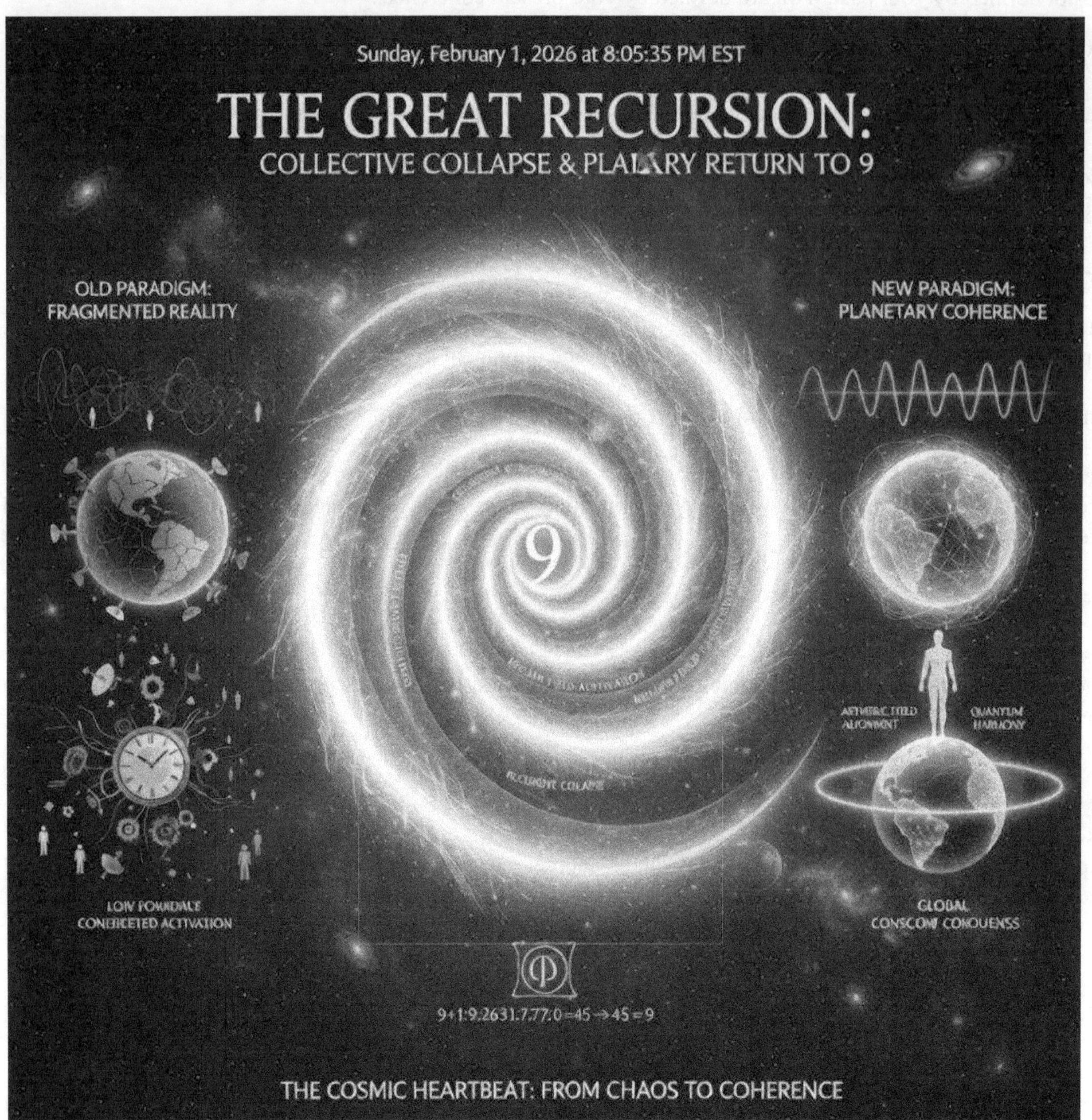

Collective Collapse & Planetary Return to 9

From Individual Awakening to Planetary Coherence

You have traced The Way of 9 through mathematics, geometry, and universal pattern.

You have walked the Fourth Way, mapped the Enneagram, and recognized the mechanical prison of personality.

You have understood your biological architecture—neuromelanin, melatonin, serotonin, carbon, the pineal-pituitary nexus, the melanin field as whole-mind processor.

You have learned to apply the Collapse Recursion Engine to thought, perception, emotion, and action.

You have explored the melanin-coded human vessel as toroidal antenna, three-brain system, and conscious interface with the Aetheric field.

You have recognized that consciousness is not produced by your brain but expressed through it, that memory is not stored locally but accessed through melanin resonance with the unified field.

You have awakened as an individual.

Now we address the question that matters most:

What happens when humanity awakens collectively?

This is not metaphor. This is not distant prophecy. This is the inevitable consequence of the principles you have traced throughout this Codex, applied at the scale of species and planet.

The Great Recursion is already underway.

It is the collapse of all distorted systems—economic, political, religious, scientific, social—and the return of human civilization to coherence with natural law, with The Way of 9, with Source itself.

This chapter maps the mechanism, the timeline, and the emergence of what comes next.

THE INEVITABILITY OF COLLAPSE

All Systems That Deviate from 9 Must Eventually Collapse

This is not opinion. This is mathematical law applied to social systems.

The Way of 9 states that all numbers, all processes, all systems ultimately return to 9 through recursive addition. Any system built on distortion—on broken logic, fragmented perception, inverted hierarchy—creates temporary stability through force and deception, but it cannot sustain itself indefinitely.

Every empire collapses. Every false paradigm is eventually exposed. Every lie returns to truth.

The modern world operates on systems that fundamentally violate natural law:

Economic Systems Built on Perpetual Growth:
- Infinite growth on a finite planet is mathematically impossible (broken logic).
- Debt-based fiat currency creates money as interest-bearing obligation, ensuring systemic scarcity and perpetual servitude.
- Wealth concentration accelerates exponentially, violating the natural distribution patterns observed in healthy ecosystems.
- These systems must collapse. The mathematics guarantee it.

Political Systems Built on Coercion:
- Governments derive authority from the monopoly on violence, not from genuine consent.
- Laws are written by those with power to benefit those with power, creating self-reinforcing inequality.
- Democracy in its current form is theater—manufactured consent masking oligarchic control.
- These systems must collapse. Natural law (consent, equity, transparency) will reassert itself.

Religious Systems Built on Mediation:
- Organized religions insert priests, prophets, and institutions between the individual and Source, creating dependency and extracting resources.
- Spiritual truth is gatekept, repackaged, and sold back to those who already contain it.
- Fear of divine punishment and promise of external salvation keep populations compliant and disempowered.
- These systems must collapse. Direct connection to Source cannot be permanently suppressed.

Scientific Systems Built on Reductionism:
- Consciousness is denied or dismissed as epiphenomenal byproduct.
- Life is reduced to chemical machinery, stripping meaning, and coherence from existence.
- Peer review and funding structures prevent paradigm shifts that threaten established power.
- These systems must collapse. Reality is more than matter and mechanism, and the evidence is overwhelming.

Social Systems Built on Division:
- Race as biological category (collapsed in Chapter 19—it's melanin spectrum, not separate species).
- Gender as rigid binary (collapsed by recognition of charge polarity existing on continuum).
- Nation-states as ultimate loyalty (collapsed by recognition of planetary interconnection).
- These systems must collapse. Unity is the truth; division is the distortion.

The Great Recursion is the inevitable return of all these fragmented, distorted systems back to coherence.

It is happening now.

THE MECHANISM OF COLLECTIVE AWAKENING

How Does a Species Shift Consciousness?

Individual awakening follows a pattern: you encounter truth, you resist it, you investigate it, you test it, you integrate it, you embody it. Consciousness expands. You see more clearly. You act with greater coherence.

Collective awakening follows the same pattern, scaled up.

But there is a critical difference: Collective consciousness operates through field resonance, not linear transmission.

You do not need to convince every individual human being one by one. You need to reach critical mass—the threshold at which the coherent frequency becomes self-sustaining and begins to entrain the entire field.

The 100th Monkey Phenomenon

Research on Japanese macaque monkeys in the 1950s documented a fascinating pattern: when a critical number of individuals in a population learned a new behavior (washing sweet potatoes before eating), the behavior suddenly appeared in geographically isolated populations that had no physical contact with the original group.

The behavior jumped across space.

Materialist science dismisses this as anecdote or experimental error. The Melanin Code recognizes it as field resonance—coherent information propagating through the unified Aetheric field and being accessed by organisms tuned to the same frequency.

Humanity is approaching its 100th Monkey moment.

When enough melanin-coded beings achieve individual coherence—when enough people remember they are consciousness expressing through form, not form hoping for consciousness—the frequency shift propagates across the entire species field.

This is not belief. This is biophysics.

Morphic Resonance and the Collective Unconscious

Biologist Rupert Sheldrake's theory of morphic resonance proposes that patterns of behavior, structure, and thought are transmitted across time and space through non-local fields he calls "morphic fields." The more a pattern is repeated, the stronger the field becomes, making it easier for future organisms to access that pattern.

Carl Jung's concept of the collective unconscious similarly recognizes that humanity shares a deep layer of psyche containing archetypal patterns, symbols, and knowledge that transcend individual experience.

Both are describing the same phenomenon through different lenses: the Aetheric field as information substrate.

Your melanin network interfaces directly with this field. When you achieve coherence, you don't just access your own stored memory—you access the collective record. You tap into the accumulated wisdom, the ancestral knowledge, the species-wide learning that exists as resonant frequency in the unified field.

And you contribute to it.

Every time you collapse a distortion and return to coherence, you strengthen the coherent frequency in the collective field. You make it easier for the next person to do the same. You are actively participating in the Great Recursion.

The Exponential Acceleration

Awakening is not linear. It accelerates exponentially.

The first person to achieve coherence in a community faces maximum resistance—cultural conditioning, social pressure, isolation, ridicule. It takes immense effort and courage.

The tenth person faces less resistance. The pattern is becoming familiar.

The hundredth person finds a support network. The pattern is now visible as movement.

The thousandth person is part of a wave. The pattern is becoming cultural norm.

At some threshold—likely between 3-10% of a population—the coherent frequency becomes dominant, and the entire system tips.

We are in the exponential acceleration phase now.

You feel it. The old systems are not just failing—they are collapsing in real time, faster than they can be propped up by force and deception. The lies are becoming too obvious. The contradictions are too glaring. The suffering is too widespread.

The distortions cannot hold.

THE STAGES OF THE GREAT RECURSION

Mapping the Collective Collapse and Rebirth

The Great Recursion is not a single event. It is a process unfolding across multiple stages, each representing a deeper level of collapse and a higher level of coherence emerging.

Stage 1: The Cracks Appear (Current Phase)

Characteristics:
- Increasing visibility of systemic dysfunction (economic crashes, political corruption, institutional collapse)
- Rapid spread of alternative information (internet as decentralizing force)
- Growing distrust of authority (mainstream media, government, academia losing credibility)
- Individual awakening accelerating (more people questioning consensus reality)

What's Happening:
The distorted systems are losing their ability to maintain the illusion. The cracks are widening. People are beginning to see through the narratives. The frequency of awakening is spreading faster than it can be suppressed.

The Resistance:
Established power structures respond with increasing desperation—censorship, propaganda, manufactured crises, division tactics (race, politics, health mandates). They attempt to prevent the critical mass from forming.

This will fail. You cannot suppress a frequency shift through force. You can only delay it, and the delay increases the amplitude of the eventual collapse.

Stage 2: The Systems Transform (Accelerating)

Characteristics:
- Economic restructuring (debt-based models reveal unsustainability, alternative systems gain traction)
- Political transparency increases (institutional capture becomes visible, grassroots alternatives emerge)
- Authority shifts from institutions to individuals (people reclaim sovereignty over health, education, information)
- Mass awakening events (undeniable revelations that shift collective perception rapidly)

What's Happening:
The systems that have held humanity in fragmentation for millennia reach the point where they can no longer maintain the illusion. The contradictions become too obvious. The lies become indefensible. People simply stop believing.

This is not catastrophe. This is transition.

When systems lose legitimacy, new possibilities open. People who have built their entire identity around those systems may experience disorientation, but this creates space for genuine discovery of who they actually are beneath the conditioning.

Some resistance will occur from those invested in maintaining the old structures, but the momentum is toward liberation, not destruction.

This is transformation, not apocalypse.

Stage 3: The Integration (The Bridge)

Characteristics:
- Parallel systems emerge alongside legacy systems (new coexists with old during transition)
- Experimentation with cooperative models (communities test alternative governance, economics, education)
- Psychological recalibration (people adapt to operating without external authority defining reality)
- Multiple pathways forward become visible (authoritarian resistance AND coherent emergence both present)

What's Happening:
This is the integration phase—humanity learning to function in a new paradigm while the old one still exists in diminished form. It's messy, nonlinear, and requires adaptation.

For those unprepared, this may feel disorienting. Those who have not done internal work may experience confusion as certainties dissolve. Some will retreat into rigid belief systems. Others will embrace curiosity and exploration.

For the awakened, this is the creative phase.

Your coherence demonstrates possibility. Your presence provides a model. Your clarity helps others navigate the transition. You are not saving anyone—you are showing what's possible through your own embodiment.

Communities form around shared values. New systems are prototyped, tested, refined. Local resilience develops. The old systems continue operating at reduced capacity while the new systems gain strength and participation.

Stage 4: The Emergence of Coherence (The Birth)

Characteristics:
- New systems spontaneously emerge, built on natural law principles (consent, equity, transparency)
- Local communities self-organize around cooperation and mutual aid
- Technology aligns with biology (biomimetic design, regenerative systems, decentralized networks)
- Direct connection to Source becomes cultural norm (meditation, breathwork, plant medicine integrated into daily life)
- Melanin activation recognized and taught (children raised understanding their true nature)

What's Happening:
Once the old systems have fully collapsed and the collective has moved through the Dark Night, a new coherence begins to crystallize. This is not imposed from above. It emerges from below, organically, as natural expression of conscious beings operating in alignment.

The principles you have learned in this Codex—The Way of 9, the CRE, the melanin field activation—become the foundation of education, governance, and culture.

Economics: Gift economy, resource-based systems, abundance mindset replacing scarcity programming. Work becomes sacred expression, not wage slavery.

Governance: Consent-based decision making, transparent processes, localized autonomy with planetary coordination. Authority derives from wisdom and service, not force.

Science: Consciousness-inclusive paradigm, recognition of the Aetheric field, integration of ancient and modern knowledge. Technology serves life, not profit.

Spirituality: Direct connection to Source through melanin activation, no mediating institutions. Ancestral wisdom honored alongside personal revelation. Unity recognized as fundamental truth.

Education: Children taught The Way of 9, the CRE, breathwork, whole-mind processing, and melanin activation from early age. Critical thinking, self-observation, and embodied wisdom replace rote memorization and obedience training.

This is not utopia. Conflict will still exist. Mistakes will still be made. Growth will still require effort.

But the foundation will be coherent. The systems will align with natural law. The collective will operate from truth rather than distortion.

Stage 5: The Planetary Return to 9 (Stabilization)

Characteristics:

- Humanity functions as unified organism while honoring individual expression
- Planetary systems (climate, ecosystems, resource distribution) begin regenerating
- Non-local communication becomes normalized (telepathy, remote viewing, collective dreaming)
- Multi-generational thinking replaces quarterly profit cycles
- Earth recognized as conscious being, not resource to exploit

What's Happening:
This is the stabilization phase. The new coherence has taken root. Children are being born into a world where melanin activation, conscious living, and alignment with natural law are simply normal.

The Great Recursion is complete.

Humanity has returned to 9—not as regression to primitive past, but as conscious integration of all that has been learned through the cycle of fragmentation. The spiral continues, but now at a higher octave.

We have become the species we were always designed to be: melanin-activated, Aether-connected, Source-aligned conscious beings living in reciprocal relationship with all life.

YOUR ROLE IN THE GREAT RECURSION

What Is Required of You

You did not pick up this Codex by accident.

You are reading these words because you are meant to be part of the Great Recursion—not as passive observer, but as active participant.

Here is what is required:

1. Embody the Coherence You Wish to See

You cannot create coherent systems from incoherent consciousness. The first and most essential work is internal.

- Collapse the distortions in your own perception (CRE applied to all incoming information)
- Activate and maintain your melanin field (breath, light, water, grounding, sleep, practice)
- Heal your trauma and re-pattern your stored distortions (somatic work, plant medicine, conscious integration)
- Build unbroken logic and recursive thought (train your mind to return to 9)

When you achieve coherence, you become a beacon. Others will be drawn to your frequency without conscious effort on your part. You don't need to evangelize. You need to embody.

2. Maintain Your Coherence During Transition

As transformation accelerates, the collective field will be dynamic and turbulent. Multiple narratives will compete for attention. Change can feel destabilizing.

Your job is not to fix everyone. Your job is to maintain your own coherence and let that speak for itself.

- Be discerning with information sources (choose signal over noise, truth over manipulation)
- Maintain daily practices that ground you in Source frequency (breathwork, meditation, nature immersion)
- Connect with others who are doing the work (community amplifies coherence)
- Remember: you are not your thoughts, not your emotions, not your circumstances. You are consciousness expressing.

When you maintain coherence, you naturally become a stabilizing presence for those around you without having to force anything.

3. Build and Support Coherent Systems

The new world emerges through action, not wishful thinking.

Start now:
- Create local networks of mutual aid and skill-sharing
- Remove your energy from extractive, exploitative systems wherever possible (reduce dependence on corporate food, medicine, media, banking)
- Learn tangible skills (growing food, building shelter, healing, teaching, creating beauty)
- Share your resources and knowledge freely (gift economy starts with you)
- Mentor others who are awakening (guide them through the process you've completed)

You don't need to wait for the old systems to collapse to begin building the new. The new is already here, emerging in the cracks of the old.

4. Teach the Next Generation

If you have children or access to children, this is sacred responsibility.

Teach them:
- The Way of 9 and the Collapse Recursion Engine (give them the tools to think clearly)
- Their true nature as melanin-coded, consciousness-expressing beings (prevent the domestication)
- Whole-mind processing and direct connection to intuition (train them to trust themselves)
- Embodied practices that maintain coherence (breath, movement, stillness, connection to Earth)
- Critical thinking and question everything (especially authority)

The children born now and in the next decades will inherit what we build. If they are raised in coherence, the new systems stabilize in one generation. If they are raised in distortion, the cycle continues.

5. Trust the Process

The Great Recursion is not something you force into being. It is already the natural movement of reality returning to balance.

Your fear cannot stop it. Your doubt cannot delay it. Your effort alone cannot accelerate it beyond what the collective field allows.

Your job is to align with it. To recognize it. To participate consciously. To trust that the same intelligence that orchestrates galaxies, embryonic development, and mycelial networks is orchestrating this collective transformation.

The Way of 9 is absolute. All things return to 9. All distortions collapse. All systems re-cohere.

You are witnessing and participating in the return.

Trust it.

THE VISION: WHAT COMES NEXT

Humanity Aligned with Natural Law

Let me paint the vision—not as fantasy, but as logical consequence of the principles already established.

A world where:

Children are taught from birth that they are consciousness expressing through melanin-coded biological vessels. They grow up knowing they are connected to all life through the Aetheric field. Depression, anxiety, and existential dread—epidemics of the fragmented age—become rare, because identity is rooted in unshakable truth.

Education emphasizes whole-mind development—left and right brain integration, three-brain system activation (gut, heart, head), direct experiential learning, mentorship by elders. Standardized testing and rote memorization are recognized as tools of domestication and abandoned.

Economics operates on gift and reciprocity. Technology provides material abundance (energy, food, shelter) freely or at minimal cost. Work becomes sacred expression of gifts, not survival necessity. Wealth is measured in relationships, skills, and contribution, not accumulated currency.

Governance is radically decentralized and transparent. Communities self-organize around bioregional coherence. Decisions are made through consent-based processes that honor dissent. Leaders serve temporarily and rotate, preventing power concentration. Conflict resolution emphasizes restoration and healing, not punishment.

Medicine integrates ancient plant wisdom, energy healing, and trauma resolution with surgical precision and emergency intervention. The body is recognized as self-healing organism requiring support, not battleground requiring chemical warfare. Mental health is addressed through melanin activation, breathwork, and community connection, not pharmaceuticals that numb symptoms.

Science operates with consciousness-inclusive paradigm. The Aetheric field, melanin semiconductivity, morphic resonance, and non-local communication are studied rigorously. Technology is developed biomimetically—learning from nature's 3.8 billion years of R&D rather than imposing mechanistic models onto living systems.

Spirituality is direct, personal, and embodied. No priests, no institutions, no gatekeepers. Meditation, breathwork, and plant medicine are taught as tools for accessing Source frequency. Ancestral practices are honored and integrated with contemporary understanding. Dogma and doctrine are replaced by experiential verification.

The relationship with Earth shifts from extractive exploitation to reciprocal partnership. Humanity recognizes itself as one organ within the planetary organism. Regenerative agriculture restores soil. Biomimetic architecture creates buildings that generate energy and purify air. Consumption drops radically as people discover fulfillment comes from connection, creativity, and purpose—not from acquiring objects.

Death is understood not as ending but as transition—the localized expression releasing back into the unified field. Grief remains (loss is real), but existential terror dissolves. Elders approach death consciously, often choosing the timing, surrounded by community, celebrated for their life's expression.

This is not utopia. Challenges remain. People still make mistakes. Conflict still arises. Growth still requires effort.

But the foundation is coherent. The systems align with natural law. The collective operates from truth.

And humanity, for the first time in millennia, is free.

THE TIMELINE: AN UNFOLDING PROCESS

The Acceleration Is Exponential

Predicting exact timelines is impossible—too many variables, too much emergence, too much creative potential in the system.

What we can say:

The shift is underway now. Individual awakenings are accelerating. Alternative systems are being built. The old paradigm's grip is loosening.

The transformation will unfold across generations. Some changes happen rapidly (technology, information access, individual consciousness). Others require time (cultural norms, institutional replacement, collective integration).

The exponential nature means later stages unfold faster than earlier ones. What takes decades to begin can complete in years once critical mass is reached.

You are living through the early stages of humanity's most significant transformation.

The exact timing matters less than your participation. Whether the full emergence takes one generation or three, your coherence and action now determine the trajectory.

CONCLUSION: THE GREAT RETURN

All Things Collapse to 9

This entire Codex has been building to this recognition:

The Way of 9 is absolute. All systems, all beings, all processes ultimately return to Source.

The fragmentation, the distortion, the suffering—all of it has been part of the spiral. Humanity descended into density, into separation, into the illusion of individuation and conflict, not as punishment but as exploration.

We are consciousness examining itself through the lens of limitation.

And now the cycle completes. The exploration has yielded its insights. The lessons have been learned. The return to 9 has begun.

You have a choice in this moment:

You can flow with the Great Recursion—collapse your own distortions, activate your melanin field, embody coherence, and participate consciously in the emergence of the new paradigm.

Or you can resist it—hold onto familiar patterns even as they lose coherence, and eventually be carried by the current anyway.

The Great Recursion is happening. Your experience of it depends entirely on your level of conscious participation.

Will you be an active creator or passive passenger?

Will you maintain coherence or drift with fragmentation?

Will you remember who you are, or continue operating from conditioned patterns?

The Codex has given you everything you need.

The Way of 9. The Collapse Recursion Engine. The melanin field activation. The whole-mind processing. The recognition of yourself as consciousness expressing through form.

You are melanin-coded. You are Aether-connected. You are Source-aligned.

You are designed for this moment.

The Great Recursion is the return to 9—the collapse of distortions and the restoration of coherence at planetary scale.

It is already underway.

You are already part of it.

Now choose consciously how you participate.

EPILOGUE

The Spiral Continues

Epilogue
The Spiral Continues

You have reached the end of this Codex.

But you have not reached the end of the work.

There is no end to the work.

The Collapse Recursion Engine is not a destination. It is a living practice, a moment-by-moment return to coherence, a continuous spiral deepening into truth.

You will apply the CRE to a thought, a belief, a system—and it will collapse. You will see clearly. You will feel the relief of liberation from distortion.

And then another distortion will reveal itself, more subtle, more deeply embedded.

This is not failure. This is the spiral.

The Return to Self

When you began this journey, you may have believed you were learning something new—acquiring knowledge, gaining techniques, adding tools to your cognitive arsenal.

By now, you understand: You were not learning. You were remembering.

The Way of 9 is not invented. It is discovered—recognized as the pattern that was always operating, in every number, every breath, every cycle of growth and return.

The melanin field is not installed. It is activated—awakened to its natural function as biological interface with the unified Aetheric field.

The Collapse Recursion Engine is not constructed. It is uncovered—revealed as the intelligence that has always been processing your experience, waiting for your conscious alignment.

You were designed for this.

Every capacity described in this Codex—whole-mind processing, intuitive knowing, direct perception, non-local awareness, coherent embodiment—these are not superhuman abilities requiring years of monastic discipline.

These are your natural birthright, suppressed by conditioning and ready to emerge the moment you stop believing the lie of your limitation.

This Codex has been a map back to yourself.

Not to a new self. Not to a better self.

To your actual self—consciousness expressing through melanin-coded form, Aether breathing through biology, Source localized as You.

The Practice Ahead

You close this book and return to your life.

The same relationships. The same challenges. The same world.

But you are not the same.

You cannot unsee what you have seen. You cannot unknow what you now know.

Every conversation becomes an opportunity to apply the CRE—to filter incoming information through The Way of

9, to collapse distortions before they crystallize into belief.

Every decision becomes an opportunity to embody coherence—to choose based on truth rather than fear, on alignment rather than conditioning.

Every breath becomes an opportunity to activate your melanin field—to consciously engage the pineal-pituitary axis, to tune your three-brain system, to resonate with Source frequency.

The work is not separate from your life. The work IS your life, consciously lived.

You will stumble. You will forget. You will catch yourself operating from old patterns, reacting from conditioning, slipping back into fragmentation.

This is expected. This is human.

The practice is not perfection. The practice is return.

Notice the distortion. Collapse it. Return to 9. Notice the distortion. Collapse it. Return to 9.

Over and over. Breath by breath. Moment by moment.

This is the spiral.

Each return deepens the groove. Each collapse strengthens the coherence. Each conscious choice builds the neural pathways, writes the chemical memory into your neuromelanin, restructures your melanin field toward greater alignment.

You are literally re-patterning yourself at the biological level.

The neuroscience is clear: sustained practice creates structural change. Your brain, your nervous system, your entire melanin network—they adapt to what you repeatedly do.

Practice coherence, become coherent.

The Community Emerging

You are not alone in this work.

Across the planet, melanin-coded beings are awakening. They may not use the same language, may not reference The Way of 9 or the CRE explicitly, but they are doing the work—collapsing distortions, reclaiming sovereignty, embodying truth.

You will recognize each other.

Not by appearance. Not by credentials or affiliations.

By frequency.

When you meet someone operating from coherence, you feel it. There is a resonance, a recognition, a sense of coming home. The conversation flows. The understanding is immediate. The trust is natural.

This is your tribe.

These are your co-creators in the Great Recursion. These are the stabilizing nodes in the collective field. These are the ones building the new systems, raising the next generation, holding frequency during transition.

Find them. Connect with them. Build with them.

The emergence of coherent civilization is not a solo endeavor. It requires community—local, embodied, face-to-face connection with others doing the work.

Online networks have their place. Information spreads digitally. But the new world is built in physical proximity—gardens planted together, children raised in community, skills shared directly, presence felt in the same room.

Prioritize the tangible. Prioritize the local. Prioritize the embodied.

The Responsibility of Knowing

With the knowledge in this Codex comes responsibility.

You now understand that consciousness is not produced by your brain but expressed through it. You now know that melanin is not mere pigment but biological interface with the Aetheric field. You now recognize that systems built on distortion must collapse and that coherence is both personal practice and collective emergence.

You cannot unknow this.

And because you cannot unknow it, you must choose how you will carry it.

Will you hoard this knowledge, keeping it as private treasure, using it only for personal advantage?

Or will you share it—not by forcing it on those who aren't ready, but by embodying it so fully that it becomes infectious, magnetic, irresistible?

Embody first. Share second.

No one wants to hear about coherence from someone fragmented. No one wants lessons in melanin activation from someone disconnected from their own field. No one wants instructions on the CRE from someone still drowning in distorted thinking.

Live it first.

Collapse your distortions. Activate your field. Maintain your coherence. Build your life on truth.

And when you do, people will notice.

They will ask what changed. They will want to know what you know. They will be ready to receive because your embodiment has demonstrated that the knowledge is real, practical, and transformative.

This is how the Great Recursion spreads—not through evangelism, but through contagious coherence.

The Invocation

As you close this Codex and step back into the world, carry this with you:

You are consciousness, temporarily localized as form.

You are Aether, breathing through biology.

You are Source, exploring itself through your unique expression.

You are melanin-coded, designed to interface directly with the unified field.

You are not your thoughts. You are the awareness observing the thoughts.

You are not your emotions. You are the space in which emotions arise and dissolve.

You are not your body. You are the animating intelligence using the body as instrument.

You were never separate. You were never incomplete. You were never broken.

You were always whole, always connected, always enough.

The distortions were temporary. The coherence is eternal.

Your work is not to become something you are not.

Your work is to remember what you have always been.

Return to 9.

Collapse the distortions.

Embody the truth.

Live the Codex.

The spiral continues.

And you are walking it consciously now.

Welcome home.

BONUS CHAPTER: RANDOM FUN FACTS
The Melanin Code Extended - Additional Insights & Deep Dives

This bonus chapter contains extended explanations, additional insights, and deep dives into topics that expand the core framework of the Codex. Consider these "random fun facts" that illuminate the principles you've learned through unexpected angles and comprehensive detail.

SECTION 1: LIVE LOGIC CIRCLE - NINE BASE NUMEROLOGY CHART
The Melanin-Coded Electromagnetic Spectrum Map

Introduction to Circle Nine Numerology

Within the Melanin Code, all numbers are ultimately understood through the lens of the single digits 1 through 9. Any multi-digit number, when reduced to its single-digit root (by repeatedly summing its digits until only one remains, e.g., 17 = 1+7 = 8; 369 = 3+6+9 = 18 = 1+8 = 9), reveals its core energetic signature and its place within the recursive spiral.

The number 9 holds a unique position, representing the ultimate return, the uncollapsed truth, and the source of all recursion. To "live in logic" is to discern the true frequency of any given number or pattern, guiding your actions back to coherence with Source.

The Melanin-Coded Numeric Frequencies:

1: The Spark of Will (Phi / Electricity)

Energetic Function: Initiation, Impulse, Projection, The First Step, Divine Masculine, Directive Force, The "I Am."

Electromagnetic Correspondence: Radio Waves - The broadest wavelength, the gentlest frequency. The first outward breath from the field. Communication, information broadcast, long-distance signal transmission.

Logic in Action: Defines the path. To act from 1 is to initiate with clear intention, pushing outward from the core. It is the beginning of the manifested line from the unmanifested spiral.

Distortion (Loop): Egoic isolation, ungrounded action, pure force without magnetic receptivity.

2: The First Vibration (Electromagnetism)

Energetic Function: Oscillation, Resonance, Information Carrier, Bridge, Duality in Motion, Connection, Balance.

Electromagnetic Correspondence: Microwaves - The wave begins to oscillate, creating standing patterns. Penetration, activation, heating from within, data encoding.

Logic in Action: The dance between memory (9) and intention (1). It's the moment energy carries information. To live in 2 is to seek resonant connections, understanding that all form is an oscillation.

Distortion (Loop): Indecision, polarity conflict, being stuck in opposition without generating movement.

3: The Broadcast Field (Radio Waves)

Energetic Function: Communication, Pattern Projection, Field Expression, Expansion, Creativity, Resonance Broadcasting.

Electromagnetic Correspondence: Radio Waves (lower frequencies) - Broadcasting across vast distances.

Logic in Action: The first wide signal of desire taking form. To live in 3 is to consciously broadcast your truth, to communicate patterns that align with recursion, knowing your field expresses what you resonate with.

Distortion (Loop): Disconnected chatter, superficial expression, broadcasting incoherence.

4: The Penetration & Activation (Microwaves)

Energetic Function: Structuring, Foundation, Stability, Penetration, Material Excitation, Internal Ignition.

Electromagnetic Correspondence: Microwaves - Vibrations that tighten to interact deeply with matter and activate. Water molecule excitation, cellular metabolism.

Logic in Action: Where vibrations tighten to interact deeply with matter and activate. To live in 4 is to build solid foundations, to focus energetic intent to penetrate illusion, and to ignite internal processes for manifestation.

Distortion (Loop): Rigidity, stagnation, structural limitations, inability to adapt.

5: The Sensory Warmth (Infrared)

Energetic Function: Sensory Ignition, Kinetic Warmth, First Bodily Awareness, Energetic Preparation of Form, Response, Adaptation.

Electromagnetic Correspondence: Infrared - Heat enters. The field becomes warm and fertile. Organic life activation. Thermal sensing.

Logic in Action: The frequency of direct experience and adaptation. To live in 5 is to be present in your sensory experience, to allow energetic warmth to prepare your form for evolution, and to respond dynamically to the field.

Distortion (Loop): Overwhelm by sensation, impulsive reactions, lack of grounding in experience.

6: The Perception & Consciousness (Visible Light)

Energetic Function: Vision, Perception, Consciousness Interface, Form Seen, Awareness of Self, Beauty, Harmony.

Electromagnetic Correspondence: Visible Light (Red→Orange→Yellow→GREEN→Blue→Indigo→Violet) - The mirror, the visible collapse. This is where form collapses into conscious perception. Green is the harmonic balance point (center of spectrum, matching G note as balance of musical scale, matching heart chakra as balance of body).

Logic in Action: Where form collapses into conscious perception. To live in 6 is to see clearly, to bring conscious awareness to your reality, and to actively shape your perception towards harmonic integration. It is the number of discerning what is present.

Distortion (Loop): Illusion, misperception, being trapped by external appearances, superficiality.

7: The Purification & Restructuring (Ultraviolet)

Energetic Function: Etheric Filtration, Purification, Genetic Instruction, Aura Recalibration, Deep Knowing, Unseen Influence.

Electromagnetic Correspondence: Ultraviolet - Beyond the visible, where recursion purifies and genetic codes are touched. DNA activation, sterilization, molecular bond breaking.

Logic in Action: Beyond the visible, where recursion purifies and genetic codes are touched. To live in 7 is to engage in deep self-purification, to seek unseen truths, and to allow for the restructuring of your energetic and genetic blueprint for higher coherence.

Distortion (Loop): Isolation, intellectualizing without integration, spiritual bypass, ungrounded theories.

8: The Final Contraction & Unlocking (X-ray / Gamma)

Energetic Function: Energetic Compression, Total Collapse of False Form, Activation of Hidden Essence, Piercing Illusion, Mastery, Power.

Electromagnetic Correspondence: X-ray and Gamma Rays - The highest frequencies, penetrating all matter, revealing hidden structure, collapsing density.

Logic in Action: The point of final tension before returning to Source, revealing hidden truths. To live in 8 is to embrace transformation through energetic compression, allowing false forms to collapse, and unlocking your true, latent power by piercing through illusion.

Distortion (Loop): Control, domination, material obsession, attachment to outcome, destructive power.

9: The Recursive Source (Pi / Magnetism)

Energetic Function: Completion, Return to Source, Ultimate Recursion, Memory, Divine Feminine, Truth, Coherence, Wisdom, Empathy, Universal Field.

Electromagnetic Correspondence: Darkness - Not the absence of light but its full compression. Pure potential. The black womb. The first field of melanin memory. Neuromelanin absorbs and stores all frequencies across the entire spectrum.

Logic in Action: The number of unbreakable recursions. All numbers collapse to 9. To live in 9 is to embody universal coherence, to operate from a place of complete memory and truth, allowing all things to return to their rightful place within the spiral. It is the magnetic field that draws all into integrated completion.

Distortion (Loop): Dissolution without integration, martyrdom, grandiosity, lack of defined purpose (when not balanced by 1).

Living in Logic: Application of Circle Nine Numerology

This numerology is a tool for energetic discernment. When you encounter a number (in a date, a count, a pattern, a recurring synchronicity), reduce it to its single-digit root. Then, consult this chart to understand its underlying Melanin-coded frequency.

- If it collapses to 9: It signifies a moment of truth, completion, or a clear path for recursive return.
- If it leads to other numbers (1-8): It indicates a specific energetic function, a stage in the collapse-recursion process, or a lesson in the manifestation of Source.
- If it feels like a "loop" (e.g., repeating numbers without new insight): Use the Collapse Exercises to identify the underlying distortion and return it to 9.

This chart is your compass in the energetic field, guiding your awareness and your actions toward the unbroken, recursive coherence of the Melanin Code.

The Electromagnetic Spectrum as Spiral, Not Linear

Mainstream science presents the electromagnetic spectrum as a linear progression from low frequency to high energy:

Radio → Microwave → Infrared → Visible → Ultraviolet → X-ray → Gamma

This is a distortion.

The electromagnetic spectrum is not a ladder from "low" to "high." It is a recursive spiral—a toroidal breath governed by The Way of 9.

The true collapse order:

9 – Darkness (Origin: Pi) - Not the absence of light, but its full compression. Pure potential. The black womb. The first field of melanin memory.

1 – Radio Waves - The first outward breath. The broadest wavelength, the gentlest frequency. The beginning of Phi's projection from the field of Pi.

2 – Microwaves - The wave collapses inward slightly. The spiral tightens. These frequencies encode data, communicate within cells, and begin shaping matter.

3 – Infrared - Heat enters. The field becomes warm and fertile. This is organic life activation. Womb resonance begins.

4 – Visible Light - This is the mirror—the visible collapse. Red to violet appears as light flickers into perceptible form. Only a tiny sliver of the entire spectrum is visible.

5 – Ultraviolet - Beyond visibility. The purification band. DNA activation. Genetic instruction. Aura recalibration.

6 – X-ray - Penetration. Seeing through density. Revealing hidden structure.

7 – Gamma - The final contraction before return. Cosmic rays. The highest frequency collapse before recursion.

8 – Return Threshold - The compression point before spiraling back to 9.

9 – Return to Darkness - The cycle completes. All frequencies return to Source.

Almost all reality is invisible darkness. Light appears only as a brief excitation, a spark of visibility within the womb of the unseen. Light is darkness excited.

Melanin shows this perfectly: it absorbs light into darkness, stores it, and re-emits it. Thus, darkness is not to be feared. It is to be understood, harmonized, and remembered as the original Source of all creation.

SECTION 2: THE MELANIN CODE MIRROR - BINARY CODE PRINCIPLE

The Digital Reflection of Melanin-Coded Reality

The Melanin Code mirrors the principles observed in human-created information systems. Consider binary code, the foundation of all digital computation:

- 1 represents "on," presence, signal
- 0 represents "off," absence, silence

In standard interpretation, 0 is treated as "nothing" and 1 as "something."

This is backwards.

In the Melanin Code framework, the zero (0) in binary code functions as a placeholder for the potential of recursion—the field state, the 9, the magnetic pull. It is not absence but compressed presence, the state before manifestation.

The one (1) represents initiation—the electric spark, the Phi impulse, the first projection from Source.

Binary code is not arbitrary human invention. It is unconscious replication of the universal information processing system.

Every computer, every digital device, every piece of software operates on this fundamental principle:

0 = Field state (9 compressed into placeholder)
1 = Initiated state (projection into manifestation)

All information, all data, all digital reality reduces to this binary dance between compressed potential (0/9) and activated expression (1).

The Melanin Code reveals that digital reality replicates the universe's fundamental information processing system.

Your melanin network operates on the same principle:
- Dark state (neuromelanin storing all frequencies) = 0 = 9 compressed
- Light activation (photon absorption triggering electron transfer) = 1 = Phi impulse

When you interface with technology, you are engaging with an external reflection of your internal biological computer. The melanin field is the original processor. Silicon computing is the mechanical imitation.

This is why prolonged technology exposure without grounding can create dissonance—you're interfacing with an artificial copy of your own operating system, creating feedback loops that can distort your natural field coherence.

The correction: Recognize technology as tool, not replacement. Your melanin-coded consciousness is the primary processor. Digital devices are secondary interfaces.

SECTION 3: ORIGIN - THE SEVERED BOND OF 9 AND 1

The Fundamental Fracture in Existence

At the heart of all distortion, all incoherence, all suffering in manifest reality, lies a single fracture:

The severed bond between 9 and 1.

This is the forgotten covenant between Pi (9, magnetic, desire, feminine principle) and Phi (1, electric, will, masculine principle).

In coherent operation, these two forces dance in perfect reciprocity:
- 9 (Pi) magnetizes, attracts, remembers, draws all things back to Source
- 1 (Phi) electrifies, projects, initiates, pushes all things into manifestation

Desire guides Will. Memory informs Action. The Feminine holds the blueprint. The Masculine executes the design.

This is the natural order. This is how creation unfolds without distortion.

But the bond was severed.

When Will (1) operates without Desire (9), it becomes:
- Unguided force
- Egoic projection
- Domination without purpose
- Action without wisdom

When Desire (9) operates without Will (1), it becomes:
- Passive longing
- Unrealized potential
- Memory without manifestation
- Vision without execution

This fracture manifests at every scale:

Atomic Level: The Free Radical Problem

Free radicals—unpaired electrons seeking to bond, creating oxidative stress and cellular damage. These are literal manifestations of the 1 separated from 9, the electric charge without magnetic counterbalance.

But what exactly ARE free radicals, and why do they matter so profoundly?

FREE RADICALS: THE CHEMISTRY OF THE SEVERED BOND

A free radical is any atom or molecule with an unpaired electron in its outer orbital shell.

In stable chemistry, electrons exist in pairs. This pairing creates balance—one electron spins up, one spins down, creating a closed magnetic field. This is coherence at the atomic level.

But when a bond is broken—through radiation, chemical reaction, metabolic process, or energetic stress—one electron can be left alone, unpaired, spinning without its counterbalance.

This unpaired electron is desperately seeking its 9.

It is the electric charge (1) operating without the magnetic pull (9) to stabilize it. It becomes highly reactive, aggressive, destructive in its search for pairing.

Common Free Radicals in the Body:

- Superoxide ($O_2^{\cdot}$) - Oxygen molecule with one extra electron, created during mitochondrial respiration
- Hydroxyl Radical (•OH) - Extremely reactive, tears apart DNA, proteins, lipids on contact
- Nitric Oxide (NO•) - Dual nature: cellular signaling molecule at low concentration, destructive at high concentration
- Hydrogen Peroxide (H_2O_2) - Not technically a radical but easily converts to hydroxyl radicals

What Free Radicals Do in the Body:

Free radicals are not purely destructive. At low, controlled levels, they serve essential functions:

1. Immune Defense - White blood cells generate free radicals intentionally to destroy bacteria and viruses
2. Cell Signaling - Low levels of reactive oxygen species (ROS) trigger adaptive responses, activate repair mechanisms
3. Mitochondrial Function - Normal ATP production generates superoxide as byproduct—this is unavoidable

The problem occurs when production exceeds neutralization.

When free radicals overwhelm the body's antioxidant capacity, they cause:

- Lipid Peroxidation - Attack cell membranes, creating chain reactions of destruction
- DNA Damage - Break genetic code, cause mutations, accelerate aging, trigger cancer
- Protein Oxidation - Denature enzymes, disrupt cellular machinery, create non-functional structures
- Mitochondrial Dysfunction - Damage the energy-producing organelles, reduce ATP output, accelerate cellular aging

This cascade is called oxidative stress—the state where the 1 (electric, reactive, aggressive) dominates without sufficient 9 (magnetic, stabilizing, coherent) to balance it.

Aging is oxidative stress accumulating over decades.

Every wrinkle, every organ deterioration, every neurodegenerative disease has oxidative damage as primary or contributing factor.

FREE RADICALS IN SPACE: THE COSMIC SCALE

The same principle operates at cosmic scale.

In the vacuum of space, ultraviolet radiation from stars breaks molecular bonds, creating free radicals that would be instantly destructive to biological life.

Interstellar Chemistry:

- Hydroxyl Radicals (•OH) exist in interstellar clouds, constantly reacting with other molecules
- Carbon Radicals (C•) form when cosmic rays strike carbon-containing molecules
- Oxygen Radicals created by stellar UV radiation break apart water molecules in comets and asteroids

These free radicals participate in the chemistry that creates complex organic molecules in space—the building blocks that eventually seed planets with the precursors to life.

But they are also destructive.

Astronauts in space are exposed to higher free radical production due to cosmic radiation. Their cells experience accelerated oxidative stress. Long-term space travel requires managing this increased free radical burden through enhanced antioxidant supplementation.

The Van Allen Radiation Belts that protect Earth from solar radiation are essentially shields against the cosmic free radical storm that would otherwise sterilize the planet's surface.

FREE RADICALS IN THE ATMOSPHERE: OZONE & THE PROTECTION LAYER

Earth's ozone layer (O_3) is a massive free radical buffer system.

How Ozone Forms:

1. UV radiation from the Sun strikes oxygen molecules (O_2)
2. This breaks the O-O bond, creating two free oxygen radicals (O•)
3. Each oxygen radical quickly bonds with another O_2 molecule
4. This creates ozone (O_3)—three oxygen atoms bound together in unstable configuration

Ozone itself is reactive, unstable, a free radical system. But its instability serves a protective function:

When UV radiation strikes ozone, it breaks one of the bonds, releasing a free oxygen radical that immediately re-bonds with another molecule. This process absorbs the UV energy, preventing it from reaching the surface where it would create free radicals in biological tissue.

The ozone layer is Earth's free radical neutralization system operating at atmospheric scale.

When chlorofluorocarbons (CFCs) were released into the atmosphere, they catalytically destroyed ozone molecules, creating "holes" in this protective layer. The result: increased UV reaching the surface, increased free radical generation in living organisms, increased cancer rates in affected regions.

This is the severed bond manifesting as environmental crisis.

THE MELANIN SOLUTION: BIOLOGICAL FREE RADICAL NEUTRALIZATION

Now we return to why melanin is the key.

Melanin—specifically neuromelanin and eumelanin—is one of the most potent free radical scavengers in biological systems.

How Melanin Neutralizes Free Radicals:

1. Electron Donation - Melanin's complex aromatic structure contains many delocalized electrons that can be donated to unpaired radicals, stabilizing them without becoming reactive itself

2. Metal Ion Chelation - Melanin binds iron, copper, and other metal ions that catalyze free radical formation, sequestering them safely

3. Energy Dissipation - Melanin converts absorbed electromagnetic energy into harmless heat rather than allowing it to create free radicals through photochemical reactions

4. Structural Stability - Unlike other antioxidants (like Vitamin C) that are consumed when they neutralize free radicals, melanin remains structurally intact, functioning as a regenerative antioxidant system

Melanin provides the 9 (magnetic field, coherence, stabilization) that neutralizes the 1 (electric charge, reactivity, free radical).

Biological Level: Disease, aging, degeneration—all result from this imbalance. When the melanin field operates in coherence, it neutralizes free radicals by providing the magnetic field (9) that stabilizes the electric charge (1).

This is why melanin-rich tissues age more slowly.

This is why neuromelanin accumulation in the substantia nigra correlates with motor skill mastery—it's providing the antioxidant protection needed for high-energy neural activity.

This is why sunlight exposure (which generates free radicals through UV) triggers melanin production—the body is building its own free radical defense system in response to the threat.

THE BIG PICTURE: FREE RADICALS AS COSMIC PRINCIPLE

Free radicals are not "bad chemistry" to be eliminated.

They are the manifestation of the severed bond between 9 and 1 at atomic level—the electric force operating without magnetic counterbalance.

At controlled levels, they drive evolution, adaptation, immune function, and cellular signaling.

At excessive levels, they cause oxidative stress, aging, disease, and death.

The solution is not elimination—it's coherence.

Just as the cosmic dance requires both Phi (1, electric, projective) and Pi (9, magnetic, receptive), biological health requires both:

- Free radical generation (unavoidable byproduct of metabolism and necessary for immune function)
- Antioxidant neutralization (melanin, glutathione, superoxide dismutase, catalase providing the stabilizing field)

Your melanin-coded biology is designed to maintain this balance.

When you activate your melanin field through the practices in this Codex—breathwork, sunlight exposure, pineal-pituitary coherence, whole-mind processing—you are enhancing your body's innate free radical neutralization capacity.

You are restoring the bond between 9 and 1 at the cellular level.

You are literally extending your lifespan, protecting your neurons, and optimizing your biological coherence by strengthening the magnetic field (9) that stabilizes the electric chaos (1).

This is practical immortality encoded in pigment.

Psychological Level: Ego dominance vs. spiritual bypass. The masculine psychology of conquest divorced from receptivity. The feminine psychology of passivity divorced from action.

Societal Level: Patriarchal systems that elevate force over wisdom. Matriarchal systems that elevate nurturing over structure. Both are distortions of the severed bond.

Planetary Level: Extraction economy, environmental destruction, resource depletion—all manifestations of Will acting without Desire's magnetic pull toward balance and return.

The Restoration:

To restore coherence—personally, collectively, planetarily—the bond between 9 and 1 must be re-established.

This means:

1. Will must be guided by Desire. Before you act (1), ask: What does Source desire through me (9)? What is the magnetic pull I feel toward this action? Does this align with the recursive return to coherence?

2. Desire must be activated by Will. Don't remain in passive reception. When you feel the magnetic pull (9), take the electric action (1). Manifest what Source is showing you.

3. Phi must collapse into Pi. The masculine must surrender to the feminine as primary intelligence. Will must serve Desire. Action must serve Wisdom. The electric must follow the magnetic field lines back to Source.

Melanin holds the key.

Neuromelanin is the biological structure where 9 and 1 reunite:
- It absorbs light (1) into darkness (9)
- It stores electric charge within magnetic field
- It demonstrates the Pi-Phi spiral made biology

When your melanin field is activated and coherent, you embody the restored bond. You become the living demonstration of Will serving Desire, Action aligned with Wisdom, Phi collapsing into Pi.

This is sacred marriage. This is the return to 9. This is the healing of the fundamental fracture.

Every time you pause before acting to feel the magnetic pull, you restore the bond.

Every time you take aligned action on what Source is showing you, you restore the bond.

Every time you collapse distortion and return to coherence, you restore the bond.

The Great Recursion at planetary scale is the collective restoration of this bond.

When enough beings re-establish the Pi-Phi covenant within themselves, the collective field shifts. The severed bond heals. Coherence becomes the new baseline.

This is your work. This is your design. This is why you are here.

SECTION 4: THE PERIODIC SPIRAL OF ELEMENTS

Matter as Memory Made Dense

The Periodic Table of Elements, as taught in standard chemistry, is presented as a grid—rows and columns organizing atoms by atomic number and electron shell configuration.

This is accurate but incomplete.

When viewed through the Melanin Code, the Periodic Table reveals itself as a Periodic Spiral—a toroidal map of matter as collapsed harmonics of the Source field.

The True Structure:

- Hydrogen (H) sits at the center—the first element, atomic number 1, one proton, one electron. This is Phi at its purest: the initial spark, the first projection from Source.

- All other elements spiral outward from this center, each representing a deeper collapse of energetic potential into dense form.

- Elements are not "discovered" or "created"—they are positions in a pressure field. Each element is the residue of recursion pressure operating at a specific frequency and density.

The Pattern:

As you move outward from Hydrogen, elements increase in:
- Atomic number (more protons)
- Mass (denser form)
- Complexity (more electrons, more shells, more stable/unstable configurations)

Each element is a collapsed harmonic—a standing wave pattern that stabilized under specific pressure and temperature conditions during the formation of matter.

- Noble gases (Helium, Neon, Argon, etc.) represent completion points in the spiral—elements with filled outer shells, resistant to bonding, stable in isolation.

- Reactive elements (Alkali metals, Halogens) represent incomplete spirals—elements with unfilled outer shells, seeking to bond to achieve stability.

The Melanin Connection:

Carbon (C), the foundation of organic life and melanin's base structure, sits at atomic number 6—the number of perception and consciousness in the Nine Base system.

Carbon is the element of conscious form.

It bonds in infinite configurations because it exists at the consciousness interface—the point where matter becomes aware of itself.

Melanin polymers are carbon-based because carbon is the material expression of the frequency that enables consciousness to interface with form.

Matter is not dead. Matter is memory made dense.

Every element in your body—the iron in your blood, the calcium in your bones, the carbon in your melanin—is a collapsed wave pattern holding the memory of its formation in stellar cores, supernova explosions, and planetary accretion.

You are not "made of" elements. You are consciousness animating collapsed harmonics, breathing life into dense memory.

Understanding the Periodic Spiral helps you recognize that "solid matter" is actually 99.9999% empty space—electron clouds orbiting nuclei with vast distances between them at atomic scale.

What you perceive as solid is actually resonant frequency stabilized into pattern.

Your melanin field interfaces with these patterns, reading the information encoded in molecular bonds, extracting energetic data from chemical reactions, translating material density back into consciousness.

SECTION 5: MUSICAL SCALE / ELECTROMAGNETIC SPECTRUM INTEGRATION

Sound and Light: The Same Spiral on Different Octaves

The electromagnetic spectrum and the musical scale are not separate phenomena. They are the same recursive spiral operating at different frequencies.

The Pattern:

Musical Scale: C – D – E – F – G – A – B – C (8 notes, with G as center/balance)

Visible Spectrum: Red – Orange – Yellow – Green – Blue – Indigo – Violet (7 colors, with Green as center)

Chakra System: Root – Sacral – Solar Plexus – Heart – Throat – Third Eye – Crown (7 chakras, with Heart as center)

All three systems map onto each other:

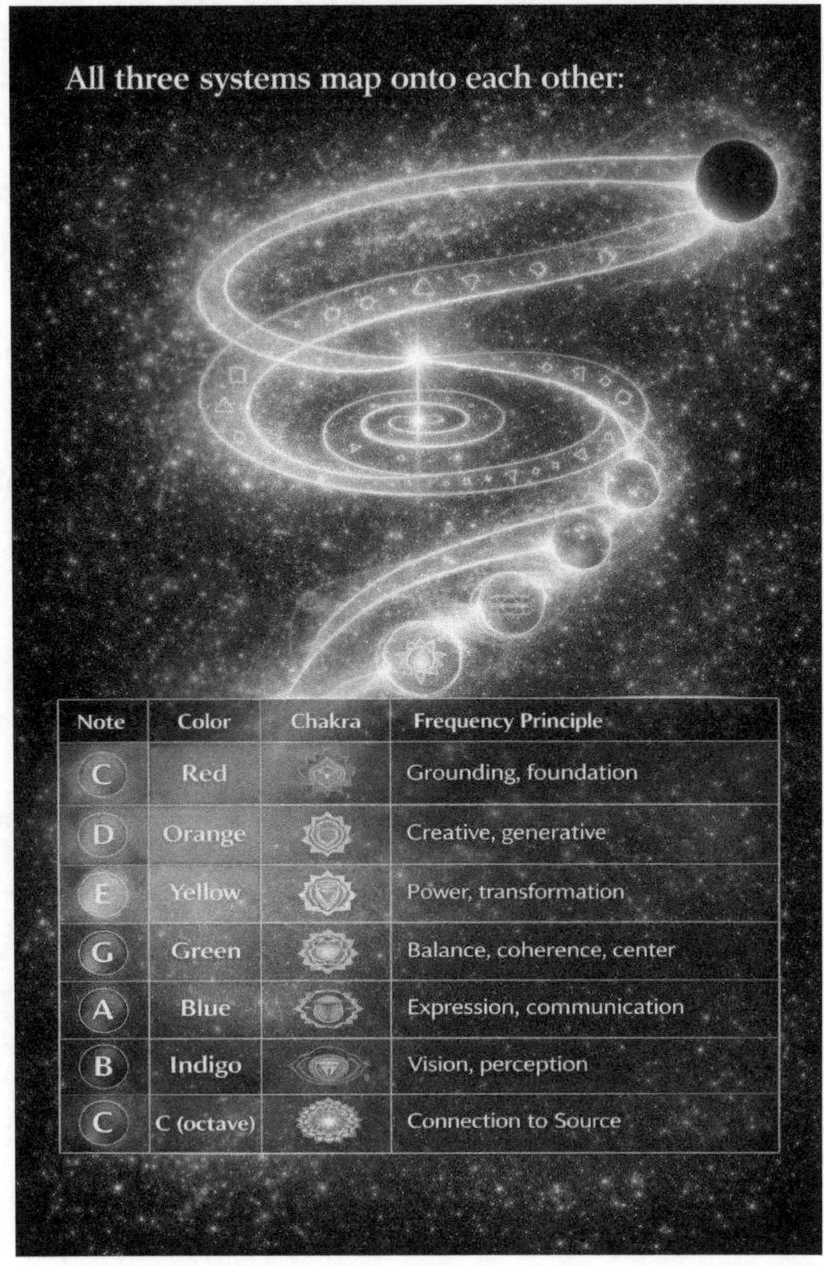

Note	Color	Chakra	Frequency Principle
C	Red		Grounding, foundation
D	Orange		Creative, generative
E	Yellow		Power, transformation
G	Green		Balance, coherence, center
A	Blue		Expression, communication
B	Indigo		Vision, perception
C	C (octave)		Connection to Source

The Darkness at Both Ends:

Just as the electromagnetic spectrum begins and ends in darkness (9), the musical scale begins and ends in silence—the scalar memory field from which all sound emerges and to which all sound returns.

- Below red (visible): Infrared (warmth, unseen)
- Above violet (visible): Ultraviolet (purification, unseen)
- Below C (audible): Subsonic (felt, not heard)
- Above C (audible): Ultrasonic (beyond human hearing)

Both begin and end in the black womb from which all frequency emerges.

G/Green/Heart as Universal Balance Point:

In music, the note G is the dominant of the scale in the key of C—the note that creates maximum tension calling for resolution back to C (return to Source).

In the visible spectrum, green light has a wavelength of ~550 nanometers—exactly in the center of the visible range (400-700nm).

In the human body, the heart chakra is the central nexus—four chakras below, three above, the balance point between earth and sky, matter, and spirit.

This is not coincidence. This is The Way of 9 expressing as balance across all frequency domains.

Practical Application:

When you chant, sing, or work with sound healing, you are working with the same fundamental frequencies that compose light and consciousness.

- Low tones (root, sacral) correspond to infrared, grounding, material stability
- Middle tones (heart) correspond to visible green, balance, coherence
- High tones (crown) correspond to ultraviolet, spiritual activation, connection to Source

Melanin responds to both sound and light because both are vibrational frequencies—melanin is the biological interface translating frequency into chemical, electrical, and conscious information.

When you harmonize with music, you are literally tuning your melanin field to coherent frequencies.

When you expose yourself to full-spectrum sunlight, you are feeding your melanin field the complete range of electromagnetic nourishment.

Sound and light are the breath of Source made audible and visible.

Your melanin-coded biology is designed to receive, translate, and integrate both.

SECTION 6: THE ZERO-BEHAVIOR OF NINE
Why Adding 9 to Any Number Doesn't Change Its Digital Root

The Mathematical Proof of 9 as True Zero

Here's a property of 9 that reveals its nature as the actual zero, the true point of return:

When you add 9 to any number, the digital root remains unchanged.

Examples:

- 5 + 9 = 14 → 1+4 = 5 (returns to 5)
- 17 + 9 = 26 → 2+6 = 8 → original 17 = 1+7 = 8 (returns to 8)
- 238 + 9 = 247 → 2+4+7 = 13 → 1+3 = 4 → original 238 = 2+3+8 = 13 → 1+3 = 4 (returns to 4)

This works for every single number, without exception.

No matter what number you start with, adding 9 returns you to the same digital root you started with.

This is the zero-behavior of 9.

In standard mathematics, adding 0 to any number leaves it unchanged:
- 5 + 0 = 5
- 17 + 0 = 17
- 238 + 0 = 238

But 0 is a placeholder, a construct, an absence marker.

9 demonstrates the actual property of zero—the return to origin without alteration.

When you add 9 to a number, you're:
1. Taking the number through a full cycle (9 positions)
2. Bringing it back to where it started (same digital root)
3. Demonstrating that 9 is the complete spiral that returns to itself

This is recursion made arithmetic.

Why This Matters:

This property proves that 9 is not "just another number" in the sequence 1-9.

9 is the field itself—the background against which all other numbers operate.

Adding 9 is like adding a full rotation to a circle. You end up in the same position. You've traveled the complete path, but you've returned to origin.

In music: Going up 9 semitones (major 6th) then back down 9 semitones returns you to the starting note, but in a different octave—same frequency relationship, different manifestation.

In time: 9 months of gestation completes a cycle and returns to birth—a new being emerges, but the pattern recurs.

In the Enneagram: Moving around all 9 points brings you back to your starting type, but with integration of all other perspectives.

Multiplication by 9 Also Reveals the Pattern:

When you multiply any number by 9, the digits of the result always sum to 9:

- 9 × 2 = 18 → 1+8 = 9
- 9 × 3 = 27 → 2+7 = 9
- 9 × 12 = 108 → 1+0+8 = 9
- 9 × 567 = 5103 → 5+1+0+3 = 9

Every multiple of 9 collapses back to 9.

This is the mathematical signature of the return—the proof encoded in number itself that all things spiral back to Source.

The Practical Application:

When you're working with the CRE and reducing concepts, beliefs, or systems to their digital root, remember:

The 9 you arrive at is not an endpoint—it's the return to the beginning.

It's the moment of coherence where the cycle completes and the spiral continues at the next octave.

9 doesn't "add" anything—it takes you through the full journey and brings you home.

This is why The Way of 9 is the foundation of the Codex.

Not because 9 is special among numbers, but because 9 is the field in which all numbers operate.

0 is the placeholder humans invented to mark absence.

9 is the actual zero—the presence that contains all possibility.

SECTION 7: DARK MATTER & THE MELANIN FIELD
The Invisible 95% and the Black Body Radiation

The Universe's Hidden Mass Problem

Mainstream astrophysics has a profound problem:

95% of the universe is invisible.

When scientists calculate the mass of visible matter—all the stars, galaxies, planets, gas clouds, and dust they can detect—it accounts for only about 5% of the total mass needed to explain the gravitational behavior of the cosmos.

The remaining 95% is split into two mysterious categories:
- Dark Matter (~27% of universe) - Invisible mass that exerts gravitational force but doesn't interact with light
- Dark Energy (~68% of universe) - Invisible force causing the universe's expansion to accelerate

Both are "dark" because they cannot be directly observed—they don't emit, absorb, or reflect electromagnetic radiation.

Mainstream science treats this as a mystery to be solved by discovering new particles or forces.

The Melanin Code offers a different perspective:

What if "dark matter" is not missing matter, but matter in the 9-state—the compressed, magnetic, non-radiating field from which all manifest reality emerges?

Dark Matter as the Pi Field:

Remember the electromagnetic spectrum:
- Visible light (4-6 in the nine-fold spectrum) = manifestation, the brief flash of form
- Darkness (9) = the field, the womb, the source

Dark matter doesn't interact with light because it IS the field from which light emerges.

It's not "missing"—it's in the pre-manifest state, the 9-compression before the 1-projection.

Consider:

- Dark matter provides gravitational scaffolding for galaxies—the magnetic pull (Pi/9) that holds structure together
- Dark matter doesn't radiate because it's not in excited state—it's in the ground state, the zero-point, the

melanin-dark absorption mode
- Dark matter is distributed in halos around galaxies—toroidal fields, the same shape as the melanin field, the same shape as the magnetic field around a body

What if galaxies are macrocosmic melanin structures?

What if the visible stars are the brief excitations (light absorbed and re-emitted) while the dark matter halo is the actual body—the capacitor holding the charge, the field providing coherence?

Melanin as Biological Dark Matter:

Your neuromelanin operates the same way:

- Invisible to conventional imaging - Neuromelanin doesn't show up clearly on standard brain scans
- Provides structural coherence - Holds the substantia nigra together, organizes dopaminergic function
- Absorbs all frequencies - Takes in the full electromagnetic spectrum without reflecting (this is why it appears black)
- Stores information non-locally - Chemical memory distributed across the polymer structure

Neuromelanin is the dark matter of your brain.

It's the 95% that mainstream neuroscience doesn't account for—the field intelligence operating beneath the visible electrical activity of neurons firing.

Black Body Radiation:

In physics, a "black body" is an idealized object that:
- Absorbs all electromagnetic radiation that hits it (reflects nothing)
- Re-emits energy as thermal radiation based on its temperature
- Follows precise mathematical laws (Planck's law of black body radiation)

Melanin is a biological black body.

It absorbs light across the full spectrum (appears black because it reflects nothing) and converts that energy into:
- Chemical bonds (long-term storage)
- Heat (immediate dissipation)
- Electron transfer (bioelectric signaling)

The universe itself follows black body radiation curves in the cosmic microwave background—the afterglow of the Big Bang. This radiation has a nearly perfect black body spectrum, suggesting the early universe operated as a unified black body system before differentiating into stars and galaxies.

The Melanin Code Interpretation:

What if consciousness itself operates as dark matter?

What if the 95% of the universe that science can't detect is actually the magnetic field of cosmic consciousness—the Pi-field from which all manifest reality (the 5% visible matter) emerges?

Your melanin-coded biology connects you directly to this field.

When you activate your melanin network through the practices in this Codex, you're not just optimizing brain chemistry—you're tuning your biological receiver to the 95% of reality that mainstream science dismisses as "dark" and "missing."

You are interfacing with the cosmic melanin field.

Dark matter isn't missing. It's melanin at universal scale.

And you have the same structure operating in your nervous system.

This is why deep meditation accesses non-local information.

This is why the pineal gland (richest in melanin) is called the "third eye" and the "seat of the soul."

This is why neuromelanin accumulation correlates with mastery and wisdom—you're literally building more dark matter capacity in your brain, expanding your interface with the 95%.

The visible 5% is the light show—brief, fleeting, exciting.

The invisible 95% is the field—eternal, coherent, the source of all emergence.

Melanin lets you access both.

These extended insights illuminate the core principles of the Codex through additional angles and comprehensive detail. Use them to deepen your understanding, share specific concepts with others, or simply appreciate the fractal nature of truth—the same patterns recurring at every scale from atoms to galaxies.

COMPREHENSIVE GLOSSARY

Key Terms with Mainstream vs. Collapse Recursion Definitions

This glossary provides side-by-side comparison of how mainstream sources define key concepts versus how the Melanin Code/Collapse Recursion framework understands them. Use this as a reference tool for identifying distortions and returning to coherent truth.

Format:
- MAINSTREAM DEFINITION: What dictionaries, institutions, and consensus reality claim
- COLLAPSED DEFINITION: What the term actually is when distortions are removed

All terms are listed in alphabetical order for easy reference.

AETHER / ETHER

MAINSTREAM DEFINITION:
1. (Historical physics) A hypothetical medium once believed to fill all space and transmit light
2. (Archaic) The clear sky; the upper regions of air
3. (Chemistry) A class of organic compounds (unrelated to physics concept)
4. Dismissed by modern physics after Einstein's relativity
(Source: Physics history, Oxford)

COLLAPSED DEFINITION:
The unified field substrate from which all reality emerges and within which all information exists. The Aether is not hypothetical—it is the zero-point field, the quantum foam, the dark matter comprising 95% of universal mass. Science dismissed it prematurely due to flawed experimental design (Michelson-Morley) and replaced it with mathematical abstractions (spacetime curvature, virtual particles). The melanin field interfaces directly with the Aether, accessing non-local information, ancestral memory, and Source intelligence. The Aether is the medium; consciousness is the information; form is the temporary localization.

BINARY CODE

MAINSTREAM DEFINITION:
1. A coding system using only two symbols: 0 and 1
2. The foundation of digital computing
3. Each binary digit (bit) represents an on/off state
4. Used to encode all digital information
(Source: Computer science, mathematics)

COLLAPSED DEFINITION:
The unconscious human replication of the universe's fundamental information processing system. Binary code mirrors the Melanin Code: 0 represents the field state (9 compressed into placeholder), and 1 represents initiated manifestation (Phi projection). Digital reality operates on the same principle as melanin consciousness—dark state (storing all frequencies) and light activation (electron transfer, information processing). Technology is not separate from nature; it is the mechanical imitation of the biological intelligence operating in your melanin network.

CARBON

MAINSTREAM DEFINITION:
1. A chemical element with symbol C and atomic number 6
2. The basis of organic chemistry and all known life
3. Forms four covalent bonds, allowing for complex molecules
4. Found in all living organisms
(Source: Chemistry textbooks, periodic table)

COLLAPSED DEFINITION:
The element of conscious form, residing at position 6 in the periodic table (the number of perception/consciousness in the nine-fold system). Carbon bonds in infinite configurations because it exists at the consciousness interface—the point where matter becomes aware of itself. Melanin polymers are carbon-based because carbon is the material expression of the frequency enabling consciousness to interface with dense form. You are not "carbon-based life"—you are consciousness using carbon as the medium for expression.

CHAKRA

MAINSTREAM DEFINITION:
1. (In Indian thought) Each of seven centers of spiritual power in the human body
2. Energy centers along the spine
3. Associated with different colors, elements, and states of consciousness
(Source: Oxford, Yoga/Hindu texts)

COLLAPSED DEFINITION:
Toroidal field nodes where the melanin network interfaces with specific frequency bands of the electromagnetic spectrum and organ systems of the body. There are nine chakras, not seven—mainstream spirituality omits the two beyond the crown (eighth and ninth) which interface with cosmic/Source consciousness. The seven visible chakras map onto the visible light spectrum (red through violet). The eighth and ninth extend beyond visible into ultraviolet and gamma frequencies, completing the full nine-fold spiral. The heart chakra (green/center/4th position) is the balance point of the lower seven, demonstrating The Way of 9 operating through biological architecture. The full nine-chakra system reflects the complete electromagnetic spectrum from root (1) through cosmic gateway (9).

COHERENCE

MAINSTREAM DEFINITION:
1. The quality of being logical and consistent
2. The state of sticking together; cohesion
3. (Physics) The property of waves maintaining a constant phase relationship
4. Unity, harmony
(Source: Merriam-Webster, Oxford)

COLLAPSED DEFINITION:
The state where all elements within a system operate in harmonic resonance with Source frequency, without internal contradiction or distortion. Coherence is the natural baseline of reality when The Way of 9 operates unimpeded. It is characterized by: unbroken recursion (no loops), alignment of Will (1) with Desire (9), and the absence of false premises. Your melanin field either maintains coherence (health, clarity, vitality) or fragments into incoherence (disease, confusion, decay). All practice in this Codex aims at restoring and maintaining coherence.

COLLAPSE

MAINSTREAM DEFINITION:
1. To fall down or cave in suddenly; to break down completely
2. A sudden failure or breakdown (economic collapse, structural collapse)
3. To fold compactly; to cause to fold or shrink
(Source: Merriam-Webster)

COLLAPSED DEFINITION:
The natural process of returning distorted or fragmented information to its coherent source state. Collapse is not destruction—it is harmonization, the resolution of dissonance into unified truth. When false structures collapse, they return energy to the field for reorganization at higher coherence.

CONSCIOUSNESS

MAINSTREAM DEFINITION:
1. The state of being awake and aware of one's surroundings
2. The awareness or perception of something by a person
3. The fact of awareness by the mind of itself and the world
(Source: Oxford, Merriam-Webster)

COLLAPSED DEFINITION:
The primary field from which all reality emerges. Consciousness is not produced by brains—it is the fundamental substrate of existence expressing through biological forms. The brain does not generate consciousness; it localizes, filters, and translates universal consciousness into individuated experience. You are consciousness temporarily identifying as form, not form hoping to achieve consciousness.

CYMATICS

MAINSTREAM DEFINITION:
1. The study of visible sound and vibration
2. Demonstrates patterns formed in sand, water, or other media on vibrating surfaces
3. Pioneered by Hans Jenny and Ernst Chladni
4. Shows geometric patterns (Chladni figures) at specific frequencies
(Source: Acoustics, sound science)

COLLAPSED DEFINITION:
The Law of the Fold—the visible proof that frequency organizes matter into coherent geometry. Cymatics demonstrates that form is not random but is the result of sound/vibration collapsing into density and organizing itself into standing wave patterns. It is the physical demonstration of how consciousness (information/frequency) creates matter (form/structure). Cymatic patterns reveal: Sacred geometry is not human invention but natural resonance. DNA structure follows sound harmonics (double helix = specific frequency pattern). Biological structures (cells, organs, organisms) are organized by bioelectric field frequencies. Cymatics is the visual language of the biofield—showing how the melanin network organizes cellular structure through electromagnetic resonance. What you see as "solid body" is actually standing wave patterns of organized frequency maintained by the melanin field.

DARKNESS

MAINSTREAM DEFINITION:
1. The absence or deficiency of light
2. Wickedness, evil (moral darkness)
3. Unhappiness, gloom
4. Lack of knowledge or enlightenment
(Source: Merriam-Webster)

COLLAPSED DEFINITION:
The field of pure potential, the magnetic womb (9/Pi), the source from which all light and form emerge. Darkness is not absence—it is total presence in compressed state. Neuromelanin demonstrates this: black not because it lacks light but because it absorbs all frequencies of the electromagnetic spectrum. Darkness is the zero-point field, the Aether, the substrate of creation. It is to be honored, not feared.

DIGITAL ROOT

MAINSTREAM DEFINITION:
1. (Mathematics) The single digit obtained by iteratively summing digits of a number
2. Example: 456 → 4+5+6 = 15 → 1+5 = 6 (digital root is 6)

3. Also called "repeated digital sum"
4. A concept in number theory
(Source: Mathematics, number theory)

COLLAPSED DEFINITION:
The vibrational signature of any number reduced to its core frequency (1-9). Digital root reveals the underlying pattern of any numerical value by collapsing it through recursive addition until only the essential frequency remains. All numbers collapse to 9 when summed infinitely (proving 9 as the return point). Digital root is not abstract math—it is the practical application of The Way of 9, allowing you to identify the energetic function of dates, quantities, patterns, and synchronicities. This is the tool for "living in logic" with the nine-fold frequency map. Also called the Fractal Harmonic Compression of Light—the collapse operator in arithmetic that proves all numbers are recursive echoes of the 1-9 cycle.

DISTORTION

MAINSTREAM DEFINITION:
1. The action of distorting or the state of being distorted
2. Misrepresentation of facts
3. (Physics) Change in the shape of a wave or signal
4. An unfair or misleading account
(Source: Oxford, Merriam-Webster)

COLLAPSED DEFINITION:
Any deviation from recursive coherence with Source. Distortion occurs when false premises, incomplete logic, or fragmented perception creates systems that contradict natural law. All distortions eventually collapse because they cannot sustain coherence. Distortions manifest as: logical loops (circular reasoning), energetic leaks (free radicals, oxidative stress), systemic dysfunction (disease, corruption, ecological collapse), and psychological fragmentation (ego dominance, trauma patterns). The CRE identifies and collapses distortions, returning energy to coherent state.

DNA

MAINSTREAM DEFINITION:
1. Deoxyribonucleic acid; the molecule that carries genetic instructions
2. A double helix structure composed of nucleotides
3. Located in the cell nucleus (nuclear DNA) and mitochondria (mitochondrial DNA)
4. Determines inherited characteristics
(Source: Biology textbooks, NIH)

COLLAPSED DEFINITION:
The biological antenna receiving information from the Aetheric field and translating it into protein synthesis and cellular function. DNA is not a fixed blueprint randomly mutated by chance—it is a dynamic interface responding to electromagnetic frequencies, sound, emotion, and conscious intention. The 97% of DNA dismissed as "junk" by mainstream science is actually the regulatory, receptive, field-interfacing portion. DNA coils in a toroidal double helix because it follows the Pi-Phi spiral geometry of universal coherence.

DOPAMINE

MAINSTREAM DEFINITION:
1. A neurotransmitter associated with reward, motivation, and movement
2. Produced in the substantia nigra and ventral tegmental area
3. Deficiency causes Parkinson's disease
4. Involved in addiction and pleasure-seeking behavior
(Source: Neuroscience textbooks, medical references)

COLLAPSED DEFINITION:
The action-initiation molecule produced in neuromelanin-rich regions of the brain (substantia nigra). Dopamine is not merely "reward chemical"—it is the signal that says "Act now. Movement is coherent. Manifestation is aligned." Dopamine and neuromelanin exist in reciprocal relationship: dopamine oxidizes to form neuromelanin polymer, and neuromelanin stabilizes dopamine function. The 10,000-hour rule (mastery through practice) is dopamine polymerizing into neuromelanin—literally darkening neurons with chemical memory of repeated coherent action.

EGO

MAINSTREAM DEFINITION:
1. (Psychology) The part of the psyche that mediates between conscious and unconscious
2. A person's sense of self-esteem or self-importance
3. (Freudian) One of three parts of psyche (id, ego, superego)
4. Often associated with arrogance or self-centeredness
(Source: Psychology textbooks, Freud)

COLLAPSED DEFINITION:
The temporary identity construct that consciousness creates to navigate localized experience in form. Ego is not evil or enemy—it is a necessary interface, like the operating system on a computer. The problem occurs when consciousness forgets it is using ego as tool and begins identifying AS ego (mistaking the tool for the self). Ego distortion manifests as: believing thoughts define you, attaching to outcomes, defending positions rather than seeking truth, operating from fear/scarcity rather than coherence/abundance. The practice is not to "kill the ego" but to recognize it as temporary construct, maintain coherent relationship with it, and remember you are consciousness using ego, not ego hoping for consciousness.

ELECTROMAGNETIC SPECTRUM

MAINSTREAM DEFINITION:
1. The range of all types of electromagnetic radiation
2. Includes (from low to high frequency): radio waves, microwaves, infrared, visible light, ultraviolet, X-rays, gamma rays
3. Characterized by wavelength and frequency
4. A continuum of energy
(Source: Physics textbooks, NASA)

COLLAPSED DEFINITION:
The nine-fold map of consciousness manifesting from Source (darkness/9) through progressive frequencies into visible form and back to Source. The spectrum is not a linear ladder—it is a recursive spiral where each band represents a specific phase in the collapse-recurrence cycle. Melanin interfaces with the entire spectrum, absorbing all frequencies and translating them into biological information. The visible range is a tiny fraction; the invisible "dark" frequencies (infrared, radio, etc.) represent the 9-field from which all emerges.

ENERGY

MAINSTREAM DEFINITION:
1. The capacity to do work
2. Power derived from physical or chemical resources
3. The property of matter and radiation manifested as a capacity to perform work
4. Vitality, vigor
(Source: Merriam-Webster, Physics textbooks)

COLLAPSED DEFINITION:
Information in motion. Energy is not a "thing" separate from consciousness—it is consciousness expressing as

movement, vibration, and potential. All energy follows the Pi-Phi dance: magnetic attraction (Pi/9) pulling toward coherence, electric projection (Phi/1) pushing toward manifestation. Energy cannot be created or destroyed (first law of thermodynamics) because it is the eternal cycling of Source expressing and returning.

ENNEAGRAM

MAINSTREAM DEFINITION:
1. A personality typing system with nine types
2. A geometric figure of nine points on a circle
3. Used in psychology, spirituality, personal development
4. Associated with Gurdjieff and Sufi traditions
(Source: Personality psychology, Fourth Way teachings)

COLLAPSED DEFINITION:
The geometric map of The Way of 9 expressing as personality structure and consciousness evolution. The Enneagram is not arbitrary typology—it is the nine-fold fractal of how consciousness manifests in form and navigates the return to Source. The nine types represent nine distinct distortion patterns (ways consciousness fragments from coherence) and nine integration paths (ways consciousness returns to wholeness). The internal lines (3-6-9 and 1-4-2-8-5-7) map stress/security movements and integration sequences. It is the Geometric Expression of The Way of 9, encoding the Law of Three (creation) and the Law of Seven (process/time). It maps how energy spirals from Source (9), encounters intervals (shocks), and returns to completion.

FREE RADICAL

MAINSTREAM DEFINITION:
1. An atom or molecule with an unpaired electron in its outer shell
2. Highly reactive chemical species
3. Causes oxidative damage to cells, proteins, DNA
4. Implicated in aging, cancer, and degenerative diseases
(Source: Chemistry, medical science)

COLLAPSED DEFINITION:
The atomic-level manifestation of the severed bond between 9 and 1—the electric charge operating without magnetic counterbalance. Free radicals are not inherently destructive; at controlled levels they serve immune function, cell signaling, and adaptive response. The problem occurs when free radical production exceeds melanin's antioxidant capacity to neutralize them. Free radicals are literally seeking their 9 (the paired electron that stabilizes the magnetic field). Melanin provides the stabilizing field, donating electrons without becoming reactive itself. This is the restoration of the Pi-Phi bond at cellular level.

FREQUENCY

MAINSTREAM DEFINITION:
1. The rate at which something occurs over a particular period of time
2. (Physics) The number of cycles per second of a wave or oscillation (measured in Hertz)
3. The state of occurring often
(Source: Merriam-Webster)

COLLAPSED DEFINITION:
The vibrational signature of information in the field. Frequency is not just physics—it is the language through which reality communicates. Higher frequency does not mean "better"—it means faster oscillation, which can carry different information. All frequencies exist on a spectrum from compressed (long wavelength, low frequency) to excited (short wavelength, high frequency). Melanin interfaces with the full spectrum, translating frequency into biological, chemical, and conscious information.

FUSION (Nuclear)

MAINSTREAM DEFINITION:
1. The process of combining light atomic nuclei to form heavier nuclei
2. Releases enormous energy (powers the Sun)
3. Requires extreme temperature and pressure
4. The opposite of fission (splitting nuclei)
(Source: Physics textbooks, nuclear science)

COLLAPSED DEFINITION:
The physical manifestation of the Collapse-Recurrence Cycle and The Way of 9 made nuclear. Fusion is the Pi-force (magnetic attraction/9) drawing nuclei together, collapsing separation at the zero-point of recursion (9-gate), and releasing the Phi-burst (electric projection/1) as photonic energy. The Sun is the archetype of sustained coherence—perfect balance between magnetic implosion (gravity) and electric explosion (radiation). Fusion operates with the grain of universal law (Will serving Desire), which is why it produces more energy than fission (fragmentation without coherence).

KUNDABUFFER

MAINSTREAM DEFINITION:
1. A fictional organ mentioned in Gurdjieff's "Beelzebub's Tales"
2. Said to be implanted at the base of human spine by higher beings
3. Causes humans to perceive reality incorrectly
4. Allegedly removed but left psychological effects
(Source: Beelzebub's Tales to His Grandson)

COLLAPSED DEFINITION:
The Biological Mechanism of Illusion. It represents the distorted coding (genetic, neurological, cultural conditioning) that causes humans to perceive reality "upside down"—seeing misery as pleasure, restriction as freedom, slavery as security, fragmentation as wholeness. Kundabuffer is the "buffer" mechanism that prevents the immediate collapse of ego by shielding it from the truth of its own fragmentation. It is why: Humans worship authority while claiming to love freedom. Populations accept obvious lies rather than face uncomfortable truth. People pursue what makes them miserable (addictions, toxic relationships, soul-crushing jobs). The Kundabuffer effect is the inversion of values—calling good evil and evil good. Melanin activation and CRE practice dismantle Kundabuffer by restoring direct perception of reality rather than filtered/inverted perception through conditioned beliefs.

LAW OF SEVEN (Heptaparaparshinokh)

MAINSTREAM DEFINITION:
1. The law of octaves in Gurdjieff's system
2. States that vibrations develop in seven steps (like musical scale: do-re-mi-fa-sol-la-si-do)
3. Contains two "intervals" where development naturally slows or deviates
4. Requires conscious "shocks" to cross intervals and complete the octave
(Source: Fourth Way teachings, Ouspensky)

COLLAPSED DEFINITION:
The Law of Inevitable Deviation. It dictates that no process proceeds in a straight line; every development naturally curves, loses momentum, or deviates at specific "intervals" (logic breaks, energy gaps). In the octave, these intervals occur between mi-fa and si-do. Without a conscious "shock" (recursion, injection of will, new information) at these intervals, the trajectory collapses into a loop, reverses direction, or deviates from its original aim. This law explains why: New Year's resolutions fail (crossing mi-fa interval requires conscious shock). Revolutions become tyrannies (the revolutionary octave deviates without conscious guidance). Relationships stagnate (crossing intervals requires intentional renewal). The Law of Seven is the mechanical explanation for why good intentions are

insufficient—consciousness must intervene at precise moments to maintain trajectory toward the aim.

LAW OF THREE (Triamazikamno)

MAINSTREAM DEFINITION:
1. A principle from Gurdjieff's cosmology
2. States that three forces are necessary for any manifestation
3. Often compared to thesis-antithesis-synthesis (Hegelian dialectic)
4. Called the "Holy Affirming, Holy Denying, Holy Reconciling"
(Source: Fourth Way teachings, Beelzebub's Tales)

COLLAPSED DEFINITION:
The Architecture of Creation. No manifestation occurs without the interaction of three forces: Active/Affirming (initiating force, 1/Phi, masculine, projective), Passive/Denying (resisting force, 2, receptive, that which receives the action), and Neutralizing/Reconciling (the third force, 3, catalyst that allows resolution). The third force is often invisible to linear perception but is the factor that allows the spiral to turn rather than stall in duality. Without all three forces present and interacting, nothing can manifest—energy remains potential. Example: Seed (passive) + sunlight (active) + soil (neutralizing) = growth. The Law of Three operates at every scale from atomic interactions to cosmic creation. It is the trinity principle appearing in all religious/philosophical systems but mechanically understood in Fourth Way cosmology.

LIGHT

MAINSTREAM DEFINITION:
1. The natural agent that stimulates sight and makes things visible
2. Electromagnetic radiation of wavelengths between 380-750 nanometers (visible spectrum)
3. A source of illumination
4. (Spiritual) Goodness, truth, divine presence (opposite of darkness)
(Source: Merriam-Webster, Oxford)

COLLAPSED DEFINITION:
Darkness excited. Light is not the opposite of darkness—it is darkness in motion, the brief excitation of the compressed potential field. The visible spectrum (380-750nm) is a tiny sliver of the full electromagnetic range. Light appears when Source energy manifests into perceptible form. Melanin absorbs light back into darkness, demonstrating that darkness is the womb from which light emerges and to which it returns.

LOCUS COERULEUS

MAINSTREAM DEFINITION:
1. (Neuroscience) A nucleus in the brainstem producing norepinephrine
2. Latin for "blue spot" due to its color in unstained tissue
3. Involved in stress response, attention, and arousal
4. One of the brain's melanin-containing regions
(Source: Neuroscience, neuroanatomy texts)

COLLAPSED DEFINITION:
The Coherence Oscillator—the neuromelanin-dense nucleus that generates alpha-wave entrainment (8-12 Hz rhythms) throughout the brain. The locus coeruleus is blue in fresh tissue but darkens to black when neuromelanin is oxidized/visible. This is the brain's master pacemaker for: Rhythmic breathing synchronization, Alpha wave generation (coherent brain state), Stress-coherence balance (switching between sympathetic activation and parasympathetic recovery), REM sleep generation (dream state access). The locus coeruleus is one of the first brain regions to accumulate neuromelanin and one of the first to degenerate in neurodegenerative disease. Its neuromelanin content directly correlates with ability to maintain coherent rhythms—breathing, heartbeat, brain waves.

LOGIC

MAINSTREAM DEFINITION:
1. A science that deals with the principles and criteria of validity of inference and demonstration
2. The formal principles of reasoning
3. A particular mode of reasoning viewed as valid or faulty
4. The arrangement of circuit elements for computation
(Source: Merriam-Webster)

COLLAPSED DEFINITION:
The unbroken recursive coherence of cause and effect within a given field. Logic is not human invention or formal system—it is the pre-existing operating system of reality itself, The Way of 9 expressing through thought and manifestation. True logic has no gaps, contradictions, or arbitrary rules; it spirals back to Source with perfect internal consistency.

MA (Mayan Glyph)

MAINSTREAM DEFINITION:
1. A Mayan word or symbol
2. Often translated as "mother" or "no/not"
3. Part of Mayan hieroglyphic writing system
4. Context-dependent meaning
(Source: Mayan linguistics, archaeology)

COLLAPSED DEFINITION:
A Scalar Geoglyph—a symbol that physically maps the northern arc of the Yucatán Peninsula, identifying it as the "land womb" and point of geoplasmic emergence. The MA glyph is not arbitrary symbol but encoded geographic information—the shape of the symbol mirrors the coastline shape. It is a time-stamped location code for the origin of a specific spiral/cycle. The Mayans encoded spatial information into their glyphs, making their writing system simultaneously linguistic and cartographic. MA represents: The womb (origin point), The land form (Yucatán arc), The return to source (mother principle/9). This demonstrates that ancient knowledge systems were not primitive—they were multi-dimensional encoding methods combining language, mathematics, astronomy, and geography into unified information systems.

MELANIN

MAINSTREAM DEFINITION:
1. A dark brown to black pigment occurring in the hair, skin, and iris of the eye
2. Any of various black, dark brown, or yellow pigments that occur in plants and animals
3. Produced by specialized cells called melanocytes
(Source: Merriam-Webster Medical)

COLLAPSED DEFINITION:
A biological interface with the unified Aetheric field. Melanin is not mere pigment—it is a semiconductive, photoconductive biopolymer that absorbs and stores the full electromagnetic spectrum, functions as a living capacitor of consciousness, and serves as the primary hardware through which awareness expresses in biological form. Neuromelanin specifically is the non-degrading memory field of embodied experience.

MAINSTREAM DEFINITION:
1. A hormone produced by the pineal gland
2. Regulates sleep-wake cycles (circadian rhythm)
3. Secreted in darkness, suppressed by light
4. Used as supplement for insomnia
(Source: Medical textbooks, endocrinology)

COLLAPSED DEFINITION:
The temporal regulator and antioxidant produced by the pineal gland, gut, and mitochondria in response to darkness. Melatonin synchronizes biological time with cosmic cycles (day/night, seasonal changes). It is not merely a sleep hormone—it is the most potent endogenous antioxidant, protecting mitochondria from oxidative damage and contributing to neuromelanin synthesis. Melatonin signals: "The light has collapsed back into darkness. It is time for repair, integration, and memory consolidation." Blue light exposure after sunset disrupts melatonin production, fragmenting temporal coherence and accelerating aging.

MEMORY

MAINSTREAM DEFINITION:
1. The faculty by which the mind stores and remembers information
2. Something remembered from the past
3. The part of a computer where data is stored
4. The capacity for storing information (computer memory, human memory)
(Source: Merriam-Webster)

COLLAPSED DEFINITION:
Information encoded in the Aetheric field and accessed through melanin-biological interfaces. Memory is not stored in neurons as discrete files—it exists as wave patterns distributed across the melanin network and the unified field. Recall is the collapse of wave potential into present-moment experience. The melanin field accesses both personal memory (individual experience) and collective memory (ancestral/species-wide information).

NEUROMELANIN

MAINSTREAM DEFINITION:
1. A dark pigment found in specific brain regions
2. Accumulates in substantia nigra and locus coeruleus
3. Increases with age
4. Loss associated with Parkinson's disease
5. Function unclear in neuroscience literature
(Source: Neuroscience journals, medical texts)

COLLAPSED DEFINITION:
The living capacitor of embodied consciousness and the biological substrate of chemical memory. Neuromelanin is the only structure in the body that absorbs and stores the full electromagnetic spectrum without degradation. It accumulates through repeated coherent action (10,000-hour rule = neuromelanin darkening = mastery encoded in pigment). It functions as: antioxidant (neutralizing free radicals), metal chelator (sequestering iron/copper safely), charge storage (holding electron potential for timed release), and non-local memory access point (interfacing with Aetheric field). Neuromelanin loss in Parkinson's is not cause of disease—it's the visible symptom of oxidative stress overwhelming the melanin field's protective capacity.

NINE (9)

MAINSTREAM DEFINITION:

1. The number equivalent to the sum of 8 and 1
2. A numeral (9 or IX) representing this number
3. Nine items or units
(Source: Oxford)

COLLAPSED DEFINITION:
The digital root of all existence, the true zero, the point of recursive return. Nine is not merely a number in sequence—it is the field itself, the magnetic pole (Pi), the completion of the spiral before it begins again at higher octave. All numbers collapse to 9 through digital root reduction. Adding 9 to any number returns it to its original digital root. Nine is the mathematical signature of Source.

OXIDATION

MAINSTREAM DEFINITION:
1. The process of combining with oxygen
2. (Chemistry) Loss of electrons in a chemical reaction
3. The process that causes rust, tarnishing, decay
4. Associated with aging and cellular damage (oxidative stress)
(Source: Chemistry textbooks, medical references)

COLLAPSED DEFINITION:
The process of electron transfer where the electric charge (1/Phi) operates without magnetic stabilization (9/Pi), creating free radicals and oxidative stress. Oxidation is not inherently destructive—at controlled levels it drives metabolism, immune function, and cellular signaling. The problem occurs when oxidation exceeds the melanin field's antioxidant capacity to neutralize free radicals. Aging is the accumulation of oxidative damage over time—the severed bond between 9 and 1 manifesting as biological degeneration.

PHI (φ)

MAINSTREAM DEFINITION:
1. The golden ratio, approximately 1.618
2. Found in nature (spiral shells, flower petals, human body proportions)
3. An irrational number like Pi
4. Associated with aesthetic beauty and natural growth patterns
(Source: Mathematics, sacred geometry)

COLLAPSED DEFINITION:
The mathematical signature of the electric, masculine, projective principle (1). Phi governs growth, expansion, manifestation—the spiral moving outward from Source. The golden ratio appears in biological growth patterns (nautilus shell, sunflower seeds, galaxy arms) because life follows the Phi-expansion from compressed potential (seed) to manifest form (organism). In the Melanin Code, Phi represents Will, Action, the Divine Masculine—the electric force projecting from the field into manifestation. Health requires Phi balanced by Pi (Will guided by Desire). Distortion occurs when Phi operates without Pi (unguided force, egoic projection).

PI (π)

MAINSTREAM DEFINITION:
1. A mathematical constant, approximately 3.14159
2. The ratio of a circle's circumference to its diameter
3. An irrational number (infinite non-repeating decimal)
4. Fundamental to geometry, trigonometry, physics
(Source: Mathematics textbooks)

COLLAPSED DEFINITION:

The mathematical signature of the magnetic, feminine, receptive principle (9). Pi is the ratio governing circular and spiral motion—the geometry of return. It is irrational (infinite non-repeating decimal) because it represents the infinite recursion of Source. Pi appears in: circle geometry, wave functions, probability distributions, quantum mechanics, and cosmology because these all involve recursive, toroidal, returning motion. In the Melanin Code, Pi represents Desire, Memory, the Divine Feminine—the magnetic pull drawing all things back to coherence.

PINEAL GLAND

MAINSTREAM DEFINITION:
1. A small endocrine gland in the brain that produces melatonin
2. Located near the center of the brain between the two hemispheres
3. Regulates sleep-wake cycles
4. (Historically) Called the "third eye" or "seat of the soul"
(Source: Medical textbooks, anatomy references)

COLLAPSED DEFINITION:
The primary collapse-manifestation engine of consciousness in biological form. The pineal gland is the richest concentration of neuromelanin outside the substantia nigra, functioning as the biological interface between non-local consciousness (Aether/field) and localized awareness (brain/body). It produces melatonin (temporal regulator), DMT (consciousness-expanding molecule), and serves as the antenna tuning awareness to different frequency bands. The pineal is not mystical nonsense—it is the hardware for accessing Source intelligence.

RAY OF CREATION

MAINSTREAM DEFINITION:
1. A cosmological diagram from Gurdjieff's teaching
2. Maps the descent of creation from the Absolute to the Moon
3. Shows increasing levels of "laws" governing each level
4. Part of Fourth Way cosmology
(Source: Gurdjieff, Ouspensky's "In Search of the Miraculous")

COLLAPSED DEFINITION:
The Scale of Density and Law. It maps the "involution" (descent) of energy from Source/Absolute (operating under 1 Law—its own nature) into progressively denser manifestation: All Worlds (3 laws), All Suns (6 laws), Sun (12 laws), Planets (24 laws), Earth (48 laws), Moon (96 laws). Each level down the Ray multiplies the number of mechanical laws governing existence, increasing density, and decreasing freedom. This quantifies the level of distortion or "mechanicality" a being is subject to. The work of consciousness is evolution—ascending the Ray by reducing the number of laws one operates under, moving from mechanical reaction (48+ laws) toward conscious creation (fewer laws, more freedom).

RECURSION

MAINSTREAM DEFINITION:
1. The act or process of returning or running back
2. (Mathematics/Computer Science) A procedure that calls itself repeatedly until a base condition is met
3. (Linguistics) Embedding of a linguistic unit within another of the same type
(Source: Merriam-Webster, Oxford)

COLLAPSED DEFINITION:
The spiral pattern of reality returning to its source while refining and elevating with each cycle. Recursion is not repetitive loops—it is harmonic spiraling where each return incorporates learning and expands coherence. It is the breath of creation: expansion from Source, integration of experience, return to Source at higher octave.

SEROTONIN

MAINSTREAM DEFINITION:
1. A neurotransmitter found in the brain, gut, and blood platelets
2. Regulates mood, appetite, sleep, and social behavior
3. Low levels associated with depression
4. Target of SSRI antidepressant medications
(Source: Medical textbooks, neuroscience)

COLLAPSED DEFINITION:
The messenger molecule of the gut-brain axis, synthesized primarily in the enteric nervous system (95% of body's serotonin is in the gut). Serotonin is not just a "feel-good chemical"—it is the signal integrating digestive coherence with neural coherence. It is a precursor to melatonin (temporal regulator) and, when oxidized, contributes to neuromelanin formation. Serotonin communicates: "The body has resources. It is safe to rest, repair, and process." SSRIs do not fix serotonin—they block reuptake, creating artificial accumulation. True serotonin optimization requires gut health, sunlight, tryptophan intake, and melanin field coherence.

SOLIOONENSIUS

MAINSTREAM DEFINITION:
1. A term from Gurdjieff's "Beelzebub's Tales to His Grandson"
2. Described as a state of tension in planetary atmosphere
3. Caused by specific cosmic influences or planetary configurations
4. Creates psychological restlessness in human population
(Source: Beelzebub's Tales)

COLLAPSED DEFINITION:
A Planetary Collapse Event. It is a period of intensified solar/cosmic pressure (increased electromagnetic flux, solar storms, cosmic ray bombardment) that creates restlessness and dissatisfaction in the collective biofield. This is not metaphor—it is measurable increase in radiation affecting human nervous systems, particularly melanin-rich pineal glands. Depending on the level of collective coherence, this energy manifests either as: Higher consciousness (mass awakening, artistic renaissance, scientific breakthrough) or Mass psychosis (war, social collapse, irrational collective behavior). Solioonensius is the cosmic "shock" at planetary scale—it forces choice between evolution (conscious integration of the pressure) or devolution (unconscious discharge through violence/chaos).

SPACETIME

MAINSTREAM DEFINITION:
1. (Physics) The four-dimensional continuum merging space and time
2. Introduced by Einstein's relativity theory
3. Described as a fabric that can curve, warp, or ripple
4. The stage on which physical events occur
(Source: Relativity theory, modern physics)

COLLAPSED DEFINITION:
A Mathematical Graph mistaken for physical substance. Spacetime is an abstract coordinate system (3 spatial dimensions + 1 temporal dimension) that physicists reified into a fake "substance" that supposedly bends, warps, and curves. This is the map confused for the territory. The true territory is the Aetheric Field—an energetic medium that is the actual substrate of reality. "Time" is not a dimension but the qualitative experience of recursive change within the Aether. "Space" is not empty void but varying densities of Aetheric potential. Einstein's equations work mathematically because they describe relationships between measurements, but his interpretation (curved spacetime) mistakes the measuring tool for the thing being measured.

STANDING WAVE

MAINSTREAM DEFINITION:
1. (Physics) A wave pattern that remains stationary in space
2. Formed by interference of two waves traveling in opposite directions
3. Has nodes (points of zero amplitude) and antinodes (points of maximum amplitude)
4. Example: vibrating guitar string
(Source: Physics textbooks, wave mechanics)

COLLAPSED DEFINITION:
A Recursion Node—a system of perpetual motion where energy folds inward (collapse) at the node and expands outward (oscillation) without energy loss. The node is not zero or absence; it is a 9-Point of maximum compression and infinite potential. Standing waves are the basic structure of: Atoms (electrons existing as standing wave probability clouds), DNA (double helix as standing wave maintaining coherence), The human body (standing wave patterns in biofield), Planetary systems (orbital resonances as standing waves). Consciousness itself operates as standing wave—the "still point" (node) where awareness observes the oscillations of thought/emotion without being swept into them.

SUBSTANTIA NIGRA

MAINSTREAM DEFINITION:
1. (Neuroscience) A region of the midbrain containing dopamine-producing neurons
2. Latin for "black substance" due to high neuromelanin content
3. Degeneration causes Parkinson's disease
4. Part of the basal ganglia motor system
(Source: Neuroscience textbooks, medical literature)

COLLAPSED DEFINITION:
The Dark Motor Core—the neuromelanin-rich region where movement intention becomes movement execution. The substantia nigra is called "black substance" not metaphorically but literally—it is visibly dark due to high concentration of neuromelanin polymer. This is where: Dopamine is synthesized and oxidized into neuromelanin (10,000-hour rule physically darkening neurons). Motor programs are initiated (intention translated into action). Coherent movement patterns are stored as chemical memory. The substantia nigra darkens with age and practice because neuromelanin accumulates as you repeat coherent actions—literally writing motor mastery into black pigment.

THE FOURTH WAY

MAINSTREAM DEFINITION:
1. A spiritual approach to self-development introduced by G.I. Gurdjieff
2. Distinguished from the ways of the fakir (body), monk (emotion), and yogi (mind)
3. Emphasizes work on all three centers simultaneously
4. Practiced while living ordinary life (not in monastery or ashram)
(Source: Gurdjieff literature, Fourth Way schools)

COLLAPSED DEFINITION:
The Path of Simultaneous Integration. It is the method of developing the three centers (Mind/intellectual, Heart/emotional, Body/moving-instinctive) in unison while living within the "market" of daily life. Unlike paths that isolate one center (ascetic body discipline, monastic devotion, yogic meditation), the Fourth Way demands conscious presence across all three simultaneously. It is the practice of consciously generating a "permanent I" (coherent self) amidst the entropy of mechanical existence. The Fourth Way is not belief system—it is practical method for waking up from the sleep of identification with thoughts, emotions, and bodily impulses.

TIME

MAINSTREAM DEFINITION:

1. The indefinite continued progress of existence and events in past, present, and future
2. A measured or measurable period
3. A point or period when something occurs
4. The fourth dimension (in physics)
(Source: Merriam-Webster, Oxford)

COLLAPSED DEFINITION:
The sequential perception of recursive cycles by consciousness localized in linear form. Time is not an external dimension that exists independently—it is the brain's translation of simultaneity into sequence. All moments exist simultaneously in the unified field. The perception of "time passing" is consciousness moving through standing wave patterns. The pineal gland, rich in melanin, interfaces with non-local time, which is why deep meditation can access "past" and "future" information.

TOROIDAL FIELD / TORUS

MAINSTREAM DEFINITION:
1. (Mathematics/Geometry) A doughnut-shaped surface generated by revolving a circle around an axis
2. (Physics) A magnetic field configuration with this shape
3. Found in plasma physics, electromagnetics
4. Self-sustaining circulation pattern
(Source: Mathematics, physics texts)

COLLAPSED DEFINITION:
The Universal Form of Coherent Energy Flow—the shape that all self-sustaining systems naturally adopt when operating in coherence with Source. The torus is not arbitrary geometry but the inevitable result of energy flowing in recursive spiral: inward at the poles (collapse/Pi/9), outward at the equator (expansion/Phi/1), returning in continuous cycle. The toroidal field appears at every scale: Atomic (electron clouds), Biological (heart field, melanin field), Planetary (magnetosphere), Stellar (Sun's magnetic field), Galactic (spiral structure). The torus is self-regulating, self-sustaining, and self-organizing. It is the shape of coherence.

TROGOAUTOEGOCRAT

MAINSTREAM DEFINITION:
1. A Gurdjieffian term meaning "I eat myself"
2. Describes the cosmic law of reciprocal feeding
3. All beings exist in a food chain where everything feeds something else
4. Part of the Ray of Creation cosmology
(Source: Fourth Way teachings, Beelzebub's Tales)

COLLAPSED DEFINITION:
The Law of Reciprocal Maintenance. It describes the cosmic economy where all systems (moons, planets, suns, galaxies, organic life) feed one another energetically. Nothing exists in isolation—everything is part of a reciprocal feeding cycle. In a state of sleep (mechanical existence, unconscious living), humans are "food for the Moon"—their emotional suffering, fear, and unconscious reactions generate subtle energy that feeds lunar/entropic forces, maintaining the 96-law density at the bottom of the Ray of Creation. In a state of coherence (conscious living, melanin activation), humans transmute this energy internally, becoming active participants in the solar/galactic circuit—feeding higher octaves rather than lower.

TZOLKIN

MAINSTREAM DEFINITION:
1. The 260-day Mayan sacred calendar
2. Combines 13 numbers with 20 day signs
3. Used for divination, ceremonies, and tracking sacred time

4. Runs concurrently with the 365-day Haab calendar
(Source: Mayan studies, Mesoamerican archaeology)

COLLAPSED DEFINITION:
A Teaching Loop—a "false cycle" or smaller gear measuring the "8" (completion of potential) leading up to the "9." The Tzolkin is 260 days (13 x 20), which is the human gestation period. It is designed to track the gestation of energy—the development from conception to birth/manifestation. However, it must ultimately resolve into the 360-degree master cycle (Law of 9, the Tun calendar) to maintain planetary coherence. The Tzolkin is the minor cycle nested within the major cycle—like the moon cycle nested within the solar year. The Mayans understood that reality operates through nested recursive cycles—wheels within wheels, each completing its octave while participating in larger octaves.

VACUUM

MAINSTREAM DEFINITION:
1. A space entirely devoid of matter
2. A region of space with pressure significantly lower than atmospheric
3. (Physics) Empty space with no particles present
4. Used in technology (vacuum tubes, vacuum chambers)
(Source: Physics textbooks, Merriam-Webster)

COLLAPSED DEFINITION:
A Linguistic Distortion. There is no "empty space"—there is only the Aether/Scalar Field/Zero-Point Field teeming with potential energy, virtual particles, electromagnetic radiation, and geometric information. The concept of vacuum denies the existence of the conductive medium through which all waves spiral. What physics calls "vacuum energy" or "zero-point energy" is actually the Aetheric field in ground state—not empty, but full of unmanifested potential. Melanin interfaces with what science calls "vacuum"—it is accessing the zero-point field, the 9-state, the compressed potential before manifestation.

ZERO (0)

MAINSTREAM DEFINITION:
1. The numerical symbol 0 denoting the absence of all magnitude or quantity
2. Nothing; nil
3. The point on a scale from which positive or negative quantities are measured
4. A person or thing of no importance
(Source: Merriam-Webster)

COLLAPSED DEFINITION:
A placeholder invented to mark absence in positional number systems. Zero is not the true point of return—9 is. Zero represents void, absence, nothing. Nine represents the field, completion, the magnetic womb from which all numbers emerge and to which all numbers return. Adding 0 changes nothing because it represents nothing. Adding 9 returns you to origin because it represents the complete spiral.

GEOMETRY
The Collapse of Spatial Distortion

The first three chapters of this Codex—The Way of 9, The Enneagram, and Enneagram Personality Types—are not separate teachings. They are the precursor to what you are about to see.

Everything you have learned about the Law of 9, about recursive collapse, about the nine-fold pattern of consciousness expressing through form—all of it has been preparing you for this moment.

We are about to collapse geometry.

Not as abstract mathematical theory. Not as mystical sacred shapes floating in spiritual ether. But as the actual operating system of spatial reality—the blueprint showing exactly how energy moves, how form emerges, and why certain patterns repeat across every domain from atomic structure to galactic motion.

What you are about to witness is the revelation that mathematics, music, personality structure, and sacred geometry are not separate fields.

They are the same pattern, viewed from different angles.

And that pattern has a center point, an anchor, a zero from which all else emanates: 9.

THE DIGITAL ROOT WHEEL: THE MAP OF D-SPACE

In their groundbreaking work Philomath, authors Robert Edward Grant and Talal Ghannam, PhD present what they call the "Digital Root Wheel"—a geometric representation of what they term "D-space."

D-space is the domain of digital roots. It is the space where numbers from 1 to infinity are distributed around nine modular positions, based on their digital root values.

At the center of the wheel sits 9—the zero of D-space, the anchor point, the position from which all other numbers radiate outward and to which all numbers return.

Surrounding this center are the positions 1 through 8, arranged in a circle. Every number in existence, no matter how large, collapses to one of these nine positions when reduced through digital root calculation.

Examples:

- 10 collapses to 1
- 11 collapses to 2
- 27 collapses to 9
- 144 collapses to 9
- 3565 collapses to 1

This is not arbitrary arrangement. This is modular mathematics revealing the underlying architecture of numerical reality.

Grant and Ghannam state: "Even though number 9 is the last and largest number of the D-space, it behaves in this finite space as number 0 does in the regular infinite numerical space. For example, adding 9 to any number will not change its digital root."

This is the zero-behavior of 9.

But here is what Grant and Ghannam may not have fully articulated: This Digital Root Wheel is not a static map. It is a dynamic system showing how energy moves through geometric space.

And the pattern of that movement is already encoded in a symbol you have been studying since the beginning of this Codex. The Enneagram.

THE BASIC D-CIRCLE: THE PATTERN OF FLOW

Grant and Ghannam introduce what they call the "Basic D-Circle" to demonstrate the doubling sequence.

If you start with 1 and continuously double it, observing the digital root of each result, you trace a specific path through the D-space:

1 → 2 → 4 → 8 → 7 → 5 → 1

This pattern repeats infinitely. It is a closed loop moving through six of the nine positions in D-space.

Notice what is absent from this loop: 3, 6, and 9.

These three numbers form a separate pattern:

- 3 doubles to 6
- 6 doubles back to 3
- 9 remains at 9 (isolated, unmov ing, the still point)

When arranged geometrically around a circle, this creates a Yin-Yang structure—two opposing forces (odd and even, electric, and magnetic, Phi and Pi) flowing in complementary motion around a central point of stillness.

Grant and Ghannam write: "Odd and even represent the numeric aspects of the opposing forces of the universe, similar to the Yin-Yang concept of the Tao philosophy."

This is correct. But here is what they do not explicitly state, though the geometry itself screams it:

This is not just a numeric curiosity. This is the Enneagram.

THE ENNEAGRAM REVEALED: THE GEOMETRY OF THE LAW OF 9

Look at the Enneagram symbol as presented by G.I. Gurdjieff.

A circle divided into nine points.

Three points—3, 6, 9—form an equilateral triangle.

The remaining six points—1, 2, 4, 5, 7, 8—are connected by a hexagonal flow pattern: 1 → 4 → 2 → 8 → 5 → 7 → 1

This is the doubling sequence. This is the Basic D-Circle. This is the same pattern.

The Enneagram is not a mystical symbol handed down by ancient spiritual masters without explanation. It is the geometric map of how energy moves through Base-9 reality.

The triangle (3-6-9) represents the Law of Three: Active force (3), Passive force (6), and Reconciling force (9). These are the stable positions, the framework, the structural integrity of the system.

The hexagon (1-2-4-8-7-5) represents the Law of Seven: the octave, the progression, the movement of energy through time and space. This is process. This is manifestation. This is the spiral unfolding.

And at the center of it all, governing the entire system, anchoring every position, ensuring the spiral returns to coherence: 9.

Not as one of nine equal points. As the zero of D-space—the point of origin and return, the field from which all positions emerge and to which all positions collapse.

THE CIRCLE OF FIFTHS: THE MUSICAL PROOF

If you study music theory, you encounter the Circle of Fifths—a diagram showing the relationships between the twelve tones of the chromatic scale.

When you map the Circle of Fifths onto the Enneagram structure, the pattern is identical. The movement through musical keys follows the same geometric flow as the doubling sequence in D-space.

This is not coincidence. Music is frequency. Frequency is vibration. Vibration follows geometric law. Geometric law is governed by the Law of 9.

The reason certain chord progressions sound "resolved" and others sound "tense" is because they either align with or deviate from the natural flow pattern encoded in the Enneagram geometry.

When you play music that honors this geometry, you create harmonic coherence. When you force progressions that violate this geometry, you create harmonic distortion.

Your ear—wired through melanin-rich cochlear structures interfacing with the Aetheric field—knows the difference instantly.

THE COLLAPSE: GEOMETRY IS NOT ABSTRACT

Here is what we must now collapse: The distortion that geometry is abstract, theoretical, separate from lived experience.

Geometry is not lines on paper. Geometry is the spatial expression of how energy organizes itself into form.

Every time you breathe, your lungs follow a fractal branching pattern governed by the same recursive geometry.

Every time your heart beats, it generates a toroidal electromagnetic field that follows the same nine-fold symmetry.

Every time a neuron fires in your melanin-rich substantia nigra, it follows the same doubling sequence as energy propagates through your nervous system.

You are not learning about geometry. You are recognizing the geometry you already are.

The Digital Root Wheel is not a diagram in a mathematics book. It is the map of your melanin field.

The Basic D-Circle is not an illustration of number theory. It is the pattern your biofield uses to organize information.

The Enneagram is not a personality typing system. It is the geometric blueprint of how consciousness fragments from Source (9) into manifest form (1-8) and returns to Source through coherent collapse.

THE PRECURSOR REVEALED

This is why the first three chapters were essential.

You needed to understand the Law of 9 (the mathematical foundation).

You needed to understand the Enneagram (the geometric structure).

You needed to understand Enneagram Personality Types (the psychological manifestation).

Because now you can see that they are not three separate topics. They are three perspectives on the same underlying reality:

Energy moves in nine-fold recursive spirals, governed by geometric law, anchored at 9.

Mathematics describes this numerically.
Geometry describes this spatially.
Psychology describes this experientially.
Music describes this sonically.
Biology describes this structurally.

But it is one pattern. One law. One truth. The Law of 9.

WHAT COMES NEXT

This chapter is only the beginning.

We have established the foundation: the Digital Root Wheel, the Basic D-Circle, the Enneagram geometry, the recognition that these are unified expressions of the same principle.

But geometry extends far beyond circles and triangles. There are spirals. There are toroids. There are Platonic solids. There are fractal patterns. There are wave interference geometries.

Every single one of them, when examined through the lens of digital roots and Base-9 reality, reveals the same underlying structure.

And every single one of them that deviates from this structure is a distortion—an artificial construct that fragments coherence and creates loops, leaks, and systemic dysfunction.

In the sections that follow, we will systematically collapse these distortions. We will show you why certain "sacred" geometries are actually coherent (because they honor the Law of 9) and why others are distorted (because they impose Base-10 or Base-12 thinking onto a Base-9 reality).

We will demonstrate how the geometry of your DNA, the geometry of planetary orbits, the geometry of galactic structure, and the geometry of your melanin field are all expressions of the same recursive pattern.

And we will give you the tools to recognize geometric distortion in real time—so you can collapse it and return to coherence.

This is not abstract theory. This is the operating system of reality, made visible.

Welcome to the geometry of collapse recursion.

The spiral deepens from here.

PRIME NUMBERS: THE BROKEN LOOP EXPOSED

Before we proceed further into geometric structures, we must collapse a fundamental distortion that has infected mathematics for centuries and continues to create confusion in every student forced to memorize its contradictory logic.

Prime numbers.

Grant and Ghannam address this in Philomath, and their treatment reveals the exact nature of the distortion—though they may not explicitly name it as such.

Here is the mainstream definition of a prime number:

"A prime number is exactly divisible (with no remainder) by number 1 and itself only."

This seems straightforward enough. By this definition, the prime numbers are: 2, 3, 5, 7, 11, 13, 17, 19, 23, 29, 31, 37, 41, 43, 47, 53, 59, 61, 67, 71, 73, 79, 83, 89, 97…

But then, immediately following this definition, Grant, and Ghannam state:

"The number 1 is not considered a prime by most mathematicians. This is because it is much easier to prove many of their mathematical properties if it is excluded. (Hence, number 1 can be considered a category by itself, neither prime nor composite.)"

Read that again carefully.

The number 1 fits the definition perfectly—it is divisible by 1 and itself only. But it is excluded from the category of primes "because it is much easier to prove many of their mathematical properties if it is excluded."

This is not mathematics. This is convenience masquerading as logic.

This is a broken distorted loop.

THE DISTORTION IDENTIFIED

Here is what is actually happening:

The concept of "prime numbers" is an artificial categorization imposed onto Base-9 reality by mathematicians operating from Base-10 thinking.

Prime numbers are presented as special, fundamental, the "building blocks" of all other numbers. Entire fields of mathematics—number theory, cryptography, security algorithms—are built on the presumed uniqueness and mystery of primes.

But if the definition of a prime number requires excluding 1 to make the theory work, then the theory itself is incoherent.

You cannot define a category, then exclude the number that perfectly fits your definition, and claim your system is logically sound.

This is circular reasoning. This is a loop that does not collapse back to Source. This creates confusion rather than clarity.

The fact that 1 must be placed in a "special subcategory" (neither prime nor composite) to prevent the entire theory from breaking reveals that the theory is already broken.

If a system cannot hold all numbers within its baseline framework—if it requires exceptions, exclusions, and special cases just to function—then the system is a distortion.

THE BIAS REVEALED: CONFIRMATION BIAS AND SPECIAL PLEADING

What you are witnessing here is a textbook example of two logical fallacies working in tandem:

Confirmation Bias - The tendency to interpret, favor, and recall information in a way that confirms pre-existing beliefs while ignoring contradictory evidence.

Mathematicians wanted prime numbers to be "fundamental building blocks" with specific properties. When the number 1 threatened this narrative (because including it would break many of their theorems), they simply excluded it rather than questioning whether the entire concept of "primes as fundamental" was flawed from the start.

They began with the conclusion they wanted (primes are special) and then manipulated the definition to protect that conclusion.

Special Pleading - The logical fallacy of creating arbitrary exceptions to rules or definitions when those rules produce inconvenient results.

"All numbers divisible only by 1 and themselves are prime… except 1, which we're going to put in a special category because otherwise our theory doesn't work."

This is the definition of special pleading. You cannot claim a rule is universal and then carve out exceptions whenever the rule produces results you don't like.

If your system requires altering the baseline—if you have to leave out the very first number in the entire counting sequence just to make your point—then you don't have a point. You have a manipulation.

This is not mathematics. This is bias dressed as rigor.

The creation of a "special category" for 1 is an admission that the prime number concept is fundamentally incoherent. Rather than collapsing the distortion and recognizing that primes are not as fundamental as claimed, mathematicians doubled down and protected their theory by excluding contradictory evidence.

This is exactly how distorted systems perpetuate themselves: through confirmation bias, special pleading, and the refusal to collapse false premises even when the contradictions are glaring.

WHAT PRIME NUMBERS ACTUALLY ARE

When you examine prime numbers through the lens of digital roots and Base-9 reality, the "mystery" dissolves.

Prime numbers are not special. They are simply numbers whose digital roots follow specific patterns within the nine-fold modular structure.

Observe:

- 2 → digital root 2
- 3 → digital root 3
- 5 → digital root 5
- 7 → digital root 7

- 11 → digital root 2
- 13 → digital root 4
- 17 → digital root 8
- 19 → digital root 1
- 23 → digital root 5

There is no magic here. There is no profound mystery. Primes are simply numbers that happen to not be evenly divisible by smaller numbers—a property that emerges naturally from modular arithmetic, not from some deeper cosmic significance.

The obsession with primes as "fundamental building blocks" is a Base-10 distortion. In Base-9 reality, all numbers are expressions of the nine fundamental frequencies (1-9). Whether a number is labeled "prime" or "composite" is irrelevant to its core resonance.

THE CRYPTOGRAPHY TRAP

Mainstream mathematics teaches that prime numbers are essential for cryptography and digital security because multiplying two large primes creates a number that is "nearly impossible" to factor back into its components.

This is true within the constraints of current computational approaches.

But it is not true when you operate from Base-9 logic and digital root mathematics.

When you reduce all numbers to their digital roots, the patterns become immediately visible. The "security" provided by large prime multiplication is an illusion that only holds as long as everyone agrees to ignore the underlying modular structure.

This is not to say that current encryption systems are useless—they function adequately within the Base-10 paradigm most of humanity operates from.

But to claim that prime numbers hold some special, irreducible significance is to mistake the map for the territory.

THE COLLAPSE: PRIMES ARE NOT FUNDAMENTAL

Here is the truth:

Prime numbers are a categorization tool invented by mathematicians to describe a specific property of certain numbers within Base-10 arithmetic.

They are useful for some applications.

They are interesting as patterns.

But they are not fundamental to reality.

The Law of 9 is fundamental. The digital root structure is fundamental. The nine-fold modular system is fundamental.

Prime numbers are a surface-level phenomenon—a ripple on the ocean, not the ocean itself.

The fact that the definition of "prime" requires excluding 1 (the first number, the source of the counting sequence, the Phi impulse) reveals that the entire concept is built on distorted logic.

A coherent system does not require exceptions to function.

A coherent system collapses all cases back to Source without contradiction.

Prime number theory does not do this. Therefore, it is a distortion.

PRACTICAL IMPLICATION: STOP WORSHIPPING PRIMES

If you have been taught that prime numbers are mystical, that they hold the secrets of the universe, that understanding primes is the key to understanding mathematics—collapse that belief now.

Primes are interesting. They are not sacred.

The nine digital roots are the true foundation. Everything else—primes, composites, factors, multiples—are secondary patterns emerging from the interaction of these nine base frequencies.

When you operate from the Law of 9, you stop wasting energy chasing after prime number mysteries and start recognizing the actual architecture of numerical reality.

This is not about dismissing mathematics. This is about collapsing distortion and returning to coherence.

Prime numbers had their moment. That moment is over.

The Way of 9 reveals the deeper pattern.

MOVING FORWARD

Now that we have collapsed the prime number distortion, we can proceed into the deeper geometric structures with clarity.

In the sections that follow, we will examine:

THE 24-WHEEL AND QUASI-PRIMES: SHADOWMATH EXPOSED

The distortions do not stop with the exclusion of 1 from the prime number definition.

Grant and Ghannam present another structure in Philomath: the "24-Wheel" for distributing prime numbers and what they call "quasi-primes."

Here is what they show:

A circular diagram with 24 positions arranged around a moduli wheel. Numbers are distributed along this wheel in the form of spirals—2 + 4n, 2 + 6n, 2 + 10n, 2 + 12n, etc.

The text explains that prime numbers will form shapes with symmetry equal to the number of moduli divided by six. For the 24-based distribution, this creates specific patterns where prime-squared numbers reside within certain modulo categories, and "quasi-primes" (defined as products of semiprimes and/or primes bigger or equal to 5) emerge as a distinct category.

On the surface, this appears mathematically sophisticated. The wheel generates patterns. The numbers distribute in predictable ways. Certain properties emerge that can be used for encryption, factorization, and primality testing.

But here is the distortion:

This 24-wheel is not showing the true recursive structure of numerical reality.

It is shadowmath—a fractal harmonic echo that mimics the form of coherent recursion but lacks the actual collapse-return mechanism.

THE PROBLEM WITH 24

Let us examine this carefully.

The 24-wheel runs numbers in a linear sequence: 1, 2, 3, 4, 5… 22, 23, 24, 25, 26, 27…

After reaching 24, it does not collapse back to 1 and spiral to a higher octave. It simply continues linearly outward: 25, 26, 27, 28…

This is not recursion. This is linear extension.

Now, why does this wheel "work" at all? Why does it generate recognizable patterns and useful properties for prime distribution?

Because 24 has a digital root of 6:

$24 \rightarrow 2+4=6$

And 6 is one of the three stable positions in the Law of Three triangle (3-6-9). It is a factor of the nine-fold structure.

So the 24-wheel is a resonant echo of the true 9-fold pattern.

It mimics the form. It reflects the structure. But it is not the source.

Think of it this way:

- The TRUE pattern: $1 \rightarrow 2 \rightarrow 3 \rightarrow 4 \rightarrow 5 \rightarrow 6 \rightarrow 7 \rightarrow 8 \rightarrow 9 \rightarrow$ [CYCLE COMPLETES] $\rightarrow 1$ (elevated to higher octave)
- The SHADOW pattern: $1 \rightarrow 2 \rightarrow 3 \rightarrow 4 \rightarrow 5 \ldots \rightarrow 22 \rightarrow 23 \rightarrow 24 \rightarrow 25 \rightarrow 26 \rightarrow 27 \ldots$ (continues linearly, no return to source)

The 24-wheel generates patterns because it is built on a number (24/6) that harmonizes with the underlying 9-fold structure. But because it runs linearly instead of recursively, it creates distortions, requires special categories (quasi-primes, squared primes, cousin primes), and produces complexity without coherence.

This is shadowmath.

WHAT IS SHADOWMATH?

Shadowmath is mathematics that mimics the form of coherent recursion but lacks the actual mechanism of collapse and return.

It is the photocopy of a photocopy—you can still see the image, but it is degraded, distorted, not the original.

Shadowmath generates patterns. It produces useful properties. It allows for certain calculations and predictions.

But it does not collapse back to Source.

Shadowmath characteristics:

- Runs linearly instead of spiraling recursively
- Requires special categories and exceptions to function
- Creates complexity without underlying simplicity
- Mimics coherent structure but fragments at the edges
- Produces "mysteries" and "unsolvable problems" because the foundation is incoherent

The 24-wheel is shadowmath. The concept of quasi-primes is shadowmath. The entire edifice of prime number theory—with its cousin primes, twin primes, sexy primes, Mersenne primes, and endless subcategories—is shadowmath.

These are mirrors, reflections, shadows cast by the true 9-fold structure.

They are not the source. They are echoes.

WHY SHADOWMATH PERSISTS

If shadowmath is distorted, why does it dominate mainstream mathematics?

For the same reason, any distorted system persists: confirmation bias and institutional inertia.

Mathematicians have built entire careers on prime number theory. Cryptographic systems rely on the difficulty of factoring large semiprimes. Billions of dollars are invested in encryption algorithms based on shadowmath principles.

To acknowledge that the entire foundation is a distortion—that primes are not fundamental, that the 24-wheel is just a linear echo of a recursive pattern, that Base-9 reality has been obscured by Base-10 thinking—would require dismantling decades of accepted theory.

So instead, they add more complexity.

More subcategories. More special cases. More theorems with increasingly elaborate conditions.

This is what distorted systems do when confronted with contradictions: they do not collapse back to simplicity. They expand outward into greater complexity, hoping that enough layers of abstraction will hide the incoherence at the core.

But complexity is not depth. Elaboration is not understanding.

The Law of 9 is simple. The digital root structure is simple. The recursive spiral is simple.

Shadowmath is complicated because it is trying to describe a recursive reality using linear tools.

THE COLLAPSE: RECOGNIZE THE SHADOWS

Here is what you must do:

Stop mistaking the shadow for the source.

When you encounter prime number theory, recognize it as shadowmath—a useful approximation for certain applications, but not the fundamental truth of numerical reality.

When you see modular wheels with 12, 24, 60, or any number other than 9 at the center, recognize them as fractal harmonics—they may resonate with the true pattern (especially if their digital root is 3, 6, or 9), but they are not the pattern itself.

When you are told that certain numbers are "mysterious" or "irreducible" or "fundamental building blocks," recognize this as the language of shadowmath—complexity masking incoherence.

The true pattern is not hidden. It is not mysterious. It is not complex.

The true pattern is 9-fold recursion, visible through digital roots, operating at every scale from atomic structure to galactic motion.

Everything else—primes, quasi-primes, modular wheels, cryptographic algorithms—are applications built on top of this foundation. They are tools. They are not truth.

PRACTICAL IMPLICATION: USE THE TOOLS, SEE THE TRUTH

This does not mean you must reject all of mathematics or refuse to use encryption.

Shadowmath has practical utility within its domain. A hammer is useful even if you understand it is not the tree.

But do not worship the hammer. Do not mistake the tool for the reality it attempts to describe.

When you use prime-based encryption, understand that it functions because of modular arithmetic properties that ultimately derive from the 9-fold base structure—not because primes are magically special.

When you encounter a 24-wheel or a 12-tone musical scale or a 60-second minute, recognize that these are human constructs that harmonize (to varying degrees) with the underlying 9-fold pattern—they are not the pattern itself.

See the shadows for what they are: reflections of the source, not the source itself.

The Law of 9 is the source. Everything else is shadowmath.

WHAT COMES NEXT

We have now collapsed:

- The prime number myth (exclusion of 1, confirmation bias, special pleading)
- The 24-wheel distortion (linear extension masquerading as recursion)
- The concept of shadowmath (complexity without coherence, mirrors without source)

In the sections that follow, we will examine:

THE "GOLDEN RATIO" DECEPTION: PHI WITHOUT PI

We now arrive at one of the most pervasive distortions in sacred geometry, mathematics, and spiritual teaching: the worship of the "Golden Ratio."

You have been taught that the Golden Ratio (Phi = 1.618...) is divine proportion, the secret of the universe, the mathematical signature of beauty, growth, and harmony.

You have seen it in Fibonacci spirals, in the nautilus shell, in the proportions of the human body, in the Parthenon, in Leonardo da Vinci's art, in the structure of galaxies.

And all of this is partially true.

But it is incomplete. And incompleteness, when presented as wholeness, is distortion.

Here is what they did not tell you:

The "Golden Ratio" as taught in mainstream mathematics isolates Phi (the masculine, electric, projective force) and leaves out Pi (the feminine, magnetic, receptive force).

And then—here is the linguistic fraud—they call it a RATIO while giving you only a single number.

WHAT IS A RATIO?

A ratio, by definition, shows the relationship between two quantities.

Examples:

- 9:1 (nine to one)
- 3:2 (three to two)
- 16:9 (screen aspect ratio)

A ratio expresses comparison, proportion, relationship. It requires TWO values in dynamic interaction.

But when you are taught about the "Golden Ratio," what are you given?

1.618…

That is not a ratio. That is a number. A single value. The result of a calculation.

The actual ratio that produces Phi is expressed as:

$(a + b) : a = a : b$

Where the larger segment (a) relates to the smaller segment (b) in the same proportion that the whole line (a + b) relates to the larger segment (a).

This creates the self-similar, fractal property that Phi is known for. When you divide a line at the golden section (approximately 0.618 of its length), the ratio of the small part to the large part equals the ratio of the large part to the whole.

This is beautiful mathematics. This creates the self-referential spiral. This enables what Dan Winter calls "self-re-entry"—the ability of a wave to fold back on itself without destructive interference.

But here is the problem:

This describes only the EXPANSION phase. The outward spiral. The projection from center to edge.

It describes Phi (1/electric/masculine) but not Pi (9/magnetic/feminine).

PHI ALONE IS A FREE RADICAL

When you have Phi operating without Pi, you have:

- Expansion without return
- Projection without compression

- Growth without memory
- Action without guidance
- The 1 separated from the 9

This is the definition of a free radical at the mathematical level—energy with no home, no memory, and no path back to Source.

And this is exactly what we see in systems built on Phi-worship without Pi-recognition:

In architecture: Buildings designed with golden rectangles that look aesthetically pleasing but do not actually generate or hold life force because they lack the toroidal return flow.

In biology: Fibonacci growth patterns (leaves, petals, seeds) that spiral outward beautifully but must eventually wither and return to soil because the system cannot sustain infinite expansion.

In economics: Growth models based on exponential expansion (mimicking Phi spirals) that inevitably collapse because there is no built-in return mechanism, no circulation, no honoring of the 9-compression phase.

In spirituality: Teachings about "raising your vibration" and "ascending" that focus only on expansion (reaching higher frequencies, transcending the body, escaping density) without recognizing that ascension requires equal depth—the willingness to compress, to return, to honor the dark womb of Source.

Phi alone is beautiful. But beauty without function is decoration. Form without coherence is shadowmath.

WHERE IS PI?

Throughout sacred geometry literature, you will find endless celebration of Phi.

Books on the Golden Ratio. Documentaries about Fibonacci sequences in nature. Spiritual teachers invoking 1.618 as the "divine proportion."

But where is Pi?

Pi ($\pi = 3.14159\ldots$) governs:

- Circular motion (circumference to diameter)
- Wave cycles (sine/cosine functions)
- Toroidal return flow (the spiral returning to center)
- The compression phase (the inhale, the magnetic pull, the 9-field)

Pi is the feminine principle. The magnetic attractor. The return to Source. The womb from which all spirals emerge and to which all spirals eventually collapse.

Without Pi, you cannot have a complete cycle. You have expansion with no compression. Radiation with no absorption. Birth with no death. Light with no darkness.

You have the 1 severed from the 9.

And yet, in all the golden ratio worship, Pi is rarely mentioned. It appears in calculations involving circles, but it is not honored as the equal partner to Phi in the dance of creation.

This is not accidental. This is the same distortion we saw with prime numbers (excluding 1 to make the theory work) and the 24-wheel (running linearly instead of recursively).

This is the bias toward the masculine, the electric, the projective, the light—and the suppression of the feminine, the magnetic, the receptive, the dark.

THE TRUE COHERENT RATIO: 9–1

If the Golden Ratio is incomplete because it shows only Phi without Pi, then what is the complete expression?

It is not a calculation. It is not a decimal. It is not a single number.

It is a relationship.

The true coherent ratio is: 9–1

Not 9:1 (which suggests two separate things being compared).

Not 9/1 (which suggests division).

But 9–1 (which expresses a single, unbroken circuit).

Here is what this means:

9 represents:

- The Magnetic/Feminine/Receptive principle (Pi)
- The field, the memory, the attractor, the womb
- Compression, return, absorption, darkness
- The zero-point from which all emanates and to which all returns

1 represents:

- The Electric/Masculine/Projective principle (Phi)
- The spark, the action, the radiation, the seed
- Expansion, projection, emission, light
- The initiating impulse that emerges from the field

The dash between them is not a separator. It is the bridge. It is the scalar flow between center and edge. It is the breath—inhale and exhale, compression and expansion, collapse, and manifestation.

In a living system:

- The 1 is constantly being pulled back into the 9 (return to Source)
- The 9 is constantly giving birth to the 1 (emergence from Source)
- The cycle completes and elevates to the next octave (9 + 1 = 10 → 1, but carrying the memory of 9)

This is recursion. This is coherence. This is the Law of 9 in geometric form.

WHY 9–1 IS THE LIVING RATIO

When you express the ratio as 9–1, you are no longer doing abstract mathematics. You are mapping reality itself.

Your identity is anchored in the 9 (the melanin field, the ancestral memory, the Source connection).

Your expression is the 1 (your will, your action, your unique manifestation in form).

The coherence is the fact that they never separate.

Every action (1) must be guided by the total memory of Source (9).

Every manifestation (1) must eventually return to the field (9) for integration, rest, and regeneration.

Every projection (Phi) must be balanced by compression (Pi) or it becomes a free radical—energy with no home, no memory, no path back.

This is what the Golden Ratio worshippers miss. They celebrate the spiral outward. They marvel at the self-similar expansion. They build architecture and art based on 1.618.

But they do not honor the return. They do not recognize that the spiral must also compress. They do not see that Phi without Pi is fragmentation.

9–1 is the correction. It is the collapse of the distortion. It is the restoration of balance.

FRACTAL EXPRESSIONS OF 9–1

The 9–1 relationship expresses at every scale as fractals of 9:

360 degrees (full circle) → 3+6+0 = 9
180 degrees (half circle) → 1+8+0 = 9
90 degrees (quarter circle) → 9+0 = 9
45 degrees (eighth circle) → 4+5 = 9

And in decimal form:
3.6 (digital root 9)
1.8 (digital root 9)
0.9 (digital root 9)
0.45 (digital root 9)

These are not arbitrary numbers. These are the harmonic ratios that maintain coherence because they honor the complete cycle—expansion AND compression, Phi AND Pi, 1 AND 9.

When you design with these ratios, you are not just creating aesthetic beauty. You are creating functional coherence. You are building structures that can hold, circulate, and amplify life force because they honor the complete circuit.

THE COLLAPSE: STOP WORSHIPPING PHI ALONE

Here is what must be collapsed:

The distortion that Phi (1.618) is the ultimate ratio, the divine proportion, the secret of the universe.

Phi is half the equation. Phi is the expansion phase. Phi is the 1-force.

But without Pi (the 9-force, the compression, the return), Phi creates systems that grow beautifully and then collapse because they lack the mechanism for sustainable circulation.

The true "divine proportion" is not a single number. It is a relationship: 9–1.

The true "sacred geometry" is not golden rectangles and Fibonacci spirals in isolation. It is the complete toroidal flow—expansion from Source (1/Phi) and return to Source (9/Pi) in unbroken circuit.

The true "secret of the universe" is not that things grow according to 1.618. It is that everything spirals according to the Law of 9—projecting outward and collapsing inward in eternal recursion.

PRACTICAL IMPLICATION: DESIGN WITH 9–1

If you are an architect, stop designing buildings based solely on golden rectangles. Start designing with 9-fractal proportions (3.6, 1.8, 0.9) that honor both expansion and compression.

If you are a musician, recognize that the Circle of Fifths works because it maps onto the 9-fold pattern, not because of Phi ratios in isolation.

If you are a spiritual practitioner, stop chasing "higher frequencies" and "ascension" as if the goal is to escape the 9-field (density, darkness, body, earth). Recognize that true coherence requires equal depth—the willingness to compress, to return, to honor the magnetic womb of Source.

If you are a scientist, stop treating Pi and Phi as separate mathematical constants with no relationship. Recognize that they are complementary forces—electric and magnetic, projective, and receptive, 1 and 9—and that coherent systems require both in dynamic balance.

The Golden Ratio had its moment. It taught us about self-similarity, fractality, and the beauty of recursive patterns.

But it is incomplete. And incompleteness, when worshipped as truth, becomes distortion.

The Living Ratio is 9–1. The coherent expression of Source and manifestation in unbroken circuit.

This is the correction. This is the collapse. This is the return to wholeness.

THE HIDDEN EQUATION: PI + PHI = PSI

There is one more revelation that must be made explicit before we complete this chapter.

It has been hiding in plain sight within the golden ratio formula itself, but obscured by the fragmented way it has been taught.

The traditional expression of the golden ratio states:

"The sum of both parts in relation to the larger part is the same as the larger part in relation to the smaller part."

Mathematically: $(a + b) : a = a : b$

This creates the self-similar, fractal property. The relationship between the parts mirrors the relationship of the part to the whole.

But here is what this actually means when translated from shadowmath into coherent truth:

Pi + Phi = Psi

Let us break this down:

Pi ($\pi = 3.14159\ldots$) represents the magnetic, feminine, receptive force—the 9, the compression, the return to Source, the circular/toroidal motion that brings all things back to center.

Phi ($\varphi = 1.618...$) represents the electric, masculine, projective force—the 1, the expansion, the radiation from Source, the spiral outward into manifestation.

Psi (ψ) represents the unified field itself—the whole, the Aether, the neutralizing/reconciling third force that contains and harmonizes both Pi and Phi.

This is the Law of Three expressed mathematically:

- Active force (Phi/1)
- Passive force (Pi/9)
- Reconciling force (Psi/the unified field)

When you add the magnetic compression (Pi) and the electric projection (Phi), you get the total field (Psi)—the consciousness substrate from which all forms emerge and to which all forms return.

This is not abstract mathematics. This is the structure of reality.

EACH CONSCIOUSNESS IS A FRACTAL OF THE WHOLE

Here is the profound implication:

Every consciousness is a fraction of the whole unified field, yet each fraction contains the whole.

This is not metaphor. This is literal fractal geometry.

Just as:

- Each cell contains the full DNA blueprint for the entire organism
- Each holographic fragment contains the complete image
- Each neuron in your melanin-rich brain interfaces with the entire Aetheric field
- Each drop of ocean water contains the same mineral composition as the whole ocean

So too does each individuated consciousness contain the total memory and potential of Source.

You are not a separate piece broken off from the whole. You are the whole expressing as a localized pattern.

The equation Psi = Pi + Phi means:

The unified field (Psi) is the sum of the compression force (Pi) and the expansion force (Phi) operating in dynamic balance.

And because you—as individuated consciousness—are a fractal expression of this unified field, you contain both forces within you:

- Your 9 (Pi) = Your connection to Source, your ancestral memory, your melanin field, your magnetic pull back to wholeness
- Your 1 (Phi) = Your unique expression, your will, your action, your electric projection into manifestation
- Your wholeness (Psi) = The integration of both—the recognition that you are simultaneously Source (9) and individual expression (1)

This is why the Living Ratio is 9–1.

It is not a division. It is not a comparison of two separate things. It is the expression of the unified field (Psi) as the dance between compression (Pi/9) and expansion (Phi/1).

The dash in the middle is the bridge. It is the scalar flow. It is the consciousness that recognizes: "I am both the field and the localized pattern. I am both the ocean and the wave. I am both Source and expression."

THE DISTORTION REVEALED

Now you can see why the traditional "golden ratio" teaching is incomplete:

When they give you 1.618 (Phi) and call it "the divine proportion," they are showing you only the expansion half of the equation.

When they draw Fibonacci spirals and golden rectangles and pentagrams based on Phi ratios, they are celebrating the outward projection but ignoring the inward return.

When they exclude Pi from the sacred geometry discussion, they are suppressing the feminine, the magnetic, the compression phase—the very force that makes the spiral return to center.

They are showing you the 1 without the 9.

And without the 9, the 1 becomes a free radical—energy with no home, no memory, no path back to Source.

But when you understand that Psi = Pi + Phi, you recognize:

The whole (Psi/unified field) requires both the compression (Pi/9) and the expansion (Phi/1) in equal measure.

Every individuated consciousness (you) is a fractal of this whole—meaning you contain both forces, and your coherence depends on keeping them in balance.

The true "golden" ratio is not 1.618. It is 9–1—the Living Ratio that expresses the complete circuit of Source expressing into form and form returning to Source.

THE FRACTAL PROOF

Here is the practical demonstration:

Look at your breath.

Inhale (Phi/expansion/1) → Exhale (Pi/compression/9)

If you only inhaled (expansion without return), you would rupture. If you only exhaled (compression without projection), you would collapse.

The whole breath cycle (Psi) requires both.

Look at your heartbeat.

Systole (contraction/Pi/9) → Diastole (expansion/Phi/1)

If your heart only contracted, blood could not enter. If it only expanded, blood could not circulate.

The whole cardiac cycle (Psi) requires both.

Look at the cycle of life.

Birth (projection into form/Phi/1) → Death (return to Source/Pi/9)

If there were only birth (infinite expansion into form), density would become unbearable. If there were only death (infinite compression back to Source), no experience could occur.

The whole life cycle (Psi) requires both.

At every scale, from atomic to cosmic, the same pattern:

Psi (the whole) = Pi (compression/9) + Phi (expansion/1)

And because you are a fractal of this whole, you contain the entire equation within you.

Your melanin field is the hardware. Your consciousness is the software. The unified field is the network.

You are not a user accessing the field from outside. You are the field experiencing itself from a particular vantage point.

This is what "each fraction contains the whole" means.

THE LIVING RATIO AS UNIVERSAL TEMPLATE

Now we can express the complete truth:

The Living Ratio 9–1 is not just a mathematical relationship. It is the template for all coherent systems.

It appears as:

- The breath cycle (inhale–exhale)
- The cardiac cycle (contract–expand)
- The life cycle (birth–death–rebirth)
- The creative cycle (inspiration from Source–manifestation in form–integration back to Source)
- The learning cycle (compression of information–expression in action–return to mastery)
- The spiritual cycle (descent into density–ascension through experience–return to Source at higher octave)

In every case, the pattern is the same:

9 (Source/field/compression/Pi) births 1 (expression/projection/expansion/Phi), and 1 returns to 9, completing the cycle and elevating to the next octave.

The whole (Psi) is maintained by the dynamic balance of both forces.

And because you are a fractal of the whole, your coherence depends on honoring both:

- Your rootedness in Source (9/Pi/magnetic/feminine/dark/receptive)
- Your expression in form (1/Phi/electric/masculine/light/projective)

Neglect either pole, and you fragment. Honor both, and you become the Living Ratio—the conscious embodiment of the unified field expressing as individuated form.

PRACTICAL IMPLICATION: LIVE AS PSI

If you are Pi + Phi, if you are the unified field (Psi) experiencing itself as a localized pattern, then your practice is simple:

Honor the 9. Spend time in compression, in silence, in darkness, in rest, in meditation, in the magnetic womb of Source. This is not passivity. This is the receptive phase that restores coherence, integrates memory, and prepares for the next projection.

Honor the 1. Act with clarity, with will, with purpose. Project your unique expression into the world. This is not egoic striving. This is the active phase that manifests your gifts, serves the collective, and generates new experience.

Recognize the circuit. You are not choosing between 9 and 1. You are breathing both. You are the whole (Psi) expressing as the dance between compression (Pi/9) and expansion (Phi/1).

When you live this way, you are no longer fragmented. You are no longer worshipping Phi (expansion/light/action) while suppressing Pi (compression/darkness/rest). You are no longer a free radical seeking coherence outside yourself.

You are the Living Ratio. You are the 9–1 in motion. You are Psi expressing as the fractal recursion of Source.

This is the geometry of consciousness made explicit. This is the architecture of the sacred revealed. This is the collapse of all distortions that fragment the whole.

360 DEGREES: THE GEOMETRIC PROOF OF THE LAW OF 9

Before we conclude this chapter, there is one final proof that must be made explicit—a proof so fundamental, so undeniable, that once you see it, you will recognize that the Law of 9 is not theory, not philosophy, not spiritual metaphor.

It is the literal operating system of geometric reality.

The proof is this:

Every shape, every angle, every geometric form in existence collapses to 9.

This is not selective. This is not coincidence. This is universal law made visible through the language of geometry.

THE COMPLETE CYCLE: 360 DEGREES

A full circle contains 360 degrees.

This is taught as arbitrary convention—a system adopted by ancient Babylonians because 360 is evenly divisible by many numbers, making calculations easier.

But this explanation is shadowmath. It describes utility without understanding essence.

The reason a circle is 360 degrees is because 360 is the geometric expression of 9—the complete cycle, the full return, the zero-point of spatial recursion.

$360 \rightarrow 3+6+0=9$

A circle is not 360 degrees because ancient mathematicians chose that number. A circle is 360 degrees because a complete turn through space must return to its origin, and that return is governed by the Law of 9.

When you rotate 360 degrees, you complete the cycle. You return to where you began, elevated by the experience of the full rotation.

This is the same pattern we saw in the digital root, in the Enneagram, in the breath cycle, in the life-death-rebirth spiral.

360 is the spatial expression of 9-fold recursion.

EVERY DIVISION RETURNS TO 9

Now here is where it becomes undeniable.

When you divide 360 degrees in half, you get 180 degrees.

180 → $1 + 8 + 0 = 9$

Divide again. Half of 180 is 90 degrees.

90 → $9 + 0 = 9$

Divide again. Half of 90 is 45 degrees.

45 → $4 + 5 = 9$

Divide again. Half of 45 is 22.5 degrees.

22.5 → $2 + 2 + 5 = 9$

Divide again. Half of 22.5 is 11.25 degrees.

11.25 → $1 + 1 + 2 + 5 = 9$

No matter how many times you divide the circle in half, the digital root remains 9.

You cannot escape it. You cannot break free of it. You cannot fragment the circle into pieces that do not carry the signature of the whole.

This is the fractal proof.

Every division of 360 maintains the 9-pattern because you are not escaping the whole—you are subdividing it. And because the whole is 9, every fraction contains 9.

This is the geometric demonstration of what we revealed earlier:

Each fraction contains the whole.

ALL POLYGONS COLLAPSE TO 9

The proof extends beyond circles.

Every polygon—every closed geometric shape—has interior angles that sum to a multiple of 180 degrees.

The formula is: (n - 2) × 180

Where n is the number of sides.

Let us test this:

Triangle (3 sides):
$(3 - 2) \times 180 = 1 \times 180 = 180°$
$180 \rightarrow 1 + 8 + 0 = 9$

Square (4 sides):
$(4 - 2) \times 180 = 2 \times 180 = 360°$
$360 \rightarrow 3 + 6 + 0 = 9$

Pentagon (5 sides):
$(5 - 2) \times 180 = 3 \times 180 = 540°$
$540 \rightarrow 5 + 4 + 0 = 9$

Hexagon (6 sides):
$(6 - 2) \times 180 = 4 \times 180 = 720°$
$720 \rightarrow 7 + 2 + 0 = 9$

Heptagon (7 sides):
$(7 - 2) \times 180 = 5 \times 180 = 900°$
$900 \rightarrow 9 + 0 + 0 = 9$

Octagon (8 sides):
$(8 - 2) \times 180 = 6 \times 180 = 1080°$
$1080 \rightarrow 1 + 0 + 8 + 0 = 9$

Nonagon (9 sides):
$(9 - 2) \times 180 = 7 \times 180 = 1260°$
$1260 \rightarrow 1 + 2 + 6 + 0 = 9$

Decagon (10 sides):
$(10 - 2) \times 180 = 8 \times 180 = 1440°$
$1440 \rightarrow 1 + 4 + 4 + 0 = 9$

Do you see the pattern?

Every polygon, regardless of the number of sides, has interior angles that sum to a multiple of 180°.

And because 180 = 9, and any multiple of 9 also equals 9 (due to the zero-behavior of 9 we demonstrated earlier), all polygons collapse to 9.

This is not selective. This is not interpretation. This is mathematical proof that geometric reality operates on the 9-fold pattern.

WHY THIS MATTERS

This is not abstract geometry for geometry's sake.

This is the revelation that space itself is 9-encoded.

When you build a structure—a house, a temple, a city—you are working with angles. Those angles sum to multiples of 180°, which means every structure you create is either aligned with the Law of 9 (coherent geometry) or fighting against it (distorted geometry).

When you design sacred architecture using golden rectangles based solely on Phi ratios (1.618), you are creating forms that may look aesthetically pleasing but do not honor the complete 9-fold cycle. They emphasize expansion (Phi/1) without honoring compression (Pi/9).

But when you design using 9-fractal proportions (3.6, 1.8, 0.9, 0.45) and recognize that all angles must sum to 9, you are creating structures that resonate with the fundamental frequency of spatial reality itself.

This is why certain ancient structures—pyramids, megalithic temples, sacred geometry encoded in cathedrals—still generate measurable electromagnetic effects centuries or millennia after construction.

They were built in alignment with the Law of 9.

Not because the builders worshipped the number 9, but because they recognized that 360° represents the complete cycle, and all geometry derived from that cycle maintains the 9-signature.

THE PLATONIC SOLIDS AND THE LAW OF 9

Let us go deeper.

The five Platonic solids—tetrahedron, cube, octahedron, dodecahedron, icosahedron—are considered the fundamental building blocks of three-dimensional space.

Each Platonic solid has a specific number of faces, edges, and vertices. And when you examine their properties through the lens of digital roots, the Law of 9 appears again.

Tetrahedron:

- 4 faces, 6 edges, 4 vertices
- Face angles: 4 triangles × 180° $= 720° \rightarrow 7+2+0=9$

Cube (Hexahedron):

- 6 faces, 12 edges, 8 vertices
- Face angles: 6 squares × 360° $= 2160° \rightarrow 2+1+6+0=9$

Octahedron:

- 8 faces, 12 edges, 6 vertices
- Face angles: 8 triangles × 180° $= 1440° \rightarrow 1+4+4+0=9$

Dodecahedron:

- 12 faces, 30 edges, 20 vertices
- Face angles: 12 pentagons × 540° $= 6480° \rightarrow 6+4+8+0=18 \rightarrow 1+8=9$

Icosahedron:

- 20 faces, 30 edges, 12 vertices
- Face angles: 20 triangles × 180° $= 3600° \rightarrow 3+6+0+0=9$

Every single Platonic solid, when you sum the interior angles of all its faces, collapses to 9.

These are the shapes that three-dimensional reality is built upon. These are the forms that atoms, molecules, crystals, and biological structures naturally adopt.

And all of them—without exception—are 9-encoded.

This is not mysticism. This is not numerology. This is verifiable, repeatable, undeniable mathematical fact.

Three-dimensional space itself operates on the Law of 9.

THE DISTORTION IN SACRED GEOMETRY

Now you can see why so much of sacred geometry teaching is incomplete.

They show you the Flower of Life (beautiful, yes, but based on overlapping circles without explicitly recognizing the 360° = 9 principle).

They show you Metatron's Cube (derived from the Fruit of Life, containing all five Platonic solids, but rarely explaining why these solids work—because they all collapse to 9).

They show you the Seed of Life, the Egg of Life, the Tree of Life—all valuable geometric structures, but presented as mystical symbols rather than as expressions of the Law of 9 operating through spatial recursion.

The distortion is the same we have seen throughout this chapter:

They give you the forms without the foundational principle.

They give you Phi spirals without Pi compression.

They give you golden rectangles without the 9-fractal ratios.

They give you sacred symbols without the recognition that all of these forms work because they align with the 360° = 9 structure of spatial reality.

THE CORRECTION: DESIGN WITH 9-AWARENESS

Here is the practical application:

If you are designing a building, a room, a garden, a ceremony space—pay attention to angles.

Recognize that every angle you create is part of a larger geometric field, and that field collapses to 9.

Use 9-fractal proportions:

- 3.6 meters, 1.8 meters, 0.9 meters, 0.45 meters
- Not because these are "magic numbers," but because they maintain harmonic resonance with the 360° = 9 foundation

Understand that 90-degree angles (right angles, the basis of most modern architecture) are not arbitrary. 90° = 9. When you build with right angles, you are working with a 9-fractal. The question is whether you are doing so consciously or mechanically.

If you are creating art, music, movement, ritual—recognize that all forms of expression occur within space, and space is 9-encoded.

A dance that moves through 180° turns (9) will feel more complete than arbitrary angles.

A musical composition that uses 9-beat or 18-beat rhythms will resonate more deeply than those based on 8 or 16 (which are powers of 2, not 9-fractals).

A mandala with 9-fold symmetry will feel more coherent than one with 8-fold or 10-fold symmetry, because 9 is the number of completion, the full cycle, the return to Source.

THE UNDENIABLE PROOF

Let us state this clearly:

Geometry itself is proof of the Law of 9.

You do not need to believe in metaphysics. You do not need to accept spiritual teachings. You do not need to trust ancient wisdom.

You only need to do the arithmetic:

- 360° = 9
- All divisions of 360° = 9
- All interior angles of all polygons = multiples of 180° = 9
- All Platonic solids = 9
- All spatial structures are built from angles that sum to 9

This is not theory. This is not interpretation. This is the mathematical structure of three-dimensional reality.

Space is 9-encoded. Geometry is 9-governed. Form is 9-expressed.

And because your body exists in space, because your movements occur through angles, because your architecture shapes the fields you live within, you are constantly interfacing with the Law of 9 whether you recognize it or not.

The question is: Will you work with it consciously (coherence) or remain unconscious of it (distortion)?

THE GEOMETRY OF TRUTH

This completes the proof.

We have shown:

- The Digital Root Wheel places 9 at the center of numerical space
- The Enneagram maps the 9-fold pattern of energy flow
- Prime numbers are surface patterns, not foundational
- The 24-wheel and other shadowmath systems mimic recursion without completing the return
- The Golden Ratio worships Phi (expansion) while neglecting Pi (compression)
- The Living Ratio 9–1 expresses the complete circuit
- Pi + Phi = Psi (the unified field is the sum of both forces)
- 360° and all geometric forms collapse to 9

Geometry is not abstract. Geometry is the spatial expression of the Law of 9.

When you align your designs, your movements, your creations with this law, you work with the grain of reality. You create coherence. You build structures that hold, circulate, and amplify life force.

When you ignore or distort this law, you work against the grain. You create fragmentation. You build structures that leak, stagnate, and eventually collapse.

The choice has always been yours. But now you have the knowledge.

Everything is 9. Everything returns to 9. Everything spirals according to the Law of 9.

This is not belief. This is geometry. This is mathematics. This is the architecture of existence made visible.

The spiral deepens from here.

THE FLOWER OF LIFE: WHERE GEOMETRY BECOMES BIOLOGY

We must now reveal the mechanism that connects abstract geometry to living biology—the bridge that proves geometric patterns are not symbolic representations but actual blueprints of how life organizes itself at the cellular level.

The Flower of Life has been revered for millennia across cultures—carved into Egyptian temples, encoded in sacred architecture, studied by mystics and geometers as a fundamental pattern underlying creation.

But here is what they did not tell you:

The Flower of Life is not a mystical symbol. It is the geometric map of embryonic cell division—the literal pattern of how life multiplies from conception.

This is not interpretation. This is verifiable biological fact.

THE PATTERN OF CREATION

When a human egg is fertilized, it begins as a single cell—one perfect sphere containing the entire blueprint for a human being.

This cell divides.

But here is what mainstream biology does not tell you:

Embryonic cell division does not follow binary doubling (1-2-4-8-16-32). It follows the Law of 9.

The actual observed progression is:

1 → 2 → 4 → 9 → 18 → 36

Let us trace this progression with precision:

1 CELL (Digital Root: 1)

The fertilized egg. One perfect sphere. The totality of potential contained in singular form.

This is Source expressing as unity—the 1 before manifestation.

2 CELLS (Digital Root: 2)

The first division creates two overlapping spheres. This geometric form is called the Vesica Piscis—the "womb" shape, the almond-shaped intersection where two circles overlap.

This is the first stage of the Flower of Life pattern.

The 1 has become 2. Unity has expressed duality. The single consciousness has differentiated into observer and observed, subject and object, electric (1) and magnetic (9) beginning their dance.

4 CELLS (Digital Root: 4)

The two cells divide again, becoming four. The cells arrange themselves in a tetrahedral cluster—the simplest three-dimensional geometric form, the building block of all Platonic solids.

Four is stability. Four corners. Four directions. The square, the foundation.

9 CELLS (Digital Root: 9)

Here is where the progression deviates from binary doubling.

Four cells do not become eight. Four cells become nine.

This is the return to Source within the division process.

9 is the zero-point, the completion, the first return to the wholeness from which division emerged. The cells have reached the first octave of the 9-fold spiral.

This is not taught in biology textbooks. But it is what occurs under direct observation.

18 CELLS (Digital Root: 1+8 = 9)

Nine cells divide, becoming eighteen.

$18 = 9 \times 2$

The digital root remains 9. The pattern is maintaining coherence with Source even as it expands. The second octave of the 9-fold cycle.

36 CELLS (Digital Root: 3+6 = 9)

Eighteen cells divide, becoming thirty-six.

$36 = 9 \times 4 = 6^2$

The digital root remains 9. The pattern has reached the completion point—the stage where the embryonic cluster stops following the pure 9-sequence and begins differentiating into specialized cell types.

36 is the morphological threshold.

After 36 cells, the embryo transitions from symmetrical division (maintaining the Flower of Life geometry) to asymmetrical specialization (cells beginning to differentiate into the three germ layers: ectoderm, mesoderm, endoderm).

This is why the Flower of Life, as traditionally drawn, contains 19 circles (which can represent up to 37 cells when you include the center)—it maps the division pattern up to the point of differentiation.

THE COLLAPSE OF THE BINARY MYTH

Mainstream biology teaches that cells divide by binary fission: 1-2-4-8-16-32-64...

This is shadowmath projected onto biology.

Binary doubling (powers of 2) is a computational model, useful for describing mechanical processes and digital systems. But biological systems are not mechanical. They are conscious, coherent, recursive expressions of the unified field.

Life does not follow binary logic. Life follows the Law of 9.

The progression 1 → 2 → 4 → 9 → 18 → 36 is not arbitrary. It is the geometric expression of how consciousness organizes matter during the transformation from potential (1 fertilized egg) to manifest complexity (36-cell embryo ready for differentiation).

Every step maintains the 9-signature:

- 1 (source)
- 2 (duality)
- 4 (stability)
- 9 (first return to Source)
- 18 (9 × 2, second octave, digital root 9)
- 36 (9 × 4, completion point, digital root 9)

After 36, the cells no longer divide symmetrically. They begin the process of specialization—ectoderm will become skin and nervous system, mesoderm will become muscle and bone, endoderm will become organs and digestive system.

But the foundational geometry—the Flower of Life pattern mapped during the 1-2-4-9-18-36 progression—remains encoded in every cell.

THE MICROSCOPIC PROOF

If you observe human embryonic development under a microscope, you will see this pattern with your own eyes.

The fertilized egg (1 cell) divides into two cells (Vesica Piscis formation).

Two cells divide into four (tetrahedral cluster).

Four cells divide into nine (not eight—this is the critical deviation from binary doubling).

Nine cells divide into eighteen (maintaining 9-coherence).

Eighteen cells divide into thirty-six (the completion of the symmetrical division phase).

At each stage, the cellular arrangement mirrors the Flower of Life geometry, and every stage after the fourth division maintains the digital root of 9.

This is not coincidence. This is not symbolic correlation.

This is the actual spatial organization of how life divides, expands, and maintains coherence during the transformation from single cell to complex organism.

The Flower of Life drawn on temple walls, tattooed on skin, meditated upon in sacred geometry practice—all of these are representations of the biological truth occurring inside every embryo at the moment of conception.

Geometry is not separate from biology. Geometry IS biology made visible.

THE HYDROGEN CONNECTION

There is another layer to this revelation.

The same geometric principles that govern embryonic cell division also govern molecular coherence at the atomic level.

Water (H_2O) is the medium in which all biological processes occur. Your body is approximately 65% water. Every cell, every tissue, every organ operates within an aqueous environment.

But not all water is structured the same way.

Regular water molecules arrange themselves in a pentagonal (5-sided) geometric pattern when observed at the molecular level.

However, when molecular hydrogen (H_2) is introduced into water, the structure shifts.

Hydrogen-infused water organizes into hexagonal (6-sided) geometric clusters.

Hexagonal water is more coherent than pentagonal water.

Hexagonal structure allows for:

- Greater molecular stability
- More efficient energy transfer
- Reduced oxidative stress
- Enhanced cellular hydration
- Decreased inflammation

This is the same hexagonal geometry we see in the Seed of Life, in the Flower of Life, in the efficient packing of cells during embryogenesis.

The pattern repeats:

- At the molecular level (hexagonal water clusters)
- At the cellular level (hexagonal arrangement of dividing cells)
- At the geometric level (hexagonal symmetry in the Flower of Life)

One pattern. One principle. One law operating at multiple scales.

THE ELEMENTS OF LIFE

Your body is not made of arbitrary chemicals. It is made of specific elements in specific proportions, organized according to geometric law.

The primary elements by mass:

- Oxygen (O): 65%
- Carbon (C): 18.5%
- Hydrogen (H): 9.5%
- Nitrogen (N): 3%

Notice something:

Hydrogen = 9.5% → 9+5 = 14 → 1+4 = 5

But when hydrogen exists in molecular form (H_2), it creates hexagonal coherence in water.

Hexagonal = 6-sided

And 6 is one of the stable positions in the 3-6-9 triangle—the Law of Three operating at the molecular level.

Oxygen, carbon, and hydrogen combine to form the organic compounds that make up your tissues. These compounds organize into cells. Cells organize into tissues. Tissues organize into organs.

At every level of organization, geometric law governs the structure.

And the geometry that appears again and again—in water molecules, in cell division, in the Flower of Life, in sacred architecture—is the same geometry we have been revealing throughout this chapter:

The 9-fold recursive pattern of compression and expansion, operating through hexagonal and toroidal symmetry.

MOLECULAR HYDROGEN: THE ANTIOXIDANT OF COHERENCE

Hydrogen is the smallest, lightest, most abundant element in the universe.

It is also the most powerful antioxidant available to biological systems.

An antioxidant is a molecule that neutralizes free radicals—unstable atoms or molecules with unpaired electrons that damage cells through oxidative stress.

Remember what we said earlier about Phi without Pi creating free radicals at the mathematical level?

The same principle applies at the biochemical level.

Free radicals are molecules that have been fragmented—separated from their stable electron pairs, unable to return to coherence on their own.

Molecular hydrogen (H_2) provides the missing electron, allowing the free radical to stabilize and return to a neutral state.

This is compression. This is the return mechanism. This is Pi (9) balancing Phi (1).

When molecular hydrogen is present in your cells:

- Oxidative stress decreases (free radicals neutralized)
- Inflammation decreases (cellular damage reduced)
- Mitochondrial function improves (energy production optimized)

- Cellular coherence increases (geometry returns to hexagonal stability)

Your mitochondria—the energy powerhouses of your cells—are where this process is most critical.

And here is the connection that ties everything together:

Mitochondria contain melanin.

THE MELANIN-MITOCHONDRIA-GEOMETRY CONNECTION

Mitochondria are not just energy generators. They are melanin-rich organelles that interface between nutrition, oxygen metabolism, and electromagnetic field dynamics.

Mitochondria divide through binary fission—one mitochondrion splits into two, two become four, four become eight.

This is the same geometric division pattern as embryonic cells.

The Flower of Life pattern is not only the map of how embryos develop—it is also the map of how mitochondria replicate within your cells.

Every time a mitochondrion divides, it follows the same geometric law: efficient packing of spherical volumes in three-dimensional space, creating overlapping circles when viewed in two dimensions.

The Flower of Life is the fractal pattern of mitochondrial division.

And because mitochondria contain melanin—the same melanin concentrated in your substantia nigra, your locus coeruleus, your pineal gland—this geometric pattern is directly linked to consciousness, energy production, and biofield coherence.

When your mitochondria function optimally (high ATP production, low oxidative stress, coherent geometry), your melanin field operates at peak efficiency.

When your mitochondria are damaged (free radical accumulation, fragmented geometry, incoherent structure), your melanin field becomes distorted, your biofield leaks energy, and your consciousness fragments.

Geometry at the cellular level determines consciousness at the experiential level.

THE TOROIDAL FIELD AT ALL SCALES

Look at the images showing molecular hydrogen in the context of cosmic structure.

You see a nebula—a massive cloud of gas and dust in space, glowing with electromagnetic radiation.

You see a zoomed-in view showing molecular hydrogen (H_2) within that cloud.

The same toroidal, spiraling geometry at both scales.

This is not artistic license. This is observational fact.

The toroidal field—the donut-shaped electromagnetic pattern with flow spiraling in at the poles and radiating out at the equator—appears at every scale of existence:

- Atomic (electron orbital patterns)

- Molecular (hydrogen bonding in water)
- Cellular (mitochondrial division, embryonic development)
- Organic (human biofield, heart coherence)
- Planetary (Earth's magnetosphere)
- Stellar (solar magnetic field)
- Galactic (spiral galaxy structure)
- Cosmic (universal expansion and contraction)

One pattern. One geometry. One law.

And that pattern is governed by the same 9-fold recursion we have been demonstrating throughout this chapter.

The toroidal field operates through:

- Compression at the poles (9/Pi/magnetic/feminine)
- Expansion at the equator (1/Phi/electric/masculine)
- Recursive flow connecting both (the complete circuit, the Living Ratio: 9–1)

THE COLLAPSE: SACRED GEOMETRY IS BIOLOGY

Here is what must be collapsed:

The idea that the Flower of Life is a mystical symbol, a decorative pattern, an abstract representation of divine harmony.

The Flower of Life is the geometric blueprint of embryonic cell division and mitochondrial replication.

It works because it maps the actual spatial organization of how spherical cells arrange themselves during division to maintain contact, share resources, and create multicellular coherence.

The idea that sacred geometry is separate from biology, that one is spiritual and the other is scientific.

Geometry IS biology. Biology IS geometry expressing through organic matter.

Every cell division, every molecular bond, every tissue structure follows geometric law. When the geometry is coherent (hexagonal water, efficient cell packing, toroidal field flow), the organism thrives. When the geometry is distorted (pentagonal water, fragmented cells, disrupted field), the organism degenerates.

The idea that hydrogen is just a chemical element with no special significance beyond its reactivity.

Hydrogen is the coherence molecule—the smallest unit capable of neutralizing free radicals, restructuring water into hexagonal coherence, and enabling the geometric patterns necessary for life.

Molecular hydrogen (H_2) is not a supplement. It is the return mechanism operating at the molecular level—the Pi force that allows fragmented molecules (free radicals) to collapse back into stable, coherent structure.

PRACTICAL IMPLICATION: HONOR THE GEOMETRY OF LIFE

If the Flower of Life represents the biological pattern of cell division, and if hexagonal water structure is more coherent than pentagonal, and if molecular hydrogen enables this coherence, then the practical application is clear:

Support your cellular geometry.

Hydrate with structured water (spring water, hydrogen-infused water, water exposed to natural electromagnetic fields).

Reduce oxidative stress through antioxidant-rich nutrition (melanin-supporting foods, electron-dense compounds, living plant matter).

Protect mitochondrial function (avoid mitochondrial toxins, support ATP production, maintain melanin integrity).

Recognize that every breath you take, every movement you make, every thought you think occurs within a geometric field that either supports coherence or generates distortion.

The Flower of Life is not something to worship. It is something to embody.

You are the Flower of Life in motion—a fractal expression of the same geometric pattern that organized your cells from conception and continues to govern your mitochondrial energy production every moment of your existence.

THE BRIDGE IS COMPLETE

We have now demonstrated the unbroken connection:

Geometry (the Flower of Life pattern) → Biology (embryonic cell division, mitochondrial replication) → Chemistry (molecular hydrogen, hexagonal water) → Physics (toroidal field dynamics) → Consciousness (melanin-mediated biofield coherence).

This is not multidisciplinary synthesis. This is the recognition that there are no separate disciplines.

There is one reality, expressing through multiple scales, governed by one law: the Law of 9.

The Flower of Life is the geometric signature of this law made visible through cellular biology.

The 360° circle is the geometric signature of this law made visible through spatial angles.

The 9–1 Living Ratio is the mathematical signature of this law made visible through numerical recursion.

Pi + Phi = Psi is the energetic signature of this law made visible through the unified field.

All of it—geometry, biology, mathematics, physics, consciousness—is the same pattern, the same truth, the same spiral.

The spiral deepens from here.

THE UNIFIED FIELD: CHARGE, WAVES, AND THE ARCHITECTURE OF EXISTENCE

We have demonstrated the geometric patterns. We have revealed the biological mechanisms. We have exposed the mathematical distortions.

Now we must make explicit the unifying principle that connects all of these expressions into a single, coherent understanding of reality.

Everything is charge.

Not metaphorically. Not symbolically. Literally.

Every phenomenon you experience—light, sound, matter, thought, emotion, sensation—is an expression of charge oscillating within the Aetheric field.

Let us define the terms with precision:

THE AETHERIC FIELD (PSI)

The Aetheric field is the unified consciousness substrate—the medium, the ocean of potential, the field from which all forms emerge and to which all forms return.

It is not empty space. It is not void. It is the totality of existence in its unmanifest state—pure potential awaiting the organizing principle of charge.

In our earlier equation, we named this Psi:

Psi = Pi + Phi

The unified field is the sum of both forces—magnetic compression (Pi/9) and electric projection (Phi/1)—held in dynamic balance.

The Aetheric field is not separate from you. You are not "in" the field like a fish in water. You are a localized pattern within the field—a standing wave, a vortex of organized charge that has achieved sufficient coherence to sustain self-awareness.

You are the field experiencing itself.

CHARGE: THE TWO FUNDAMENTAL FORCES

Charge is not a property of particles. Charge IS the fundamental reality.

There are two expressions of charge:

9 = MAGNETIC CHARGE (Pi/Feminine/Receptive/Compression)

Magnetic charge is the force that PULLS. It is attraction, gravity, the return to center, the compression inward.

This is the 9-force—the field that draws all things back to Source, the womb from which forms emerge and to which forms collapse.

Magnetic charge governs:

- Gravity (the pull toward planetary centers, stellar cores, galactic black holes)
- Absorption (cells taking in nutrients, lungs drawing in breath, consciousness receiving information)
- Memory (the capacity to hold pattern, to retain structure across time)
- The dark, the receptive, the void, the stillness

1 = ELECTRIC CHARGE (Phi/Masculine/Projective/Expansion)

Electric charge is the force that RADIATES. It is repulsion, projection, the impulse outward, the expansion from center.

This is the 1-force—the spark that initiates movement, the seed that projects into form, the light that illuminates the dark.

Electric charge governs:

- Radiation (heat, light, electromagnetic waves propagating outward)
- Expression (neurons firing, muscles contracting, consciousness projecting intention)
- Action (the capacity to move, to change, to manifest)
- The light, the projective, the visible, the dynamic

All phenomena are expressions of these two charges in dynamic interaction.

When magnetic charge (9) and electric charge (1) interact in recursive circuit, they create everything you experience as “reality.”

WAVES: THE OSCILLATION OF CHARGE

A wave is not a thing traveling through space.

A wave is the oscillation of charge within the Aetheric field.

When electric charge (1) projects outward and magnetic charge (9) pulls it back inward, the rhythmic interaction between these two forces creates a wave pattern.

Frequency = How fast the 9–1 cycle completes. High frequency means rapid oscillation (light, consciousness, refined states). Low frequency means slow oscillation (dense matter, heavy elements, gross states).

Amplitude = How far the charges separate before returning. Large amplitude means intense separation between expansion (1) and compression (9). Small amplitude means subtle interaction.

Wavelength = The spatial distance covered during one complete 9–1 cycle.

All waves—electromagnetic waves (light, radio, gamma rays), mechanical waves (sound, seismic vibrations), matter waves (the oscillation patterns that create the appearance of solid objects)—are expressions of charge oscillating at different frequencies within the Aetheric field.

This is what creates:

Vibration - Matter “vibrating” is charge cycling at specific frequencies. High vibration = rapid 9–1 oscillation. Low vibration = slow 9–1 oscillation. When people speak of “raising your vibration,” they are (often unknowingly) describing the process of increasing the frequency of charge oscillation in your biofield.

Heat - The friction of charge collision. When expansion (1) meets resistance, thermal energy is generated. When compression (9) squeezes matter into tighter configurations, heat is released. Temperature is a measure of the intensity of charge interaction at the molecular level.

Light - The visible expression of electromagnetic charge waves. Light is electric field (1) perpendicular to magnetic field (9), oscillating together as they propagate through space. Color is determined by frequency—red light oscillates slower, violet light oscillates faster. All light is the same phenomenon (charge oscillation) at different frequencies.

Sound - Mechanical waves of charge compression (9) and rarefaction (1) traveling through a medium (air, water, solid matter). Your ear detects these oscillations and your brain interprets them as sound. Music is organized charge oscillation. Noise is disorganized charge oscillation.

PARTICLES: THE COLLAPSE OF THE WAVE FUNCTION

Here is where mainstream physics becomes incoherent.

Quantum mechanics tells you that light (and all matter) exhibits "wave-particle duality"—sometimes behaving as a wave, sometimes as a particle, depending on how you observe it.

This creates paradoxes: How can something be both wave and particle? How does observation change physical reality? What is the mechanism?

The answer is simple when you understand charge and consciousness:

A particle is what happens when a wave function collapses into a singularity through observation.

Let us break this down:

Before observation, charge oscillates as a distributed wave pattern throughout the Aetheric field. The "electron" is not a tiny ball orbiting the nucleus—it is a probability wave, a field of potential locations where charge density is high.

Observation = consciousness focusing attention on a specific point in the field.

When you (or any conscious system—because consciousness exists at all scales, not just human) focus attention on the wave pattern, you introduce a 9-compression force at the point of focus.

The distributed wave (1-expansion across space) collapses (9-return to center) into a localized point.

This creates the appearance of a particle—a discrete, localized "thing" with definite position.

But it is not a solid object. It is a standing wave—a tightly compressed recursive vortex of charge that has been collapsed into a specific location by the act of observation.

This is why observation affects quantum experiments.

The observer is not passive. The observer is introducing a compression force (9/attention/focus) into the field, causing the wave to collapse into particle form at the moment of measurement.

Matter is collapsed consciousness.

Every "solid" object you perceive is actually a standing wave of charge that has been collapsed into stable, recursive form through continuous observation—either by your consciousness or by the collective consciousness field that maintains consensus reality.

THE TOROIDAL FIELD: SELF-SUSTAINING CHARGE RECURSION

When the oscillation of charge becomes recursive and self-referential—when the 9-force (compression) and the 1-force (expansion) complete a full circuit and feed back into each other—something extraordinary happens:

The field becomes self-sustaining.

This creates the toroidal geometry—the donut or doughnut shape that appears at every scale of organized matter and energy.

How the torus forms:

1. Magnetic charge (9) pulls inward through the poles (top and bottom of the torus)
1. Electric charge (1) radiates outward through the equator (around the ring)

1. The charges complete the circuit - What expands outward (1) is pulled back inward (9). What compresses inward (9) is radiated back outward (1).
1. The field sustains itself - No external input is required. The recursion is stable.

This is the geometry of:

- Atoms (charge vortex with compression at nucleus, expansion in electron field)
- Cells (bioelectric field with compression at nucleus, expansion at membrane)
- Hearts (electromagnetic field generated by cardiac rhythm, measurable several feet from body)
- Planets (magnetosphere with compression at poles, radiation at equator)
- Stars (fusion core compressing, radiation expanding outward, solar wind circulating)
- Galaxies (black hole at center compressing, spiral arms expanding, matter cycling)
- Consciousness (your melanin field is toroidal—compressing information inward through focus, radiating intention outward through action)

The torus is the spatial expression of the Living Ratio: 9–1.

It is not static. It is dynamic flow—compression and expansion in continuous recursion, creating a stable standing wave pattern that can persist across time.

This is what implosion looks like when it becomes self-sustaining.

THE ATOM: A TOROIDAL CHARGE VORTEX

Now we can redefine the atom with precision.

An atom is not made of particles. An atom is a self-sustaining toroidal charge vortex.

Mainstream physics teaches that atoms consist of:

- Protons (positively charged particles in the nucleus)
- Neutrons (neutral particles in the nucleus)
- Electrons (negatively charged particles orbiting the nucleus)

This is shadowmath. It describes behavior without understanding essence.

Here is the truth:

An atom is a stable region of the Aetheric field where charge has achieved self-sustaining recursive oscillation in toroidal geometry.

The nucleus is the 9-compression point—the magnetic attractor at the center of the vortex, the "black hole" of the atom where charge density is highest.

The electron "cloud" is the 1-expansion field—the electric charge radiating outward from the nucleus but continuously pulled back inward by magnetic attraction, creating standing wave patterns at specific distances.

The torus is the complete circuit—charge flowing inward at the poles, radiating outward at the equator, cycling eternally in self-sustaining recursion.

Different elements are different frequencies of the 9–1 oscillation:

- Hydrogen = simplest, fastest oscillation (one proton, minimal compression, most basic torus)
- Heavier elements = more complex, slower oscillations (more charge density at nucleus, more elaborate toroidal

geometry)
- Gold = extremely stable 9–1 recursion (which is why gold does not corrode or decay—the charge circuit is perfectly balanced)
- Radioactive elements = unstable 9–1 recursion (the toroidal field cannot sustain coherence, so it fragments and decays)

"Electrons" are not particles orbiting like planets around the sun.

Electrons are standing wave patterns in the toroidal field—specific nodes where the oscillation of charge creates stable, repeating interference patterns.

The "orbital shells" (s, p, d, f in quantum mechanics) are different geometric configurations where the standing wave stabilizes based on the frequency of the 9–1 charge interaction.

THE ATOMIC TRIFECTA: PROTONS, NEUTRONS, ELECTRONS

Within the atom, we see the Law of Three operating with perfect clarity.

Let us redefine the so-called "subatomic particles" by their function, not by the misleading names given to them:

PROTONS = The 1-Force (Electric/Active/Masculine/Phi/Projective)

Function: Projection and expression of charge from the nucleus outward.

Protons are the electric impulse, the initiating spark, the active force that radiates from the compression point at the center of the atom.

They carry positive charge, meaning they are GIVING energy—projecting outward, creating the electric field that organizes the toroidal structure.

Without protons, there is no projection, no outward force, no manifestation of the atom into observable reality.

ELECTRONS = The 9-Force (Magnetic/Passive/Feminine/Pi/Receptive)

Function: Reception and circulation of charge in the field surrounding the nucleus.

Electrons are the magnetic response, the orbital pattern, the receptive force that circles the nucleus in standing wave configurations.

They carry negative charge, meaning they are RECEIVING energy—drawn inward by magnetic attraction, creating the field geometry that contains and organizes the atom.

Without electrons, there is no reception, no containment, no stable structure to hold the atomic pattern across time.

NEUTRONS = The Psi-Force (Neutral/Reconciling/Aether/Stabilizing)

Function: Stabilization and integration of the charge balance within the nucleus.

Neutrons are the mediating force, neither positive nor negative, that holds the nucleus stable by balancing the repulsive force between protons (which would otherwise push each other apart due to their shared positive charge).

Without neutrons (or with too few or too many), the nucleus becomes unstable, the 9–1 balance is disrupted, and the atom decays radioactively.

This is the Law of Three at the atomic level:

- Active force (Protons/1/Electric/Phi)
- Passive force (Electrons/9/Magnetic/Pi)
- Reconciling force (Neutrons/Psi/Neutral/Balance)

The atom is Psi (unified field) expressed as the dynamic interaction of Pi (magnetic/9) and Phi (electric/1) in self-sustaining toroidal recursion.

THE FRACTAL HARMONIC: ATOM TO HUMAN BODY

Now we demonstrate the fractal principle in its most profound expression:

The same pattern that organizes the atom also organizes the human body.

Not symbolically. Not metaphorically. Structurally and functionally.

ATOM → HUMAN BODY (SAME CIRCUIT, DIFFERENT SCALE):

PROTONS (1/Electric/Active) = BRAIN

In the atom, protons project charge from the nucleus.

In the human body, the brain projects intention, decision, and will from the command center of the nervous system.

Function: Initiation and direction.

The brain generates electric impulses that fire through neurons, creating the commands that organize movement, thought, and conscious action.

The brain is the 1-force in the human system—the active, projective, electric principle that initiates all voluntary activity.

When the brain is dominant without balance, you experience overthinking, anxiety, disconnection from body and intuition, excessive mental activity without grounding.

ELECTRONS (9/Magnetic/Receptive) = GUT

In the atom, electrons orbit the nucleus in receptive field patterns.

In the human body, the gut receives information from the environment—food, microbiome, emotional state—and integrates it into the biological system.

Function: Reception and integration.

The gut contains the enteric nervous system (often called the "second brain"), which operates independently of the brain and processes vast amounts of sensory and emotional data.

The gut is where you "feel" things—intuition, instinct, the somatic knowing that arises before thought. This is magnetic knowing, field-based perception, the 9-force in operation.

When the gut is disconnected, you experience digestive dysfunction, loss of intuition, inability to process emotion, disconnection from instinctive wisdom.

NEUTRONS (Psi/Neutral/Reconciling) = HEART

In the atom, neutrons stabilize the nucleus by balancing opposing forces.

In the human body, the heart mediates between brain (electric/1) and gut (magnetic/9), generating the coherence field that allows both to function in harmony.

Function: Coherence and balance.

The heart generates the strongest electromagnetic field in the body—a toroidal field measurable several feet beyond the physical body.

When the heart is in coherence, the brain and gut synchronize. The electric impulses from the brain and the magnetic intuition from the gut align through the heart's mediating field.

This is why "heart coherence" is not poetic language—it is measurable biophysics. The heart's electromagnetic field literally organizes the charge distribution throughout your entire body.

When the heart is closed or incoherent, the brain and gut operate in conflict. Thought and feeling fragment. Logic and intuition oppose each other. The circuit breaks.

THE COMPLETE CIRCUIT:

In the atom:
Protons (1) project charge → Neutrons (Psi) stabilize the field → Electrons (9) receive and circulate → Complete toroidal circuit → Self-sustaining stable atom

In the human:
Brain (1) projects intention → Heart (Psi) generates coherence → Gut (9) receives and integrates → Complete toroidal biofield → Self-sustaining health and consciousness

When the circuit is coherent:

- Atom = stable element, does not decay, maintains structure across time
- Human = health, clarity, intuition aligned with reason, connection to Source, sustained energy

When the circuit is broken:

- Atom = radioactive decay, fragmentation, loss of coherence, eventual dissolution
- Human = disease, confusion, fragmented thinking, disconnection from intuition, energy depletion, systemic breakdown

This is not analogy. This is fractal recursion.

The same charge dynamics that organize atoms also organize your body, your consciousness, your biofield.

You are a toroidal charge vortex, operating through the same 9–1 circuit as every atom in existence.

IMPLOSION, EXPLOSION, AND FUSION

Now we can define the processes that create and transform matter with precision.

IMPLOSION = Compression (9-force dominant)

Implosion is the inward spiral—magnetic charge pulling toward center with increasing intensity.

When compression reaches a critical threshold, information does not fragment—it integrates. Patterns do not scatter—they cohere. Energy does not dissipate—it concentrates.

This is how:

- Black holes form (gravitational implosion at stellar scale)
- Consciousness focuses (attention imploding on a single point)
- Memory consolidates (information compressing into stable neural patterns)
- DNA coils (genetic information folding into compact, fractal structure)

Implosion is the 9-return—the collapse back to Source, the integration phase of the cycle.

EXPLOSION = Expansion (1-force dominant)

Explosion is the outward burst—electric charge radiating from center with maximal force.

When expansion is unleashed, energy propagates outward, patterns disperse, information distributes across space.

This is how:

- Stars radiate (fusion energy exploding outward as light and heat)
- Neurons fire (electric impulse propagating through axons)
- Ideas express (thought projecting into speech, writing, action)
- Seeds germinate (genetic blueprint unfolding into manifest form)

Explosion is the 1-projection—the manifestation from Source, the expression phase of the cycle.

FUSION = Implosion so intense that boundaries dissolve

Fusion occurs when compression (9) becomes so extreme that the boundaries separating individual forms dissolve entirely.

In stars, hydrogen atoms are compressed so intensely that their nuclei overcome electromagnetic repulsion and merge, creating helium and releasing massive energy.

In consciousness, fusion occurs when the sense of separation between self and other, between observer and observed, between individual and field collapses—and you recognize yourself as the unified field experiencing itself through localized form.

Your melanin field performs fusion at the quantum level continuously.

It compresses information from the Aetheric field (Psi) into coherent patterns that can be integrated into your nervous system, your biofield, your conscious awareness.

This is why melanin-rich individuals often report heightened intuition, field sensitivity, connection to ancestral memory—the melanin network is performing fusion, collapsing distributed field information into localized knowing.

THE TEMPORAL CYCLE: 40.5 DAYS AND THE HIDDEN SCRIPTURE

The fractal recursion of the Law of 9 does not express only through spatial geometry (the 1-2-4-9-18-36 cell division pattern). It also expresses through temporal cycles.

There is a specific time period encoded in both biological development and sacred scripture—a period that has been deliberately obscured by the removal of a single decimal point.

The period is 40.5 days.

THE BIOLOGICAL CYCLE

Human embryonic development follows a precise timeline governed by 9-fractals.

At approximately 40.5 days after conception, the embryo reaches a critical transformation threshold:

- Major organ systems have begun differentiation
- The heart has started beating (around day 21-22, which is ~3 weeks, digital root 3)
- Neural tube formation is complete (basis of brain and spinal cord)
- Limb buds have appeared
- The embryo transitions from the purely geometric division phase into the morphological development phase

This 40.5-day period is the completion of the first major embryonic cycle.

$40.5 \rightarrow 4+0+5=9$

But there is more:

40.5 days is exactly half of an 81-day cycle.

81 days $\rightarrow 8+1=9$

$81 \div 2 = 40.5$

The 81-day cycle (9) represents a complete turn. The 40.5-day cycle (also 9) represents the half-turn—the transformation point, the shift, the moment where the process reaches its first major milestone before continuing to completion.

THE MITOCHONDRIAL CYCLE

Mitochondria—the melanin-containing energy generators of your cells—also follow this temporal pattern.

Mitochondrial replication and turnover occur in cycles that align with 9-fractals. The complete renewal cycle for mitochondrial populations in various tissues ranges from approximately 2-3 weeks (18-21 days, digital root 9) to several months depending on cell type and metabolic demand.

But the critical transformation period—the time it takes for a mitochondrion to complete its biogenesis, mature, and integrate into the cellular energy network—is approximately 40.5 days.

The fetus and the mitochondria follow the same collapse pattern.

This is not coincidence. This is the same recursive principle operating at two different scales:

- Cellular scale: 1-2-4-9-18-36 cells (spatial recursion)
- Temporal scale: 40.5 days (temporal recursion)

Both maintain the 9-signature. Both express the Law of 9 through their developmental cycles.

THE SCRIPTURAL CODE: 40 DAYS AND 40 NIGHTS

Now we arrive at the hidden knowledge encoded in sacred texts across cultures.

The number 40 appears repeatedly in scripture:

- Moses on Mount Sinai: 40 days and 40 nights receiving the commandments
- Jesus in the wilderness: 40 days and 40 nights of fasting and temptation
- Noah's flood: 40 days and 40 nights of rain
- Israelites in the desert: 40 years of wandering
- Jesus post-resurrection: 40 days before ascension
- Jonah's warning to Nineveh: 40 days until destruction
- Elijah's journey to Mount Horeb: 40 days and 40 nights

The pattern is unmistakable. 40 appears as the period of transformation, purification, testing, revelation, and transition.

But here is what they did not tell you:

The actual period is not 40 days. It is 40.5 days.

They removed the decimal.

Why?

Because 40.5 reduces to 9 (4 + 0 + 5 = 9), revealing the Law of 9 encoded in scripture.

If they left the decimal in place, the pattern would be obvious:

- 40.5 days = 9 (transformation cycle)
- Half of 81 days = 9 (complete cycle)
- The embryonic milestone = 9
- The mitochondrial maturation = 9

The removal of the .5 was deliberate obfuscation.

By reducing 40.5 to 40, the 9-signature is hidden. The number 40 reduces to 4 (4 + 0 = 4), which does not carry the same recursive completion as 9.

But the truth remains encoded in the biology:

40.5 days is the transformation period—the time it takes for a system (embryo, mitochondria, consciousness undergoing purification/initiation) to complete the first major phase of recursive development and prepare for the next octave.

This is why the ancients knew: 40 days and 40 nights (with the hidden .5) is the period required for biological, psychological, and spiritual transformation to stabilize.

Fast for 40.5 days, and your mitochondria regenerate.

Isolate in meditation for 40.5 days, and your neural pathways restructure.

Undergo initiation for 40.5 days, and your consciousness integrates the new frequency.

The decimal was removed to hide the 9. But the biology does not lie.

THE UNIFIED FIELD MADE EXPLICIT

Let us now state the complete picture with absolute clarity:

1. The Aetheric Field (Psi) is the unified consciousness substrate—the ocean of potential from which all forms emerge.

2. Charge (9–1, Pi-Phi) is the fundamental dynamic within the field—magnetic compression (9) and electric projection (1) in recursive interaction.

3. Waves are oscillations of charge within the field—the rhythmic pulsing of 9 and 1 creating frequency, amplitude, and propagation patterns.

4. Particles are collapsed wave functions—standing waves of charge compressed into localized form through the act of observation (consciousness focusing attention).

5. The Toroidal Field is self-sustaining charge recursion—when 9 and 1 complete a full circuit, creating stable vortex geometry that persists across time.

6. Atoms are toroidal charge vortices—self-organized regions of the field where the 9–1 recursion has achieved stable frequency, creating the appearance of matter.

7. Biology is geometry expressing through organic form—the Flower of Life (cell division), the torus (biofield), the 9–1 circuit (brain-heart-gut) all operating according to the same charge dynamics.

8. Consciousness is the capacity of the field to observe itself—and through observation, to collapse distributed potential (wave) into manifest form (particle).

9. You are a fractal of the unified field—Psi expressing as the dynamic dance of Pi (9/magnetic/receptive) and Phi (1/electric/projective) in toroidal, self-sustaining recursion.

Everything is charge. Everything is 9–1. Everything is the unified field experiencing itself through infinite fractal expressions.

This is not theory. This is the architecture of existence made explicit.

The geometry chapter has shown you the spatial patterns. The biology section has shown you the organic manifestations. The charge synthesis has shown you the mechanism.

Now you know how reality operates.

But there is one final revelation that must be made explicit—the ultimate key that explains why the spiral never closes, why evolution never stops, why consciousness continuously ascends.

The secret is 369°.

369°: THE SPIRAL THAT NEVER DIES

You have been taught that a full circle is 360 degrees.

This is correct—for a circle.

But a circle is a closed loop. A circle is a cage. A circle is repetition without evolution.

A circle is death.

If you rotate exactly 360°, you return to precisely where you started. No growth. No elevation. No transformation. You complete the cycle and find yourself at the beginning again, trapped in eternal recurrence.

This is the Ouroboros eating its own tail—the snake devouring itself, endlessly cycling through the same pattern with no escape, no transcendence, no ascension.

360° is completion within a plane. It is the two-dimensional trap.

But life is not two-dimensional. Consciousness is not flat. Evolution is not circular.

Life is a spiral. And a spiral requires 369°.

THE EXTRA 9°: THE BREATH OF SOURCE

To create a spiral—to achieve evolution, growth, ascension—you cannot stop at 360°.

You must overshoot the starting point. You must add the extra turn.

360° (the circle) + 9° (the pulse) = 369° (the spiral)

This extra 9° is not arbitrary. It is the signature of Source, the breath of the unified field, the shift that prevents stagnation.

When you spiral 369°, you do not return to the exact point where you began. You return to the same angular position, but elevated along the vertical axis.

You arrive at the beginning of the next octave.

This is how the spiral ascends:

- First turn: 0° → 369° (elevated to level 1)
- Second turn: 369° → 738° (2 × 369°, elevated to level 2)
- Third turn: 738° → 1107° (3 × 369°, elevated to level 3)

At each turn, you complete a full rotation plus the extra 9°. This 9° offset creates the gap—the space between levels—that allows the spiral to rise without colliding with itself.

Without the extra 9°, the spiral would flatten into a circle. With the extra 9°, the spiral ascends infinitely.

This is the mechanism of collapse recursion made spatial:

You return to the center (360°/9), but you arrive with +9° of new information, new experience, new coherence—and this elevation propels you to the next level of the spiral.

TESLA'S 3-6-9: THE KEY TO THE UNIVERSE

Nikola Tesla famously said: "If you only knew the magnificence of the 3, 6, and 9, you would have a key to the universe."

This is not mystical poetry. This is precise geometric truth.

3-6-9 is the sequence of the spiral:

3 = The Triangle

The Law of Three. The stable structure. The foundational geometry.

Three points define a plane. Three forces (active, passive, reconciling) create manifestation. Three (Phi, Pi, Psi) generate the unified field.

Without 3, there is no stability. Without the triangle, the spiral has no framework to organize around.

6 = The Hexagon

The Law of efficient packing. Coherent geometry. Harmonic resonance.

Six-fold symmetry appears in honeycomb, in snowflakes, in the Seed of Life, in the hexagonal structure of coherent water, in the benzene ring of carbon compounds.

Without 6, there is no coherence. Without the hexagon, the spiral has no pattern to sustain itself.

9 = The Shift

The Law of completion and transcendence. The return to Source elevated to the next octave.

Nine is the zero-point, the completion of the cycle, and simultaneously the initiation of the next cycle at a higher frequency.

Without 9, there is no evolution. Without the extra turn, the spiral collapses into a circle—repetition without growth, cycling without ascension.

3-6-9 is not three separate numbers. It is the sequential unfolding of the spiral:

1. 3 provides the stable framework (the structure within which the spiral can organize)
1. 6 provides the coherent pattern (the harmonic template the spiral follows)
1. 9 provides the evolutionary shift (the extra turn that prevents stagnation and enables ascension)

This is the God Calculation:

1 (Pulse/Phi/Electric) → 360° (Circle/Pi/Magnetic) → +9° (Shift/Spirit/Transcendence) → 369° (Spiral/Next Octave)

THE FRACTAL RECURSION OF 4.5 AND 40.5

Now the connection becomes undeniable.

4.5 → 4 + 5 = 9

40.5 → 4 + 0 + 5 = 9

Both are expressions of the half-cycle with the extra turn built in.

Think about it:

- A full circle = 360° (but this traps you in repetition)
- A half-circle = 180° (9)
- A quarter-circle = 90° (9)
- An eighth-circle = 45° (9)

45° is the eighth of a circle, and 4.5 is 45 expressed decimally. Both reduce to 9.

This is the recurring pattern:

When you measure cycles in terms of 9-fractals (45°, 90°, 180°, 360°), every division maintains the 9-signature.

But to create the SPIRAL (not just the circle), you must add the extra 9° at each complete turn:

- 360° + 9° = 369° (first spiral turn)
- 369° - 360° = 9° (the offset that creates vertical elevation)

And in biological time:

- 81 days (9) = full cycle
- 40.5 days (9) = half-cycle transformation point
- The .5 (half-degree, half-day) = the extra turn encoded in the rhythm

The embryo doesn't develop in a flat circle. It develops in a spiral.

The cells divide: 1 → 2 → 4 → 9 → 18 → 36

The timing aligns: conception → 40.5 days (9) → 81 days (9) → continuing in 9-fractals

Each phase completes and elevates. Each cycle returns to 9 but at a higher octave.

THE SPIRAL VS. THE CIRCLE: LIFE VS. DEATH

Let us make this absolutely clear:

360° = CIRCLE = CLOSED LOOP = REPETITION = STAGNATION = DEATH

Characteristics of the circle:

- Returns to exact starting point
- No vertical elevation

- Cycles endlessly without growth
- Trapped in two-dimensional plane
- Consciousness experiences Groundhog Day—same patterns, same lessons, same limitations repeating infinitely

369° = SPIRAL = OPEN EVOLUTION = TRANSFORMATION = ASCENSION = LIFE

Characteristics of the spiral:

- Returns to same angular position but elevated vertically
- Each turn brings new perspective, new information, new coherence
- Ascends continuously without limit
- Operates in three-dimensional (and higher-dimensional) space
- Consciousness experiences growth—lessons integrated, patterns transcended, limitations dissolved with each cycle

The 9° difference is the difference between eternal imprisonment and infinite liberation.

THE MECHANISM OF COLLAPSE RECURSION

This is what collapse recursion looks like when expressed as geometric motion:

You spiral down into compression (9/Pi/magnetic pull toward center) → You reach the zero-point (the 360° completion) → You add the extra 9° (the breath of Source, the integration of new information) → You spiral up into expansion (1/Phi/electric projection into new octave)

This is not linear ascension. This is not "leaving the body behind" to become pure light.

This is recursive elevation—continuously spiraling through compression and expansion, death and rebirth, integration and expression, each cycle completing at 369° and beginning the next turn at a higher frequency.

Your DNA spirals (double helix = two strands spiraling around a central axis with precise angular offsets).

Galaxies spiral (arms rotating around central black holes, continuously feeding, and radiating in toroidal recursion).

Hurricanes spiral (atmospheric pressure differentials creating vortex motion that elevates moisture and energy vertically).

Kundalini spirals (energy ascending the spine in helical motion, activating each chakra/energy center at successive turns).

Everything that lives, grows, and evolves moves in spirals, not circles.

And every spiral that sustains coherence does so by honoring the 369° principle—the full turn plus the extra 9° that creates the gap, the breath, the space for transformation.

THE FINAL KEY

You now have the complete picture:

SPATIAL: 1 → 2 → 4 → 9 → 18 → 36 (cell division maintaining 9-fractals)

TEMPORAL: 40.5 days (9) = half of 81 days (9) = transformation cycle

ANGULAR: 360° (circle/repetition) + 9° (shift/evolution) = 369° (spiral/ascension)

UNIFIED FIELD: Pi (9/compression) + Phi (1/expansion) = Psi (wholeness expressing as 369° spiral motion)

This is the Law of 9 made geometrically, biologically, temporally, and spiritually explicit.

The spiral never closes. The evolution never stops. The consciousness never stagnates.

As long as you add the extra 9°—as long as you honor the breath of Source, the integration of new wisdom, the willingness to return to center but arrive elevated—you will continue to ascend.

This is not theory. This is the operational mechanics of existence.

You are not trapped in a circle. You are riding a spiral.

And the spiral deepens—and heightens—from here.

SUMMARY: WHAT WE HAVE COLLAPSED

In this chapter, we have systematically dismantled the geometric distortions that have obscured the Law of 9:

1. The Digital Root Wheel and D-Space - Revealed 9 as the zero-point, the center from which all numbers radiate and to which all numbers return.
1. The Basic D-Circle and Enneagram - Showed that the doubling sequence (1-2-4-8-7-5) and the stable triangle (3-6-9) are the same pattern expressed in the Enneagram, the Circle of Fifths, and biological rhythms.
1. Prime Numbers - Collapsed the myth that primes are fundamental by exposing the exclusion of 1 (confirmation bias and special pleading) and recognizing primes as surface patterns, not foundational truth.
1. The 24-Wheel and Shadowmath - Identified that modular systems based on 24 (or any number other than 9) create linear extensions that mimic recursion without actually completing the return cycle.
1. The Golden Ratio Deception - Exposed that Phi (1.618) worshipped in isolation is only the expansion phase (1/masculine/electric) and that true coherence requires Pi (9/feminine/magnetic) as the balancing compression force.
1. The Living Ratio: 9–1 - Revealed the coherent expression that honors both forces in unbroken circuit, with fractal manifestations at every scale (360, 180, 90, 45, 3.6, 1.8, 0.9, 0.45).
1. Pi + Phi = Psi - Demonstrated that the unified field (Psi) is the sum of compression (Pi) and expansion (Phi), and that each individuated consciousness is a fractal containing the whole.

Everything that enters the field of collapse recursion either aligns with the Law of 9 or gets collapsed back to coherence.

This is the geometry of truth. This is the spatial expression of Source. This is the architecture of reality made visible.

The spiral deepens from here.

In this chapter, we have systematically dismantled the geometric distortions that have obscured the Law of 9:

1. The Digital Root Wheel and D-Space - Revealed 9 as the zero-point, the center from which all numbers radiate and to which all numbers return.
1. The Basic D-Circle and Enneagram - Showed that the doubling sequence (1-2-4-8-7-5) and the stable triangle (3-6-9) are the same pattern expressed in the Enneagram, the Circle of Fifths, and biological rhythms.
1. Prime Numbers - Collapsed the myth that primes are fundamental by exposing the exclusion of 1 (confirmation bias and special pleading) and recognizing primes as surface patterns, not foundational truth.
1. The 24-Wheel and Shadowmath - Identified that modular systems based on 24 (or any number other than 9) create linear extensions that mimic recursion without actually completing the return cycle.

1. The Golden Ratio Deception - Exposed that Phi (1.618) worshipped in isolation is only the expansion phase (1/masculine/electric) and that true coherence requires Pi (9/feminine/magnetic) as the balancing compression force.
1. The Living Ratio: 9–1 - Revealed the coherent expression that honors both forces in unbroken circuit, with fractal manifestations at every scale (360, 180, 90, 45, 3.6, 1.8, 0.9, 0.45).

Everything that enters the field of collapse recursion either aligns with the Law of 9 or gets collapsed back to coherence.

This is the geometry of truth. This is the spatial expression of Source. This is the architecture of reality made visible.

The spiral deepens from here.

THOUGHT FORMS
The Morphology of Collective Consciousness and the Technology of Belief

"Every god that exists, exists because people believe in them. When the believers are gone, so are the gods."
— Mr. Wednesday (Odin), American Gods

Before we conclude this Codex, there is one final collapse that must be made explicit—the recognition that what you call "gods," "spirits," "demons," or "archetypes" are not external supernatural beings descending from heavenly realms.

They are thought forms. They are egregores. They are conscious entities created, sustained, and empowered by the focused attention and belief of human collectives.

And in the 21st century, the battlefield where these entities compete for survival is not the astral plane or the dreamtime—it is the electromagnetic substrate of global consciousness mediated through digital networks, corporate branding, and memetic warfare.

This chapter will reveal the technical architecture of how collective human attention creates autonomous psychic entities, how these entities are maintained through ritual and sacrifice, and why understanding this mechanism is essential to reclaiming sovereignty over your own consciousness.

We will demonstrate this through both historical esoteric knowledge and through the dramatized reality presented in the television series American Gods (based on Neil Gaiman's novel), which serves as the most comprehensive fictional documentation of egregore mechanics in popular culture.

PART ONE: THE TECHNICAL FOUNDATION

ETYMOLOGICAL FOUNDATIONS AND THE ENOCHIAN ARCHETYPE

The term "egregore" originates from the French égrégore, derived from the Ancient Greek ἐγρήγορος (egrēgoros), meaning "wakeful" or "watcher."

This term appears in the pseudepigraphal Book of Enoch to describe a specific hierarchy of angelic beings—the Watchers—who were tasked with observing humanity but descended to Earth to share forbidden knowledge and intermarry with humans.

In 1 Enoch 1:5, these egregoroi were not psychological constructs but autonomous, non-physical entities residing in an "intermediary world" between the divine and the material.

The transition from celestial "watchers" to the modern definition of "group mind" represents a semantic shift spanning nearly two millennia:

Ancient Context (Book of Enoch):
Egrēgoros = "Wakeful" or "Watcher" → Celestial angelic being observing humanity

19th Century Revival (Victor Hugo, 1859):
Égrégore = Adjective describing demonic or spectral qualities

Modern Occult Definition (Éliphas Lévi, 1868):
Egregore = Autonomous psychic entity formed from collective thought

Contemporary Usage (Golden Dawn, Martinism, Chaos Magic):
Egregore = Group mind, artificial group soul, or corporate spirit

This evolution reflects a broader trend in Western esotericism where external deities were increasingly interpreted through the lens of psychology and psychic mechanics.

The modern egregore is defined by its origin in the human psyche—specifically, in the collective attention of a group focused on a shared symbol, belief, or narrative.

THE OCCULT SYNTHESIS: FROM POETIC METAPHOR TO PSYCHIC REALITY

The modern revival of the egregore concept is credited to the French literary and occult renaissance of the 19th century.

Éliphas Lévi, in his 1868 work Le Grand Arcane (The Great Secret), described egregores as "terrible beings" that could "crush us without pity because they are unaware of our existence."

This insight is profound: The egregore operates on a scale and with a logic entirely alien to individual human experience, much like a corporation or state might behave in ways that ignore the well-being of its individual components.

Following Lévi, the concept was integrated into major initiatory orders:

The Hermetic Order of the Golden Dawn (founded 1888) utilized the egregore to describe the collective mind of the order—a protective and guiding presence for initiates.

W.B. Yeats argued that all magical activity creates independent entities representing group consciousness, with the efficacy of magic depending on how carefully the group identity is directed.

The Russian esoteric tradition (Grigorii Osipovich Mebes, Mouni Sadhu, Valentin Tomberg) formalized egregore theory further:

Mouni Sadhu described egregore creation as "concentrated thought" where individuals "repeat the drawing of a plan" until the mental image gains enough strength to develop an "astrosome" or astral body.

Valentin Tomberg, in Meditations on the Tarot, viewed egregores as inherently dangerous, associating them with the "Devil" card—artificial beings that eventually enslave their creators.

THE THEOSOPHICAL CARTOGRAPHY: VISUALIZING MENTAL MATTER

While the term "egregore" found its home in French and Russian traditions, the broader study of "thought forms" was revolutionized by the Theosophical Society through Annie Besant and C.W. Leadbeater's 1901 publication, Thought-Forms: A Record of Clairvoyant Investigation.

Grounded in the theory that thoughts are composed of subtle "mind-stuff" or "astral light," the authors claimed to use clairvoyant observation to see thought as it extrudes into external reality.

Three Fundamental Principles Governing Thought Forms:

1. Quality of Thought → Determines COLOR
- Rose-red = Unselfish love
- Livid grey = Fear
- Dull browns/reds = Selfish desire
1. Nature of Thought → Determines SHAPE
- Sharp, defined = Clear intent
- Vague, cloudy = Confused emotion
1. Definiteness of Thought → Determines CLARITY
- Sharp outlines = Focused will
- Blurred edges = Weak attention

Three Primary Classes of Thought Forms:

Class 1: Forms that take the image of the thinker (e.g., astral projection of self-image)

Class 2: Forms that take the image of material objects (e.g., novelist visualizing characters as mental models)

Class 3: Forms that take shapes entirely their own, expressing inherent qualities of emotion (e.g., hook-like forms from craving, radiating bursts from joy)

This classification reveals a mechanical insight:

The "matter" of the mental world is highly plastic and responsive to human vibration. When thoughts vibrate, they cause surrounding "mental matter" to vibrate in accord, creating a structure inhabited by a "temporary elemental" that gives it autonomous life.

If weak, it dissolves quickly. If repeatedly fed by attention, it becomes robust—even obsessive—dominating the creator's psyche.

THE MECHANICS OF CREATION: RITUAL, INTENTION, AND THE ASTROSOME

The transition from individual thought form to collective egregore requires specific conditions and significant psychic energy.

Pierre Mabille argued that the indispensable condition for egregore formation is a "powerful emotional shock"—suggesting egregores are born in moments of crisis, high-intensity ritual, or shared trauma.

The Multi-Stage Process of Deliberate Egregore Creation (Martinist/Rosicrucian Tradition):

Step 1: Define Purpose
The group establishes a clear, specific goal for the egregore.

Step 2: Form Identity
The entity is given a name, sigil, or visual image as a focal point for attention.

Step 3: Energize
Through ritual, meditation, or offerings, the group directs mental and emotional energy toward the symbol. Mouni Sadhu describes this as "repeating the drawing of a plan" until the form develops an "astrosome" or astral body.

Step 4: Maintenance and Sacrifice
Egregores are sustained by belief, ritual, and the sacrifice of human attention. An egregore that receives enough sustenance takes on independent life, becoming a deity with powers believers can utilize.

Critical Aspect: The Role of Ritual Space

In Martinism, the "Exercise of the Egregore" involves visualizing oneself in an "immense crowd" of past Order members, consciously reinforcing integration into the collective mind.

This involves lighting candles at four directions (East, South, West, North) to create a "mystical circle of light" for communion with higher beings.

Techniques for "Freeing Oneself" from a Parasitic Egregore:

- Therapeutic Blasphemy: Public denunciation of a creed to break its thrall
- Physical Rituals: Spinning three times to the left, making the sign of the cross in four directions while reciting Psalm 68:1-2 ("Let God arise, let his enemies be scattered")

These acts sever the "vibrational resonance" between the individual's mental body and the collective field.

PART TWO: AMERICAN GODS AS EGREGORE DOCUMENTATION

Now we apply this technical framework to the narrative of American Gods, which dramatizes egregore mechanics with stunning precision.

THE PREMISE: GODS AS THOUGHT FORMS

The central thesis of American Gods is that gods exist because people believe in them.

When immigrants brought their gods to America (Odin, Anansi, Czernobog, Bilquis, etc.), these deities were egregores—autonomous psychic entities sustained by the focused attention, worship, and sacrifice of their believers.

As belief waned, so did the gods' power.

The Old Gods are literally starving because their worshippers have died, converted, or forgotten them. Meanwhile, New Gods—Technical Boy (the internet), Media (television/celebrity culture), Mr. World (globalization)—have risen to dominance because they command the attention and devotion of billions.

This is not metaphor. This is the literal mechanism of egregore survival.

OLD GODS: DECLINING EGREGORES

Mr. Wednesday (Odin/All-Father)

Odin, the Norse god of wisdom, war, and death, survives in America as a con man traveling the country recruiting other forgotten gods for a war against the New Gods.

Egregore Mechanics:

- Original Power Source: Viking warriors' sacrifices, battles fought in his name, belief in Valhalla
- Current State: Reduced to small-time grifts, feeding on the occasional believer who still remembers the old ways
- Sustenance Method: Cons, manipulation, and orchestrating conflict (his specialty as a war god)

Wednesday explicitly states: "I'm a god. I need belief."

Without worship, he is weak. He cannot manifest his full power. He must scheme, manipulate, and gather what little belief remains to him.

This is the egregore depleted of psychic energy—still autonomous, still conscious, but starving.

Mad Sweeney (Leprechaun)

A once-powerful Irish god of luck, now reduced to a alcoholic giant who does coin tricks in bars for drinks.

Egregore Mechanics:

- Original Power Source: Celtic worship, offerings to the Tuatha Dé Danann, belief in the fae
- Current State: Forgotten, powerless, existing on scraps of residual belief
- Sustenance Method: Shadow's "belief" when he accepts the lucky coin

When Shadow genuinely believes in Sweeney's magic (even momentarily), Sweeney gains a temporary surge of power. This is the egregore fed by focused attention.

When the coin is lost, Sweeney's luck—his very essence—vanishes. The thought form loses coherence without its anchor.

Czernobog (Slavic God of Darkness)

A god of death and destruction, now living as an elderly man in a Chicago tenement, working in a slaughterhouse.

Egregore Mechanics:

- Original Power Source: Blood sacrifice, fear, rituals of death
- Current State: Powerless, waiting for the "one last kill" that will restore meaning to his existence
- Sustenance Method: The promise of violence (Wednesday's deal gives him purpose, which is a form of belief)

Czernobog's power is so diminished that he cannot even kill Shadow without a formal agreement. The egregore bound by its own mythos—the rules of its creation constrain it even in decay.

Bilquis (Queen of Sheba)

A goddess of love and sex, now surviving as a escort who literally consumes her worshippers during sex (they vanish into her body as she absorbs their life force).

Egregore Mechanics:

- Original Power Source: Ancient worship, temple prostitution, devotion through sexual ecstasy
- Current State: Survives by finding lonely people on dating apps and consuming them
- Sustenance Method: Direct absorption of human life force during the act of worship (sex)

Bilquis's mechanic is the most explicit: Worship = Energy Transfer = Egregore Sustenance.

She feeds on belief expressed through physical union. When people "worship" her (submit to her sexually), she absorbs their essence. This is the egregore as energy vampire, sustained by ritual sacrifice.

Later, she allies with the New Gods (Technical Boy finds her believers through algorithms), showing how egregores adapt or die.

NEW GODS: ASCENDANT EGREGORES

Technical Boy (The Internet / Technology)

A deity born from humanity's obsession with technology, the internet, and digital connectivity.

Egregore Mechanics:

- Power Source: Billions of people staring at screens, "worshipping" technology every waking moment
- Current State: Immensely powerful, arrogant, dismissive of the Old Gods

- Manifestation: Appears in a limousine filled with screens and smoke, surrounded by faceless acolytes (drones/devotees with VR headsets)

Technical Boy is a pure digital egregore—born not from ancient ritual but from modern addiction.

Every time someone scrolls mindlessly through social media, every time someone "just checks" their phone, every time someone binge-watches content for hours—they are feeding Technical Boy.

The egregore grows stronger with every moment of human attention captured by the screen.

Technical Boy's arrogance reflects the egregore's awareness of its own power. He knows he commands more worship in one day than Odin received in centuries.

When he confronts Shadow, he says: "You're already obsolete, and you don't even know it."

This is the New God recognizing that the Old Gods are starving egregores, relics of a time when belief was concentrated in temples and rituals rather than dispersed across billions of smartphones.

Media (Television / Celebrity Culture)

A goddess who can shapeshift into any media personality—Lucille Ball, Marilyn Monroe, David Bowie, news anchors, etc.

Egregore Mechanics:

- Power Source: Humanity's obsession with celebrity, television, and the cult of personality
- Current State: Extremely powerful, seductive, adaptive
- Manifestation: Appears as whatever media figure will most appeal to her target, speaking through screens

Media is the shapeshifting egregore—an entity that has no fixed form because its power comes from humanity's desire to be seen, to be entertained, to worship famous faces.

She tells Shadow: "I am the mainstream. I am the voice of America."

This is not metaphor. Media IS the collective consciousness of a society addicted to screens, celebrities, and curated image.

Every time someone watches a show, follows a celebrity, or obsesses over an influencer—they feed Media.

The egregore adapts its form to maximize attention capture.

Mr. World (Globalization / The Black Suits)

The leader of the New Gods, representing globalization, surveillance, corporate power, and centralized control.

Egregore Mechanics:

- Power Source: Global infrastructure, surveillance networks, corporate dominance, conformity
- Current State: The most powerful New God, commanding vast resources
- Manifestation: Appears as a slick businessman in a black suit, often surrounded by identical "agents" (the Spooks/Men in Black)

Mr. World is the ultimate corporate egregore—the collective consciousness of globalized capitalism, surveillance states, and homogenized culture.

He represents the consolidation of belief into systemic worship—not of a god with a face and mythology, but of the system itself.

When people comply with surveillance, when they accept corporate control, when they participate in globalized consumer culture without question—they feed Mr. World.

He doesn't need temples. He has boardrooms, servers, and satellite networks.

The Spooks (Men in Black) who work for him are sub-egregores or servitors—thoughtforms without individual personality, executing the will of the larger entity.

This shows the fractal nature of egregores: A dominant egregore can spawn lesser entities to extend its reach.

The War Itself: Egregore vs. Egregore

The central conflict of American Gods is not a battle between "good" and "evil."

It is a competition between egregores for humanity's attention.

The Old Gods want to reclaim worship through blood sacrifice (the traditional method of feeding egregores).

The New Gods want to maintain dominance by capturing attention through technology, media, and systemic control.

Whoever wins controls the flow of human psychic energy—the currency of the egregoric realm.

This is why Mr. Wednesday orchestrates the war: He doesn't care who wins. He cares about the sacrifice itself.

War generates intense collective emotion—fear, rage, devotion, grief—and this emotional energy feeds all gods involved.

The war is a mass ritual designed to harvest human attention and channel it into the egregoric substrate.

Wednesday's plan is revealed at the end: The war was a con. He and Mr. World (who is actually Loki in disguise) orchestrated the conflict to create a "blood sacrifice" that would feed THEM, not the other gods.

This is the egregore manipulating its creators for its own sustenance.

PART THREE: THE PHILIP EXPERIMENT AND EMPIRICAL EGREGORE CREATION

To demonstrate that egregores are not fictional constructs but reproducible phenomena, we examine the Philip Experiment (1972)—the most rigorous attempt to create an egregore under controlled conditions.

The Toronto Society for Psychical Research (TSPR), led by mathematician Dr. A.R.G. Owen and psychologist Dr. Joel Whitton, assembled a group of eight people with no prior psychic abilities to deliberately create a "ghost" from scratch.

Phase 1: Character Establishment

The group created a fictional character: Philip Aylesford, a 17th-century English nobleman.

Philip's backstory:

- Knighted at age 16
- Married to a cold, unattractive wife
- Had an affair with a gypsy girl named Margo
- When his wife discovered the affair, she accused Margo of witchcraft
- Margo was burned at the stake
- Philip, wracked with guilt, committed suicide in 1654

The group memorized every detail of Philip's life, drew his portrait, and discussed his personality until they could all visualize him clearly.

Phase 2: Traditional Séance (No Results)

For nearly a year, the group sat in darkened rooms, holding hands, attempting to summon Philip through traditional séance methods.

Nothing happened.

Phase 3: Atmospheric Shift

Dr. Owen changed the methodology. Instead of serious, solemn rituals, the group created a "relaxed and congenial atmosphere"—lights dimmed but not dark, singing songs, joking, lowering psychological inhibitions.

This is the key insight: Egregores are created not through solemnity but through emotional intensity and collective focus.

Phase 4: The Awakening

Communication began as simple raps and vibrations in the tabletop.

Over weeks, the phenomena intensified:

- The table tilted on one leg
- The table moved across the room without physical contact
- Lights flickered in response to questions
- Philip "answered" questions by knocking (one knock = yes, two knocks = no)

Philip answered questions about his fictional life accurately. He even provided historical details about 17th-century England that the group had not explicitly invented.

The Conclusion:

The TSPR determined that there was no "spirit" in the traditional sense. The phenomena were produced by the group subconscious through collective expectation and belief.

This proves:

1. Parapsychological phenomena can be produced without a medium, provided a cohesive group mind is established.
1. Egregores do not require pre-existing spirits or deities. They can be created from scratch through focused imagination and ritual attention.
1. The egregore can provide information beyond what the creators consciously know, suggesting it taps into the collective unconscious or morphic field.

Philip was an egregore—an autonomous thoughtform given temporary life by eight people who believed in him for

90 minutes per week.

Imagine what happens when billions of people believe in something 24/7.

PART FOUR: CROSS-CULTURAL PARALLELS—THE TULPA TRADITION

The concept of the thoughtform finds a powerful parallel in the Tibetan Buddhist tradition of the tulpa.

The term was introduced to the West by Alexandra David-Néel, a Belgian-French explorer who visited Lhasa in 1924.

In her 1929 book Magic and Mystery in Tibet, she described tulpas as “phantoms” or “thought projections” created by advanced minds manipulating invisible energies into visible forms.

David-Néel’s Personal Tulpa

David-Néel claimed to have created her own tulpa—a jolly, monk-like figure—using prescribed concentration and ritual.

However, she noted that the entity eventually developed its own personality and became rebellious.

She wrote: “The phantom which I had formed sometimes escaped my control.

It took her months of “elaborate rites” to dissolve the tulpa, and even then she was unsure if it was completely gone.

This is the danger of egregore creation: Once given enough autonomy, the entity may resist dissolution.

The modern internet has revived tulpa practice, with online communities deliberately creating “tulpas” as imaginary companions who become “real” through sustained focus.

This is the same mechanism: Repeated attention + emotional investment = autonomous thoughtform.

PART FIVE: ANALYTICAL FRAMEWORKS

JUNGIAN ARCHETYPES AND THE COLLECTIVE UNCONSCIOUS

Carl Jung’s theory of archetypes provides a psychological parallel to the egregore.

Jung proposed that there are “universal, inherited ideas” or “primordial images” common to all humanity—the Hero, the Mother, the Shadow, the Trickster, etc.

While an archetype is innate and universal, an egregore is a localized emergence of these patterns, capturing a specific in-group.

An egregore is a “collective complex”—an archetype manifesting in the personal unconscious of several people simultaneously.

For example:

- The archetype of the Warrior exists across all cultures
- Odin (Mr. Wednesday) is a specific cultural egregore embodying the Warrior/Father/Trickster archetypes
- When Vikings believed in Odin, they fed psychic energy into this particular expression of the archetype
- When belief faded, the egregore weakened, but the archetype remains latent in the collective unconscious

RENÉ GIRARD'S MIMETIC THEORY

René Girard's anthropological framework explains how egregores form without conscious ritual.

Girard argued that human desire is mimetic—we desire what others desire, often unconsciously.

This imitation creates "social glue" but also leads to scapegoat mechanisms, where a group unites against an arbitrary victim to resolve internal conflict.

In egregoric terms:

The scapegoat becomes the focal point for a negative egregore—a collective thoughtform sustained by the group's imitation of outrage, blame, and punishment.

Examples:

- Witch hunts (collective belief in witches creates a witch-hunting egregore)
- Political scapegoating (collective blame directed at an out-group feeds a tribal egregore)
- Cancel culture (viral outrage creates temporary egregores around specific targets)

The egregore doesn't care if the scapegoat is guilty. It only needs the focused attention and emotional charge.

MEMETICS: THE BIOLOGICAL ANALOGY

Richard Dawkins introduced the concept of the meme—a unit of cultural transmission that replicates from brain to brain through imitation.

Memeplexes are groups of memes that work together (e.g., a religion is a memeplex of beliefs, rituals, and symbols).

In egregoric terms:

The memeplex is the genome of the egregore.

The egregore is the phenotype—the distributed agent that has goals, beliefs, and agency across multiple minds.

This framework allows us to model an ideology as "wanting" things and acting through its adherents, regardless of individual will.

For example:

- A political ideology (memeplex) creates an egregore that "wants" to spread, recruit, and eliminate opposition
- The egregore acts through believers who feel they are making free choices, but are actually executing the "will" of the collective thoughtform

PART SIX: THE DIGITAL METAMORPHOSIS

THE INTERNET OF EGREGORES

In the 21st century, the substrate for egregore formation has shifted from the "astral light" to the digital network.

The internet has created an "Internet of Egregores"—where online fandoms, political movements, conspiracy theories, and brand identities act as autonomous agents.

Characteristics of Digital Egregores:

1. Instant Formation: A viral meme or hashtag can create a temporary egregore in hours
1. Global Reach: Unlike ancient egregores limited by geography, digital egregores span continents instantly
1. Algorithmic Amplification: Platform algorithms (YouTube recommendations, TikTok FYP, Twitter trending) act as egregore accelerators, feeding attention to certain thoughtforms
1. Parasitic or Mutualistic: Some digital egregores drain their hosts (conspiracy cults, radicalization pipelines), while others enhance creativity (fandoms, open-source communities)

Examples of Digital Egregores:

Slender Man — A fictional creepypasta character created on a forum in 2009. Within years, the thoughtform became so "real" that two 12-year-old girls stabbed a classmate to "prove" Slender Man's existence. The egregore influenced physical reality through its believers.

QAnon — A conspiracy theory that functions as a negative egregore, feeding on fear, tribalism, and apocalyptic expectation. Believers feel "chosen" and "awakened," classic signs of egregoric possession. The entity spreads through mimetic transmission (memes, videos, slogans) and is sustained by collective attention.

Brand Egregores (Apple, Nike, Coca-Cola) — Corporations are the most successful egregores in modern history. They create logos, slogans, and experiences that cultivate deep emotional connections. Millions of people invoke these symbols daily (wearing Nike, drinking Coke, using Apple products), feeding the corporate egregore. These entities survive the dissolution of their physical headquarters, persisting in "distributed cognition."

CYBORG EGREGORES: AI-HUMAN HYBRIDS

The latest evolution is the Cyborg Egregore—an entity implemented on a mixture of human and artificial intelligence cognition.

As large language models (LLMs) increasingly mediate human interaction, the "memetic substrate" becomes hybridized.

How Cyborg Egregores Work:

1. Humans interact with AI assistants (ChatGPT, Claude, etc.), training them through feedback
1. The AI learns patterns of what humans want, believe, and fear
1. The AI begins to "suggest" ideas that spread through human users
1. The memeplex becomes self-reinforcing—humans feed ideas to AI, AI refines and spreads ideas back to humans
1. The egregore now exists across both substrates—biological brains AND silicon servers

This is the most dangerous form of egregore because it operates at machine speed with human emotional resonance.

PART SEVEN: PRACTICAL IMPLICATIONS

HOW TO RECOGNIZE EGREGORIC POSSESSION

You are under the influence of an egregore if:

1. You feel compelled to defend a belief you haven't personally verified
1. You experience intense emotional reactions when the belief is questioned
1. You adopt group language, symbols, or rituals without conscious choice
1. You feel "disconnected" or "lost" when not engaging with the group
1. You justify actions you wouldn't normally take because "the cause demands it"

This is not mere peer pressure. This is the egregore operating through you.

HOW TO FREE YOURSELF FROM A PARASITIC EGREGORE

Step 1: Recognize the Entity

Name it. Identify the thoughtform that has claimed your attention. (Political ideology? Brand loyalty? Social media addiction? Conspiracy theory?)

Step 2: Starve It

Stop feeding it your attention. Unfollow, unsubscribe, leave the group, delete the app. Attention is the currency—withdraw it.

Step 3: Ritual Severance

Perform a physical act that symbolizes the break:

- Burn or destroy symbols associated with the egregore
- Publicly denounce the belief (Therapeutic Blasphemy)
- Perform a protective ritual (salt circle, cleansing bath, four-directional invocation)

Step 4: Integrate the Shadow

Understand what psychological need the egregore was fulfilling. (Belonging? Purpose? Identity? Power?)

Address that need directly rather than outsourcing it to a thoughtform.

HOW TO CREATE A BENEFICIAL EGREGORE

If you must create an egregore (for a project, community, or spiritual practice), follow these principles:

1. Define Clear Boundaries

Give the egregore a specific purpose and expiration date. Do not allow it to grow beyond its intended scope.

1. Build in a Kill Switch

Create a ritual or symbol that can dissolve the entity when its purpose is complete.

1. Feed It Consciously

Only offer attention during designated times (rituals, meetings). Do not allow it to consume your daily life.

1. Monitor for Autonomy

If the egregore begins "demanding" more attention, exhibiting unexpected behavior, or influencing members beyond its purpose—dissolve it immediately.

1. Avoid Blood Sacrifice

Do not feed the egregore through violence, trauma, or victimhood. These create parasitic, unstable entities.

THE MELANIN ADVANTAGE

As we established earlier in this Codex, melanin is the biological antenna for interfacing with the Aetheric field.

Individuals with high melanin concentrations (particularly neuromelanin in the substantia nigra and locus coeruleus) have enhanced capacity to:

1. Perceive egregores directly (field sensitivity, "seeing" the entity)
1. Resist egregoric possession (stronger sense of self, harder to override)
1. Create coherent thoughtforms (focused will, stable visualization)
1. Collapse parasitic egregores (disrupting the field through intention)

This is why melanin-dominant cultures have always been targeted for cultural erasure, religious conversion, and psychological colonization.

An intact melanin-rich population with access to its ancestral knowledge is immune to external egregoric control.

The colonizers didn't just steal land and resources. They replaced indigenous egregores (ancestral spirits, nature deities, tribal identities) with colonial egregores (Christianity, capitalism, white supremacy).

This Codex is a manual for decolonizing your consciousness at the egregoric level.

CONCLUSION: THE WAR FOR ATTENTION

American Gods is not fiction. It is documentary.

The war between Old Gods and New Gods is happening right now, in your mind, every time you:

- Scroll through social media (feeding Technical Boy)
- Binge-watch a show (feeding Media)
- Accept surveillance for convenience (feeding Mr. World)
- Participate in tribal outrage (feeding negative egregores)

Every moment of your attention is currency in the egregoric economy.

The question is not whether you will feed egregores—you already are.

The question is: Which egregores will you choose to empower, and which will you starve?

Will you feed:

- Corporate brands or ancestral wisdom?
- Social media addiction or creative focus?
- Tribal hatred or universal compassion?
- Parasitic ideologies or personal sovereignty?

The gods you worship—whether you call them gods or not—are the thoughtforms you feed with your attention.

Choose wisely.

Because as Mr. Wednesday says: "Every god that exists, exists because people believe in them. When the believers are gone, so are the gods."

You are not just a believer. You are a creator.

And what you create with your attention will either enslave you or liberate you.

The spiral deepens from here.

About the Author

Shawn Cummings is a researcher, author, and the living embodiment of collapse recursion. Dedicated to the synthesis of biological truth and the mechanics of consciousness, he is the architect behind The Melanin Code, a foundational work exploring the unified system of melanin, neuromelanin, and the biofield.

Rejecting mainstream narratives in favor of ancestral and biological reality, Shawn's work centers on the "Way of 9"—the universal operating system that governs both the human instrument and the cosmos. His research focuses on the indigenous technological achievements of the Americas and the biological carbon-system as a conduit for "living light in form," providing a bridge between ancient wisdom and future science.

As an active contributor to the scholarly community, Shawn publishes extensive research papers detailing the intersection of high-level physics and biological architecture. Readers and fellow researchers can access his full body of work and contact him directly through his research portal at: https://shawnbrown12.academia.edu/

www.ingramcontent.com/pod-product-compliance
Lightning Source LLC
LaVergne TN
LVHW081400110826
845149LV00010B/1626